Analytical Techniques in Biotechnology

Analytical Techniques in Biotechnology

Edited by **Suzy Hill**

SYRAWOOD
PUBLISHING HOUSE
New York

Published by Syrawood Publishing House,
750 Third Avenue, 9th Floor,
New York, NY 10017, USA
www.syrawoodpublishinghouse.com

Analytical Techniques in Biotechnology
Edited by Suzy Hill

International Standard Book Number: 978-1-68286-238-4(Hardback)

Printed in China.

Contents

Preface

This book has been a concerted effort by a group of academicians, researchers and scientists, who have contributed their research works for the realization of the book. This book has materialized in the wake of emerging advancements and innovations in this field. Therefore, the need of the hour was to compile all the required researches and disseminate the knowledge to a broad spectrum of people comprising of students, researchers and specialists of the field.

Biotechnology is a rapidly progressing field of science and technology. It integrates principles and applications from diverse fields such as genomics, immunology, and many more. This book strives to equip the reader with thorough knowledge about various techniques of applied biotechnology, its applications in the areas of medicine, agriculture, environment, etc., along with details about the latest equipment and analytical tools which would help the reader gain an in-depth perspective about the subject. This book elucidates new techniques and their applications in a multidisciplinary approach and is apt for students of biotechnology and associated disciplines of study.

At the end of the preface, I would like to thank the authors for their brilliant chapters and the publisher for guiding us all-through the making of the book till its final stage. Also, I would like to thank my family for providing the support and encouragement throughout my academic career and research projects.

Editor

1

Experimental and modeling studies on microwave-assisted extraction of mangiferin from *Curcuma amada*

Jeke Kullu · Abhishek Dutta · Denis Constales · Surabhi Chaudhuri · Debjani Dutta

Abstract Mangiferin, a bioactive compound having potent nutraceutical, strong antioxidant and pharmacological significance has been extracted using microwave-assisted extraction (MAE) technique from *Curcuma amada*, commonly known as mango ginger. The extraction solvent ethanol is eco-friendly, nontoxic and reduces the risk of environmental hazards. The influence of several independent variables such as microwave power, ethanol concentration, extraction (irradiation) time and pre-leaching time has been studied under varying conditions using one-factor-at-a-time analysis to obtain an optimal extraction ratio. The maximum mangiferin content of 1.1156 mg/g is obtained at microwave power of 550 W and extraction time of 50 s with 80 % ethanol as a solvent and pre-leaching time of 20 min. The results indicate that microwave power and ethanol concentration have the most significant effect on the yield of mangiferin content. The presence of mangiferin in final *Curcuma amada* extract is confirmed through high-performance liquid chromatography and the functional groups are identified through Fourier transform infrared spectroscopy analyses using standard mangiferin. The experimental profiles are fitted into a two-parameter modified first-order kinetic model and a three-parameter modified logistic model and checked using the goodness-of-fit criterion. The *Curcuma amada* retained its antioxidant activity after MAE treatment and the antioxidant activity of mangiferin obtained after extraction using DPPH free radical scavenging assay is studied.

Keywords Microwave-assisted extraction · Mangiferin · Antioxidant activity · ANOVA · Mathematical modeling

Abbreviations

ANOVA	Analysis of variations
MAE	Microwave-assisted extraction
OFAT	One-factor-at-a-time
HPLC	High performance liquid chromatography
FTIR	Fourier transform infrared spectroscopy
SCFE	Supercritical fluid extraction
UAE	Ultrasound assisted extraction
HRE	Heat reflux extraction

Introduction

Traditional plant spices, similar to fruits and vegetables, are known to contain health-promoting components such as vitamins, minerals, antioxidants and prebiotics (Omenn et al. 1996). In particular, plant spices are used in foods because they impart desirable flavours and may fulfil more than the one function for which they are added. Extensive research is being conducted on traditional medicines, on different plant species and their therapeutic applications all over the world. *Curcuma amada*, commonly known as mango ginger, is an important member of the

J. Kullu · S. Chaudhuri · D. Dutta (✉)
Department of Biotechnology, National Institute of Technology, Durgapur 713209, India
e-mail: debs_2000in@yahoo.com; debjani.dutta@bt.nitdgp.ac.in

A. Dutta
Department of Chemical Engineering, Groep T-Leuven Engineering College (Associatie KU Leuven), A. Vesaliusstraat 13, 3000 Leuven, Belgium

D. Constales
Vakgroep Wiskundige Analyse, Ghent University, Galglaan 2, Blok S22, 9000 Ghent, Belgium

Zingiberaceae family. It has an Indo-Malayan origin and is distributed widely in the tropics from Asia to Africa and Australia (Sasikumar 2005). *Curcuma amada* is named mango ginger because it is morphologically similar to ginger and imparts a mango flavour and is typically used in the manufacture of pickles, culinary preparations and salads for flavour, candy and sauce (Shankaracharya 1982). *Curcuma amada* has pharmacological significance for a variety of ailments. Therapeutically, mango ginger is used to treat a range of mood and medical disorders in traditional and ayurvedic medicine. *Curcuma amada* is credited with diverse bioactive molecules demonstrating antibacterial, antifungal, anti-inflammatory, anti-hypercholesterolemic, insecticidal, aphrodisiac, antipyretic and antioxidant properties (Singh et al. 2010). Mangiferin is an important bioactive constituent of mango ginger containing xanthone-C-glycoside, which has numerous pharmacological properties and is an important phytochemical. It has antidiabetic, cardioprotective, immunomodulatory, antioxidant, antitumour, hepatoprotective and vasorelaxant properties and is useful in the treatment of biliousness, skin diseases, bronchitis, asthma and inflammation (Jatoi et al. 2007). Extraction forms the first basic step in medicinal plant research because the preparation of crude extracts from plants is the starting point for the isolation and purification of chemical constituents (Romanik et al. 2007). Keeping in mind the requirements such as shortened extraction time, reduced solvent consumption, increased pollution prevention and the special care needed for thermolabile constituents, numerous extraction techniques have been developed for the purpose of obtaining pharmacologically active compounds from various plant sources such as supercritical fluid extraction (SCFE), microwave-assisted extraction (MAE), ultrasound-assisted extraction (UAE) and heat reflux extraction (HRE). However, because of several disadvantages with the traditional extraction techniques like sonication and Soxhlet extraction, non-conventional extraction techniques like SCFE, extraction by microwave and ultrasound sources have gained importance. The use of microwaves in analytical sciences is not new; the first reported analytical use for microwave oven was in 1986 for the extraction of organic compound (Dean 2010). In recent years, MAE has attracted growing interest as it allows rapid extraction of solutes from solid matrices, with extraction efficiency comparable to that of the classical techniques (Camel 2000). Heating occurs in a targeted and selective manner in MAE with practically no heat being lost to the environment, and the mechanism can significantly reduce the extraction time (Huie 2002). This means it requires less solvent volume and is thus time conserving with improved product recovery. Further, the extraction solvent used is usually water or ethanol, which is inexpensive, nontoxic and environmentally benign (Ferguson et al. 2012). Samples pretreated with solvents with higher microwave absorbing capacity when coupled with extracting solvents like ethanol bring about heating by at least two competing mechanisms, namely direct heating from the interaction of microwaves with ethanol and heating from the diffusion of excess heat resulting from the interaction of the microwaves with the pretreated matrix (Mandal et al. 2007). In our previous study (Padmapriya et al. 2012), MAE of mangiferin from *Curcuma amada* was studied using only two independent factors, namely microwave power and extraction (irradiation) time. However, it has been observed that several other extraction variables such as solvent concentration, ethanol concentration and pre-leaching time could also be influential factors in the optimization of the extraction protocol of a bioactive compound, which may act dependently or independently (Dhobi et al. 2009). In the present study, therefore, a more rigorous approach has been applied to understand the influence of these independent factors on mangiferin extraction using mathematical modeling. The presence of mangiferin in final *Curcuma amada* extract was confirmed using high-performance liquid chromatography (HPLC) using standard mangiferin and was further subjected to Fourier transform infrared spectroscopy (FTIR) analysis for identification of the functional groups. The antioxidant activity of mangiferin obtained after extraction using DPPH free radical scavenging assay has also been studied.

Materials and methods

Plant material

Fresh and healthy *Curcuma amada* (mango ginger) were purchased from the local market in Durgapur, West Bengal. The rhizomes were washed, peeled and cut into fine pieces and then dried in a hot-air oven (OVFU) at 70 °C until constant weight and was well blended. Mangiferin standard was purchased from Sigma-Aldrich, USA.

Microwave-assisted extraction (MAE)

Microwave-assisted extraction was performed using a microwave apparatus (Samsung Trio, Model CE117ADV; 230 V ~50 Hz) in a closed vessel system. 2.5 g of dried *Curcuma amada* powder was extracted with 25 ml solvent under different MAE conditions. After extraction, the vessels were allowed to cool at room temperature before opening. Microwave power (250, 350, 450, 500, 550 and 900 W), ethanol concentration (50–100 %, v/v), extraction time (1–120 s, with an interval of 5 s) and pre-leaching time (1–30 min, with an interval of 5 min) were evaluated for the extraction of mangiferin from *Curcuma amada*. The extraction of mangiferin was carried out using the method of Padmapriya et al. (2012). The final extract was evaporated and dissolved in DMSO before UV–vis spectrophotometric (Techcomp, UV 2310) analysis. For the estimation of mangiferin, the method described by Joubert et al. (2008) was used and the absorbance was measured at 410 nm.

High-performance liquid chromatography (HPLC) analysis

The final extract of *Curcuma amada* was analysed by HPLC (Waters 600) equipped with a UV–vis detector (Waters 2489) according to the method described by Muruganandan et al. (2002). Chromatographic separation was performed on a reverse-phase column (C18, 4.6 × 250 mm, Waters) with the temperature of the column being maintained at 25 °C. The mobile phase was acetonitrile and 3 % acetic acid in the ratio 16:84 at a flow rate of 0.5 ml/min. The sample injection volume was 10 µl. The peaks were evaluated based on their absorbance at 254 nm. Retention time and concentration of the samples were compared with pure standard of mangiferin (Sigma-Aldrich, USA).

Fourier transform infrared spectroscopy (FTIR) analysis

The mangiferin extracted after MAE at 550 W was further subjected to FTIR analysis for identification of the functional groups. Comparing the functional groups present in standard mangiferin, the damaged functional group of the extracted mangiferin can be identified. A known weight of the final sample extract was mixed with potassium bromide and loaded onto a Perkin Elmer instrument. The samples were scanned in model spectrum-100 system in range of 400–4,000 cm^{-1}. The spectral data obtained were compared with a standard mangiferin chart to identify the functional groups present in the sample.

DPPH radical scavenging activity

The DPPH assay was carried out according to the method reported by Ara and Nur (2009). DPPH solution (0.004 % w/v) was prepared in 95 % methanol. The stock solution was diluted to final concentration of 1, 5, 10, 20, 40, 60, 80 and 100 µg/ml . The freshly prepared DPPH solution was added in each of the test tubes containing the final concentrations of *Curcuma amada* methanolic extract, and after 10 min of incubation the absorbance was taken at 517 nm using a spectrophotometer. The scavenging effect (%) of DPPH free radical was measured using the following equation:

$$\text{DPPH radical-inhibition scavenging effect}\,(\%) = [(\text{absorbance of control} - \text{absorbance of test sample})/(\text{absorbance of control})] \times 100.$$

Statistical analysis

The screening of the variables has been done using one-factor-at-a-time (OFAT) analysis, which has several advantages such as run size economy, fewer level changes and providing protection against the risk of premature termination of experiments (Qu and Wu 2005). It must be noted that although processes are commonly optimized in most industrial experiments using OFAT design approach, optimal conditions or interactions between variables cannot be predicted with this methodology (Wardhani et al. 2010). However, OFAT design allows to find out more rapidly whether a factor has any effect and is therefore a sequential learning process (Morgan and Deming 1974). The statistical software Graphpad Prism v5.0.0.2 was used for the data analysis. A two-way analysis of variance (ANOVA) was implemented to calculate the significance of the differences in the content of mangiferin. Means and coefficients of variance were computed for all qualitative analysis and treatments with homogeneous means ranked using the Newman–Keuls post hoc test. The significance of the results was established at values greater than 0.05 in all the experiments performed. The parameters of the empirical models were fitted with a nonlinear least-squares (NLLS) Marquardt–Levenberg algorithm, using the device-independent plotting program Gnuplot.

Mathematical modeling of mangiferin extraction

Kinetics of MAE of mangiferin is performed at the experimental design points for the three independent variables namely, microwave power, ethanol concentration and pre-leaching time. In all these cases, the experimental data seem to follow a sigmoidal curve for which a two-parameter modified first-order kinetic model (Wardhani et al. 2010) and a three-parameter delayed logistic model with a final asymptote (Yukalov et al. 2009) are chosen to describe the evolution of microwave-assisted mangiferin extraction given as follows:

$$Y = Y_{\max}(1 - e^{-k_{\mathrm{m}}.t}), \quad (1)$$

and

$$Y = Y_{\max}/(1 + e^{-k_{\mathrm{m}}(t-\tau)}), \quad (2)$$

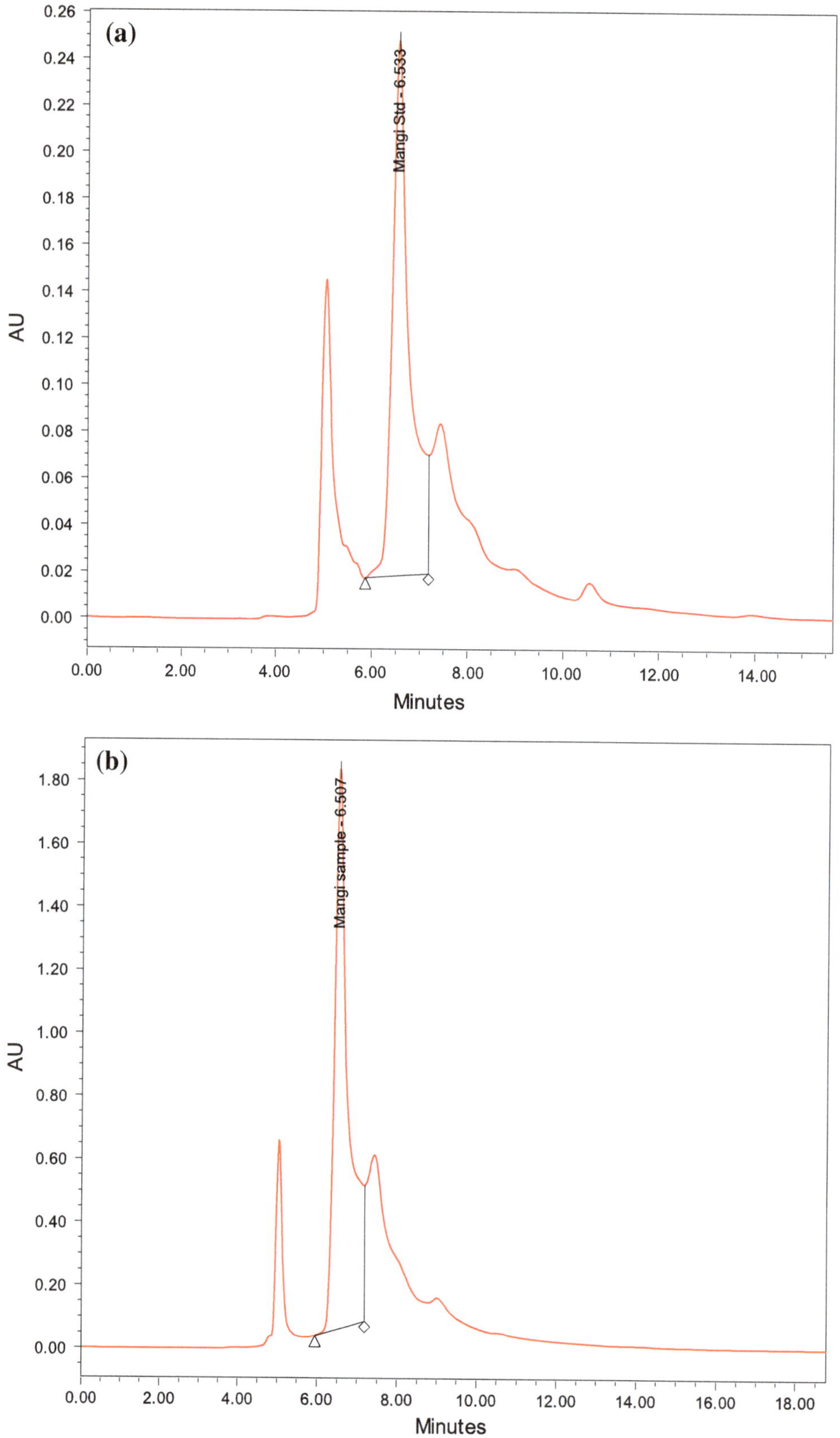

Fig. 1 **a** HPLC chromatogram of mangiferin standard. **b** HPLC chromatogram of mangiferin extracted from *Curcuma amada* by microwave-assisted extraction (MAE)

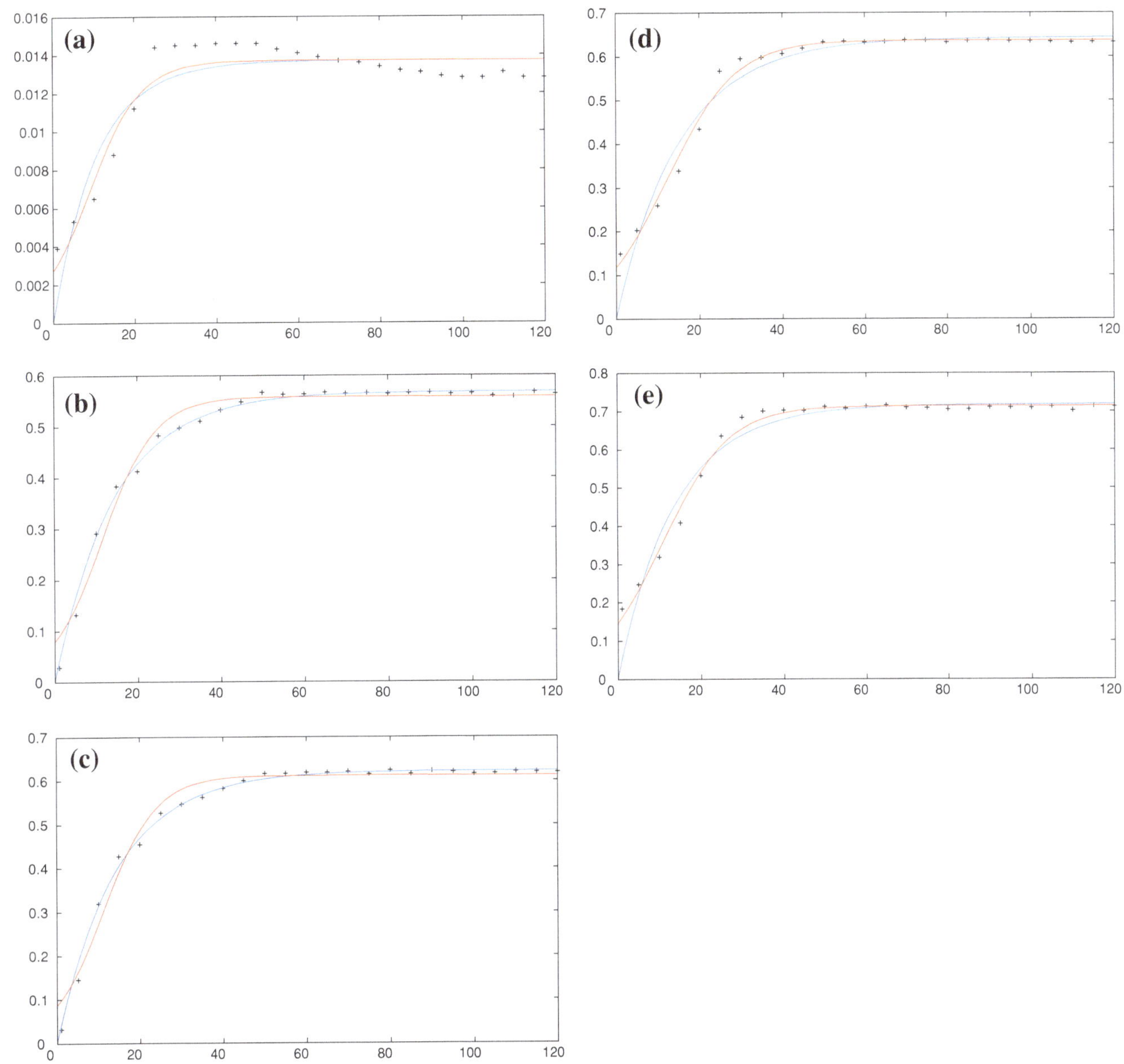

Fig. 2 Temporal evolution of the effect of microwave power (as an independent variable) on the yield of mangiferin content extracted from *Curcuma amada* at various experimental design points (x axis: time in s; y axis: yield in mg/g). The experimental data (*symbols*) fitted to the 2-parameter model (Eq. 1, *blue*) and 3-parameter logistic model (Eq. 2, *red*) are shown in Fig. 2a–e for 250, 350, 450, 500 and 550 W, respectively

where Y is the mangiferin content (mg/g) at time t, $Y_{\max}$ is the maximum mangiferin content (mg/g) when time approaches infinity, k_m is the first-order mangiferin extraction constant or the specific rate of mangiferin concentration (s^{-1}) and τ is the time delay (s); else the value at $t = 0$ would always be half the value at $t = \infty$ (and there is no sufficient reason to assume such a restriction).

A generalized expression to describe the dependency of both microwave power (P) and ethanol concentration (E) on extraction time can be written using a slightly modified delayed logistic model as follows:

$$Y = \frac{Y_{\max}\left(1 + \frac{P\,(\text{or } E)}{P_{\text{ref}}\,(\text{or } E_{\text{ref}})}\right)}{1 + e^{-k_m(t-\tau)}}, \tag{3}$$

where P_{ref} (or E_{ref}) is the parameter related to microwave power (or ethanol concentration), respectively, and where the mangiferin yield Y is twice the initial value.

It is important to clarify that the extraction process is the result of an interaction between *Curcuma amada* (mangiferin) and ethanol, causing the kinetic dependence to be of the second order. During extraction, concentration of mangiferin (solute) increases and goes to saturation

Table 1 Test of between-subjects effects: a two-way analysis of variance with microwave power and extraction time as independent factors

Source		Type III sum of squares	*df*	Mean square	*F*	Sig.	Noncent. parameter	Observed power[a]
Microwave power	Hypothesis error	10.067	5	2.013	316.937	0.0	1,584.685	1.0
		0.762	120	6.353E–03[b]				
Extraction time	Hypothesis error	1.464	24	6.101E−02	9.604	0.0	230.504	1.0
		0.762	120	6.353E–03[b]				
Microwave power × extraction time	Hypothesis error	20.906	1	20.906	342.642	0.0	342.642	1.0
		1.464	24	6.101E–02[c]				

Dependent variable: concentration of extracted mangiferin

[a] Computed using alpha = 0.05

[b] MS (Error)

[c] MS (TMIN)

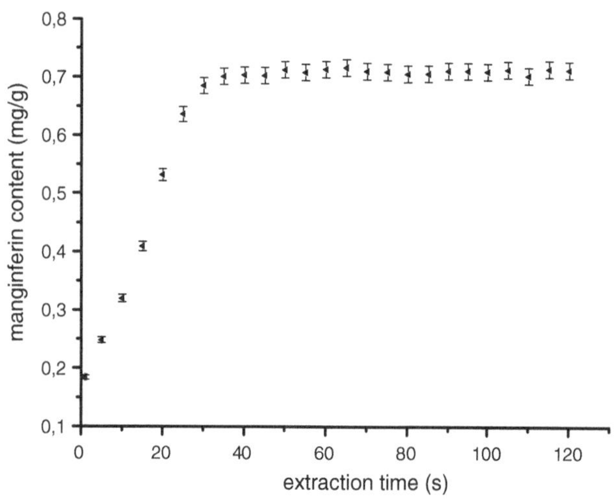

Fig. 3 Influence of extraction time on the yield of mangiferin content. Extraction condition: pre-leaching time—1 min, microwave power—550 W and ethanol concentration—100 %. The results are expressed as means of yield ± SD

although it is not yet extracted. However, if excess ethanol (solvent) is added to the solution, the extraction of mangiferin occurs. This reduces the apparent kinetic dependence from a second-order rate equation to a pseudo-first order rate equation (see Eqs. 1, 2).

Results and discussions

Figure 1a, b shows the chromatographic profile of the mangiferin from *Curcuma amada* after MAE and standard mangiferin, respectively. The retention time of 6.51 min obtained from the extract agreed well with the standard verifying the presence of mangiferin in *Curcuma amada* extract.

Effect of microwave power

Figure 2a–e shows the effect of microwave power on mangiferin content with extraction time; the symbols represent the experimental data while the continuous curves represent the model fit. It can be clearly seen in these figures that there is a steady increase in mangiferin content up to 50 s at the power range from 250 to 550 W after which it reaches a threshold value. However, at 900 W there is a significant decrease in the mangiferin content and the yield decreases drastically, as was observed by the response at 900 W (not shown). This is in accordance with the observations of our previous study (Padmapriya et al. 2012) where a similar response was obtained for 600 W. The mangiferin content of *Curcuma amada* in the control sample, i.e. before MAE, was 0.0046 mg/g. The mangiferin content at 250 and 550 W after 50 s of microwave extraction is found to be 0.0146 and 0.7161 mg/g, respectively. This manifold increase in mangiferin content is maximum (more than 150 times higher than the control sample) at 550 W after 50 s of extraction compared with 250 W (around three times higher than the control sample) for the same extraction time. This accelerated extraction of mangiferin by increasing microwave power can be correlated to the direct effects of microwave energy on molecules by ionic conduction and dipole rotation which result in power dissipated in volumetric basis inside the solvent and plant material which generate molecular movement and heating. Microwave irradiation energy disrupts the bonds because of microwave-induced dipole rotation of molecules and migration of dissolved ions. Microwave irradiation energy can enhance the penetration of solvent into the matrix and deliver efficiently to materials through molecular interaction with the electromagnetic field and thus offer a rapid transfer of energy to the solvent and

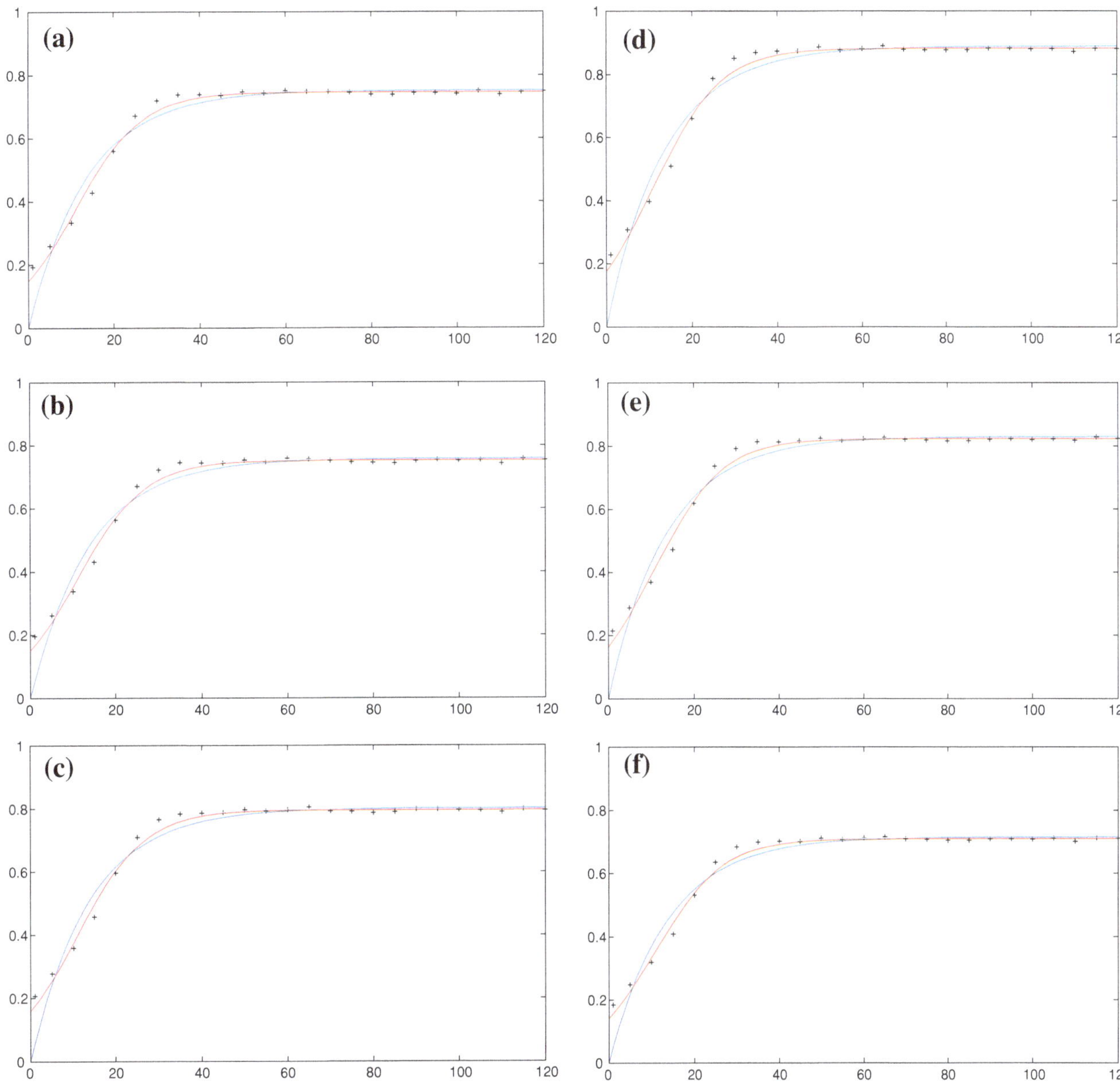

Fig. 4 Temporal evolution of the effect of ethanol concentration (as an independent variable) on the yield of mangiferin content extracted from *Curcuma amada* at various experimental design points (*x* axis: time in s; *y* axis: yield in mg/g). The experimental data (*symbols*) fitted to the 2-parameter model (Eq. 1, *blue*) and 3-parameter logistic model (Eq. 2, *red*) are shown in Fig. 4a–f for 50, 60, 70, 80, 90 and 100 %, respectively

Table 2 Test of between-subjects effects: a two-way analysis of variance with ethanol concentration and extraction time as independent factors

Source		Type III sum of squares	*df*	Mean square	*F*	Sig.	Noncent. parameter	Observed power[a]
Ethanol conc.	Hypothesis error	0.373	5	7.455E−02	392.693	0.0	1,963.466	1.0
		2.278E−02	120	1.898E−04[b]				
Extraction time	Hypothesis error	4.499	24	0.187	987.452	0.0	23,698.838	1.0
		2.278E−02	120	1.898E−04[b]				
Ethanol conc. × extraction time	Hypothesis error	72.827	1	72.827	388.509	0.0	388.509	1.0
		4.499	24	0.187[c]				

Dependent variable: concentration of extracted mangiferin

[a] Computed using alpha = 0.05

[b] MS (Error)

[c] MS (TMIN)

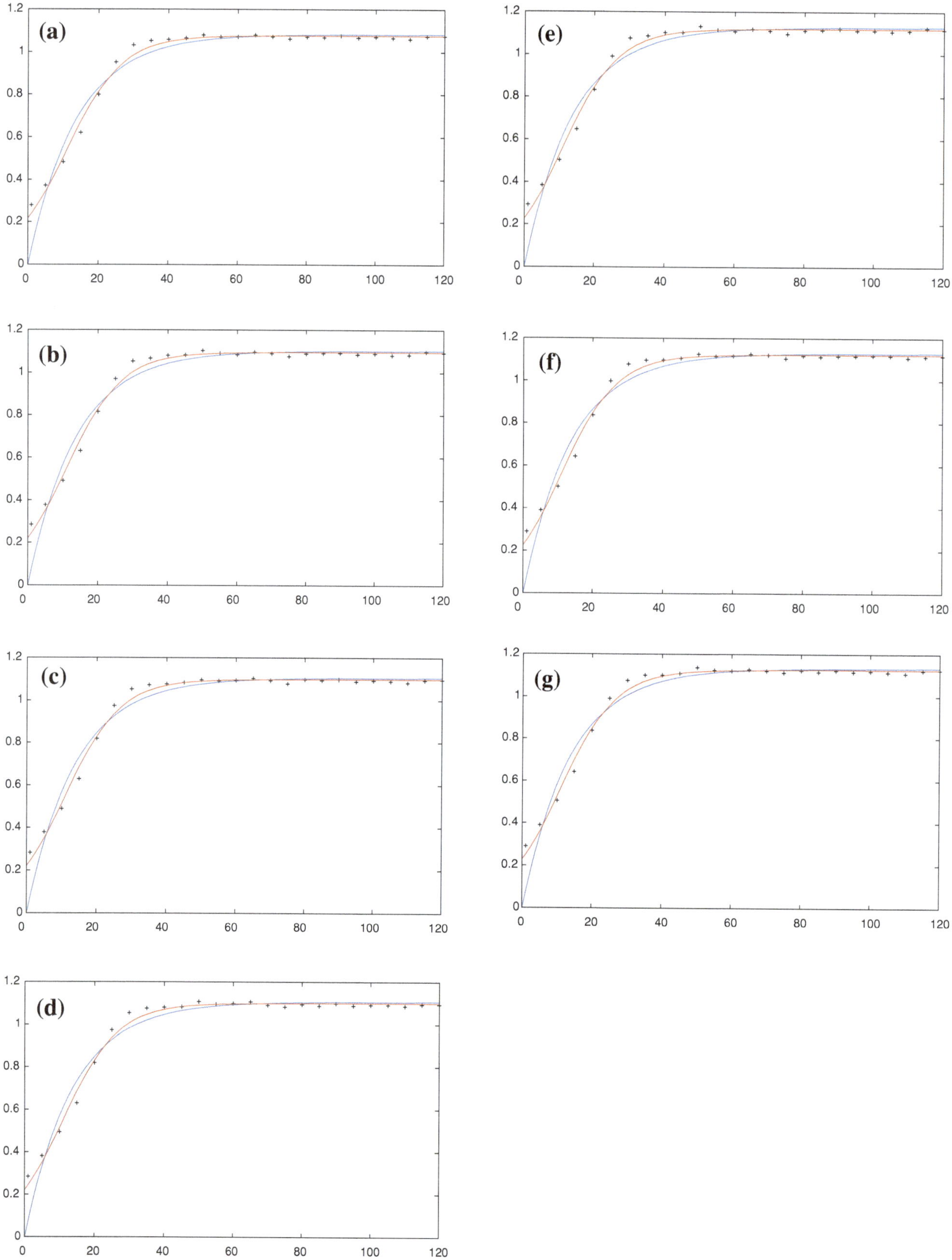
(a)
(b)
(c)
(d)
(e)
(f)
(g)
0 0.2 0.4 0.6 0.8 1 1.2
0 20 40 60 80 100 120

◀**Fig. 5** Temporal evolution of the effect of pre-leaching time (independent variable) on the yield of mangiferin content extracted from *Curcuma amada* at various experimental design points (*x* axis: time in s; *y* axis: yield in mg/g). The experimental data (*symbols*) fitted to the 2-parameter model (Eq. 1, *blue*) and 3-parameter logistic model (Eq. 2, *red*) are shown in Fig. 5a–g for 1, 5, 10, 15, 20, 25 and 30 min respectively

matrix, allowing the dissolution of components to be extracted. The steep decrease in mangiferin content at 900 W is due to the rapid degradation of mangiferin at higher microwave power range. As the experiments are conducted in dry matter, as is usually the case (Mandal et al. 2007), chances of degradation due to drying or evaporation at a higher microwave power intensity are ruled out. Similar results of decrease in extraction yield of astragalosides from *Radix astragali* at high power due to disorderly molecular interactions have been reported in the optimization study of MAE of four main astragalosides in *Radix astragali* (Yan et al. 2010).

Results of a two-way ANOVA with extraction time and microwave power as independent variables are given in Table 1. The mangiferin content in *Curcumaamada* is significantly dependent on microwave power and extraction time as well as their interaction. Newman–Keuls test suggest that the mangiferin content is significant at 550 W, validating our experimental results of extracting the highest mangiferin content at 550 W from *Curcuma amada*. Student's independent *t* test further confirms that both microwave power and extraction time have a significant effect on the mangiferin content.

Effect of extraction time

As seen in Fig. 3, mangiferin content increases significantly with the increase in extraction time from 1 to 50 s before reaching a steady state. The mangiferin content of *Curcuma amada* kept in a pre-leaching time of 1 min and extracted at 550 W for 50 s is found to be maximum around 0.7121 mg/g. Beyond 50 s of extraction time, no significant increase in mangiferin content is observed. Similar observations are also reported for MAE of artemisinin in from *Artemisia annua* (Pan et al. 2007) and tanshinones from *Salvia Miltiorrhiza* Bunge (Hao et al. 2002).

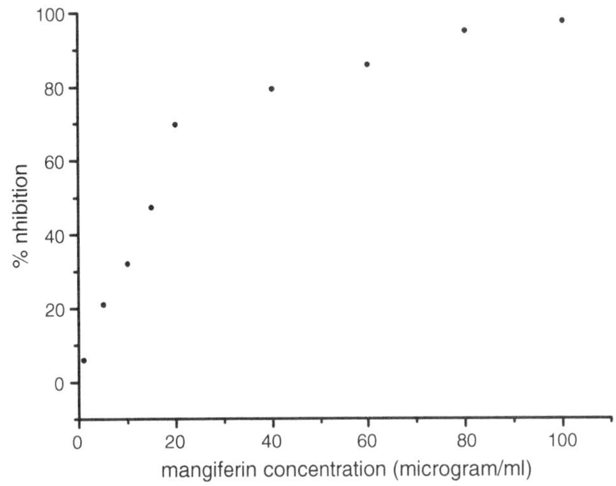

Fig. 6 DPPH radical scavenging activity of mangiferin obtained from microwave-assisted extraction of *Curcuma amada* at microwave power—550 W, pre-leaching time—20 min, extraction time—50 s and ethanol concentration—80 %

Effect of solvent concentration

Preliminary screening experiments (not reported in this study) with different organic extraction solvents such as acetone, acetonitrile, methanol and ethanol have been carried out and it was observed that ethanol yielded significant mangiferin content. Ethanol undergoes less microwave absorption than water due to its lower dielectric loss value but the overall heating efficiency for the solvent will remain higher than that of water due to increased value of the dissipation factor. Extraction with aqueous ethanol has been reported in earlier studies since it has less

Table 3 Test of between-subjects effects: a two-way analysis of variance with pre-leaching time and extraction time as independent factors

Source		Type III sum of squares	*df*	Mean square	*F*	Sig.	Noncent. parameter	Observed power[a]
Pre-leaching time	Hypothesis error	3.711E−02	6	6.186E−03	254.053	0.0	1,524.317	1.0
		3.506E−03	144	2.435E−05[b]				
Extraction time	Hypothesis error	10.311	24	0.430	17,644.599	0.0	42,347.037	1.0
		3.506E−03	144	2.435E−05[b]				
Pre-leaching time × extraction time	Hypothesis error	166.914	1	166.914	388.523	0.0	388.523	1.0
		10.311	24	0.430[c]				

Dependent variable: concentration of extracted mangiferin

[a] Computed using alpha = 0.05

[b] MS (Error)

[c] MS (TMIN)

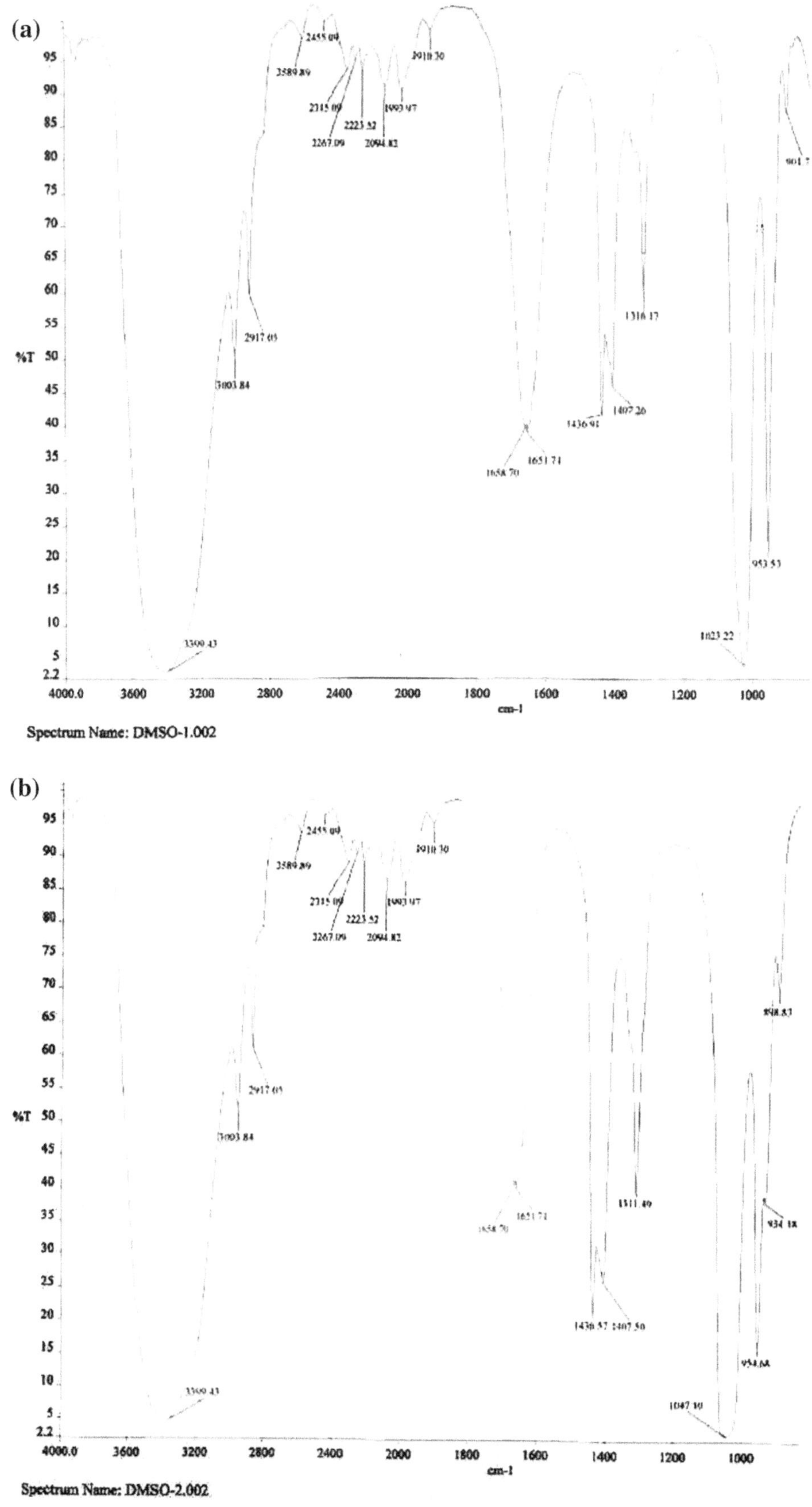

Fig. 7 FTIR spectrum of mangiferin extracted from **a** Curcuma amada by MAE at 550 W and **b** mangiferin standard (for peak values refer Table 1)

restrictions in food applications (Wardhani et al. 2010; Wang et al. 2010; Hemwimon et al. 2007). MAE of 2.5 g of dry *Curcuma amada* powder is carried out at microwave power of 550 W, pre-leaching time of 1 min and irradiation time of 1–120 s with aqueous ethanol as solvent. The effect of aqueous ethanol concentration on mangiferin content can be seen in Fig. 4a–f. The mangiferin content increases significantly with increase in ethanol concentration up to 80 % ethanol concentration; beyond 80 % ethanol concentration there is a decrease in mangiferin content. Dhobi et al. (2009) found similar results in their work related to optimization of MAE of bioactive flavonolignan-silibinin. A maximum mangiferin content of 0.8864 mg/g is obtained in 80 % ethanol concentration at extraction time of 50 s. One possible reason for the increased efficiency with 80 % ethanol might be due to the increase in swelling of plant material by presence of some amount of water, which increased the contact surface area between the plant matrix and the solvent. Presence of some amount of water can also increase the mass transfer process by increasing the relative polarity of the solvent, thus improving its solubilizing capacity. Similar results were reported by Li et al. (2004) during microwave-assisted solvent extraction and HPLC determination of effective constituents in *Eucommia ulmoides* Oliv.

Statistical results indicate that the mangiferin is positively correlated but insignificant with the ethanol concentration but significant with extraction time. The results of ANOVA are given in Table 2. Both the ethanol concentration and extraction time along with their interaction are significant with respect to the mangiferin content. Newman–Keuls test show that the mangiferin content is significant at higher ethanol concentration of 70, 80, 90 and 100 % with ethanol concentration of 80 % yielding the highest mangiferin content.

Table 4 FTIR peak values of mangiferin after microwave-assisted extraction at 550 W with standard mangiferin

Functional group	Wave number (cm^{-1}) mangiferin extracted from *Curcuma amada* by microwave-assisted extraction at 550 W	Wave number (cm^{-1}) mangiferin standard
O–H	3,399.43	3,399.43
C–H	2,917.05	2,917.05
(>C=O)	1,658.70	1,660.91
C=C	1,436.91	1,436.57
C–O	1,316.17	1,311.49
C–O–C	1,023.22	1,047

Effect of pre-leaching time

Figure 5a–g shows the effect of pre-leaching time on the yield of mangiferin content. Similar to Figs. 2a–e and 4a–f, the symbols represent the experimental data and the continuous curves represent the model fit. Pre-leaching time can be defined as the contact time between sample matrix and extracting solvent before microwave extraction. MAE

Table 5 Optimized parameter set using the two-parameter modified first order kinetic model (Eq. 1) and the corresponding statistical ratio of the experimental design points for the three independent variables: microwave power, ethanol concentration and pre-leaching time

Independent variables	Experimental design points	Model parameters		Statistical indicator
		Y_{max} (mg/g)	k_m (s^{-1})	Y_{RMS}/Y_{max} (−)
Microwave power	250 W	0.013715	0.094507	0.084504
	350 W	0.568469	0.069558	0.018957
	450 W	0.622861	0.069927	0.018408
	500 W	0.643451	0.065034	0.051907
	550 W	0.718277	0.073040	0.052717
Ethanol concentration	50 %	0.754853	0.073097	0.052993
	60 %	0.761361	0.072713	0.052911
	70 %	0.804594	0.072975	0.052548
	80 %	0.890682	0.073307	0.052608
	90 %	0.831694	0.073287	0.053215
	100 %	0.718277	0.073040	0.052717
Pre-leaching time	1 min	1.083663	0.072867	0.052288
	5 min	1.102309	0.073028	0.052639
	10 min	1.105546	0.072777	0.052213
	15 min	1.107016	0.072974	0.052566
	20 min	1.127495	0.072855	0.052249
	25 min	1.129510	0.072919	0.052386
	30 min	1.133223	0.072485	0.052057

Table 6 Optimized parameter set using the three-parameter delayed logistic model (Eq. 2) and the corresponding statistical ratio of the experimental design points for the three independent variables: microwave power, ethanol concentration and pre-leaching time

Independent variables	Experimental design points	Model parameters			Statistical indicator
		Y_{max} (mg/g)	k_m (s^{-1})	τ (s)	Y_{RMS}/Y_{max} (−)
Microwave power	250 W	0.013716	0.013716	9.078383	0.064274
	350 W	0.558402	0.156855	11.625592	0.040377
	450 W	0.611637	0.158482	11.513974	0.040776
	500 W	0.636598	0.121393	12.340408	0.023855
	550 W	0.713695	0.125821	10.917342	0.021411
Ethanol concentration	50 %	0.749934	0.126824	10.938109	0.022006
	60 %	0.756452	0.125319	10.96249	0.021600
	70 %	0.799469	0.125387	10.909826	0.021116
	80 %	0.885108	0.126128	10.886925	0.021164
	90 %	0.826444	0.126240	10.88443	0.021859
	100 %	0.713695	0.125821	10.917342	0.021411
Pre-leaching time	1 min	1.076805	0.125075	10.928752	0.020460
	5 min	1.095355	0.125485	10.911725	0.020907
	10 min	1.098343	0.125436	10.941557	0.020672
	15 min	1.100119	0.125104	10.919032	0.020930
	20 min	1.120260	0.124949	10.905220	0.020455
	25 min	1.122212	0.125592	10.921825	0.021031
	30 min	1.126029	0.124124	10.972533	0.020265

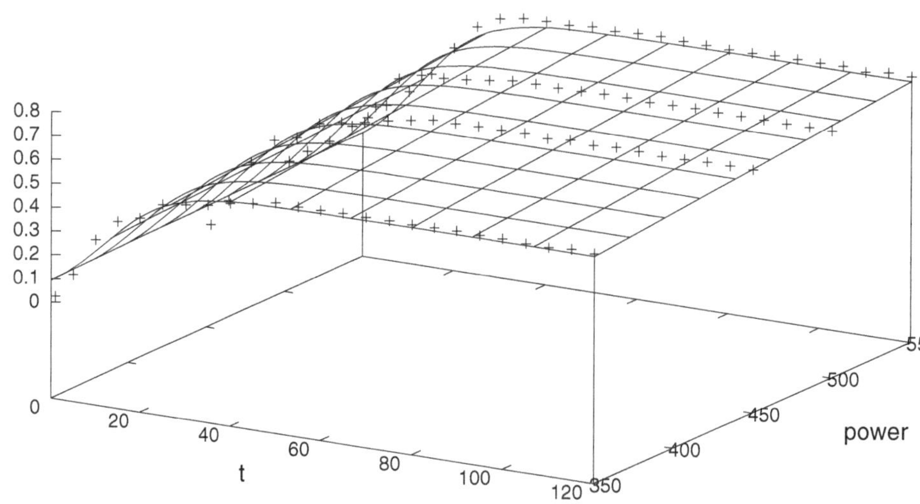

Fig. 8 Temporal evolution of yield on microwave power during the extraction process (in accordance with a modified logistic expression; see Eq. 3) for all the design points used in the experimental setup

of 2.5 g of dry *Curcuma amada* powder is carried out at microwave power of 550 W, 80 % ethanol concentration and irradiation time of 1–120 s for different pre-leaching time of 1–30 min. It is observed from these figures that with an increase in pre-leaching time from 1 to 20 min, there is an increase in mangiferin content. Beyond a pre-leaching time of 20 min, there is no noticeable increase in the yield of mangiferin content. It can be inferred that pre-leaching time of 20 min allows sufficient swelling of the plant matrix. This increased hydrated status of plant material helps in the bursting of the cell wall due to internal thermal stress and enlargement of the cellular pores, thus facilitating leaching of the target analyte. The results for ANOVA for pre-leaching time and extraction time as independent factors are given in Table 3. The Newman–Keuls test indicates that the pre-leaching time is not a significant factor contributing to the mangiferin content.

Figure 6 shows the DPPH radical scavenging activity of mangiferin extracted from *Curcuma amada* by MAE at the optimal condition of microwave power 550 W, pre-leaching time 20 min, extraction time 50 s and ethanol

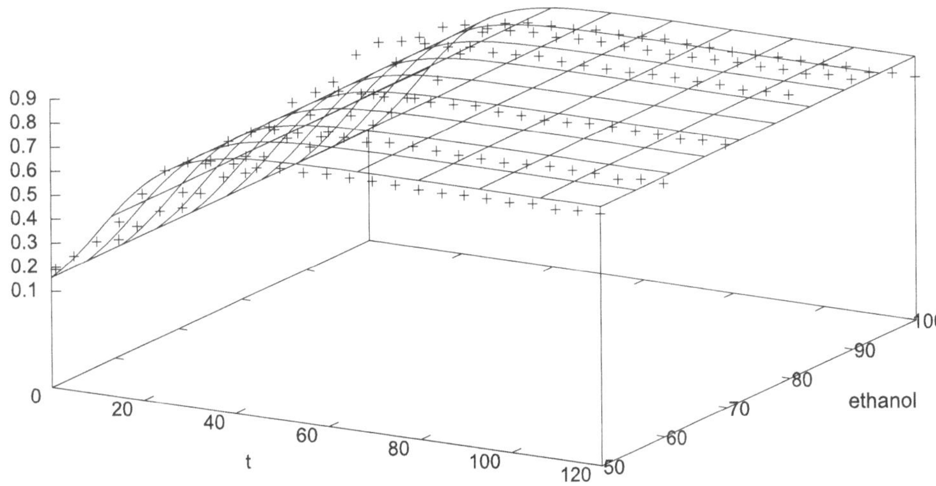

Fig. 9 Temporal evolution of yield on the ethanol concentration during the extraction process (in accordance with a modified logistic expression; see Eq. 3) for all the design points used in the experimental setup

concentration 80 %. It is observed that the IC50 value for mangiferin extracted from *Curcuma amada* was 17.04 μg/ml and the radical scavenging activity was directly proportional to the concentration of mangiferin with an inhibition of 97.65 % at 100 μg/ml. From this observation, it is clear that mangiferin obtained from MAE at 550 W, pre-leaching time of 20 min, extraction time of 50 s and 80 % ethanol concentration retained its antioxidant property. It is important to note here that Stoilova et al. (2005) had earlier established the antioxidant properties of mangiferin standard using DPPH radical scavenging activity of mangiferin.

FTIR analysis has proven to be a valuable tool for the characterization and identification of compounds or functional groups (chemical bonds) present in an unknown mixture of plant extracts (Eberhardt et al. 2007; Hazra et al. 2007). Figure 7 shows the FTIR spectrum of mangiferin extracted from *Curcuma amada* by MAE at 550 W and mangiferin standard. Six functional groups were identified: FTIR spectrum results of mangiferin after MAE showed peaks at 3,399 cm^{-1} and indicated the presence of secondary OH^- bond, peak at 2,917 cm^{-1} showed the presence of C–H anti-symmetric stretching, peak at 1,658.70 cm^{-1} indicated the presence C–O stretching, peak at 1,436.91 cm^{-1} indicated the presence of CH–CH bending and peak at 1,316.17 cm^{-1} indicated the presence of C–O bond. Peak at 1,023.22 cm^{-1} showed the presence of C–C stretching in the mangiferin structure. Comparing the FTIR analysis of mangiferin extracted by MAE and mangiferin standard (see Table 4) revealed the similarity and variation in the functional group. The absorption spectra showed that the C–O bond and C–O–C stretching of the mangiferin were affected during the extraction process.

The results of the validation using the two-parameter first-order kinetic model (Eq. 1) and using the three-parameter logistic model (Eq. 2) for microwave power (Fig. 2a–e), ethanol concentration (Fig. 4a–f) and pre-leaching time (Fig. 5a–g) are shown, respectively. As the response for 900 W was found to vary widely from the initial five responses (i.e. 250, 350, 450, 500 and 550 W) and did not follow a clear sequence, it was neglected while validating the kinetic model for microwave power with extraction time. To check the goodness of fit, the ratio of the root mean square (RMS) value to the maximum (limit) value of mangiferin content is considered. The optimized parameter set and the corresponding value of the statistical indicator Y_{RMS}/Y_{max} are summarized in Tables 5 and 6. The goodness of fit statistical indicator helps to determine how well the curve fits the data. The curve fits (based on Eq. 3) of the temporal evolution of yield on microwave power and ethanol concentration are shown in Figs. 8 and 9, respectively. The best-fit parameter values for (P_{ref}, k_m, τ) are found to be (759.42, 0.14, 11.68) and (2,315.22, 0.13, 10.91) using the NLLS Marquardt–Levenberg algorithm. The corresponding indicator Y_{RMS}/Y_{max} equals 0.054598 and 0.070991, respectively, for microwave power and ethanol concentration, indicating a good fit.

Conclusions

Mangiferin was extracted from *Curcuma amada* using microwave-assisted extraction technique. Maximum mangiferin content of 1.1156 mg/g was obtained at microwave power of 550 W and extraction time of 50 s with 80 % ethanol as a solvent and pre-leaching time of 20 min and retained its antioxidant properties. The experimental profiles fitted into a two-parameter modified first-order kinetic model and a three-parameter modified logistic model with sufficient accuracy. The MAE of mangiferin from *Curcuma amada* using ethanol can be safely employed in food and medicinal industries as it is not only efficient from the industrial point of view, but also eco-friendly since it prevents environmental hazards. This indicates the usefulness and significance of MAE as a novel extraction technique in biotechnological applications.

Conflict of interest The authors declare that they have no conflict of interest.

References

Ara N, Nur H (2009) In Vitro antioxidant activity of methanolic leaves and flowers extracts of *Lippia alba*. Res J Med Med Sci 4:107–110

Camel V (2000) Microwave-assisted solvent extraction of environmental samples. TrAC Trends Anal Chem 19:229–248

Dean JR (2010) Extraction techniques in analytical sciences. Wiley, New York, pp 167–183

Dhobi M, Mandal V, Hemalatha S (2009) Optimization of microwave-assisted extraction of bioactive flavonolignan-silybinin. J Chem Metrol 3:13–23

Eberhardt TL, Li X, Shupe TF, Hse CY (2007) Chinese tallow tree (*Sapium sebiferum*) utilization: characterization of extractives and cell-wall chemistry. Wood Fiber Sci 39:319–324

Ferguson P, Harding M, Young J (2012) Green analytical chemistry. In: Zhang W, Cue BW (eds) Green techniques for organic synthesis and medicinal chemistry. Wiley, Chichester

Hao JY, Han W, Huang S, Xue BY (2002) Microwave-assisted extraction of artemisinin from *Artemisia annua* L. Sep Purif Technol 28:191–196

Hazra KM, Roy RN, Sen SK, Laska S (2007) Isolation of antibacterial pentahydroxy flavones from the seeds of *Mimusops elengi* Linn. Afr J Biotechnol 6:1446–1449

Hemwimon S, Pavasant P, Shotipru A (2007) Microwave-assisted extraction of antioxidative anthraquinones from roots of *Morinda citrifolia*. Sep Purif Technol 54:44–50

Huie CW (2002) A review of modern sample preparation techniques for the extraction and analysis of medicinal plants. Anal Bioanal Chem 373:23–30

Jatoi SA, Kikuchi A, Gilani SA, Watanabe KN (2007) Phytochemical, pharmacological and ethnobotanical studies in mango ginger (*Curcuma amada* Roxb.; Zingiberaceae). Phytother Res 21:507–516

Joubert E, Manley M, Botha M (2008) Evaluation of spectrophotometric methods for screening of green rooibos (*Aspalathus linearis*) and green honeybush (*Cyclopia genistoides*) extracts for high levels of bio-active compounds. Phytochem Anal 19:169–178

Li H, Bo C, Zhang ZH, Yao SZ (2004) Focused microwave-assisted solvent extraction and HPLC determination of effective constituents in *Eucommia ulmoides* Oliv. (*E. ulmoides*). Talanta 63:659–665

Mandal V, Mohan Y, Hemalatha S (2007) Microwave-assisted extraction—an innovative and promising extraction tool for medicinal plant research. Pharmacogn Rev 1:7–18

Morgan SL, Deming SN (1974) Simplex optimization of analytical chemistry methods. Anal Chem 46:1170–1181

Muruganandan S, Gupta S, Kataria M, Lal J, Gupta PK (2002) Mangiferin protects the streptozotocin-induced oxidative damage to cardiac and renal tissues in rats. Toxicology 176:165–173

Omenn GS, Goodmann GE, Thornquist MD, Balmes J (1996) Effects of a combination of beta-carotene and vitamin A on lung cancer and cardiovascular disease. N Engl J Med 334:1150–1155

Padmapriya K, Dutta A, Chaudhuri S, Dutta D (2012) Microwave-assisted extraction of mangiferin from *Curcuma amada*. 3Biotech 2:27–30

Pan X, Niu G, Liu H (2007) Microwave-assisted extraction of tanshinones from *Salvia miltiorrhiza* Bunge with analysis by high-performance liquid chromatography. J Chromatogr A 922:371–375

Qu X, Wu CFJ (2005) One-factor-at-a-time designs of resolution V. J Stat Plan Inference 131:407–416

Romanik G, Gilgenast E, Przyjazny A, Namiesnik J (2007) Techniques of preparing plant material for chromatographic separation and analysis. Biochem Biophys Methods 70:253–261

Sasikumar B (2005) Genetic resources of Curcuma: diversity, characterization and utilization. Plant Genet Resour Charact Util 3:230–251

Shankaracharya NB (1982) Mango ginger. Indian Cocoa Arecanut Spices J 5:78–80

Singh S, Kumar JK, Saikia D, Shanker K (2010) A bioactive labdane diterpenoid from *Curcuma amada* and its semisynthetic analogues as antitubercular agents. Eur J Med Chem 45:4379–4382

Stoilova I, Gargova S, Stoyanova A, Ho L (2005) Antimicrobial and antioxidant activity of the polyphenol mangiferin. Herba Polonica 51:37–44

Wang YL, Xi GS, Zheng YC, Miao FS (2010) Microwave-assisted extraction of flavonoids from Chinese herb *Radix puerariae* (*Ge Gen*). J Med Plant Res 4:304–308

Wardhani DH, Vázquez JA, Pandiella SS (2010) Optimisation of antioxidants extraction from soybeans fermented by *Aspergillus oryzae*. Food Chem 118:731–739

Yan MM, Liu W, Fu YJ, Zu YG (2010) Optimisation of the microwave-assisted extraction process for four main astragalosides in *Radix astragali*. Food Chem 119:1663–1670

Yukalov VI, Yukalova EP, Sornette D (2009) Punctuated evolution due to delayed carrying capacity. Physica D 238:1752–1767

2

Identification of surfactins and iturins produced by potent fungal antagonist, *Bacillus subtilis* K1 isolated from aerial roots of banyan (*Ficus benghalensis*) tree using mass spectrometry

Khyati V. Pathak · Hareshkumar Keharia

Abstract The banyan endophyte, *Bacillus subtilis* K1, produces a complex mixture of lipopeptides exhibiting potent antifungal activity. These lipopeptides were purified by high-performance liquid chromatography and analyzed using MALDI-TOF-MS as well as liquid chromatography coupled with ESI-MS. A heterogenous mixture of lipopeptides belonging to three different families of cyclic lipopeptides, viz., fengycins, iturins and surfactins, was detected in the cell-free extracellular extract of *B. subtilis* K1. The detailed mass spectrometric characterization revealed the presence of four variants of iturin A and three variants of iturin C varying in the β-amino fatty acid chain length from C_{13} to C_{17}. The MS/MS of monovalent alkali metal ion adducts (Na and K) of iturin suggested the Glu^4 as a binding site for metal ion. The LC-ESI-MS/MS analysis of surfactins enabled the identification of seven surfactin variants with the variations in Val/Ile/Leu at position 4 and C_{13}–C_{17} β-hydroxy fatty acids. This study demonstrates the application of tandem mass spectrometry in identification of closely related lipopeptides from a heterogenous mixture obtained from a natural source. Furthermore, this is the first report of an endophytic bacillus strain co-producing so many variants of surfactins and iturins.

K. V. Pathak · H. Keharia (✉)
BRD School of Biosciences, Sardar Patel University, Sardar Patel Maidan, Satellite Campus, Vadtal Road, P.O. Box 39, Vallabh Vidyangar 388120, Gujarat, India
e-mail: haresh970@gmail.com

K. V. Pathak
e-mail: khyati835@gmail.com

Keywords Endophyte · *B. subtilis* K1 · Surfactin · Iturin · Mass spectrometry

Introduction

Endophytes are microorganisms which reside within plant tissues without causing any harm to host and they represent potential source of novel bioactive metabolites (Strobel 2006). In search for endophytes with potential biocontrol activity, we isolated bacilli strains exhibiting potent antifungal activity from hanging roots of Banyan tree. The *B. subtilis* K1, isolated from the aerial roots of banyan tree, was found to inhibit growth of *Aspergillus niger*, *Alternaria brunsii*, *Fusarium oxysporum*, *Cladosporium herbarum* 1112, *Candida albicans* and *Lasiodiplodia theobromae* ABFK1 (Pathak et al. 2012; Pathak and Keharia 2013). The mass spectrometric analysis of crude lipopeptide extract from cell-free culture supernatant of *B. subtilis* K1 revealed that it was a complex mixture of iturins, surfactins and fengycins. This prompted us to characterize these cyclic lipopeptides employing tandem mass spectrometry.

Bacilli in general are known for their ability to produce wide array of broad spectrum antibiotics synthesized either ribosomally or non-ribosomally (Katz and Demain 1977; Stein 2005; Ongena and Jacques 2007). Among these, the non-ribosomally synthesized lipopeptide antibiotics are well documented for their antifungal, antibacterial, antiviral, insecticidal, antitumor and surfactant activities (Winkelmann et al. 1983; Vanittnakom et al. 1986; Vollenbroich et al. 1997; Kracht et al. 1999; Stein 2005; Ongena and Jacques 2007; Kim et al. 2007, 2010; Geetha et al. 2010). The production of surfactins, iturins and fengycins by various strains of *B. subtilis* has been reported by several researchers (Winkelmann et al. 1983; Vanittnakom et al. 1986; Vollenbroich et al.

1997; Kracht et al. 1999; Kim et al. 2007; Nagorska et al. 2007; Ongena and Jacques 2007; Kim et al. 2010; Geetha et al. 2010). The lipopeptides belonging surfactin family are β-hydoxy hepta cyclic depsipeptides with possibilities of Ala, Val, Leu or Ile amino acid variations at positions 2, 4, and 7 in cyclic depsipeptide moiety and C_{13} to C_{16} variation in β-hydroxy fatty acid chains (Peypoux et al. 1994; Kowall et al. 1998; Hue et al. 2001). Surfactin is well known for its extraordinary surfactant activity and has also been demonstrated to possess antiviral, antitumor and insecticidal activities (Ongena and Jacques 2007, Vollenbroich et al. 1997; Kracht et al. 1999; Kim et al. 2010; Geetha et al. 2010). Iturins are broad spectrum potent antifungal heptapeptides cyclized by amide bond formed between α-COO group of seventh amino acid and β-NH_2 group of β-amino fatty acid (βAA), which in turn is peptide bonded through its α-COO group to N-terminal amino acid. The members of iturin family exhibit heterogeneity at 1, 4, 5, 6 and 7 amino acid position/s in the peptide moiety as well as in the βAA length, which varies from 14 to 17 carbons. On the basis of variation of amino acids in peptide moiety, iturins have been classified as: iturin A, iturin C, iturin D, iturin E, bacillomycin D, bacillomycin F, bacillomycin L, bacillomycin Lc and mycosubtilin (Winkelmann et al. 1983; Peypoux et al. 1981, 1984, 1986; Isogai et al. 1982; Gong et al. 2006; Romero et al. 2007; Pecci et al. 2010). Among all the iturins, iturin A has been found to be most potent antifungal lipopeptide and is secreted by most bacilli strains exhibiting strong as well as broad spectrum antifungal activity (Romero et al. 2007; Pecci et al. 2010). Different strains of bacilli exhibit diversity in production of cyclic lipopeptides, with most strains reported to produce lipopeptides belonging to only one family while few reported to be co-producers of lipopeptides belonging to two or all three families (Vater et al. 2002; Nagorska et al. 2007; Romero et al. 2007; Pecci et al. 2010). Furthermore, the fungal antagonistic activity of *Bacillus* sp. has been correlated with the amount as well heterogeneity in the production of lipopeptides (Nagorska et al. 2007; Ongena and Jacques 2007; Pecci et al. 2010).

The present study describes detailed mass spectrometric analysis of the surfactins and iturins produced by *B. subtilis* K1. To the best of our knowledge, this is a first report on mass spectrometric characterization of such a heterogenous mixture of surfactin and iturin variants co-produced by a single bacillus strain.

Materials and methods

Bacterial strain and production of lipopeptides

The *B. subtilis* K1 exhibiting broad spectrum antifungal activity was isolated from the aerial roots of banyan tree (Pathak et al. 2012). For production of lipopeptides, an inoculum was prepared by inoculating a single colony of *B. subtilis* K1 into 50 mL of Luria broth in 250-mL Erlenmeyer flask and incubating at 30 °C for 12 h. The 12-h-old inoculum was used to seed 250 mL of glucose yeast extract (GY medium: composition in g/L: glucose, 10.00; yeast extract, 1.00; $KH_2PO_4{\cdot}2H_2O$, 1.00; $K_2HPO_4{\cdot}2H_2O$, 1.00; $MgSO_4{\cdot}7H_2O$, 0.20; $CaCl_2{\cdot}2H_2O$, 0.02; $FeSO_4{\cdot}2H_2O$, 0.05) medium in 1,000-mL flask to an initial $OD_{(600\ nm)}$ of ~0.05 and incubated at 30 °C for 72 h on orbital shaker (150 rpm).

Extraction and purification of lipopeptides

The 72-h-old fermentation broth of *B. subtilis* K1 was centrifuged at ~10,000×*g* at 4 °C for 20 min. The supernatant was collected and its pH was lowered to pH 2 using 6 N HCl and incubated at 4 °C for 4 h in order to precipitate lipopeptides. The acid precipitates were recovered by centrifugation at ~10,000×*g* at 4 °C for 20 min, and pellet thus obtained was solubilized in anhydrous methanol (Pathak et al. 2012). The methanolic lipopeptide extract was further concentrated using vacuum evaporator (Buchi Switzerland).

The lipopeptides were purified by HPLC employing a semi-preparative reverse phase C-18 HPLC column (4.6 × 250 mm, 10 m particle size, 90 pore size) using methanol/water/0.1 % trifluroacetic acid solvent system. The elution of lipopeptides was done using a gradient of 80–95 % methanol (v/v) for 50 min, 95 % methanol (v/v) for 5 min and 95–80 % for 5 min (Pathak et al. 2012). The lipopeptide fractions were then analyzed by mass spectrometry.

Mass spectrometry of lipopeptides

MALDI-TOF mass spectrometry (MALDI-TOF-MS)

All the lipopeptide fractions obtained by reverse phase HPLC were subjected to MALDI-TOF-MS analysis using Ultraflex TOF/TOF spectrometer (Bruker Daltonics, Billericia, MA, USA and Bremen, Germany). The lipopeptide fractions were mixed with equal volume of α-cyano-4-hydroxy-cinnamic acid matrix and spot applied on the target plate. The mass spectrometric (MS and MS/MS) data acquisition was carried out as per the method described previously (Pathak et al. 2012).

Liquid chromatography electrospray ionization MS (LC–ESI–MS)

LC–ESI–MS spectra were acquired over the mass range of *m/z* 50–2,000 by passing the analyte through a Zorbax

300SB-C18 column (Santa Clara, CA, USA) (4.6 × 150 mm, 5 μm particle size) using MeOH/H_2O/0.1 % (v/v) formic acid on an Agilent 1100 HPLC system, coupled with HCT Ultra ETD II (Bruker Daltonics, Bremen, Germany) ion trap mass spectrometer in positive ion mode. A gradient (80–95 % MeOH in 50 min, 95 % MeOH for 5 min and 95–80 % MeOH in 5 min) pumped at a constant flow rate of 0.2 mL/min was used with a run time of 60 min. The data were acquired in an auto MS^2 mode using collision-induced dissociation method for fragmentation and data were analyzed by Data Analysis software ver. 4.0 (Bruker Daltonics, Bremen, Germany).

Minimum inhibitory concentration (MIC) of purified lipopeptides

Stock solutions (1 mg/mL) of iturin A2, iturin A2/A3/A4 and fengycin were prepared by dissolving lyophilized fractions in MeOH and analyzed for activity against *A. niger* 40211, *A. flavus*, *A. parasiticus*, *F. oxysporum* 1072, *Chrysosporium indicum*, *Candida albicans*, *Trichosporon* sp. 1110, *Alternaria brunsii* (2), and *Cladosporium herbarum* 1112. Fungal spore suspension was prepared by harvesting spores into sterile distilled water and the spore counts were determined using haemocytometer. In case of *Candida albicans* and *Trichosporon* sp. 1110, culture suspension was prepared by cultivating them in 50-mL potato dextrose broth under agitated condition (150 rpm) at 30 °C for 10–12 h. The cell numbers were determined using a hemocytometer and adjusted to 1×10^6 cells/mL by appropriate dilution with sterile distilled water. The MIC of pure lipopeptides were determined by double dilution technique against susceptible fungal cultures in sterile 96-well microtiter plates with each well containing 100 μL of potato dextrose broth. After dilution $\sim 10^2$ spores of test fungus were inoculated into each well. To control wells, corresponding aliquot of MeOH instead of sample was added. The plates were incubated for 24–48 h at 30 °C and MIC values for each fraction were determined against susceptible test fungi based on the highest dilution showing no growth.

Results and discussion

The MALDI-TOF MS analysis of HPLC purified fractions of crude antifungal extract obtained from cell-free supernatant of *B. subtilis* K1 revealed the presence of compounds with molecular mass ions in the range, *m/z* 1,028–1,109, *m/z* 994–1,065 and *m/z* 1,421–1,566, which were assigned as iturins, surfactins and fengycins, respectively (Kowall et al. 1998; Vater et al. 2002; Williams and Brodbelt 2004; Gong et al. 2006; Romero et al. 2007; Pathak et al. 2012). The characterization of iturins and surfactins produced by *B. subtilis* K1 is described below:

Identification and characterization of iturins

The iturins containing fraction with *m/z* range 1,028–1,109 eluted in HPLC peak no. P2–P6 and were further analyzed by MALDI-TOF MS/MS for peptide sequence determination. The MALDI-TOF MS of HPLC peaks P2–P6 corresponded the molecules with *m/z* 1,029.5 (P2), 1,043.5 (P3), 1,057.5 (P4), 1,071.5 (P5) and 1,072.5 (P6) (for HPLC chromatogram see publication Pathak et al. 2012). The molecules at *m/z* 1,029.5, 1,043.5, 1,057.5 and 1,071.5 differed in their masses by 14 or multiples of 14 Da while the *m/z* 1,072.5 differed from 1,071.5 by 1 Da, suggesting them to be members of a same family (Gong et al. 2006; Williams and Brodbelt 2004; Pecci et al. 2010). Two additional sets of *m/z* ions (with corresponding mass difference of 22 and 39 Da) were also detected which were putatively assigned as sodium adducts (*m/z* 1,051.5, 1,065.6, 1,079.5 and 1,093.5) and potassium adducts (*m/z* 1,067.5, 1081.5, 1,095.5 and 1,109.5) of corresponding iturins (Vater et al. 2002; Gong et al. 2006; Romero et al. 2007; Pecci et al. 2010). The appearance of sodium and potassium adducts along with their protonated species is a common feature in mass spectral studies of peptides under conventional MALDI as well as electron spray conditions (Eckart 1994; Hue et al. 2001; Rutenbach et al. 2001; Vater et al. 2002). The MS/MS of the cyclic peptides results into cleavage of either peptide bond or ester bond in the peptide backbone after gas phase protonation, which leads to formation of the protonated ring-opened forms with different linear sequences (Eckart 1994; Hue et al. 2001). In MALDI-MS/MS spectrum of protonated precursor ion at *m/z* 1,043.5, series of *a*-, *b*-, *b'*-, *y*-, *y'*-ions, immonium ions, ions with loss of H_2O or NH_3 or $H_2O + NH_3$ and internal fragment ions could be assigned (Fig. 1). These observed product ions were derived from two distinct modes of cleavage of the cyclopeptide ring, corresponding to initial fragmentation at the β-amino acid (βAA)-Asn site and Gln-Pro peptide bonds. The MS/MS spectrum of other protonated precursor ions at *m/z* 1,028.5, 1,043.5, 1,071.5 also yielded similar profile of product ions. The general sequence derived from the fragment ions formed upon cleavage at C-terminal end of βAA in MS/MS spectra of *m/z* 1,028.5, 1,043.5, 1,071.5 was N-Y–N-Q-P-N-S-βAA whereas, the sequence deduced from Pro^5 directed cleavage was P-N-S-βAA-N-Y-N-Q. A third series fragment ions at *m/z* 225.5 (βAA), *m/z* 339.6 (βAA-N), *m/z* 502.9 (βAA-N-Y-), *m/z* 616.6 (βAA-N-Y-N), *m/z* 744.9 (βAA-N-Y-N-Q) and *m/z* 201.7 (S-N), *m/z* 298.8 (S–N-P), *m/z* 426.6 (S-N-P-Q), *m/z* 541.0 (S-N-P-Q-N), *m/z* 703.7 (S-N-P-Q-N-Y-), *m/z* 817.6 (S-N-P-Q-N-Y-N) were found in the MS_2

spectrum of *m/z* 1,043.5. This fragment series suggested yet another ring cleavage site, i.e. amide bond between βAA and Ser[1] of iturin homolog. The ion at *m/z* 183.7 seemed to have formed due to loss of 41 Da $(–CH_2–C=O)^+$ from βNC_{14} and was assigned as βNC_{12}. The lipopeptide sequences derived from the MS/MS data for the molecules at *m/z* 1,029.5, 1,043.5, 1,057.5 and 1,071.5 were assigned as iturin A homologs, based on published literature (Vater et al. 2002; Isogai et al. 1982; Gong et al. 2006; Romero et al. 2007). The Pro-directed cleavage has been reported by Vater et al. (2002) and Gong et al. (2006). The cleavage of peptidyl-prolyl bond is ubiquitously observed in the tandem mass spectra of peptide containing proline (Papayannopoulos 1995; Rutenbach et al. 2001; Vater et al. 2002; Sabareesh et al. 2007). The basic nature of tertiary amide nitrogen of proline in its N-terminal peptide bond makes it susceptible towards low-energy fission and therefore it readily undergoes dissociation under gas phase (Papayannopoulos 1995; Sabareesh et al. 2007).

Gong et al. (2006) reported the ring opening in iturins due to cleavage of peptide bond between Tyr[2] and Asn[1] while, Vater et al. (2002) reported the ring opening due to cleavage of amide bond between βAA and Ser[7]. In our studies, we observed ring opening due to cleavages of amide bonds between βAA and Asn[1], βAA and Ser[7] as well as Gln[4] and Pro[5]. The variation in pattern of ring opening in iturin A homologues could be due to different methods employed for their characterization. Gong et al. (2006) used FAB-TOF/TOF and Vater et al. (2002) used PSD-MALDI, whereas we studied fragmentation of purified iturin A homologues using MALDI-LIFT. The MS/MS spectra of all the three protonated species of iturin showed the low-mass ion region containing immonium ions as well as internal fragment ions. The immonium ions of amino acids, Q (*m/z* 101.4), Y (*m/z* 136.5) and P (*m/z* 70.3) could be identified in MS/MS spectra of all these three iturins while the ion at *m/z* 339.7 could be assigned as an internal ion of tripeptide –NPQ (Fig. 1). The presence of immonium ions along with internal ions have also been reported by Gong et al. (2006).

In the *b*- and *y*-type fragmentation of *m/z* 1,029.5, 1,043.5 and 1,057.5 and 1071.5, *y*-ion series of all four $[M+H]^+$ precursor ions differed by 14 Da, whereas in case of *b′*- and *y′*-type fragmentation, $b_1'b\text{-}_3'$ ions remained same while $b_4'b\text{-}_7'$ ions differed from each other by 14 Da. The 14 Da difference observed in both cases suggested that

Fig. 1 MALDI-MS/MS spectrum of $[M+H]^+$ ion at *m/z* 1,043.5 from HPLC fraction P3

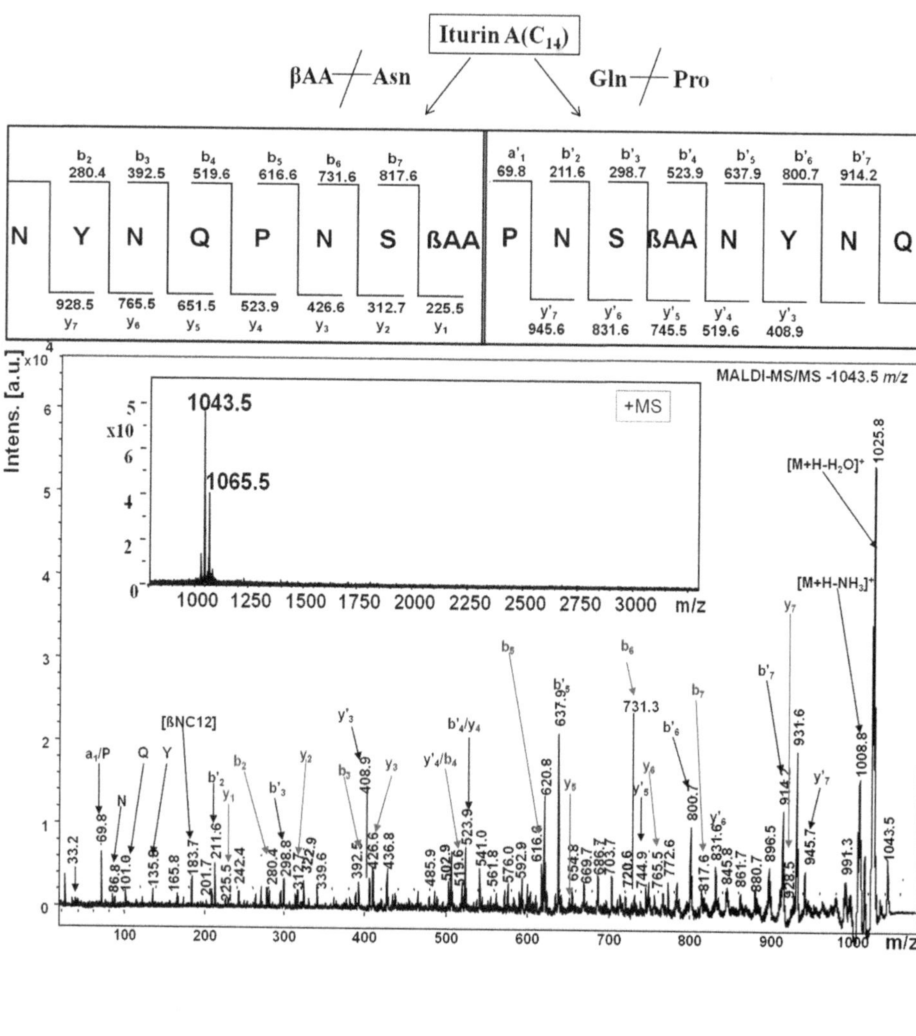

the metabolites with *m/z* 1,029.5, 1,043.5 and 1,057.5 and 1,071.5 differed only in carbon length of βAA and thus were assigned as iturin A homologues with 13-, 14- and 15- and 16-carbon βAA, respectively (Table 1). Iturin A homologues have been classified into eight different groups (iturin A_1–A_8) based on the heterogeneity of βAA (chain length as well as isoforms of βAA viz., normal, iso and anteiso) (Isogai et al. 1982; Gong et al. 2006). In present study, *B. subtilis* K1 was found to produce a mixture of iturins consisting of at least three homologues of iturin A, of which *m/z* 1,029.5, *m/z* 1,043.5 and *m/z* 1,071.5 could be assigned as iturins A1, A2 and A6, respectively. The iturins A3, A4 and A5 are isomers (normal, iso and anteiso with respect to βAA) with same molecular mass of 1,057.5 and thus, *m/z* 1,057.5 may be one of these three iturin A isomers, which cannot be distinguished by mass spectrometry. Along with these four protonated iturin A species, fifth species with *m/z* 1,072.5 of iturin was also detected. The molecular mass ion at *m/z* 1,072.5 differed from 1,071.5 by 1 Da and it is difficult to separate the species varying by 1 Da in MS spectrum. Moreover, in MALDI-MS/MS spectrum of *m/z* 1,072.5 most of the fragmentation belonging to iturin A species was at *m/z* 1,071.5 therefore, to determine the identity of iturin species at *m/z* 1,072.5, LC-ESI-MS/MS method was used. In the ESI-MS spectrum of molecular mass ions ranging within the *m/z* 1,040 to *m/z* 1,080 (eluted within the time range of 9–23 RT), three isotopic peak clusters with mono isotopic masses at *m/z* 1,043.5, 1,057.5, 1,072.5 were observed (Fig. S1). The isotopic distribution in these peak clusters seemed to be abnormal as the intensity of second isotopic peak (1,044.5, 1,058.5 and 1,072.5) in each cluster is higher than the first isotopic peak. Moreover, these compounds are falling in the *m/z* range where normal isotopic distribution is anticipated. This type of isotopic distribution in the fengycins has been previously described and has been attributed to the presence of two fengycins differing by 1 Da in same isotopic cluster (Pathak et al. 2012). Thus, the presence of iturins differing by 1 Da may be attributed to the abnormal distribution in the iturin isotopic clusters observed in present study. According to literature, iturin A and iturin C are the homologues varying from each other by 1 Da due to variation of Asn to Asp at amino acid position 1 (Park et al. 1995). To confirm 1-Da variation in the peptide sequences, each of these ions was selected for MS–MS analysis. Similarly, the sequences P-N-S-βAA-D-N-Q and Q-P-N-S-βAA-D-Y-N were deduced from the MS/MS analysis of *m/z* 1,058.5 (Fig. S2). The MS/MS spectra of ions at *m/z* 1044.5, 1,072.5 also yielded similar peptide sequences with variation of βAA fatty acids chain length. Using LC-MS/MS approach, molecular mass ions at *m/z* 1,044.5, 1,058.5 and 1,072.5 were identified as iturin C homologues with C14 to C16 βAA fatty acids (Park et al. 1995). To the best of our knowledge, this is the first report on the identification of iturin A and iturin C (differing by 1 Da) co-produced by a *B. subtilis* strain, using LC-ESI-MS/MS.

The precursor ions at *m/z* 1,051.5, 1,065.5, 1,079.5 and 1,093.5 with mass difference of 22 Da from ions at *m/z* 1,029.5 and 1,043.5, 1,057.5 and 1,071.5 were assigned as sodium adducts of corresponding iturin homologues. Each

Table 1 Iturin homologues secreted by *B. subtilis* K1

Mass (*m/z*)	Peptide sequence	Identification of iturins	References
1,029.5	Cyclo(βAA-N-Y-N-Q-P-N-S)	(C_{13}) Iturin A1$[M+H]^+$	This work (Gong et al. 2006)
1,051.5		(C_{13}) Iturin A1$[M+Na]^+$	
1,067.5		(C_{13}) Iturin A1$[M+K]^+$	
1,043.5	Cyclo(βAA-N-Y-N-Q-P-N-S)	(C_{14}) Iturin A2$[M+H]^+$	This work (Gong et al. 2006; Romero et al. 2007)
1,065.5		(C_{14}) Iturin A2$[M+Na]^+$	
1,081.5		(C_{14}) Iturin A2$[M+K]^+$	
1,057.5	Cyclo(βAA-N-Y-N-Q-P-N-S)	(C_{15}) Iturin A3/A4/A5$[M+H]^+$	This work (Romero et al. 2007)
1,079.5		(C_{15}) Iturin A3/A4/A5$[M+Na]^+$	
1,095.5		(C_{15}) Iturin A3/A4/A5$[M+K]^+$	
1,071.5	Cyclo(βAA-N-Y-N-Q-P-N-S)	(C_{16}) Iturin A6/A7$[M+H]^+$	This work (Romero et al. 2007)
1,093.5		(C_{16}) Iturin A6/A7$[M+Na]^+$	
1,109.5		(C_{16}) Iturin A6/A7$[M+K]^+$	
1,044.5	Cyclo(βAA-D-Y-N-Q-P-N-S)	(C_{14}) Iturin C1$[M+H]^+$	This work (Williams and Brodbelt 2004)
1,058.5	Cyclo(βAA-D-Y-N-Q-P-N-S)	(C_{15}) Iturin C2$[M+H]^+$	This work (Williams and Brodbelt 2004)
1,072.4	Cyclo(βAA-D-Y-N-Q-P-N-S)	(C16) Iutrin C3$[M+H]^+$	This work (Williams and Brodbelt 2004)

βAA β-amino acid

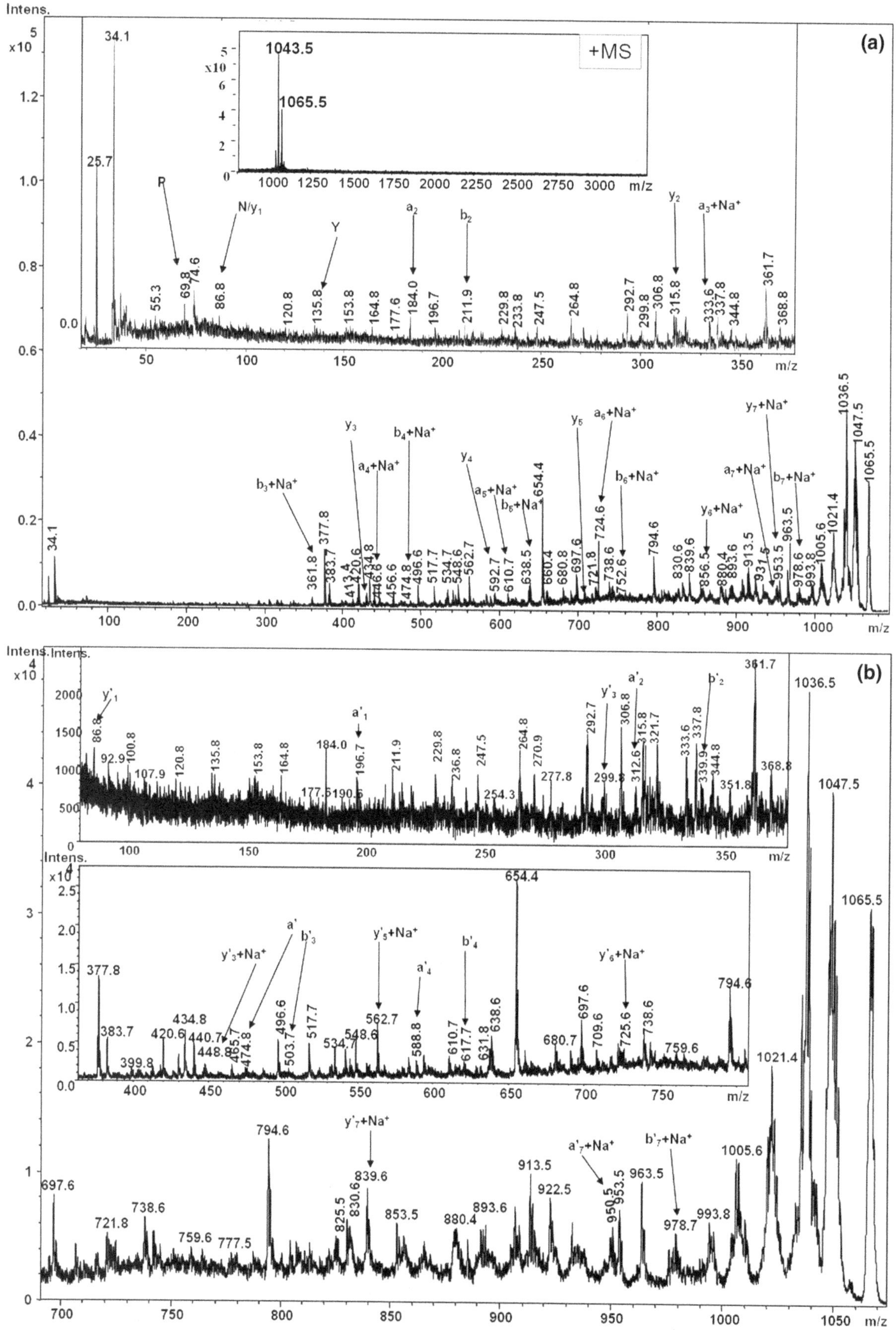
(a)
+MS
Intens.
1043.5
1065.5
34.1
25.7
P
N/y$_1$
Y
a$_2$
b$_2$
y$_2$
a$_3$+Na$^+$
361.7
b$_3$+Na$^+$
y$_3$
a$_4$+Na$^+$
b$_4$+Na$^+$
y$_4$
a$_5$+Na$^+$
b$_5$+Na$^+$
654.4
y$_5$
a$_6$+Na$^+$
b$_6$+Na$^+$
y$_6$+Na$^+$
a$_7$+Na$^+$
y$_7$+Na$^+$
b$_7$+Na$^+$
1036.5
1047.5
1065.5
1021.4
794.6
377.8
(b)
y'$_1$
a'$_1$
y'$_3$
a'$_2$
b'$_2$
y'$_3$+Na$^+$
a'
b'$_3$
y'$_5$+Na$^+$
a'$_4$
b'$_4$
y'$_6$+Na$^+$
y'$_7$+Na$^+$
a'$_7$+Na$^+$
b'$_7$+Na$^+$
m/z

◀**Fig. 2** MALDI-MS/MS spectrum of $[M+Na]^+$ ion at *m/z* 1,065.5 from HPLC fraction P3. The product ions, *a*-, *a* + Na-, *b*-, *b* + Na-, *x*-, *x* + Na-, *y*-, *y* + Na- assigned in (**a**) represents peptide sequence upon gas phase ring opening due to peptide bond cleavage between Ser_1-Asn_2 while the product ions, *a*′-, *a*′ + Na-, *b*′-, *b*′ + Na-, *x*′-, *x*′ + Na-, *y*′-, *y*′ + Na- and assigned in (**b**) shows peptide sequence upon ring opening between Ser^1-βAA

of these sodium adducts of iturin were subjected to MALDI-TOF MS/MS analysis and the fragment ions were analyzed. In MALDI-TOF MS/MS of $[M+Na]^+$ at *m/z* 1,065.5, *a*-, *a*′-, *b*-, *b*′-, *y*- and *y*′-fragment ions could be assigned (Fig. 2). The fragment ions at *m/z* 1,021.4, 1,036.8 and 1,047.8 observed in MS/MS spectrum of $[M+Na]^+$ ion at *m/z* 1,065.5 could be assigned as product ions formed upon loss of $-CONH_2$, –C=O and NH_3, respectively from the parent ion. The ions observed at *m/z* 794.6 $[CH_2$-CO-N-Y-N-Q-P-N + Na$]^+$, 839.6 [S-N-P-Q-N-Y-N +Na$]^+$ and 913.5 [N-P-Q-N-Y-N- βNC13 + Na$]^+$ were assigned as internal fragment ions. The sequences deduced from the fragmentation patterns of four sodium adducts of iturin A were N-P-Q-N-Y-N-βAA-S (Fig. 3a) and βAA-N-Y–N-Q-P–N-S (Fig. 3b), respectively. All the *b*-ions corresponding to both sodium adducts (*m/z* 1,065.5 and 1,079.5) were found to be same except b_6 and b_7 ion. The b_6 and b_7 ions of *m/z* 1,065.0 and 1,079.7 varied from each other by 14 Da, which may be attributed to the difference in the chain length of βAA by single $-CH_2$.

In MALDI-MS spectrum of metabolites eluted in HPLC peaks no. P3, the ion at *m/z* 1,081.5 was assigned as $[M+K]^+$ ion of iturin A. To confirm its identity, tandem mass spectrometric analysis was carried out. Figure 4 shows the MS/MS analysis of *m/z* 1,081.5 with *a*-, *a*′-, *b*-, *b*′-, *y*- and *y*′-fragment ions. Along with *b*- and *y*-ions some internal ions were also observed. The ions at *m/z* 337.6, 464.9, 562.8, 655.4, 677.4, 838.8, 1,046.2 and 1,064.1 could be assigned as [S-N-P+K$]^+$, [S-N-P-Q+K$]^+$, [βAA-S-N-P+K$]^+$, [N-P-Q-N-Y+K$]^+$, [N-βAA-S-N-P+K$]^+$,

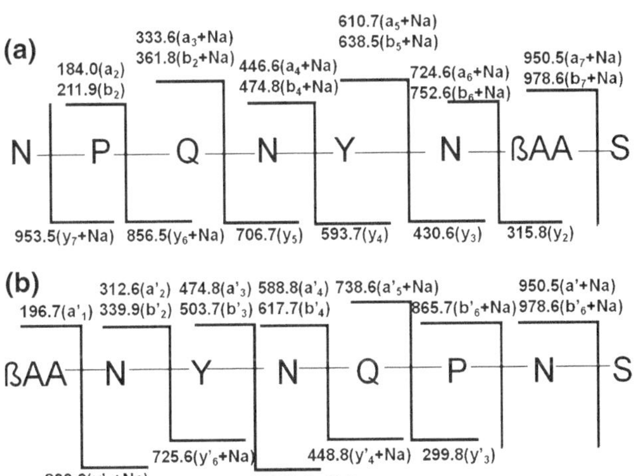

Fig. 3 Peptide sequence assignments of $[M+Na]^+$ ion at *m/z* 1,065.5

[Y-N-βAA-S-N-P+K$]^+$, $[M+K-CONH_2]^+$ and $[M+K-NH_3]^+$, respectively. On the basis of the fragmentation pattern observed, two ring cleavage sites: (1) between Asn^6-Ser^7 and (2) βAA-Ser^7 were predicted. The sequences derived from the fragment ion series corresponding to the predicted ring cleavages were N-P-Q-N-Y-N-βAA-S and βAA-N-Y-N-Q-P-N-S, respectively (Fig. 4).

The fragmentation pattern observed for sodium and potassium adducts of iturin A were different from the fragmentation pattern of protonated isoforms of iturin A. The major product ions from sodium adducts of iturin A were *a*-, *b*- and *y*-ions. This pattern of product ions from sodium adducts has also been reported by others (Papayannopoulos 1995; Rutenbach et al. 2001; Sabareesh et al. 2007). The fragment ions containing Glu^4 were detected as sodium or potassium adducts while the metal adduct form of fragment ions lacking Glu4 was not observed in respective MS/MS spectrum of sodium ion adducts (*m/z* 1,051.5, 1,065.6, 1,079.5 and 1,093.5) and potassium adducts (*m/z* 1,067.5, 1,081.5, 1,095.5 and 1,109.5), suggesting that possible site of metal interaction was Glu^4 in iturin A homologues (Papayannopoulos 1995; Sabareesh et al. 2007). The fragmentation pattern showed no cleavage at highly labile Glu^4-Pro^5 peptide bond in the sodiated and potassiated iturins. It has been reported that the complexation of iturin homologues with sodium/potassium ions confers the extra-stability to the otherwise highly susceptible bond (X-Pro bond) (Rutenbach et al. 2001). In our studies employing MALDI-TOF MS/MS, only mono-sodium/mono-potassium adducts of iturin A homologues could be detected. Rutenbach et al. (2001) also reported the predominant occurrence of mono sodium/potassium adducts of cyclic iturin A2, synthetic iturin A2 and its shorter analogues. They also proposed that the site for binding of alkali metal ion was present in one of the two β-turns on the interior of the peptide ring of natural iturin A molecule, with C=O oxygen as the chelating atom. The interaction of metal ions in the interior region of peptide ring may play a significant role in the formation of anion selective pores in the target cell membranes.

The isoforms of iturins exists which have same mass but differ in their peptide sequence for example iturin A(C-16) and mycosubtilin (C-16) differ in peptide sequence but have same mass (1,070 Da) (Isogai et al. 1982; Gong et al. 2006). The tandem mass spectrometry can be employed to easily identify such iturin isoforms, as each will yield a unique fragmentation finger print.

Identification and characterization of surfactin

The surfactin group of lipopeptides within the mass range *m/z* 994–1,065 eluted at long retention times from C18 reverse phase HPLC columns and were largely concentrated in the peaks P20–P23. Figure 5a shows the LC-ESI-MS spectrum of surfactins eluted with retention time of 4.0–4.2 min. In the LC-ESI-MS/MS spectrum of ion at *m/z* 1,036.7, series of *b*- and *y*-ions could be easily assigned (Fig. 5b), which resulted upon initial cleavage of protonated ester bond. The previous studies on cyclodepsipeptides have established the facile cleavage of protonated species at ester linkages (Sabareesh et al. 2007). The two peptide sequences of a molecular mass ion at *m/z* 1,036.7 could be deduced based on the fragmentation profile as: β-OH FA-Glu^1-Leu/Ile^2-Leu/Ile^3-Val^4-Asp^5-Leu/Ile^6-Leu/Ile^7 and β-OH FA-Glu^1-Leu/Ile^2-Leu/Ile^3-Leu/Ile^4-Asp^5-Leu/Ile^6-Leu/Ile^7. The common peaks at *m/z* 699.2 and 685.2 observed in CID spectrum of the surfactin ion at *m/z* 1,036 could be assigned as an internal protonated fragment ions [(H)Leu/Ile^2-Leu/Ile^3-Leu/Ile^4-Asp^5-Leu/Ile^6-Leu/Ile^7(OH)+H]$^+$ and [(H)Leu/Ile^2-Leu/Ile^3-Val^4-Asp^5-Leu/Ile^6-Leu/Ile^7(OH)+H]$^+$ with a gain of 18 Da, which is in agreement with observations reported by Hue et al. (2001). The presence of fragment ions with a gain of 18 Da could be due to cleavage of the peptide bond under gas phase upon double hydrogen transfer (Hue et al. 2001; Williams and Brodbelt 2004). The presence of *m/z* 685.2 and 699.2 fragment ions is ubiquitous in CID spectra of all surfactin [M+H]$^+$ homologues, thus this fragment ions represent characteristic marker ion for identification of surfactins. All these results were sufficient to establish the sequence identity of protonated surfactin homologues varying in their β-OH fatty acid chain lengths as well as exhibiting variation of Val to Leu/Ile at position 4 in cyclic depsipeptide ring.

The alkali metal ion adducts have been a common feature in the mass spectra of surfactins appearing as minor peaks along with their corresponding protonated precursor ions. The alkali metal ions bind to free carboxylic acid group in surfactin and form metal ion adducts. The sodium and potassium adducts of surfactin have been observed most frequently as these metal ions are ubiquitously present in nature (Hue et al. 2001; Vater et al. 2002). The additions of chloride salts of metal ions to these molecules

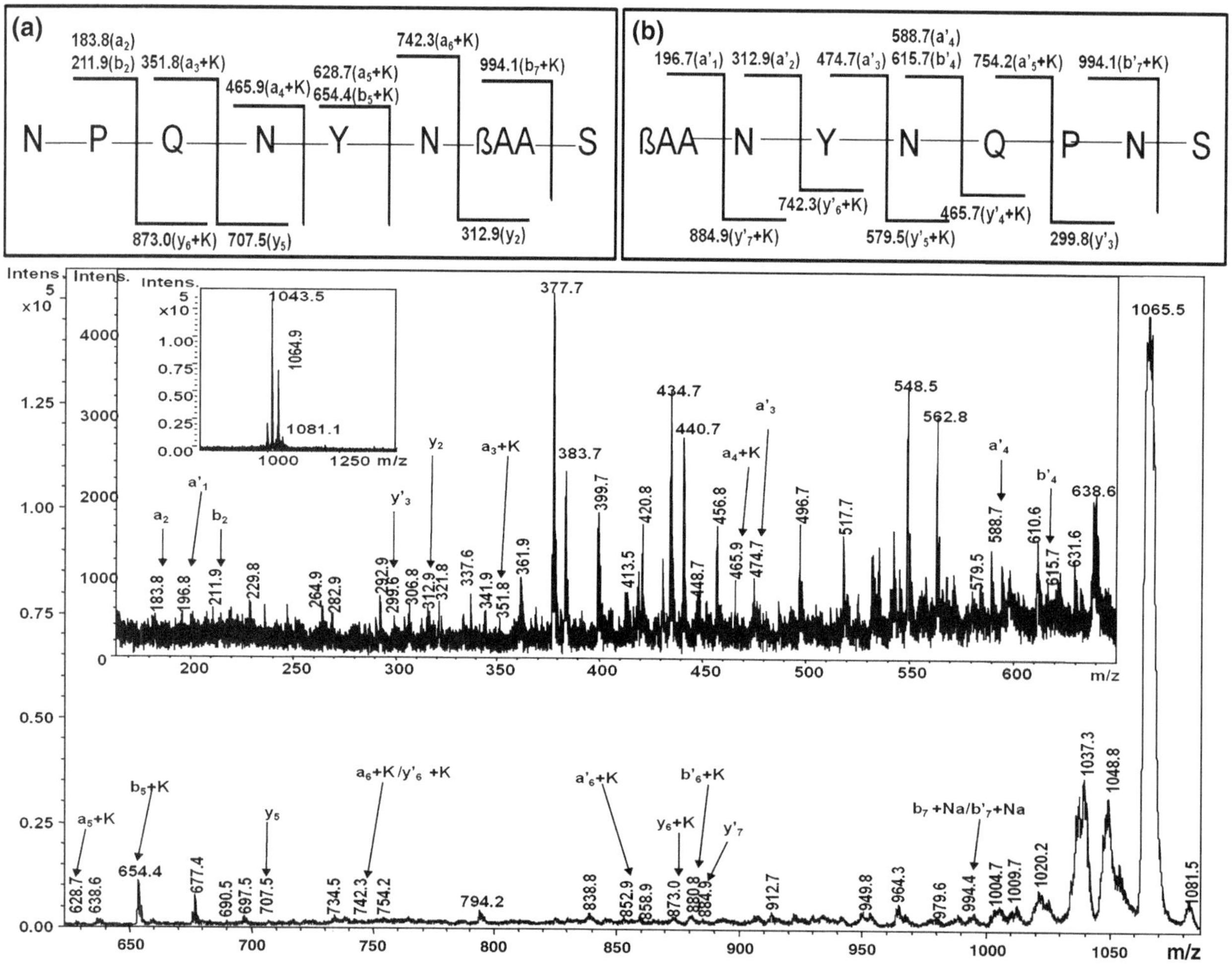

Fig. 4 MALDI-MS/MS spectrum of [M+K]$^+$ ion at *m/z* 1,081.5 from HPLC fraction P3

greatly enhance the relative intensity of corresponding metal ion species in the mass spectra (Eckart 1994; Hue et al. 2001). The precursor ions *m/z* 1,030.8, 1,044.7 and 1,073.0 (Fig. 6a) were assigned as the sodium ion adducts of surfactin homologues with 1,008.7, 1,022.7 and 1,050.7 Da mass, respectively, whereas the precursor ion at *m/z* 1,060.8 (Fig. 7) was assigned as potassium adduct of surfactin homologue with mass of 1,022.7 Da (Hue et al. 2001; Vater et al. 2002; Williams and Brodbelt 2004).

The *a*-, *b*- and *y*-series of fragments ions could be assigned in the CID spectrum of *m/z* 1,030.8 (Fig. 6b; Table 2). The peak at *m/z* 481.3 observed in spectrum of *m/z* 1,030.8 could be assigned as an internal fragment ion [Val^4-Asp^5-Leu^6-Leu^7(OH)+H+Na]$^+$, which is equivalent to y_4 + H_2O. The common peaks at *m/z* 707.5 and 594.3 were observed in this sodium adduct which could be assigned as internal peptide fragment ions [Leu^2-Leu^3-Val^4-Asp^5-Leu^6-Leu^7(OH)+H+Na]$^+$ and [Leu^3-Val^4-

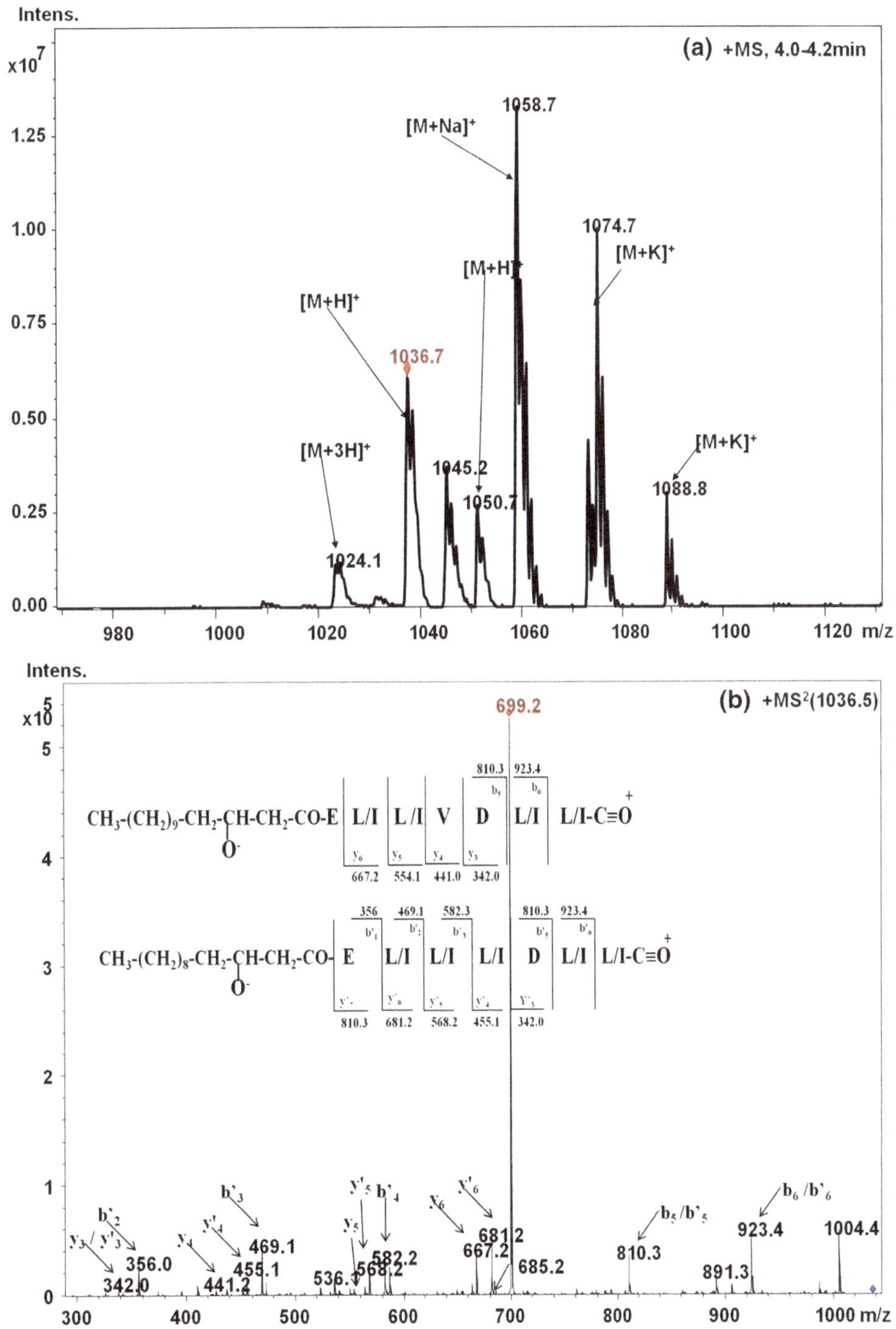

Fig. 5 **a** LC-ESI-MS spectrum of surfactin cluster separated at retention time period of 4.0–4.2 min, **b** LC-ESI-MS/MS spectrum of [M+H]$^+$ ion at *m/z* 1,036.5

Asp^{5}-Leu^{6}-Leu^{7}(OH)+H+Na]$^{+}$, respectively, which are equivalent to y_6 and y_5-ions with addition of water molecule. The presence of these intense fragment ions in ESI-MS/MS spectra of surfactin is a result of cleavage of peptide bond after double hydrogen transfer (Hue et al. 2001).

The CID analysis of 1,060.7 showed *a*- and *b*-type of fragment ions (Fig. 7). The fragment ions at *m/z* 723.4 and 610.3 could be assigned as internal ions [Leu^{2}-Leu^{3}-Val^{4}-Asp^{5}-Leu^{6}-Leu^{7}(OH)+H+K]$^{+}$ and [Leu^{3}-Val^{4}-Asp^{5}-Leu^{6}-Leu^{7}(OH)+H+K]$^{+}$, respectively. The ion at *m/z* 495.3 could be assigned as [Val^{4}-Asp^{5}-Leu^{6}-Leu^{7}(OH)+H+K]$^{+}$ (Fig. 7). The sequence of sodium and potassium adducts of surfactin homologues revealed Leu at seventh position. The sodium adducts of surfactin (*m/z* 1,030.8, 1,044.7 and 1,073.0) were assigned as homologues with 13-, 14- and 16-C β-OH FA, respectively while potassium adduct *m/z* 1,060.7 was assigned as surfactin homologue with 14-C β-OH FA. The heterogeneity in the Val/Leu/Ile at positions 2, 4 and 7 in the peptide sequence of surfactins is well established (Kowall et al. 1998; Hue et al. 2001; Vater et al. 2002; Williams and Brodbelt 2004). However, differentiation of Leu and Ile from each other is not an easy task using LC–ESI–MS/MS. Hue et al. (2001) employed LSI–MS technique to differentiate between Leu and Ile based upon detection of *d*- and *w*-ions generated upon side chain fragmentation. The *d*- and *w*-ions could not be assigned in the MS_2 spectra of surfactin homologues and thus, it was not possible to specify whether these homologues contained Leu/Ile at positions 2 and 7 in the peptide sequence.

The protonated and alkali metal ion species of surfactin isoforms characterized from the culture supernatants of *B. subtilis* K1 are summarized in Table 2 along with the peptide variants of surfactins reported by various researchers. The surfactin homologues produced by banyan

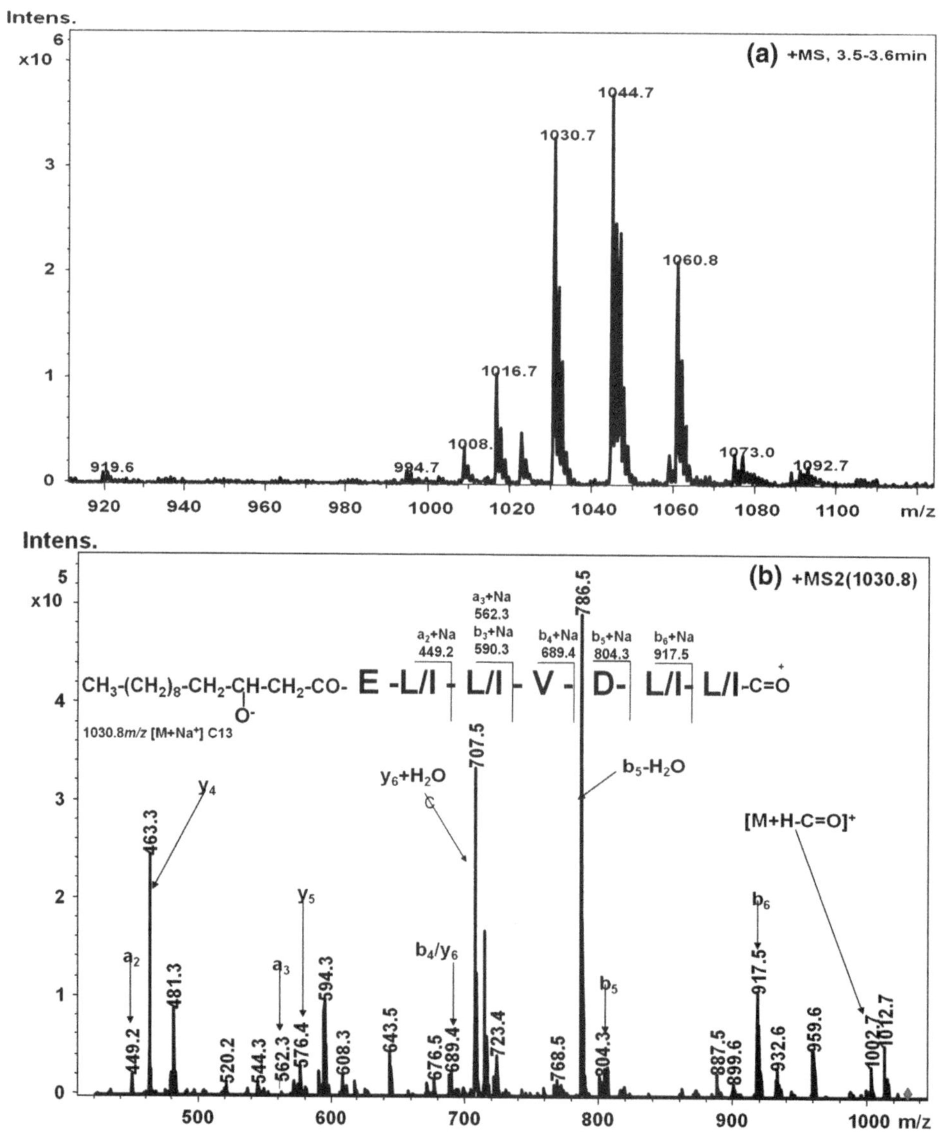

Fig. 6 **a** LC-ESI-MS spectrum of surfactin cluster separated at retention time period of 3.5–3.6 min, **b** LC-ESI-MS/MS spectrum of [M+Na]$^{+}$ ion at *m/z* 1,030.8

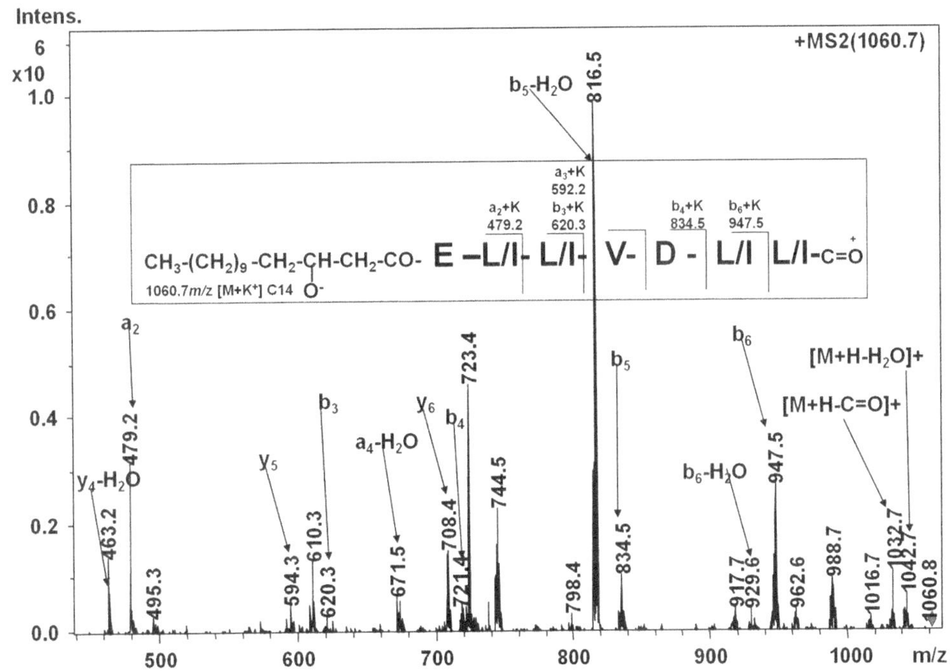

Fig. 7 LC-ESI-MS/MS spectrum of $[M+K]^+$ ion at *m/z* 1,060.7

endophyte, *B. subtilis* K1, comprise C_{13} to C_{17} isoforms which is in agreement to reports for other surfactin producers (Kowall et al. 1998; Hue et al. 2001; Vater et al. 2002; Kim et al. 2010; Williams and Brodbelt 2004).

Biological activity of lipopeptides

The minimal inhibitory concentrations (MIC) for purified iturin A homologues and crude antifungal extract were determined using double dilution method in 96-well microtitre plate. Table 3 represents MIC for purified iturin fractions and crude antifungal extract. The surfactins could not be purified as some fengycins co-eluted along with surfactins. The MIC values of iturin A1 and A2/A4/A5 against *Candida albicans* were found to be 5 and 10 μg/mL, respectively, which is in agreement to observations reported by Winkelmann et al. (1983). The iturins were found to be more potent against *A. niger* 40211, *C. indicum*, *Alt. brunsii*, and *Clad. herabarum* 1126, in comparison to *Can. albicans*, *Trichosporon* 1110 and *F. oxysporum*. Klich et al. (1991) also reported requirement of higher concentration iturin A to inhibit the growth of *A. parasiticus*, *A. flavus* and *Fusarium moliniforme*. The MIC of crude extract was found to be lower in comparison to both purified iturins. The lipopeptide crude extract is composed of iturin A, iturin C, iturin D, fengycin A1/A1′, fengycin A2/A2′, fengycin B1/B1′, fengycin B2/B2′ and fengycin C1/C1′ as well as surfactins, suggesting synergism amongst these lipopeptides towards fungal antagonism. However, it is not possible to make any conclusive comment on synergism between the

Table 2 Surfactins secreted by *B. subtilis* K1

Maas (*m/z*)	Peptide sequence	Identification of surfactins	References
1,008.6	Cyclo(β-OHFA-E-L/I-L/I-V-D-L/I)	(C_{13}) Surfactin (V^4), $[M+H]^+$	This work (Park et al. 1995)
1,030.6	Cyclo(β-OHFA-E-L/I-L/I-V-D-L/I)	(C_{13}) Surfactin (V^4), $[M+Na]^+$	
1,022.5	Cyclo(β-OHFA-E-L/I-L/I-V-D-L/I)	(C_{14}) Surfactin (V^4), $[M+H]^+$	This work (Park et al. 1995)
1,022.5	Cyclo(β-OHFA-E-L/I-L/I-L/I-D-L/I)	(C_{13}) Surfactin (I/L^4), $[M+H]^+$	This work (Kowall et al. 1998)
1,044.5	Cyclo(β-OHFA-E-L/I-L/I-V-D-L/I)	(C_{14}) Surfactin (V^4), $[M+Na]^+$	This work (Park et al. 1995)
1,060.6	Cyclo(β-OHFA-E-L/I-L/I-V-D-L/I)	(C_{14}) Surfactin (V^4), $[M+K]^+$	This work (Vater et al. 2002)
1,036.5	Cyclo(β-OHFA-E-L/I-L/I-V-D-L/I)	(C_{15}) Surfactin (V^4), $[M+H]^+$	This work (Vater et al. 2002)
1,036.5	Cyclo(β-OHFA-E-L/I-L/I-L/I-D-L/I)	(C_{14}) Surfactin (I/L^4), $[M+H]^+$	
1,050.6	Cyclo(β-OHFA-E-L/I-L/I-V-D-L/I)	(C_{16}) Surfactin (V^4), $[M+H]^+$	This work
1,073.6	Cyclo(β-OHFA-E-L/I-L/I-V-D-L/I)	(C_{16}) Surfactin (V^4), $[M+Na]^+$	
1,065.6	Cyclo(β-OHFA-E-L/I-L/I-V-D-L/I)	(C_{17}) Surfactin (V^4), $[M+H]^+$	This work

β-OHFA β-hydroxy fatty acid

Table 3 Minimum inhibitory concentration (MIC) of lipopeptides fractions and crude extract of *B. subtilis* K1

Test culture	MIC of lipopeptides containing fractions (μg/mL)		
	Iturin A2	Iturin A3/A4/A5	Crude extract
A. niger 40211	2.5 ± 0.03	1.25 ± 0.01	0.63 ± 0.008
A. paraciticus 40211	5.0 ± 0.05	5.0 ± 0.05	1.25 ± 0.05
C. indicum	5.0 ± 0.05	2.5 ± 0.05	0.063 ± 0.001
F. oxysporum 1072	5.0 ± 0.03	5.0 ± 0.04	2.5 ± 0.02
Alt. brunsii	2.5 ± 0.03	1.25 ± 0.01	0.063 ± 0.001
Ca. albicans	5.0 ± 0.05	10.0 ± 0.08	2.5 ± 0.02
Trichosporon sp. 1110	5.0 ± 0.06	10.0 ± 0.07	2.5 ± 0.02
Clad. herbarum 1126	5.0 ± 0.03	2.5 ± 0.02	0.031 ± 0.0015

lipopeptides produced by *B. subtilis* K1, unless each of them is purified in sufficient concentration and systematic study in this regards is conducted. We have not been able to separate all isoforms of these lipopeptides even after repeated cycles of purification by HPLC as well as attempts using other methods. The main reason for inability to purify them is the extent of heterogeneity amongst these lipopeptides both at the level of amino acid sequence as well as in the length of fatty acid or βAA. Nevertheless, the synergistic action of iturin and fengycin, iturin and surfactin as well as surfactin and fengycin has been documented earlier (Maget-Dana and Peypoux 1994; Ongena et al. 2005; Romero et al. 2007).

Conclusion

The paper describes antifungal activity and in detail mass spectrometric investigation of cyclic lipopeptides such as C_{13}–C_{16} β-amino fatty acid variants of iturin A as well as iturin C and C_{13}–C_{17} β-OH fatty acid isoforms of surfactin exhibiting variation in Val/Ile/Leu at position 4. Seven different variants of iturins as well as of surfactins produced by *B. subtilis* K1 could be identified by tandem mass spectrometry. The MIC of crude extract was lower than purified iturin fractions, suggesting synergism amongst cyclic lipopeptides co-produced by *B. subtilis* K1 in inhibiting the growth of fungi. Further, it would be worth exploring the role of these isoforms in biology of producing bacilli strain.

Acknowledgments This work was supported by University Grant Commission (UGC), New Delhi, India. The authors thank Prof. Padmanabhan Balaram, Molecular Biophysics Unit for reviewing mass spectral analysis. The authors acknowledge proteomics facility at Molecular Biophysics Unit, Indian Institute of Sciences, Bangalore, India for mass spectrometry facility. KP acknowledges Ms. Sunita Prakash for helping with mass spectrometers.

Conflict of interest The authors declare that they have no conflict of interest.

References

Eckart K (1994) Mass spectrometry of cyclic peptides. Mass Spectrom Rev 13:23–55

Geetha L, Manonmani AM, Paily KP (2010) Identification and characterization of a mosquito pupicidal metabolite of *Bacillus subtilis* subsp. *subtilis* strain. Appl Microbiol Biotechnol 86:1737–1744

Gong M, Wang JD, Zhang J, Yang H et al (2006) Study of the antifungal ability of *Bacillus subtilis* strain PY-1 in vitro and identification of its antifungal substance (Iturin A). Acta Biochim Biophys Sin (Shanghai) 38:233–240

Hue N, Serani L, Laprevote O (2001) Structural investigation of cyclic peptidolipids from *Bacillus subtilis* by high energy tandem mass spectrometry. Rapid Commun Mass Spectrom 15:203–209

Isogai A, Takayama S, Murakoshi S, Suzuki A (1982) Structure of β-amino acids in antibiotics Iturin A. Tetrahedron Lett 23:3065–3068

Katz E, Demain AL (1977) The peptide antibiotics of *Bacillus*: chemistry, biogenesis, and possible functions. Bacteriol Rev 41:449–474

Kim SY, Kim JM, Kim SH et al (2007) Surfactin from *Bacillus subtilis* displays anti-proliferative effect via apoptosis induction, cell cycle arrest and survival signalling suppression. FEBS Lett 581:865–871

Kim PI, Ryu J, Kim YH, Chi YT (2010) Production of biosurfactant lipopeptides iturin A, fengycin and surfactin from *Bacillus subtilis* CMB32 for control of *Colletotrichum gloeosporides*. J Microbiol Biotechnol 20:138–145

Klich MA, Lax AR, Bland JM (1991) Inhibition of some mycotoxigenic fungi by iturin A, a peptidolipid produced by *Bacillus subtilis*. Mycopathologia 116:77–80

Kowall M, Vater J, Kluge T, Stein P, Ziessow D (1998) Separation and characterization of surfactin isoforms produced by *Bacillus subtilis* OKB 105. J Colloid Interface Sci 203:1–8

Kracht M, Rokos H, Ozel M, Kowall M, Pauli G, Vater J (1999) Antiviral and haemolytic activities of surfactin isoforms and their methyl ester derivatives. J Antibiot 52:613–619

Maget-Dana R, Peypoux F (1994) Iturins, a special class of pore-forming lipopeptides: biological and physicochemical properties. Toxicology 87:1–3

Nagorska K, Bikowski M, Obuchowki M (2007) Multicellular behaviour and production of a wide variety of toxic substance support usage of *Bacillus subtilis* as powerful biocontrol agent (review). Acta Biochim Pol 54:495–508

Ongena M, Jacques P (2007) *Bacillus* lipopeptides: versatile weapons for plant disease biocontrol. Trends Microbiol 16:115–124

Ongena M, Jacques P, Toure Y, Destain J, Jabrane A, Thonart P (2005) Involvement of fengycin type lipopeptides in the multifaceted biocontrol potential of *Bacillus subtilis*. Appl Microbiol Biotechnol 69:29–38

Papayannopoulos IA (1995) The interpretation of collision-induced dissociation tandem mass spectra of peptides. Mass Spectrom Rev 14:49–73

Park JK, Hasumi K, Endo A (1995) Inhibition of the binding of oxidized low density lipoprotein to the macrophages by iturin C-related compounds. J Antibiot 48:226–232

Pathak KV, Keharia H (2013) Application of extracellular lipopeptide biosurfactant produced by endophytic *Bacillus subtilis* K1 isolated from aerial roots of banyan (*Ficus benghalensis*) in microbially enhanced oil recovery (MEOR). 3 Biotech.

Pathak KV, Keharia H, Gupta K, Thakur SS, Balaram P (2012) Lipopeptides from banyan endophyte, *Bacillus subtilis* K1: mass spectrometric characterization of a library of fengycins. J Am Soc Mass Spectrom 10:1716–1728

Pecci Y, Rivardo F, Martinotti MG, Allegrone G (2010) LC/ESI-MS/MS characterization of lipopeptide biosurfactants produced by *Bacillus licheniformis* V9T14 strain. J Mass Spectrom 45:772–778

Peypoux F, Besson F, Michel G, Delcambe L (1981) Structure of Bacillomycin D, a new antibiotic of Iturin group. Eur J Biochem 118:323–327

Peypoux F, Pmmier MT, Das BC, Besson F et al (1984) Structures of Bacillomycin D and Bacillomycin L, peptidolipid antibiotics from *Bacillus subtilis*. J Antibiot 37:1600–1604

Peypoux F, Pommier MT, Marion D, Ptak M (1986) Revised structure of Mycosubtilin, a peptidolipid antibiotic from *Bacillus subtilis*. J Antibiot 39:636–641

Peypoux F, Bonmatin JM, Labbe H, Grangemard I et al (1994) [Ala4] Surfactin, a novel isoforms from Bacillus subtilis studied by mass and N, R spectroscopies. Eur J Biochem 224:89–96

Romero D, Vicente A, Rakotoaly RH, Dufour SE et al (2007) The iturin and fengycin families of lipopeptides are key factors in antagonism of *Bacillus subtilis* towards *Podosphaera fusca*. Mol Plant Microbe Interact 118:323–327

Rutenbach M, Swart P, Van der Merwe MJ (2001) Sequence specific stabilization of a linear analog of the antifungal lipopeptide iturin A2 by sodium during low energy electrospray ionization mass spectrometry conditions. J Am So Mass Spectrom 12:505–516

Sabareesh V, Ranganayaki RS, Raghotoma S, Bopanna MP et al (2007) Identification and characterization of a library of microheterogeneous cyclohexadepsipeptides from fungus *Isaria*. J Nat Prod 70:715–729

Stein T (2005) *Bacillus subtilis* antibiotics: structures, syntheses and specific functions (micro review). Mol Microbiol 56:845–857

Strobel G (2006) Harnessing endophytes for industrial microbiology. Curr Opin Microbiol 9:240–244

Vanittnakom N, Loeffler W, Koch U, Jung G (1986) Fengycin- A novel antifungal lipopeptide antibiotic produced by *Bacillus subtilis* F-29-2. J Antibiot 7:888–901

Vater J, Barbel K, Wilde C, Frnake P et al (2002) Matrix-assisted laser desorption ionization time of flight mass spectrometry of lipopeptide biosurfactants in whole cells and culture filterates of *Bacillus subtilis* C-1 isolated from petroleum sludge. Appl Environ Microbiol 68:6210–6219

Vollenbroich D, Ozel M, Vater J, Kamp RM, Pauli G (1997) Mechanism of inactivation of enveloped viruses by the biosurfactant surfactin from *Bacillus subtilis*. Biologicals 25:289–297

Williams S, Brodbelt JS (2004) MSn characterization of protonated cyclic peptides and metal complexes. J Am So Mass Spectrom 15:1039–1054

Winkelmann H, Allgaier H, Lu R, Jung G (1983) Iturin AL- A new long chain iturin A possessing an unusual high content of C16 β amino acids. J Antibiot 11:1451–1457

3

Bioconversion of lignocellulosic waste to bioethanol by *Trichoderma* and yeast fermentation

K. Saravanakumar · K. Kathiresan

Abstract The present work aimed at producing bioethanol using lignocellulosic waste sawdust and marine yeast fermentation. Lignocellulosic waste materials were converted into monosugars through acid hydrolysis and finally treated with cellulase enzyme derived from *Trichoderma/Hypocrea*. To enhance the conversion of the glucose from sawdust, the experimental conditions were statistically optimized. The efficient conversion of sawdust to glucose of 78.56 % was achieved under the conditions of pH 6.19, temperature 29 °C, cellulase enzyme (8.16 IU ml^{-1}) and sawdust (7.95 g l^{-1}). The lignocellulosic waste-sawdust hydrolysis was used as the carbon source for the production of bioethanol. Bioethanol production of 85.6 % was achieved (55.2 g l^{-1}) under the optimized conditions of temperature of 36.5 °C, incubation time of 102 h and enzyme-treated sawdust of 45.14 ml l^{-1} and agitation of 330 rpm. This work achieved maximum bioethanol production using *H. estonica* and *S. cerevisiae* fermentation.

Keywords Bioethanol · Marine yeasts · *Trichoderma* · Lignocellulosic waste · Mangroves

Introduction

Fuel deficiency is a global issue due to exhaustion of fossil fuel and growing climate change (EC 2003; Wingren et al. 2003; Li 2003; Qureshi et al. 2006; Sveinsdottir et al. 2009). In order to overcome this issue, different types of techniques have been invented for the possible conversion of cellulosic waste materials into glucose for the ethanol production, as an alternative way for fuel conservation (Sun and Cheng 2002). The bioconversion of lignocellulosic materials (i.e., agricultural residues, woods, and residues from pulp and paper industries, solid wastes) to bioethanol produces high yield of glucose after hydrolysis (McMillan 1994). The utilization of the lignocellulosic materials for the conversion of the biofuel involves only low cost (Li et al. 2007).

Lignocellulose, the most abundant organic matter in the Earth, can be utilized to produce various renewable fuels and chemicals (Lau et al. 2010). For the digestion of these materials into glucose, many methods have been used such as thermal pretreatments, chemical pretreatments, biological pretreatments and enzymatic pretreatments (Meinita et al. 2012). Cellulolytic and hemicellulolytic enzymes are able to increase monosugars from the digestion of lignocellulose (Pakarinen et al. 2011). For this purpose, fungi in particular *Trichoderma* species that can produce cellulolytic enzymes are employed (Aro et al. 2005; Rodriguez Gomez et al. 2012). In general, yeasts are potential microorganisms for the production of bioethanol by sugar fermentation (Kathiresan et al. 2011; Senthilraja et al. 2011). However, most of the studies are restricted to microbes of terrestrial origin, but not of marine origin. Hence, the present work was undertaken for the conversion of sawdust (lignocellulosic wastes) to sugars for fuel fermentation using cellulolytic enzymes of the mangrove-derived *Trichoderma* and further conversion of sugars to bioethanol using the mangrove-derived yeast strain of *Saccharomyces cerevisiae*. Based on the data, the culture

K. Saravanakumar (✉) · K. Kathiresan
Faculty of Marine Sciences, Centre of Advanced Study in Marine Biology, Annamalai University, Parangipettai 608502, Tamil Nadu, India
e-mail: saravana732@gmail.com

K. Kathiresan
e-mail: kathirsum@rediffmail.com

conditions were optimized statistically for the enzyme hydrolysis process and bioethanol fermentation.

Materials and methods

Microorganisms and maintenance

Trichoderma (*Hypocrea*) species (*Trichoderma estonicum/ H. estonica* SKS1 JQ611722) and yeasts species (*Saccharomyces cerevisiae* JN387604) were isolated from mangrove sediment, located in the south east coast of India using selective medium (Askew and Laing 1993) and yeast peptone agar medium (Fell 2005), respectively. The stock cultures were maintained at 4 °C on slants of potato dextrose agar for *H. estonica* and yeast peptone agar for *S. cerevisiae*.

Substrate

Sawdust was obtained from local wood mills, passed through a 1.5-mm sieve to maintain uniform particle size, washed through distilled water to remove impurities present in it and finally dried at 60 °C overnight. The sawdust was pre-hydrolyzed using 0.8 % of phosphoric acid by the method of Kathiresan et al. (2011).

Cellulase enzyme production by *Hypocrea estonica*

Spore suspension of *Hypocrea estonica* SKS1 (JQ611722) was used as microbial inoculum for the fermentation. It was prepared by culturing *H. estonica* in potato dextrose agar slant cultures at 30 °C for 7 days and spore suspension was washed through the Tween-80 water (0.02 % v/v); the suspension was then assessed as final spore count of 2.3×10^3 CFU ml^{-1}. Then 10 ml of spore suspension was inoculated into a 250-ml Erlenmeyer flask containing 100 ml glucose pre-cultured medium (%, w/v) (glucose, 3; yeast extract, 0.8; K_2HPO_4, 0.4; $MgSO_4 \cdot 7H_2O$, 0.2; pH, 7.0) and was cultured at 30 °C, 200 rpm for 48 h in a rotary shaking incubator to prepare the mycelial suspension. At last, mycelial suspension was inoculated into a 250-ml Erlenmeyer flask containing the statistically optimized medium for cellulase production by *H. estonica* (Saravanakumar and Kathiresan 2013a) that consisted of 7.69 g l^{-1} of sawdust, 3.59 g l^{-1} of $(NH_4)_2$ SO_4, under the temperature of 49 °C at pH of 8.8 with inoculum size of 6.0 % (v/w) and was cultivated at 30 °C for 120 h to produce cellulose.

Determination of cellulase enzyme activity

5 ml of the fermented residues was suspended in 150 ml distilled water and shaken at 120 rpm for 2 h. Then the filtrate was centrifuged (10,000 rpm) using a high speed centrifuge for 15 min and the supernatant was used for enzyme assay using the dinitrosalicylic acid (DNS) method (Miller 1959).

Enzymatic hydrolysis

Enzymatic hydrolysis was carried out in 125 ml screw-capped Erlenmeyer flasks. Diluted acid-treated sawdust powder was suspended in distilled water in the flasks. Water was added during the pretreatment in such a way so as to maintain the substrate concentration at 15 % (w/v) after the addition of the crude cellulase enzyme derived from *H. estonica*. Hydrolysis was performed at 50 °C for 24 h at 120 rpm in an incubator shaker. 1 ml sample was periodically removed from each flask (0, 2, 8, 18, 24, 72, 120, and 168 h). Each sample was centrifuged for 10 min at 13,500*g*, and 500 μl supernatant was then removed and placed into a 1.5-ml Eppendorf tube containing the stop buffer (512 mM Na_2CO_3 and 288 mM $NaHCO_3$; pH 10.0) (Sandhu et al. 2012). The buffered samples were stored at 4 °C for subsequent glucose measurement; commercial glucose was used as the standard for the percentage calculation.

Bioethanol production by *S. cerevisiae*

Bioethanol production experiments were carried out according to the experimental setups derived from the center composite design. The enzyme-treated sawdust hydrolysis was used as the carbon source. The percentage of the bioethanol was calculated according to the method of Saravanakumar et al. (2013b).

Determination of bioethanol using gas chromatography

Concentration of bioethanol in the distillate of culture filtrate was estimated using a Hewlett Packard 5890 Series II gas chromatography with chromosorb 105 column and nitrogen as a carrier gas. The temperature of the injection port, oven and detection port was 250, 120, and 250 °C, respectively. For the analysis, 1 μl of liquid samples was injected into gas chromatography. The bioethanol concentration was determined using bioethanol standard plot and is expressed in percentage with help of instrumental default standard value of 65 g l^{-1} which is equivalent to 100 % (Saravanakumar and Kathiresan 2013a, 2013b).

Optimization of the enzyme hydrolysis process

Optimization of the environmental conditions for the maximum hydrolysis process of the sawdust to sugar using

H. estonica-derived cellulase was studied using 30 runs of statistical model assessed from central composite design of response surface methodology. In this experiment, important factors such as pH (6–8), temperature (20–50 °C) and cellulase concentration (2–10 IU ml^{-1}) on glucose conversion were studied.

$$Y = \beta_0 + \Sigma_i \beta_i X_i + \Sigma_i \beta_{ii} X_i^2 + \Sigma_{ij} \beta_{ij} X_i X_j \quad (1)$$

where Y is the predicted response (glucose), X_i, X_j the independent variables, β_0 the offset term, β_i the ith linear coefficient, β_{ii} the ith quadratic coefficient and β_{ij} is the ijth interaction coefficient. The experimental design in coded and uncoded value is presented in Table 1. In this study, the independent variables are coded as X_1, X_2, and X_3, Thus, the second-order polynomial equation can be presented as follows:

$$\begin{aligned} Y = {} & \beta_0 + \beta_1 X_1 + \beta_2 X_2 + \beta_3 X_3 + \beta_4 X_4 + \beta_{11} X_1^2 + \beta_{22} X_2^2 \\ & + \beta_{33} X_3^2 + \beta_{44} X_4^2 + \beta_{12} X_1 X_2 + \beta_{13} X_1 X_3 + \beta_{14} X_1 X_4 \\ & + \beta_{23} X_2 X_3 + \beta_{24} X_2 X_4 + \beta_{34} X_3 X_4 \end{aligned} \quad (2)$$

Statistical optimization of bioethanol production

In this experiment, *S. cerevisiae* (JN387604) was selected for optimization. The individual and interaction effects of temperature (°C), incubation time (0–120 h), and enzyme-treated sawdust (10–50 ml l^{-1}) and agitation (100–500) on bioethanol production were carried out. The maximum yield of bioethanol production was tested using 30 experimental setups derived from a statistical model-central composite design of response surface methodology. The

Table 1 Center composite design of response surface methodology for the glucose production of experimental and predicted responses

Std	(A) pH	(B) Temperature (°C)	(C) Cellulase concentration (IU ml^{-1})	(D) Sawdust (mg l^{-1})	Glucose (%)	
					Experimental	Predicted
1	6	20	2	2	65.26	63.13
2	8	20	2	2	52.26	44.33
3	6	50	2	2	42.56	34.68
4	8	50	2	2	32.26	31.08
5	6	20	10	2	53.26	52.91
6	8	20	10	2	41.23	31.96
7	6	50	10	2	23.25	19.11
8	8	50	10	2	15.26	13.36
9	6	20	2	10	45.65	52.32
10	8	20	2	10	36.56	31.85
11	6	50	2	10	35.62	36.04
12	8	50	2	10	25.65	30.77
13	6	20	10	10	78.56	70.89
14	8	20	10	10	35.62	48.27
15	6	50	10	10	36.56	49.26
16	8	50	10	10	48.56	41.84
17	5	35	6	6	68.56	67.71
18	9	35	6	6	36.56	41.49
19	7	5	6	6	45.56	49.89
20	7	65	6	6	15.26	15.01
21	7	35	2	6	35.26	39.03
22	7	35	14	6	39.56	39.87
23	7	35	6	2	12.25	27.61
24	7	35	6	14	56.56	45.28
25	7	35	6	6	65.23	65.23
26	7	35	6	6	65.23	65.23
27	7	35	6	6	65.23	65.23
28	7	35	6	6	65.23	65.23
29	7	35	6	6	65.23	65.23
30	7	35	6	6	65.23	65.23

Table 2 Center composite design of response surface methodology for the bioethanol production of experimental and predicted response

Std	(A) Temperature (°C)	(B) Incubation time	(C) Enzyme-treated sawdust (ml l^{-1})	(D) Agitation (rpm)	Bioethanol yield (%)	
					Experimental	Predicted
1	0	0	10	100	0.00	4.71
2	50	0	10	100	0.00	9.51
3	0	120	10	100	15.69	19.99
4	50	120	10	100	75.56	82.40
5	0	0	50	100	0.00	4.24
6	50	0	50	100	0.00	8.58
7	0	120	50	100	15.65	19.09
8	50	120	50	100	85.65	81.04
9	0	0	10	500	0.00	21.21
10	50	0	10	500	0.00	18.69
11	0	120	10	500	12.26	25.81
12	50	120	10	500	68.56	80.92
13	0	0	50	500	0.00	15.28
14	50	0	50	500	0.00	12.30
15	0	120	50	500	12.36	19.45
16	50	120	50	500	56.69	74.10
17	25	60	30	300	15.58	−1.97
18	75	60	30	300	78.65	57.47
19	25	60	30	300	19.65	−8.25
20	25	180	30	300	79.65	68.82
21	25	60	10	300	75.26	49.04
22	25	60	70	300	54.26	41.75
23	25	60	30	100	15.26	16.12
24	25	60	30	700	65.26	25.67
25	25	60	30	300	75.60	75.60
26	25	60	30	300	75.60	75.60
27	25	60	30	300	75.60	75.60
28	25	60	30	300	75.60	75.60
29	25	60	30	300	75.60	75.60
30	25	60	30	300	75.60	75.60

coded values of the fermentation factors were determined by the following equation:

$$Y = \beta_0 + \Sigma_i \beta_i X_i + \Sigma_i \beta_{ii} X_i^2 + \Sigma_{ij} \beta_{ij} X_i X_j \quad (3)$$

where Y is the predicted response, X_i, X_j the independent variables, β_0 the offset term, β_i the ith linear coefficient, β_{ii} the ith quadratic coefficient and β_{ij} is the ijth interaction coefficient. The experimental design in coded and uncoded value is presented in Table 2. In this study, the independent variables are coded as X_1, X_2, X_3 and X_4. Thus, the second-order polynomial equation can be presented as follows:

$$Y = \beta_0 + \beta_1 X_1 + \beta_2 X_2 + \beta_3 X_3 + \beta_4 X_4 + \beta_{11} X_1^2 + \beta_{22} X_2^2 + \beta_{33} X_3^2 + \beta_{44} X_4^2 + \beta_{12} X_1 X_2 + \beta_{13} X_1 X_3 + \beta_{14} X_1 X_4 + \beta_{23} X_2 X_3 + \beta_{24} X_2 X_4 + \beta_{34} X_3 X_4 \quad (4)$$

A statistical program package Design Expert 8.0.6 was used for regression and model fit analysis of the data was obtained, and then estimated the coefficient of the regression equation, and analyzed the variance of selected factors and model significance.

Results

Effect of the cellulase enzyme on sawdust hydrolysis process

The interaction and individual effects of the experimental condition on hydrolysis process were studied by adopting the statistical design of the center composite model of 30 experimental setups and are presented in Table 1 along with the predicted and experimental responses of glucose production. The acceptability of the statistical model was tested by the quadratic model along with the contour error plot for the predicted and experimental responses and further model fitness tested by the normal standardized residual plot. The standard error value of 0.43 showed fitness for the acceptable model and R^2 value of the model was 0.87 which also proved the 87 % of model fitness. The application of the response surface methodology based on the estimates of the parameters indicated an experimental relationship between the response and input variables with the glucose production expressed in the following quadratic mode:

$$\begin{aligned}\text{Glucose}\,(\%) = & 65.23 - 6.56X_1 - 8.72X_2 + 0.21X_3 + 4.42X_4 \\ & + 3.80X_1X_2 - 0.54X_1X_3 - 0.42X_1X_4 \\ & - 1.34X_2X_3 + 3.04X_2X_4 + 7.20X_3X_4 - 2.66X_1^2 \\ & - 8.19X_2^2 - 6.44X_3^2 - 7.20X_4^2\end{aligned} \quad (5)$$

where X_1 (pH), X_2 (temperature), X_3 (cellulase) and X_4 (agitation) are independent variables. Significance of each coefficient is presented in Eq. (5), determined by the Student's t test and p values (Montgomery 2001). The ANOVA results of the quadratic model for glucose production are given in Table 3. The value of R^2 and adjusted R^2 was close to 0.87 and it revealed a high correlation between the observed values and the predicted values. This means that regression model provides an excellent explanation of the relationship between the independent variables (factors) and the response (glucose production). The lack-of-fit term was non-significant as it was desired. The non-significant value of lack-of-fit observed (0.856) was more than probability of 0.05 and this revealed that the quadratic model was valid for the present study.

In this case, the individual and interaction effects on the hydrolysis process were tested and the data are presented in Table 3. The factors of the X_1, X_2, X_4, X_3X_4, X_2^2, X_3^2, X_4^2 were significant but other factors and their interaction effects were not significant on the glucose production ($p < 0.05$).

The optimization of the specific conditions for the glucose production was assessed. Statistically optimized conditions for the efficient sawdust hydrolysis process of glucose production were pH (6.19), temperature (29 °C), cellulase enzyme (8.16 IU ml^{-1}) and sawdust (7.95 mg l^{-1}). This hydrolysed glucose was used further for the alcohol production as the carbon source.

Table 3 Analysis of variance for the response of glucose production by *H. estonica*

Source	Sum of squares	*df*	Mean square	*F* value	*p* value prob > *F*
Model	8,011.377	14	572.2412	7.290321	0.0002***
(A) Ph	1,031.233	1	1,031.233	13.13785	0.0025**
(B) Temperature (°C)	1,824.922	1	1,824.922	23.2494	0.0002***
(C) Cellulase concentration (IU ml^{-1})	1.075267	1	1.075267	0.013699	0.9084^{NS}
(D) Sawdust (mg l^{-1})	468.6968	1	468.6968	5.971171	0.0274*
AB	231.04	1	231.04	2.943437	0.1068^{NS}
AC	4.6225	1	4.6225	0.05889	0.8115^{NS}
AD	2.7889	1	2.7889	0.03553	0.8530^{NS}
BC	28.6225	1	28.6225	0.364649	0.5550^{NS}
BD	148.1089	1	148.1089	1.886899	0.1897^{NS}
CD	828.8641	1	828.8641	10.55968	0.0054**
A^2	193.6786	1	193.6786	2.467455	0.1371^{NS}
B^2	1,841.955	1	1,841.955	23.4664	0.0002***
C^2	1,139.255	1	1,139.255	14.51405	0.0017**
D^2	1,420.334	1	1,420.334	18.09498	0.0007***
Residual	1,177.399	15	78.49328		
Lack-of-fit	1,177.399	10	117.7399		0.856^{NS}
Pure error	0	5	0		
Cor total	9,188.776	29			

NS not significant

*** $p < 0.001$; ** $p < 0.01$; * $p < 0.05$

Optimization of bioethanol production by *S. cerevisiae*

The interaction and individual effects of experimental conditions on bioethanol production process were studied by adopting the statistical design of the center composite for the bioethanol production. The acceptability of the statistical model was tested by the quadratic model along with the contour error plot for the predicted and experimental responses. The standard error value of the 0.43 showed fitness for the acceptable model and R^2 value of the model of 0.81 also proved 81 % of model fitness. The application of the response surface methodology based on the estimates of the parameters indicated an experimental relationship between the response and input variables, which is expressed in the following quadratic model:

$$\begin{aligned}\text{Bioethanol yield}\,(\%) = & 75.60 + 14.86X_1 + 19.27X_2 \\ & - 1.82X_3 + 2.39X_4 + 14.41X_1X_2 - 0.12X_1X_3 \\ & - 0.83X_1X_4 - 0.11X_2X_3 - 2.67X_2X_4 - 2.67X_3X_4 \\ & - 11.96X_1^2 - 11.33X_2^2 - 7.55X_3^2 - 13.68X_4^2\end{aligned} \quad (6)$$

where X_1 (temperature), X_2 (incubation time), X_3 (enzyme-treated sawdust) and X_4 (agitation) are independent variables. Significance of each coefficient is presented in Eq. (6), determined by the Student's t test and p values

(Montgomery 2001). The ANOVA results of the quadratic model for bioethanol production are given in Table 4. The value of R^2 and adjusted R^2 was close to 0.81 and it revealed a high correlation between the observed values and the predicted values. This means that regression model provides an excellent explanation of the relationship between the independent variables (factors) and the response (bioethanol production). The lack-of-fit term was non-significant. The non-significant value of lack-of-fit observed (0.52) was more than probability of 0.05 and this revealed that the quadratic model was valid for the present study.

In this case, the individual and interaction effects on the hydrolysis process were tested and the data are presented in Table 4. The factors of the X_1, X_2, X_1X_2, X_1^2, X_2^2, X_3^2, X_4^2 were significant but other factors and their interaction effects were not significant on the bioethanol production ($p < 0.05$).

The optimization of the specific conditions for the bioethanol production was assessed. Statistically optimized conditions for the efficient bioethanol production process were temperature of 36.5 °C, incubation time of 102 h and enzyme-treated sawdust of 45.14 ml l^{-1} and agitation of 330 rpm.

Table 4 Analysis of variance for the response of bioethanol production by *S. cerevisiae*

Source	Sum of squares	*df*	Mean square	*F* value	*p* value prob > *F*
Model	28,097.22	14	2,006.944	4.824282	0.0023**
(A) Temperature (°C)	5,299.67	1	5,299.67	12.73932	0.0028**
(B) Incubation hours	8,909.677	1	8,909.677	21.41704	0.0003***
(C) Enzyme-treated sawdust (ml l^{-1})	79.64327	1	79.64327	0.191446	0.6680^{NS}
(D) Agitation (rpm)	136.8993	1	136.8993	0.329078	0.5747^{NS}
AB	3,320.641	1	3,320.641	7.982139	0.0128*
AC	0.2116	1	0.2116	0.000509	0.9823^{NS}
AD	53.4361	1	53.4361	0.128449	0.7250^{NS}
BC	0.1849	1	0.1849	0.000444	0.9835^{NS}
BD	113.8489	1	113.8489	0.273669	0.6085^{NS}
CD	29.75703	1	29.75703	0.07153	0.7928^{NS}
A^2	3,924.794	1	3,924.794	9.4344	0.0078**
B^2	3,519.94	1	3,519.94	8.461214	0.0108*
C^2	1,563.842	1	1,563.842	3.759156	0.0716^{NS}
D^2	5,129.922	1	5,129.922	12.33128	0.0031**
Residual	6,240.133	15	416.0088		
Lack-of-fit	6,240.133	10	624.0133		0.52^{NS}
Pure error	0	5	0		
Cor total	34,337.35	29			

NS not significant

*** $p < 0.001$; ** $p < 0.01$; * $p < 0.05$

Discussion

Bioethanol production from the lignocellulosic waste materials depends on the treatment of the sawdust and removal of the hemicelluloses and lignin from the sawdust for the production of monosugars such as glucose. This is a challenge to the current researchers on the renewable fuel production (Meinita et al. 2012). Marine-derived *Trichoderma* is a highly potential source for the bioprospecting research (Saravanakumar and Kathiresan 2012). These marine *Trichoderma* strains were used in this work. The present study successfully attempted the pretreatment of the sawdust with 0.8 % phosphoric acid followed by enzyme hydrolysis using *H. estonica*-derived cellulase for the maximum production of monosugars and then conversion of bioethanol by fermentation with yeast (*S. cerevisiae*).

Statistically optimized conditions for the efficient sawdust hydrolysis process were pH (6.19), temperature (29 °C), cellulase enzyme (8.16 IU ml^{-1}) and 8 % phosphoric acid-treated sawdust (7.95 mg l^{-1}). For suitability of each pre-hydrolysis treatment for maximum solubilization of hemicelluloses and lignin, the effect of pre-hydrolysis treatment of sawdust with the enzyme was investigated using *H. estonica*-derived cellulase enzyme. It resulted in maximum yield of glucose (78.56 %). The significant conversion of the sawdust to the monosugars (glucose) was attained at temperature above 29 °C. This is due to the fact that high temperature increases hemicellulose degradation and lignin transformation, thus increasing the potential of cellulose hydrolysis (Li et al. 2009), whereas low temperature increases the solubilization of the hemicellulose (Sun and Cheng 2002). Addition of dilute acid in steam treatment can effectively improve enzymatic hydrolysis, decrease the production of inhibitory compounds, and lead to more removal of hemicelluloses (Martin et al. 2002; Kathiresan et al. 2011; Matsushita et al. 2010; Wang et al. 2009).

Saravanakumar and Kathiresan (2013a, b) have reported the potential of the marine strain over the terrestrial yeast strains and achieved the maximum yield of the bioethanol of 69.58 % under optimal conditions of temperature at 30 °C, sawdust concentration of 6.84 mg l^{-1} under the agitation speed of 360 rpm in 89 h of incubation. Similarly, the present study attained maximum bioethanol production of 85.6 % under the optimized conditions of temperature of 36.5 °C, incubation time of 102 h and enzyme-treated sawdust of 45.14 ml l^{-1} and agitation of 330 rpm; little variations of the optimized conditions were attributed due to the higher availability of monosugars in the *Trichoderma*-derived enzyme-treated sawdust hydrolysis. Thus, the temperature has a significant role in the bioethanol production by increasing the enzyme hydrolysis of the sawdust concentration and also the bioethanol yield due to increased production of fermentable monosugars (Fromanger et al. 2010; Zuroff and Curtis 2012). The present work suggested that the maximum production of the bioethanol from the sawdust could be achieved by proper method of acid and enzyme pretreatment and yeast fermentation employing *H. estonica* and *S. cerevisiae*.

Acknowledgments The authors are thankful to the authorities of Annamalai University, India and to Ministry of Human Resource Development, Govt. of India, New Delhi.

Conflict of interest The authors declare that they have no conflict of interest.

References

Aro N, Pakula T, Penttiillae M (2005) Transcriptional regulation of plant cell wall degradation by filamentous fungi. FEMS Microbiol Rev 29:719–739

Askew DJ, Laing MD (1993) An adapted selective medium for the quantitative isolation of *Trichoderma* sp. Plant Pathol 42: 686–690

Fell JW (2005) The role of nucleotide analysis in the systematics of the yeast genera *Cryptococcus* sp. and *Rhodotorula* sp. Stud Mycol 38:129–146

Fromanger R, Guillouet SE, Uribelarrea JL, Molina-Jouve C, Cameleyre X (2010) Effect of controlled oxygen limitation on *Candida shehatae* physiology for ethanol production from xylose and glucose. J Ind Microbiol Biotechnol 37:437–445

Kathiresan K, Saravanakumar K, Senthilraja P (2011) Bio-ethanol production by marine yeasts isolated from coastal mangrove sediment. Int Multidiscip Res J 1(1):19–24

Lau MJ, Lau MW, Gunawan C, Dale BE (2010) Ammonia fiber expansion (AFEX) pretreatment, enzymatic hydrolysis and fermentation on empty palm fruit bunch fiber (EPFBF) for cellulosic ethanol production. Appl Biochem Biotechnol 162: 1847–1857

Li K (2003) The role of enzymes and mediators in lignocellulose degradation. In: Goodell B, Nicholas D, Schultz T (eds) American Chemical Society Symposium Series 845. Wood deterioration and preservation, advances in our changing world. American Chemical Society, Washington, DC, pp 196–209

Li A, Antizar-Ladislao B, Khraisheh M (2007) Bioconversion of municipal solid waste to glucose for bioethanol production. Bioprocess Biosyst Eng 30(3):189–196

Li XH, Yang HJ, Roy B, Wang D, Yue WF, Jiang LJ, Park EY, Miao YG (2009) The most stirring technology in future: cellulose enzyme and biomass utilization. Afr J Biotechnol 8:2418–2422

Martín C, Galbe M, Wahlbom CF, Hagerdal BH, Jonsson LJ (2002) Ethanol production from enzymatic hydrolysates of sugarcane bagasse using recombinant xylose-utilising *Saccharomyces cerevisiae*. Enz Microb Technol 31:274–282

Matsushita Y, Yamauchi K, Takabe K, Awano T, Yoshinaga A, Kato M, Kobayashi T, Asada T, Furujyo A, Fukushima K (2010) Enzymatic saccharification of Eucalyptus bark using hydrothermal pretreatment with carbon dioxide. Biores Technol 101: 4936–4939

Mcmillan JD (1994) Fuels production In: Himmel ME, Baker JO, Overend RP (eds) ACS symposium series 566. American Chemical Society, Washington, DC, pp 292–324

Meinita MD, Hong YK, Jeong GT (2012) Comparison of sulfuric and hydrochloric acids as catalysts in hydrolysis of *Kappaphycus alvarezii* (cottonii). Bioprocess Biosyst Eng Jan 35(1–2):123

Miller GL (1959) Use of DNS reagent for the measurement of reducing sugar. Anal Chem 31:426–428

Montgomery DC (2001) Design and analysis of experiments, 5th edn. John Wiley and Sons, New York

Pakarinen A, Zhang J, Brock T, Maijala P, Viikari L (2011) Enzymatic accessibility of fiber hemp is from GH families 10 and 11 with thermostable cellulases in *lignocellulose* hydrolysis. 108(12):2823–2834

Qureshi TA, Mirbahar KB, Samo MU, Soomro NM, Solangi AA, Memon A (2006) Clinical study of experimentally induced anaphylactic shock in goats. Int J Pharmacol 2:357–361

Rodriguez Gomez D, Lehmann L, Olkjar et al (2012) Examining the potential of plasma-assisted pretreated wheat straw for enzyme production by *Trichoderma reesei*. Appl Biochem Biotechnol 166(8):2051–2063 (Data provider: Technical University of Denmark Document type)

Sandhu H, Nidumolu U, Sandhu S (2012) Assessing risks and opportunities arising from ecosystem change in primary industries using ecosystem-based business risk analysis tool. Hum Ecol Risk Assess 18(1):47–68

Saravanakumar K, Kathiresan K (2012) Bioprospecting potential of marine-derived Trichoderma. Asian Pac J Trop Biomed 1–8

Saravanakumar K, Kathiresan K (2013a) Thermostable cellulase produced by trichoderma (*Hypocrea estonica*). J Biotechnol Sci 1(1):22–41

Saravanakumar K, Senthilraja P, Kathiresan K (2013b) Bioethanol production by mangrove derived marine yeast, *Sacchromyces cerevisiae* King Saud University Journal of King Saud University–Science http://dx.doi.org/10.1016/j.jksus.2012.12.005

Senthilraja P, Kathiresan K, Saravanakumar K (2011) Comparative analysis of bioethanol production by different strains of immobilized marine yeast. J Yeast Fungal Res 2(8):113–116

Sherrard EC, Harris EE (1932) Factors influencing properties of isolated wood lignin. Ind Eng Chem 24:103

Sun Y, Cheng J (2002) Hydrolysis of lignocellulosic material from ethanol production: a review. Bioresour Technol 83:1–11

Sveinsdottir M, Baldursson SRB, Orlygsson J (2009) Ethanol production from monosugars and lignocellulosic biomass by thermophilic bacteria isolated from Icelandic hot springs. Icelandic Agric Sci 22:45–58

Wang Y, Sakamoto Y, Kamiya Y (2009) Remediation of actual groundwater polluted with nitrate by the catalytic reduction over copper-palladium supported on active carbon. Appl Catal A 361:123–129

Wingren A, Galbe M, Zacchi G (2003) Techno economic evaluation of producing ethanol from softwood: comparison of SSr and SHF and identification of bottlenecks. Biotechnol Prog 19:1109–1117

Zuroff TR, Curtis WR (2012) Development of symbiotic consortia for *lignocellulosic* biofuel production. Appl Microbiol Biotechnol 93:1423–1435

4

Enhanced production of napthoquinone metabolite (shikonin) from cell suspension culture of *Arnebia* sp. and its up-scaling through bioreactor

Komal Gupta · Shashank Garg · Joginder Singh · Manoj Kumar

Abstract Cell culture in shake flask and air-lift bioreactor was carried out to exploit the potential of *Arnebia* sp. for napthoquinone metabolite production. Cell suspension cultures of *Arnebia* were established from friable callus in liquid MS medium supplemented with 6-benzylaminopurine (BAP) (10 μM) and indole-3-butyric acid (IBA) (5 μM). Growth kinetic studies were done by using settled cell volume and fresh/dry cell weight method. Suspension cultures were maintained by sub-culturing at 10 days interval. A two-stage culture system is employed using growth medium (GM) and modified M9 medium (production medium) for cell biomass and naphthoquinone pigment production, respectively. Results showed that cultivation of cells under dark conditions at room temperature (25 ± 2 °C) enhanced the cell biomass from 100 to 625 g l^{-1}. The pigment production was also found to be increased in dark conditions at room temperature. Alkaline pH found to have positive effect on pigment yield. In case of M9 medium constituents, absence of Na_2SO_4 does not affect the pigment yield. The current approaches have the cumulative effect to meet an increased level of (25.5 μg/ml) metabolite production in air-lift bioreactor.

S. Garg · M. Kumar (✉)
Department of Biotechnology, Lovely Professional University, Punjab, India
e-mail: manojjnu@gmail.com

K. Gupta
NIT, Surathkal, Karnataka, India

J. Singh
School of Biotechnology and Bio sciences, Lovely Professional University, Punjab, India

Keywords Secondary metabolites · Plant cell culture · Air-lift bioreactor · Shikonin

Introduction

The Indian Himalayan region (IHR) is a rich reservoir of biodiversity in the world. Over 1,748 species of medicinal and aromatic plants reported from IHR are used in different systems of medicine (Kandari et al. 2008). There are two types of plants derived molecules: (1) primary metabolites, and (2) secondary metabolites. Primary metabolites such as nucleic acids, proteins, carbohydrates and fats are primary compounds because of their roles in basic cell metabolism, and are usually "high volume low value" chemicals. Secondary metabolites are biosynthetically derived from primary compounds and are more limited in distribution and produced in smaller quantities. They tend to be synthesized in specialized cell types and at distinct development stages and difficult to synchronize the extraction and purification of these metabolites. Secondary metabolites such as shikonin, anthocyanin, hyoscyamine, vincristine, etc., has no apparent physiological function, but have ecological roles, e.g., attractant and chemical defense. Human use secondary metabolites as medicines, flavorings, and recreational drugs. Secondary metabolites are "high value low volume" products (Katare et al. 2009; Verpoorte and Alfermann 2000; Babula et al. 2009).

Arnebia (family Boraginaceae), commonly known as "Ratanjot", is an important and valuable plant species found at high altitude of Western Himalayan region (4,000–4,200 m amsl). It has been used as a natural dye for coloring silk and cosmetics, food additives and medicine. Its roots possess naphthoquinone pigments or shikonin derivatives having medicinal properties such as antimicrobial, anticancer, antipyretic and anti-inflammatory (Saker et al. 2012; Verpoorte et al. 2002; Manjkhola and Dhar 2002). These properties of plant are attributed to the presence of shikonin derivatives. Owing to these properties, genus *Arnebia* is over-exploited and therefore, placed under the category of critically endangered species. The plants are very difficult to cultivate through conventional agriculture (possess third level of difficulty in seed germination) (Malik et al. 2011).

In this regard, plant cell culture seems to be a promising technique for production of naphthoquinone pigments under in vitro conditions (Onrubia et al. 2012; Verpoorte et al. 2002; Sharma et al. 2008). Several attempts have been made in recent past to enhance the production level which are very much significant for phytotherapeutical applications (Hussain et al. 2012).

In consideration of the above facts, the present study was undertaken with an aim to produce naphthoquinone pigment through cell culture of *Arnebia* sp. and its scale-up from shake-flask to air-lift bioreactor. This will not only help in meeting the industrial demand of these metabolites, but also lead to the conservation of the useful species in their natural habitat.

Materials and methods

Establishment of in vitro callus culture

Culture media, plant growth regulators and glassware

Murashige and Skoog (MS) medium (Chattopadhyay et al. 2002) was used for induction of callus from different explants. Sucrose (3 %; w/v) was added as a carbohydrate source and agar (0.8 %; w/v) was incorporated as a gelling agent. Different plant growth regulators (PGRs), i.e., kinetin (Kn) and 2,4-dichlorophenoxyacetic acid (2,4-D), 6-benzylaminopurine (BAP) and indole-3-butyric acid (IBA), BAP and 2,4-D, 2,4-D and 3, 6-dichloro-2-methoxybenzoic acid (DICAMBA) were used in factorial combinations (Table 1). PGR-free media served as control. The pH of the media was adjusted to 5.75 prior to autoclaving.

Plant material and sterilization

Leaf explants excised from the young plants growing in polyhouse were washed in running tap water for 5–10 min. The explants (5–7 cm) were rinsed in distilled water having 1–2 drops of liquid detergent (Teepol or Tween-20) for 10 min followed by 5 min washing with distilled water. The leaves were then surface sterilized with bavistin (0.02 % w/v) and streptomycin (0.02 % w/v) for 10 min followed by washing with distilled water thrice. The explants were then disinfected by 70 % ethyl alcohol treatment for 30–40 s. The leaves were then treated with 0.01 % $HgCl_2$ (having 1–2 drops of tween-20) for 5 min followed by five washings with sterile distilled water inside the laminar airflow.

Inoculation

The leaf explants were dissected into small pieces (1.0–1.5 cm) and were inoculated into petri plates placing adaxial side in contact with solid MS medium with and without growth regulators. Cultures were kept at 25 ± 2 °C under dark condition.

Callus induction and proliferation The percentage for callus induction can be calculated by the following formula:

$$\text{Callus induction } (\%) = \frac{\text{Number of calli}}{\text{Number of explants inoculated}} \times 100.$$

Establishment of cell suspension culture

The suspension cultures were established using growth medium (GM) (MS medium supplemented with BAP (10 μM) and IBA (5 μM)) as earlier standardized at the host institute. Sucrose (3 % w/v) was added as carbohydrate source. The pH of media was adjusted to 5.75 prior to autoclaving. The cultures were kept on shaker set at 100 rpm, 25 ± 2 °C under dark conditions and maintained with regular sub-culturing at 10 days interval.

Growth kinetics

The following methods were employed for studying the growth kinetics:

(a) *Settled cell volume (SCV)* SCV was determined by allowing a cell suspension to sediment in the side arm of the flask. It is reported as the percentage of the total volume of suspension occupied by the cell mass.
(b) *Fresh and dry cell weight* To measure fresh weight cell biomass was harvested by sieving suspension cultures through 40 μm mesh and weighed on a balance immediately, to reduce variations caused by water evaporation. Dry weight was estimated by drying the fresh tissue at 60 °C in a convection oven, until it reached constant weight (usually 12–14 h). Fresh weight was reported as gram, fresh weight per liter (g, FW l^{-1}), while dry weight was reported as gram, dry weight per liter (g, DW L^{-1}).

Pigment production

A two-stage culture system was adopted, i.e., cell growth in GM and pigment in modified M9 medium, i.e., production

Table 1 Combinations of PGRs used for studying the effect of salts on callus induction of *Arnebia*

PGR	2,4-D			
	0 μM	2 μM	5 μM	10 μM
Kn				
0 μM	0,0 (T1)	0,2 (T2)	0,5 (T3)	0,10 (T4)
2 μM	2,0 (T5)	2,2 (T6)	2,5 (T7)	2,10 (T8)
5 μM	5,0 (T9)	5,2 (T10)	5,5 (T11)	5,10 (T12)
10 μM	10,0 (T13)	10,2 (T14)	10,5 (T15)	10,10 (T16)
PGR	IBA			
	0 μM	2 μM	5 μM	10 μM
BAP				
0 μM	0,0 (T17)	0,2 (T18)	0,5 (T19)	0,10 (T20)
2 μM	2,0 (T21)	2,2 (T22)	2,5 (T23)	2,10 (T24)
5 μM	5,0 (T25)	5,2 (T26)	5,5 (T27)	5,10 (T28)
10 μM	10,0 (T29)	10,2 (T30)	10,5 (T31)	10,10 (T32)
PGR	2,4-D			
	0 μM	2 μM	5 μM	10 μM
BAP				
0 μM	0,0 (T33)	0,2 (T34)	0,5 (T35)	0,10 (T36)
2 μM	2,0 (T37)	2,2 (T38)	2,5 (T39)	2,10 (T40)
5 μM	5,0 (T41)	5,2 (T42)	5,5 (T43)	5,10 (T44)
10 μM	10,0 (T45)	10,2 (T46)	10,5 (T47)	10,10 (T48)
PGR	2,4-D			
	0 μM	2 μM	2.5 μM	5 μM
Dicamba				
0 μM	0.0 (T49)	0,2 (T50)	0,2.5 (T51)	0,5 (T52)
2 μM	2,0 (T53)	2,2 (T54)	2,2.5 (T55)	2,5 (T56)
2.5 μM	2.5,0 (T57)	2.5,2 (T58)	2.5,2.5 (T59)	2.5,5 (T60)
5 μM	5,0 (T61)	5,2 (T62)	5,2.5 (T63)	5,5 (T64)

medium (PM) (Katare et al. 2009). Sucrose (5 %; w/v) was added as carbohydrate source. The pH of the media was adjusted to 6.00 prior to autoclaving.

Physical factors versus cell growth and pigment production

To study the effect of different factors on cell growth and pigment production in cell suspension culture, the cells from growth medium were inoculated at 10 % inoculum density in GM and PM. The cultures were kept on shaker set at 100 rpm, temperature at 25 ± 2 °C and incubated under required illumination conditions.

Light conditioning

To study the effect of light, cell suspension cultures were incubated under continuous light (photosynthetic photon flux density (PPFD), 70 μmol m^{-2} s^{-1}) or complete darkness. Differential exposures of illumination period on pigment production from cell suspension cultures were determined (Table 2).

Effect of pH

To investigate the effect of pH on cell growth and pigment production, cells were incubated in GM and PM. with varied pH (ranging from 3.75 to 9.75). The pH of the media was adjusted with 0.1 N HCl or 0.1 M KOH prior to autoclaving. The flasks were incubated in dark condition at 25 ± 2 °C on shaker set at 100 rpm. Three replicates were used for each treatment.

M9 medium constituents and pigment production

To investigate the effect of individual component of M9 medium on pigment production, cells were inoculated in production medium having one component missing as per

Table 2 Effect of light exposure duration on shikonin production

Replicates	Shikonin (μg/ml)	
	Mean	SD
R1 = treatment under 2 days dark + 8 days light	11.1	1.249
R2 = treatment under 4 days dark + 6 days light	10.75	1.671
R3 = treatment under 6 days dark + 4 days light	24.95	2.144
R4 = treatment under 8 days dark + 2 days light	12.95	2.308
R5 = treatment under 10 days dark	11.90	1.61
R6 = treatment under 10 days light	6.00	0.655

Table 3. Treatments were kept on shaker set at 100 rpm, at 25 ± 2 °C in dark conditions.

Spectrophotometric analysis of shikonin derivatives

The quantification of shikonin derivatives was done at 2-day intervals in all the treatments. To release the pigment from cells, suspension cultures were sieved through 40 μm mesh and centrifuged for 15 min at 10,000*g* in Eppendorf tubes containing iso-amyl alcohol. The shikonin derivatives in the organic phase obtained being recovered as a blue aqueous solution by using 1 ml of 2.5 % KOH. Absorbance of samples was recorded at 620 nm using spectrophotometer (Hussain et al. 2012, 2011).

Standard shikonin solution

A total of 1 mg shikonin was dissolved in 2 ml iso-amyl alcohol to make 500 ppm solution. This solution was diluted 50 times to make 10 ppm solution (0.01 mg/ml). 200 μl of each concentration of shikonin solution was taken in 2 ml Eppendorf tube and 1 ml of KOH (2.5 %) was added to it. The solution was centrifuged at 10,000*g* for 15 min and allowed to stand for 10–15 min. The upper oily layer was removed and OD of blue colored solution thus obtained recorded at 620 nm. Blank was prepared with 200 μl iso-amyl alcohol instead of shikonin solution; all other steps being as stated above. This calibration curve was used to determine shikonin derivatives in cells and culture medium.

Scale-up in bioreactor

Shake-flask experiments

Shake-flask experiments were performed with 10 % inoculum density in 250 ml glass Erlenmeyer flasks containing

Table 3 Effect of PGRs on callus induction

Treatment	Response	Treatment	Response	Treatment	Response	Treatment	Response
T1		T17		T33		T49	
T2	*	T18		T34		T50	
T3		T19		T35		T51	
T4		T20		T36		T52	
T5	*	T21		T37		T53	
T6	**	T22		T38		T54	
T7		T23	***	T39		T55	**
T8		T24		T40		T56	
T9	**	T25		T41		T57	
T10	****	T26		T42	**	T58	
T11		T27		T43		T59	
T12		T28		T44		T60	
T13		T29		T45		T61	
T14	****	T30		T46	***	T62	
T15	****	T31		T47		T63	
T16		T32	****	T48		T64	

NIL columns represent no swelling of explants and induction up to 40 days of incubation

* Swelling in all the explants after 20 days but induction was observed only in one explant after 30 days

** Swelling in all the explants after 15 days but induction was observed only in explants after 20 days

*** Swelling in all the explants after 20 days but induction was observed only in one explant after 30 days

**** Swelling in all the explants after 20 days but induction was observed only in one explant after 30 days

50 ml of GM or PM for the pigment production. Two types of bioreactors were used as mentioned below:

Stirred-tank bioreactor

The modular bench-top, water-jacketed, glass vessel stirred-tank bioreactor (Bio Flo 110) having 2 L working volume was used for scale-up studies. The medium agitation was done at 100 rpm using top-driven motor having six-blade turbine impellers. The parameters like temperature (25 ± 2 °C) and pH (5.75 for GM and 6.00 for PM, prior to autoclaving) were set at the value as optimized during shake-flask experiments. The dissolved oxygen was supplied at 2 l min^{-1} using compressed air. Proportional integral derivative (PID) of biocontroller unit controlled the set parameters during bioreactor cultivation. The inoculation was done in sterile conditions under laminar airflow to avoid contamination.

Air-lift bioreactor

The bench-top, water-jacketed with internal concave bottom section glass vessel air-lift bioreactor (Biostat Bplus) of 2 L working volume was also used for scale-up studies. Temperature was set at 25 ± 2 °C, pH (set at 5.75 for growth medium and 6.00 for production medium before autoclaving) and DO is supplied with compressed air at 2 L min^{-1}. The culture medium is mixed by gassing, where the air is introduced near the bottom via a sparger ring and rises through an internal draft tube (Table 4). Inbuilt PID controller with the bioreactor system controlled and monitored the set parameters during cultivation.

Statistical analyses

One-way ordinary ANOVA was performed for analyzing differences in shikonin production under (1) combinatorial salt treatment (2) combination of illumination and dark conditions. ANOVA was performed using GraphPad Prism (Version 5.0) statistical software at a significance level (α) of 0.05.

Results and discussion

Establishment of in vitro callus cultures and callus induction

The leaf segments were inoculated in Petri plates containing MS medium supplemented with different PGR combinations for callus induction (Fig. 1) and their effect on callus induction was observed at different time intervals (Table 3). Highest frequency of callus induction was recorded in T10, T15 and T32 after 35–40 days. However, lowering the auxin concentration in T14 resulted in early induction, i.e., 25–30 days. In these treatments, all leaf segments showed callus induction. About 50 % callus induction frequency was observed in MS medium supplemented with Kn (2 μM) and 2,4-D (2 μM), BAP (5 μM) and 2,4-D (2 μM) and DICAMBA (2 μM) and 2,4-D (2.5 μM). However, further proliferation was very slow in all the treatments tried and failed to produce friable callus. In basal MS medium (control) only swelling of leaf segments was observed, which later on turned brown.

Growth kinetics

Settled cell volume (SCV)

The results showed that the cells take about 2 days (lag phase) after inoculation to start growing actively. The exponential growth (log phase) persist for about 8 days (2–10 days of cultivation period), as evident from the Fig. 2. Thereafter the growth starts ceasing to follow stationary phase (between 10 and 16 days of cultivation) and onset of decline phase (begins from 16th day onwards).

Table 4 Effect of initial pH of medium on shikonin production

Sr. no.	Time interval (days)	Shikonin (μg/ml)							
		pH 3.75		pH 5.75		pH 7.75		pH 9.75	
		Mean	SD	Mean	SD	Mean	SD	Mean	SD
1	0	0.5	0.096	0.525	0.082	0.525	0.035	0.5	0.072
2	2	0.75	0.075	1.25	0.121	1.05	0.095	1.275	0.197
3	4	1.025	0.109	2.25	0.252	1.825	0.175	2.6	0.319
4	6	2.375	0.218	5.025	0.229	3.05	0.353	7.05	0.321
5	8	3.2	0.229	5.8	0.409	4.675	0.303	9.3	1.029
6	10	5.9	0.377	7.75	0.804	7.5	0.750	9.85	0.721
7	12	6.15	0.242	8.375	0.533	8.8	0.393	10.025	0.801

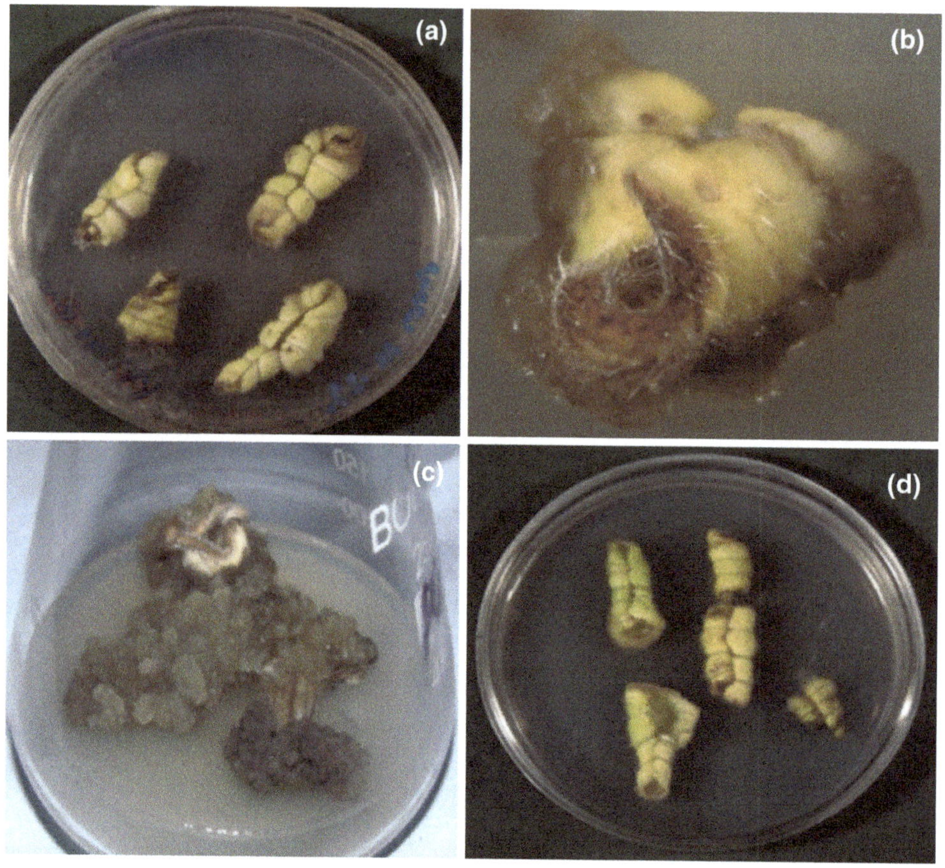

Fig. 1 Leaf explants of *Arnebia* sp. for callusing **a** swelling of leaves, **b** initiation of callus (T14), **c** proliferation of callus (T15), and **d** control

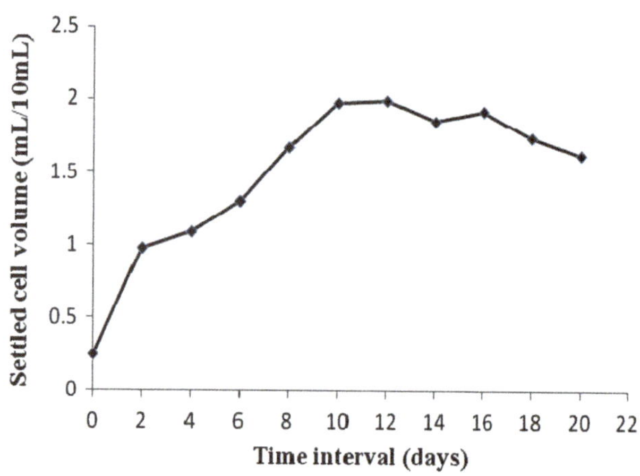

Fig. 2 Growth curve based on settled cell volume (SCV)

Hence, the results showed that the *Arnebia* cells have 10 days of cultivation period before fresh sub-culturing for further multiplication can be done in liquid medium.

Fresh and dry weight

The growth kinetics based on fresh and dry weight was also studied. This is more accurate method than SCV, as in this actual increase in weight being recorded. The results showed a lag phase up to 1 day, log (exponential) phase from 2nd to 8th day, stationary phase from 8th to 12th day and decline phase 12th day onwards (Fig. 3). The reduced lag phase and early log phase can be corroborated to the age of inoculum used, making fast cell acclimatization.

Pigment production

The already established two-stage culture system was used for pigment production. The cells grown in GM were transferred to modified M9 medium at 10 % inoculum density, as described in "Materials and methods". The cultures were kept under dark conditions at 25 ± 2 °C on shaker.

Factors affecting cell growth and pigment production in suspension culture

Effect of illumination

Cell growth The cells grown in GM under 24 h light as well as dark conditions revealed that cell biomass grows almost six-times higher under dark condition and increased from 5 to 31.58 g as compared to 20.91 g in the presence

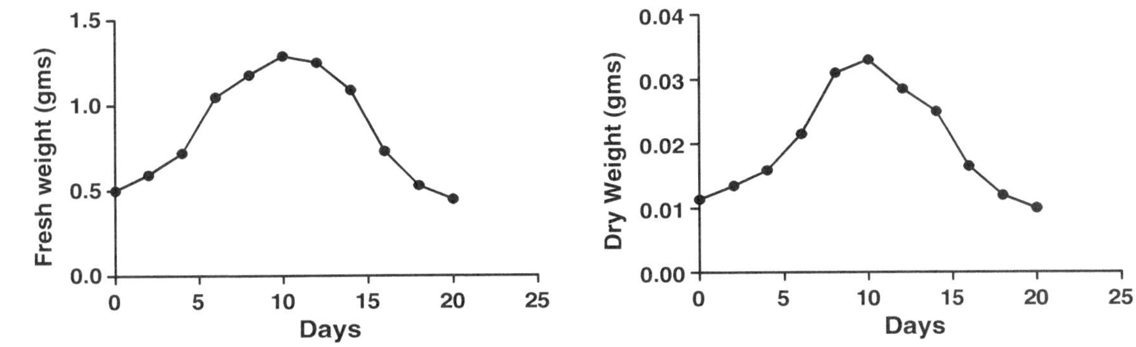

Fig. 3 Growth curve based on fresh and dry cell weight

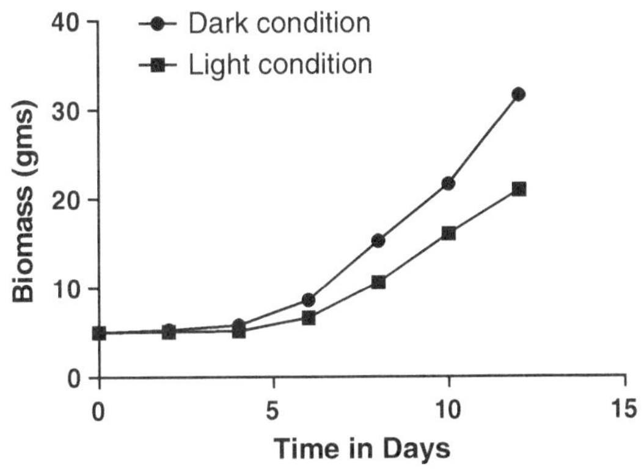

Fig. 4 Effect of dark and light conditions on biomass production

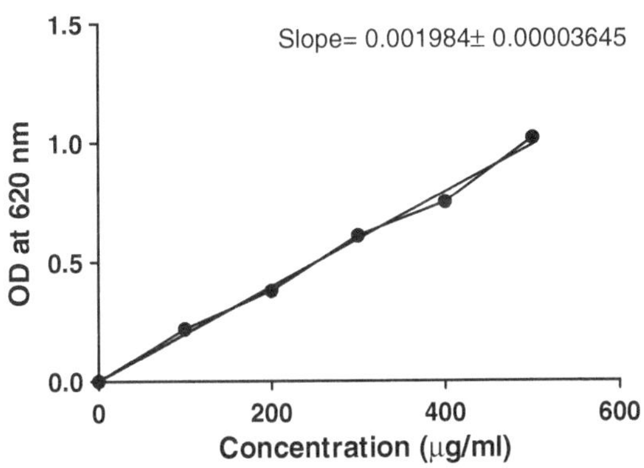

Fig. 5 Standard curve of shikonin

of light, at the end of cultivation period (Figs. 4, 5). The per cent increase in yield was calculated to be 531 %.

Pigment production The cells were inoculated in modified M9 medium under 24 h light as well as dark conditions to observe effect of illumination on pigment production. There was an increase in pigment content irrespective of light conditions till 4th day of culture but as the culture period progresses, the pigment content decreases in the presence of light, whereas in dark it continued to increase. Dark red colored pigment was observed at the end of culture period in suspension culture growing under dark conditions. However, there was very less pigment formation in cells growing under the presence of light. This shows that light completely inhibits the production of pigment during earlier phase of cultivation period (Fig. 6). Results revealed that maximum shikonin content ($OD_{620} = 0.218$) was observed in dark conditions after 12 days of culture period as compared to light ($OD_{620} = 0.120$) (Fig. 6).

Effect of illumination exposure period on pigment yield Results showed that maximum pigment production was obtained when the cell suspension culture was kept in light for 6 days and then in dark for 4 days (Fig. 7) than rest of the illumination dark period combination.

The decrease in pigment production in cultures grown under the presence of light is attributed to the inhibition of *p*-hydroxybenzoic acid geranyl transferase (PGT) activity—a key enzyme of shikonin biosynthesis Irradiation results in accumulation of *o*-glucosidic form of *p*-hydroxybenzoic acid (PHB). PHB is stored in vacuole of cells in its glucoside form in presence of light and upon transfer of cells to dark condition, this glucoside is hydrolyzed to give free hydroxybenzoic acid, which on further reaction

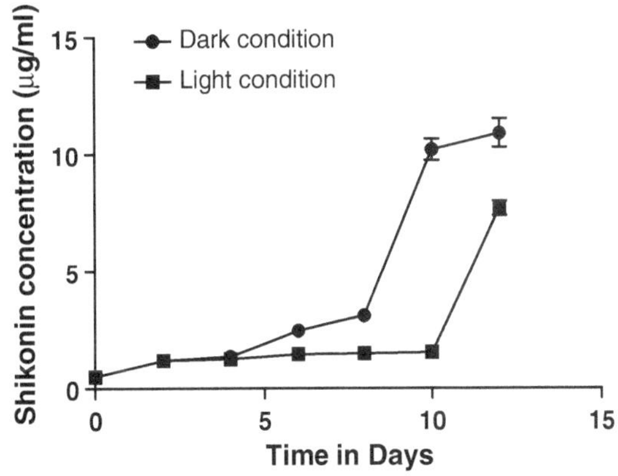

Fig. 6 Comparison of dark and light conditions on shikonin production

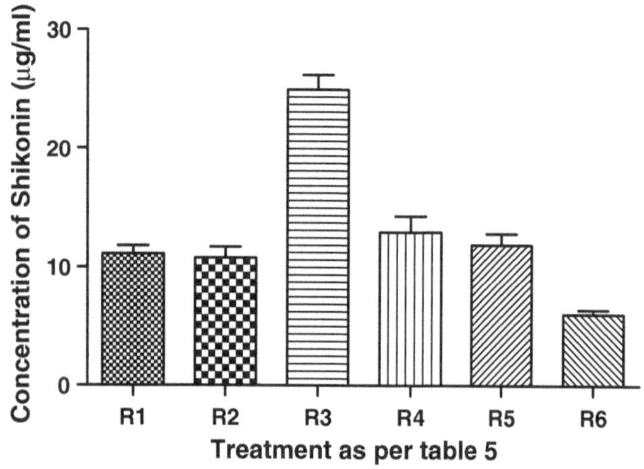

Fig. 7 Effect of light exposure period on shikonin production

metabolized to shikonin biosynthesis (Sommer et al. 1999; Xuqing and Dewei 1998).

The maximum production of naphthoquinone pigment was observed within 10 days of cultivation period, thereafter it declined. The decline was possibly due to cell death or as result of feedback inhibition caused by the accumulation of pigment. Dark condition was a pre-requisite for the biosynthesis of pigment as there was no pigment production in light. This is due to the fact that these compounds are accumulated in the underground part of the plant.

Effect of pH

The pH of the medium was found to have pronounced effect on cell biomass and pigment production. Cell growth is directly influenced by the nutritional composition of the medium, but the uptake of nutrient is mainly influenced by pH of the medium. The cell growth is directly influenced by the nutritional composition of the medium, but the uptake of nutrient is mainly influenced by pH of the medium. The various biochemical reactions occurring at the cellular level are also influenced by change in pH due to the specificity of pH-dependent enzymes.

Cell growth The *Arnebia* cells were cultured in GM under different pH regime to observe its effect on biomass growth. Results revealed that cell biomass recorded almost seven-times higher growth at pH 5.75 and increased from 5 to 35.99 g as compared to 18.97 g at pH 3.75, after 12 days of cultivation period. The increase in yield was found to be 620 %. As the pH of the medium increased from 5.75 to 9.75, the cell growth was found to decrease. At pH 5.75, an increase in fresh weight of cell biomass was observed throughout the cultivation period. However, a low biomass yield was recorded in extremes of pH, i.e., 3.75 and 9.75 (Fig. 8). The pattern of cell growth is indicative of 10 days

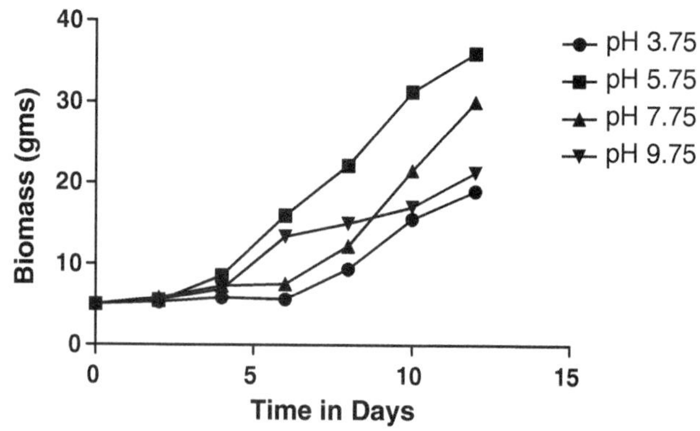

Fig. 8 Effect of initial pH of medium on biomass growth

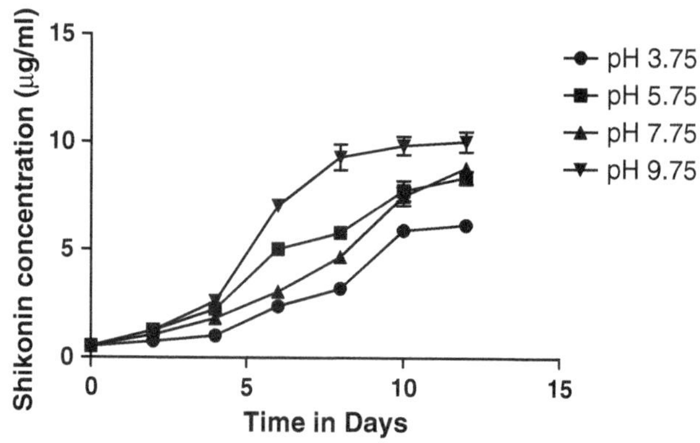

Fig. 9 Effect of initial pH of medium on shikonin production

sub-culturing period for maximum biomass production in cell suspension cultures.

Pigment production The cells were grown in M9 medium under different pH regime. Results revealed that the cells growing at high pH (9.75) exhibited high and early synthesis of pigment (6 days). The maximum shikonin content ($OD_{620} = 0.210$) was observed at pH 9.75, while the lowest value ($OD_{620} = 0.123$) was observed at pH 3.75 (Fig. 9). Cell growth and production of shikonin was observed to be greatly influenced by the pH of the respective media (Table 5). Acidic pH favors cell growth, whereas alkaline pH favors pigment production. It was observed that pH of the growth medium as well as that of production medium changed by 1–2 Units after autoclaving. The change was more pronounced in production medium because of the low buffering capacity of the medium. The change in pH after autoclaving is due to the high during sterilization temperature, which alters the composition of various nutrients and salts of the medium. The initial increase in pH value might result from the fast uptake and reduction of nitrates (formation of OH^-) while acidification of the medium at later stages of cell culture can be associated with the uptake of ammonium ion.

Table 5 Production of shikonin under dark and light conditions

Sr. no.	Time interval (days)	Shikonin (μg/ml)			
		Dark condition		Light condition	
		Mean	SD	Mean	SD
1	0	0.5	0.132	0.5	0.055
3	2	1.2	0.264	1.2	0.132
3	4	1.35	0.229	1.25	0.079
4	6	2.45	0.132	1.45	0.11
5	8	3.1	0.360	1.49	0.127
6	10	10.2	0.781	1.55	0.125
7	12	10.9	1.044	7.7	0.529

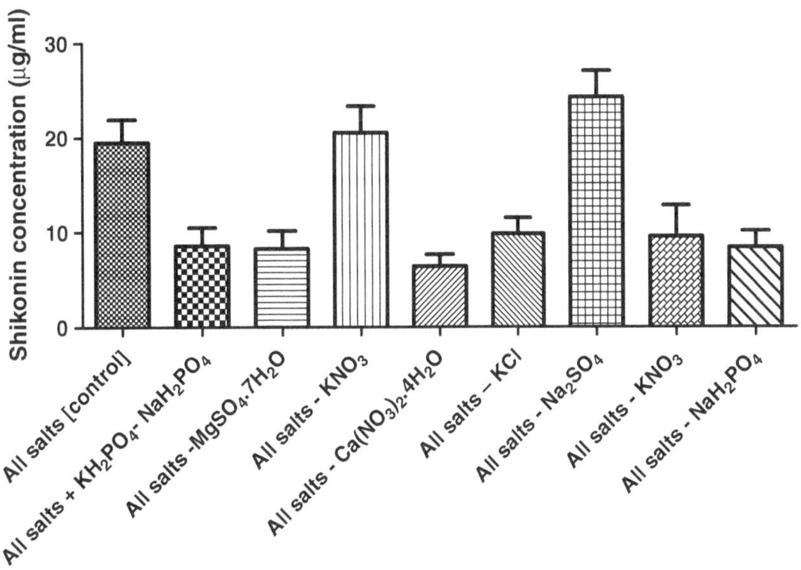

Fig. 10 Effect of salts on production of shikonin

Increased production of pigment at alkaline pH of production medium could be correlated with the enhanced activity of PGT, one of the key regulatory enzymes of shikonin biosynthetic pathway. This enzyme catalyzes the condensation of *p*-hydroxybenzoic acid and geranyl pyrophosphate (GPP) to yield *m*-gernayl-*p*-hydroxybenzoic acid (GBA) requires the optimum pH which ranges from 7.1 to 9.3 for its activity (Malik et al. 2011; Xuqing and Dewei 1998).

Effect of M9 medium constituents on pigment production

The purpose of this experiment was to study the effect of individual constituents of M9 medium (Table 3) on pigment production. Na_2SO_4 is used at higher concentration (1,480 mg l^{-1}) in M9 medium. It was interesting to notice that its exclusion of Na_2SO_4 (T7) does not affect the pigment yield and recorded a little higher OD value ($OD_{620} = 0.141$) as compared to T1, i.e., control ($OD_{620} = 0.112$). However, the exclusion of other constituents does not reveal any pronounced effect on metabolite yield. TSS and pH were found to be higher in combination where $MgSO_4{\cdot}7H_2O$ (T3) was missing (Fig. 10).

Scale-up in bioreactor

Experiments on scale-up of cell biomass production (GM) and pigment production (modified M9) from shake flask to bioreactor was done with optimized parameters as discussed in "Materials and methods" section. The data were recorded for cell biomass and pigment production in shake-flask and different bioreactors. Results showed that the cell biomass was drastically decreased with the scale-up from shake-flask (1,249.2 g) to stirred-tank bioreactor (480 g) and air-lift bioreactor (450 g). However, the pigment production as evident from Table 6 is considerably higher in air-lift reactor as well as stirred-tank bioreactor (Fig. 11).

Statistical analyses

ANOVA table for evaluating the effect of M9 media constituents (as per Table 7) is given in Table 8. From

Table 6 Production of shikonin in three different culture systems

Sr. no.	Culture system	Concentration of shikonin produced (μg/ml)	
		Mean	SD
1	Shake-flask (2 L)	9.7	3.522
2	Stirred-tank bioreactor (2 L)	21.15	3.050
3	Air-lift bioreactor (2 L)	25.5	5.286

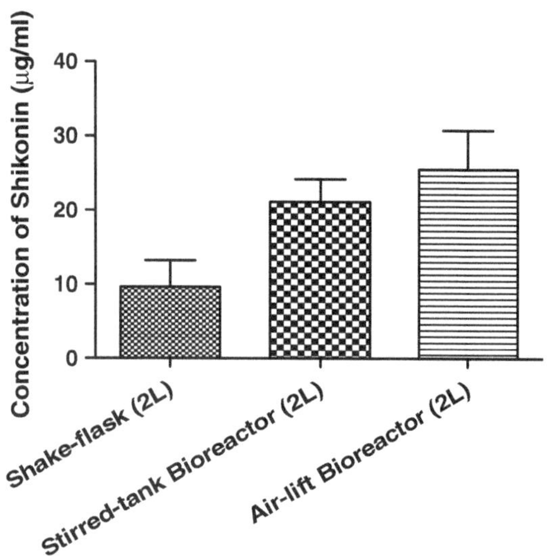

Fig. 11 Comparative histogram for produce of shikonin in 3 different culture systems

Table 8 it is revealed that there is a significant effect ($P < 0.001$) of M9 media constituents on pigment production.

Table 9 gives the ANOVA statistics for data collected after exposing the cell culture to light conditions for different duration. A very significant correlation ($P < 0.001$) is obtained between light exposure duration and pigment yield and ANOVA revealed that there is a very less probability that the variance among the means of obtained result is due to chance.

Conclusion

The present study was carried out to exploit the potential of *Arnebia* sp. for naphthoquinone pigment production through cell culture. Growth kinetics was studied by settled cell volume and fresh/dry cell weight methods. It was found that fresh/dry weight was more efficient to predict the growth phases.

Effect of factors like pH, illumination and medium constituents affecting the cell growth and pigment production was also studied. Different factors affecting the cell growth and pigment production were studied. It was concluded that dark conditions at room temperature (25 ± 2 °C) and acidic pH (5.75) favors the cell growth in GM while alkaline pH at same conditions favors the

Table 7 Combination of salts used for studying effect on pigment production

Sr. no.	No. of combinations	Shikonin concentration (μg/ml)	
		Mean	SD
T1	All salts [control]	19.5	2.438
T2	All salts + KH_2PO_4–NaH_2PO_4	8.55	1.931
T3	All salts: $MgSO_4{\cdot}7H_2O$	8.25	1.920
T4	All salts: KNO_3	20.5	2.825
T5	All salts: $Ca(NO_3)_2{\cdot}4H_2O$	6.4	1.264
T6	All salts: KCl	9.8	1.739
T7	All salts: Na_2SO_4	24.25	2.818
T8	All salts: KNO_3	9.55	3.308
T9	All salts: NaH_2PO_4	8.4	1.757

Table 8 One-way ANOVA of effect of salts combination as per Table 7

P value	$P < 0.0001$		
P value summary	***		
Are means significant different? ($P < 0.05$)	Yes		
Number of groups	9		
F	24.91		
R squared	0.9172		
ANOVA table	SS		
Treatment (between columns)	1,062	df	MS
Residual (within columns)	95.9	8	132.7
Total	1,158	18	5.328
		26	

Table 9 One-way ANOVA of effect of light exposure duration as per Table 2

P value	$P < 0.0001$		
P value summary	***		
Are means significant different? ($P < 0.05$)	Yes		
Number of groups	6		
F	42.04		
R squared	0.946		
ANOVA table	SS	df	MS
Treatment (between columns)	605	5	121
Residual (within columns)	34.54	12	2.878
Total	639.5	17	

pigment production. In case of M9 medium constituents, interestingly, Na_2SO_4 does not affect the pigment yield and was found at par with control. During scale-up, the cell biomass productivity decreases drastically; however, the pigment yield increased.

Finally, statistical analyses of the obtained data through ANOVA rule out any chance of random differences among the final production of pigment production and establish the fact that there is effect of constituents of M9 medium and illumination on the pigment yield.

Acknowledgments Authors acknowledge the National Institute of Technology, Karnataka, India, and Lovely Professional University, Punjab, India, for providing the space and other supports.

References

Babula P, Adam V, Havel L, Kizek R (2009) Noteworthy secondary metabolites naphthoquinones- their occurrence, pharmacological properties and analysis. Curr Pharm Anal 5:47–68

Chattopadhyay S, Farkya S, Ashok K, Srivastava AK, Bisaria VS (2002) Bioprocess considerations for production of secondary metabolites by plant cell suspension cultures. Biotechnol Bioprocess Eng 7:138–149

Hussain MS, Fareed S, Ansari S, Rahman MA, Ahmad IZ, Saeed M (2012) Current approaches toward production of secondary plant metabolites. J Pharm Bioallied Sci 4:10–20

Kandari LS, Rao KS, Maikhuri RK, Chauhan K (2008) Effect of presowing, temperature and light on the seed germination of *Arnebia benthamii* (Wall. ex G. Don): an endangered medicinal plant of Central Himalaya, India. Afr J Plant Sci 2(1):5–11

Katare DP, Aeri V, Bora M (2009) Secondary metabolites and metabolic engineering. J Cell Tissue Res 9(3):2027–2036

Malik S, Bhushan S, Sharma M, Ahuja PS (2011) Physico-chemical factors influencing the shikonin derivatives production in cell suspension cultures of *Arnebia euchroma* (Royle) Johnston, a medicinally important plant species. Cell Biol Int 35:153–158

Manjkhola S, Dhar U (2002) Conservation and utilization of *Arnebia benthamii* (Wall. ex G. Don) Johnston—a high value Himalayan medicinal plant. Curr Sci 83:4–25

Onrubia M, Cusidó RM, Ramirez K, Hernández-Vázquez L, Moyano E, Bonfill M, Palazon J (2012) Bio processing of plant in vitro systems for the mass production of pharmaceutically important metabolites: paclitaxel and its derivatives. Current medicinal chemistry (Online draft)

Saker MM, Farid MM, Fahmi AA, El-Mekkawy SA, Taha HS, Amin AI (2012) Large scale production of antitumor cucurbitacins from *Ecballium Elaterium* using bioreactor. Afr J Biotechnol 11:12974–12982

Sharma N, Sharma UK, Malik S, Bhushan S, Kumar V, Verma SC, Sharma N, Sharma M, Sinha AK (2008) Isolation and purification of acetylshikonin and b-acetoxyisovaleryl shikonin from cell suspension cultures of *Arnebia euchroma* (Royle) Johnston using rapid preparative HPLC. J Sep Sci 31:629–635

Sommer S, Kohle A, Yazaki K, Shimomura K, Bechthold A, Heide L (1999) Genetic engineering of shikonin biosynthesis hairy root cultures of *Lithospermum erythrorhizon* transformed with the bacterial *ubiC* gene. Plant Mol Biol 39:683–693

Verpoorte R, Alfermann AW (2000) Metabolic engineering in plant secondary metabolism, 1st edn. Kluver Academic Publisher, USA, pp 1–20

Verpoorte R, Contin A, Memelink J (2002) Biotechnology for the production of plant secondary metabolites. Phytochem Rev 1:13–25

Xuqing F, Dewei L (1998) Enhancement of shikonin production in cell suspension cultures of *Arnebia euchroma* employing two-liquid phase systems. Chin J Chem Eng 6(1):86–90

5

Quantitative trait loci underlying resistance to sudden death syndrome (SDS) in MD96-5722 by 'Spencer' recombinant inbred line population of soybean

J. Anderson · M. Akond · M. A. Kassem ·
K. Meksem · S. K. Kantartzi

Abstract The best way to protect yield loss of soybean [*Glycine max* (L.) Merr.] due to sudden death syndrome (SDS), caused by *Fusarium virguliforme* (Aoki, O'Donnel, Homma & Lattanzi), is the development and use of resistant lines. Mapping quantitative trait loci (QTL) linked to SDS help developing resistant soybean germplasm through molecular marker-assisted selection strategy. QTL for SDS presented herein are from a high-density SNP-based genetic linkage map of MD 96-5722 (a.k.a 'Monocacy') by 'Spencer' recombinant inbred line using SoySNP6K Illumina Infinium BeadChip genotyping array. Ninety-four $F_{5:7}$ lines were evaluated for 2 years (2010 and 2011) at two locations (Carbondale and Valmeyer) in southern Illinois, USA to identify QTL controlling SDS resistance using disease index (DX). Composite interval mapping identified 19 SDS controlling QTL which were mapped on 11 separate linkage group (LG) or chromosomes (Chr) out of 20 LG or Chr of soybean genome. Many of these significant QTL identified in one environment/year were confirmed in another year or environment, which suggests a common genetic effects and modes of the pathogen. These new QTL are useful sources for SDS resistance studies in soybean breeding, complementing previously reported loci.

J. Anderson · K. Meksem · S. K. Kantartzi (✉)
Department of Plant, Soil and Agricultural Systems, Southern Illinois University, Carbondale, IL 62901-4415, USA
e-mail: kantart@siu.edu

M. Akond · M. A. Kassem
Plant Genomics and Biotechnology Lab, Department of Biological Sciences, Fayetteville State University, Fayetteville, NC 28301-4298, USA

Keywords Sudden death syndrome · Inbred lines · Quantitative trait loci

Introduction

Sudden death syndrome (SDS) caused by *Fusarium virguliforme* (Aoki, O'Donnel, Homma & Lattanzi) (Aoki et al. 2003) is one of the most devastating diseases of soybean (*Glycine max* L. Merr.) responsible for severe yield loss. SDS spreads quickly across the Americas and became a major pest problem for soybean growers starting in the 1980s (Wrather et al. 2003; Roy 1997), but it was not known in Asia until 2011 (Srour et al. 2012). Yield loss ranges between 5 and 15 %, while trace losses can go up to 80 %, depending on environmental conditions, genotype, and planting date (Ohmes 2013). SDS genetics are complex because it is multi-genetic (Anderson 2012; Kazi et al. 2008) and the environmental-controlled nature of disease (Gongora-Canul 2010). *F. virguliforme* is capable of causing root rot that affects root mass and produces a toxin that is responsible for foliar symptoms (Njiti et al. 1998; Kazi et al. 2008). Some of soybean cultivars have a dual resistance to SDS leaf scorch and root infection (Njiti et al. 1997, 2001, 2003; Hartman et al. 1997) such as 'Forrest' (Hartwig and Epps 1973) and 'Ripley' (Njiti et al. 1997), but more genotypes need to be explored as sources of resistance (Hartman et al. 1997; Rupe et al. 1989; Iqbal et al. 2001). Although, the development of resistant varieties with high-yield performance requires considerable effort and time, it is considered the best approach to manage SDS (Westphal et al. 2008).

Genomic-assisted breeding using molecular techniques and quantitative trait loci (QTL) for cultivar screening shows promise. Previous reported projects screened or developed genotypes with resistance to SDS and identified important QTL closely linked to resistance genes. Stephens et al. (1993) identified a single dominant gene (*Rfs*) in soybean variety 'Ripley', while another locus for resistance to root infection (*Rfs1*) was detected on linkage group (LG) G in 'Essex' by 'Forrest' population. In the same population, Hnetkovsky et al. (1996), Chang et al. (1996) and Kassem et al. (2006) reported QTL conferring SDS on nine LG, e.g., A2, C2, D2, F, G, I, J, L and N. There are other QTL found on LG-A2 in 'Ripley' by 'Spencer' (Farias-Neto et al. 2006), on LG-L in 'Minsoy' × 'Noir 1' and on LG-H in 'Essex' by 'Forrest' populations of soybeans. However, currently SoyBase (2010): the Soybean Breeder's Toolbox includes more than 56 detections of QTL for SDS, which identified in different populations. Identification of more QTL and DNA markers linked to SDS is worthwhile for cultivar development via marker-assisted selection (MAS).

Although simple sequence repeat (SSR) markers were used successfully for localizing QTL in soybeans (Hnetkovsky et al. 1995; Kassem et al. 2006; Kazi et al. 2008), single nucleotide polymorphism (SNP) markers are the most ample genetic markers available for taking research to the next level (Nicod and Largiader 2003; Fan et al. 2003; Barbazuk et al. 2007; Hsu et al. 2008). Currently, very few SNP markers/QTL are available for SDS resistance in soybean. Recently, Kassem et al. (2012) mapped 14 significant QTL linked to SDS resistance through a high-density SNP-based genetic linkage map using PI438489B by 'Hamilton' RIL population of soybean. Another SNP-based high-density genetic linkage map (Akond et al. 2013) was constructed using MD 96-5722 × 'Spencer' RIL population to map QTL for SDS resistance. Therefore, the objective of this research was to identify QTL conferring SDS resistance in a population advanced from a cross between soybean line MD 96-5722 and variety 'Spencer'.

Materials and methods

Plant material

Ninety-four $F_{5:7}$ lines generated by crossing a resistant line, MD 96-5722 (a.k.a Monocacy), with a susceptible line, Spencer (Wilcox et al. 1989), were evaluated against SDS by growing the population at Carbondale (2010) and Valmeyer (2010 and 2011) in southern Illinois, USA. Detailed information of parental lines and methods of development of RIL population can be found in Akond et al. (2013).

Screening of SDS symptoms and data analysis

SDS leaf symptoms were rated and compared to two checks, one resistant, 'Ripley' (Cooper et al. 1990), and one susceptible, 'Spencer' (Wilcox et al. 1989), as close as possible to the R_6 stage (Fehr et al. 1971) when seeds fill the pod cavity, but are not yet entered to senescence. SDS was rated by two scores; disease incidence (DI), which is the percentage of plants with SDS symptoms in a plot, and disease severity (DS). DS is rated on a 1–9 scale with 1 describing mild symptoms and 9 being the premature death of the plant: (1) 0–10 % where 1–5 % of leaf surface chlorotic/necrotic, (2) 10–20 % where 6–10 % of leaf surface chlorotic/necrotic, (3) 20–40 % where 10–20 % of leaf surface chlorotic/necrotic, (4) 40–60 % where 20–40 % of leaf surface chlorotic/necrotic, (5) >60 % where more than 40 % of leaf surface chlorotic/necrotic, (6) up to 33 % premature defoliation, (7) up to 66 % premature defoliation, (8) >66 % premature defoliation, and (9) premature death of plant. These two scores were used to calculate disease index (DX) with the formula (DI × DS)/9 (Njiti et al. 1996) and was used for QTL analysis. Heritability estimates and variation of SDS were calculated using the linear–bilinear model, i.e., additive main effects and multiplicative interactions (AMMI) function of JMP Genomics 6 (SAS Institute, Cary, NC). Main effects that tested were environment, replication, genotype, and their interactions.

Genetic map and QTL identification

MD 96-5722 by 'Spencer' RIL population was genotyped using 5,376 SNPs through SoySNP6K Illumina Infinium BeadChip array. The genetic linkage map (Akond et al. 2013) was constructed through JoinMap 4 (Kyazma BV, Wageningen, Netherlands; Van Ooijen 2001). Composite interval mapping (CIM) was used to detect QTL from genotypic and phenotypic data using WinQTLCart 2.5 software (Wang et al. 2005). Model 6 with four parameters for forward and backward stepwise regression, 10 cM window size, 1 cM step size and five (5) control markers were chosen for running WinQTLCart (Wang et al. 2005). Threshold was determined by permutations in 1,000 times.

Results

Frequency distribution of DX was nearly normal as skewness and kurtosis values were <1.00 (Fig. 1). Mean squares of environment (E), replication (R), and genotype (G) were significant at $P \leq 0.0001$, while G × E interaction was significant at $P \leq 0.005$ for DX (Table 1). DX

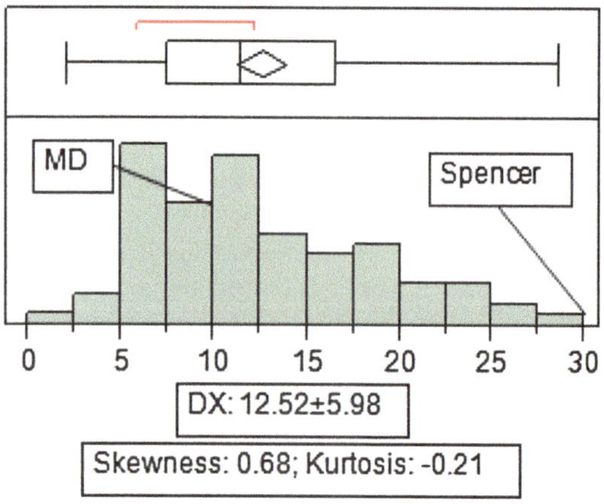

Fig. 1 Phenotypic distribution of mean values of disease index (DX) from three different growing environments (Carbondale, IL in 2010 and Valmeyer, IL in 2010 and 2011) of MD 96-5722 and 'Spencer' recombinant inbred line (RIL) population. Mean values and standard deviation (mean ± SD) of DX, skewness and kurtosis are shown at the *lower* part of figure

Table 1 G × E interaction pattern from the additive main effect and multiplicative interaction (AMMI) of RIL population of disease index (DX); RIL population was grown in three different environments i. e., Carbondale, IL in 2010, Valmeyer, IL in 2010 and 2011

Main effect	Disease index, DX (CV: 47.80; heritability: 0.200)			
	DF	MS	*F* value	Pr > *F*
Environment (E)	2	27,822.03	292.63	<0.0001
Replication (R)	3	4,082.88	42.94	<0.0001
Genotype (G)	97	312.82	3.29	<0.0001
G × E	194	138.57	1.46	0.0018

variation was 47.9 % and heritability was as low as 0.20, which is probably the consequence of high environmental variances (Table 2).

Table 2 and Fig. 2 show chromosomal locations and parameters associated with SDS QTL based on DX as calculated from data obtained from three different environments, i.e., Carbondale, IL in 2010 (C-10) and Valmeyer, IL in 2010 (V-10) and Valmeyer, IL in 2011 (V-11). QTL nomenclature was designated based on SoyBase principles and contained abbreviation of disease index (DX) then the serial number of QTL and name of each environment or year in parenthesis. CIM identified 19 QTL for SDS and platted on 11 separate chromosomes of soybean genome (Table 2; Fig. 2). Two QTL for DX [*qDX001* (*Car.10*); *qDX002* (*Val.11*)] were identified on same peak position (15.7 cM) with same marker interval ss244562583-ss244554797 (Table 2; Fig. 2) on LG D1a/Chr_1. QTL for DX had peak LOD score of 7.36 [*qDX001* (*Car.10*)] and 3.61 [*qDX002* (*Val.11*)] with additive effect of 11.77 and −0.14, respectively (Table 2). LG-N/Chr_3 had also two QTL for DX [*qDX003* (*Car.10*); *qDX004* (*Val.11*)]. DX QTL *qDX003* (*Car.10*) and *qDX003* (*Car.10*) had peak LOD score of 9.19 and 3.48 with corresponding additive effect of 13.24 and 2.96 (Table 2). CIM analysis identified two QTL for DX [*qDX005* (*Car.10*); *qDX005* (*Val.11*)] on LG-A1/Chr_5 with the peak LOD score of 9.21 and 10.10 with additive effect of −1.46 and −0.28, respectively (Table 2).

LG-C2/Chr_6 had three QTL for DX [*qDX006* (*Car.10*); *qDX007* (*Val.10*); *qDX008* (*Val.11*)]. The peak LOD scores for DX QTL were 5.55 [*qDX006* (*Car.10*)], 5.60 [*qDX007* (*Val.10*)], and 7.40 [*qDX008* (*Val.11*)] with corresponding additive values 12.71, 0.95 and 0.46. On LG-K/Chr_9, one QTL for DX [*qDX009* (*Val.11*)] was identified with LOD score of 4.13 and additive effect of 0.49. One QTL for DX [*qDX01 0* (*Car.10*)] on LG-O/Chr_10 at two different peak positions (7.3 and 2.3 cM) with marker interval ss247085505-ss247098566 was identified (Table 2; Fig. 2). QTL for DX on LG-O had peak LOD score of 7.63 with additive effect of −0.80 (Table 2). QTL analysis also identified one QTL for DX [*qDX011* (*Car.10*)] on LG-F/chr_13 with peak LOD score and additive of 5.04 and −0.80, correspondingly.

CIM analysis identified three QTL for DX [*qDX012* (*Car.10*); *qDX013* (*Car.10*); *qDX014* (*Val.11*)]. These three QTL had peak LOD scores of 5.96, 17.10 and 6.39 with corresponding additive effect of 0.87, 0.45 and 14.33 (Table 2). LG-E/Chr_15 had two for DX [*qDX015* (*Car.10*); *qDX016* (*Val.11*)] with same marker interval ss248604753-ss248616287. One QTL for DX [*qDX017* (*Car.10*)] was identified on LG-J/Chr_16 having LOD score of 10.18 with additive effect of −0.58. On LG-L/Chr_19, one QTL was identified for DX [*qDX018* (*Car.10*)] having a peak LOD score and additive effect of DX QTL as 3.12 and 0.33, respectively. In total, 19 QTL for DX were identified which explained 47.80 % of total phenotypic variance (Fig. 1) in the $F_{5:7}$ RIL population of soybean line MD 96-5722 by Spencer.

Discussion

MD 96-5722 by 'Spencer' RIL population was evaluated for its reaction to SDS in Carbondale and Valmeyer, IL in 2010 and 2011. Across environments some lines in the population had significantly lower DX than MD 96-5722, while no line had a DX higher than 'Spencer' (Fig. 1). Genotype, replication and environment (location/year) showed significance at $P < 0.0001$, while genotype × environment (G × E) was significant at $P \leq 0.005$. Heritability was calculated and displayed low values for DX, which may show that environmental effect has highly influenced the population. Small heritability may also be

Table 2 Chromosomal locations and parameters associated with quantitative trait loci (QTL) of disease index (DX) in MD 96-5722 and 'Spencer' RIL population evaluated in three different environments in Carbondale, IL in 2010, Valmeyer, IL in 2010 and Valmeyer, IL in 2011

No.	QTL (Env.)	LG/Chr	Peak position (cM)[a]	2-LOD support interval (cM)[b]	Markers interval	Peak LOD[c]	R^2 (%)[d]	Additive effects[e]
1	*qDX001 (Car.10)*	D1b/Chr_1	15.7	15.60–15.90	ss244562583-ss244554797	7.36	0.92	11.77
2	*qDX002 (Val.11)*	D1b/Chr_1	15.7	15.40–16.10	ss244562583-ss244554797	3.61	0.50	−0.14
3	*qDX003 (Car.10)*	N/Chr_3	15.80	15.70–15.90	ss245026358-ss245025977	9.19	0.80	13.24
4	*qDX004 (Val.11)*	N/Chr_3	16.1	15.90–16.10	ss245025977-ss245026227	3.48	0.10	2.96
5	*qDX005 (Car.10)*	A1/Chr_5	9.50	9.20–10.00	ss245747167-ss245786667	9.21	0.04	−1.46
6	*qDX005 (Val.11)*	A1/Chr_5	9.50	8.50–11.70	ss245747167-ss245786667	10.10	0.01	−0.28
7	*qDX006 (Car.10)*	C2/Chr_6	7.30	7.20–7.30	ss246091245-ss246092064	5.55	0.89	12.71
8	*qDX007 (Val.10)*	C2/Chr_6	7.30	7.20–7.50	ss246087580-ss246092064	3.38	0.60	12.26
9	*qDX008 (Val.11)*	C2/Chr_6	7.30	7.20–7.30	ss246091245-ss246092064	7.40	0.54	0.46
10	*qDX009 (Val.11)*	K/Chr_9	0.50	0.40–0.50	ss246870684-ss246865400	4.13	0.53	0.49
11	*qDX010 (Car.10)*	O/Chr_10	2.30	1.30–3.30	ss247085505-ss247098566	7.63	0.04	−0.80
12	*qDX011 (Car.10)*	F/Chr_13	2.70	2.50–2.80	ss247942156-ss247937719	5.04	0.11	−0.08
13	*qDX012 (Car.10)*	B2/Chr_14	3.00	1.40–4.10	ss248293401-ss248275088	5.96	0.03	0.87
14	*qDX013 (Car.10)*	B2/Chr_14	15.00	12.90–18.20	ss248293401-ss248275088	17.10	0.03	0.45
15	*qDX014 (Val.11)*	B2/Chr_14	12.00	10.30–13.00	ss248293401-ss248275088	6.39	6.37	14.33
16	*qDX015 (Car.10)*	E/Chr_15	2.50	1.50–3.00	ss248604753-ss248616287	4.30	0.05	0.09
17	*qDX016 (Val.11)*	E/Chr_15	2.50	1.40–2.70	ss248604753-ss248616287	5.89	0.60	3.11
18	*qDX017 (Car.10)*	J/Chr_16	11.90	11.50–14.00	ss248983974-ss248977568	10.18	0.85	−0.58
19	*qDX018 (Car.10)*	L/Chr_19	0.10	0.0–0.70	ss250232030-ss250233870	3.12	0.01	0.33

[a] Position of peak LOD value on composite maps described previously (Coles et al. 2010)

[b] The positions that define the two LOD intervals around the position of peak likelihood for the QTL

[c] The log of odds (LOD) value at the position of peak likelihood of the QTL

[d] R^2 estimates the proportion of RIL mean variance (%) explained by the detected QTL

[e] A positive number in additive effect of the QTL indicate that the allele for susceptibility was derived from the line indicated and a negative number mean that the allele for resistance was derived from the line indicated

the consequence of greater additive genetic variation leading to large environmental variances (Barton and Turelli 1989; Roff 1997).

QTL for SDS presented herein are from a high-density SNP-based genetic linkage map of soybeans (Akond et al. 2013) using SoySNP6K Illumina Infinium BeadChip genotyping array, and, in the best of our knowledge, few other scientific groups identified SDS QTL from high-density SNP maps published in soybeans up to date (Choi et al. 2007; Hyten et al. 2010b; Vuong et al. 2010). Among 20 LG of soybean, 11 had SDS QTL of DX at three environments. Many of QTL for DX on each LG over environment were common or on in very adjacent locations. Significant QTL identified here in one environment were confirmed in another, suggesting genetic effects and modes.

Most of SDS QTL identified in our study are on different LG or regions of previously reported ones. Two QTL for DX were identified herein on Chr 1, but on different position than those identified by Kassem et al. (2012) for the same disease in 'Essex' by 'Forrest' population. There are previous reports of SDS QTL in 'Essex' by 'Forrest' and 'Pyramid' by 'Douglas' populations on LG-N (Kassem et al. 2007; Njiti et al. 2002; Hnetkovsky et al. 1996), however, the positions were different probably because those QTL were mapped with SSR markers. Up to date, there are no reports for SDS QTL on LG A1. We also detected three QTL on LG-C2 that were different from SDS QTL on the same linkage group mapped (32.8–39.2 cM positions) by Kassem et al. (2012) in PI 438489B by 'Hamilton' RIL population. Iqbal et al. (2001) detected SDS QTL by CIM on LG-C2 on about 13 cM between two microsatellite markers, BARC_Satt277 and BARC_Satt079. SDS (*Rfs*) region on LG-C2 (Kassem et al. 2006; Chang et al. 1996; Hnetkovsky et al. 1996) is very close (4.4 cM) to QTL that is reported herein (2.60–3.40 cM and 7.20–7.30; Table 2; Fig. 2). Kassem et al. (2012) mapped an SDS QTL on LG-O from SNP derived map of PI 438489B by 'Hamilton' at 13.5–15.2 cM positions and it is not in the same region of the QTL we identified in this study. We identified several QTL on LG-F and J which were found by Kassem et al. (2007) in 'Essex'

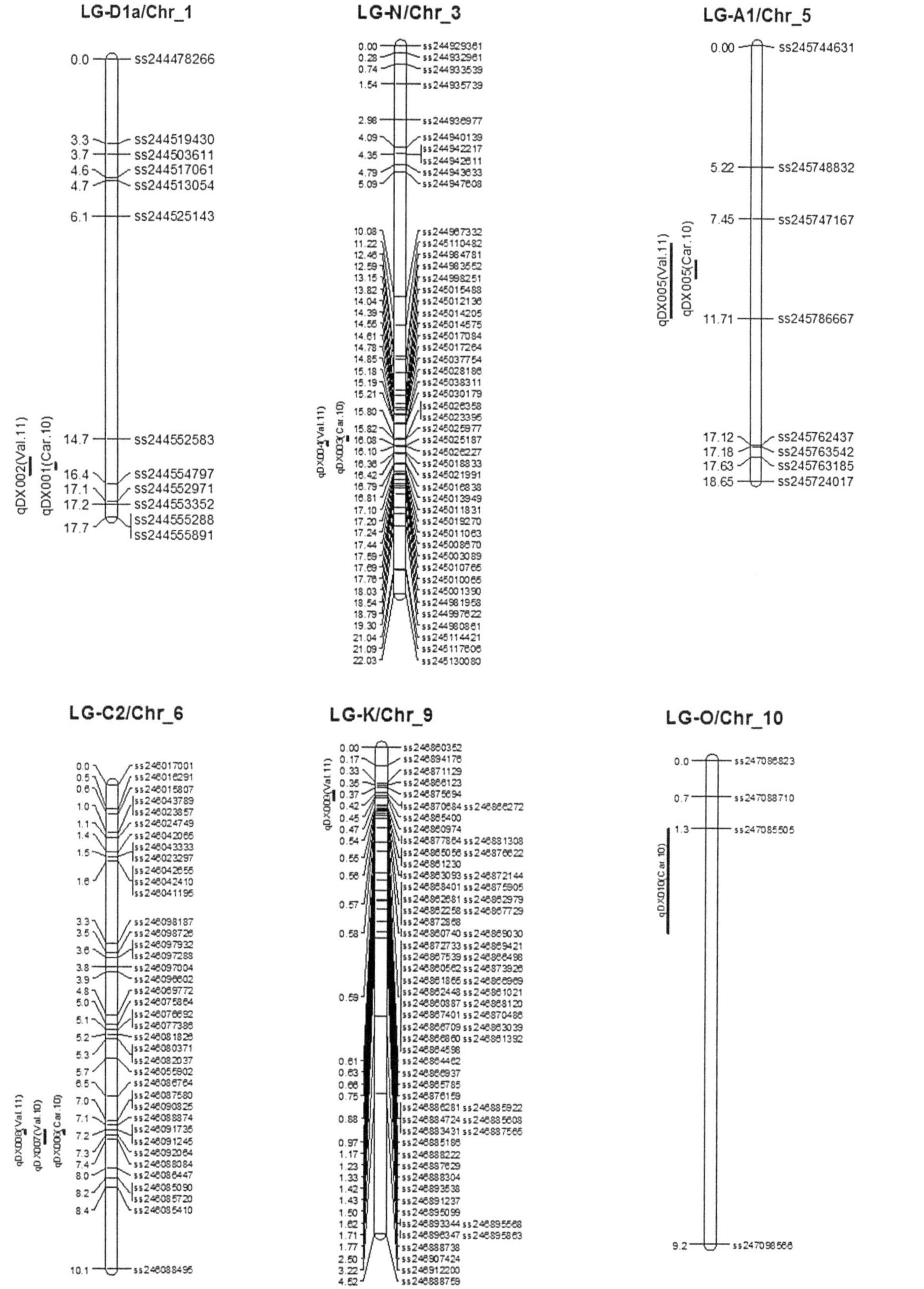

Fig. 2 Chromosomal locations of QTL of disease index (DX) in MD 96-5722 and 'Spencer' RIL population evaluated in three different environments in Carbondale, IL in 2010, Valmeyer, IL in 2010 and Valmeyer, IL in 2011

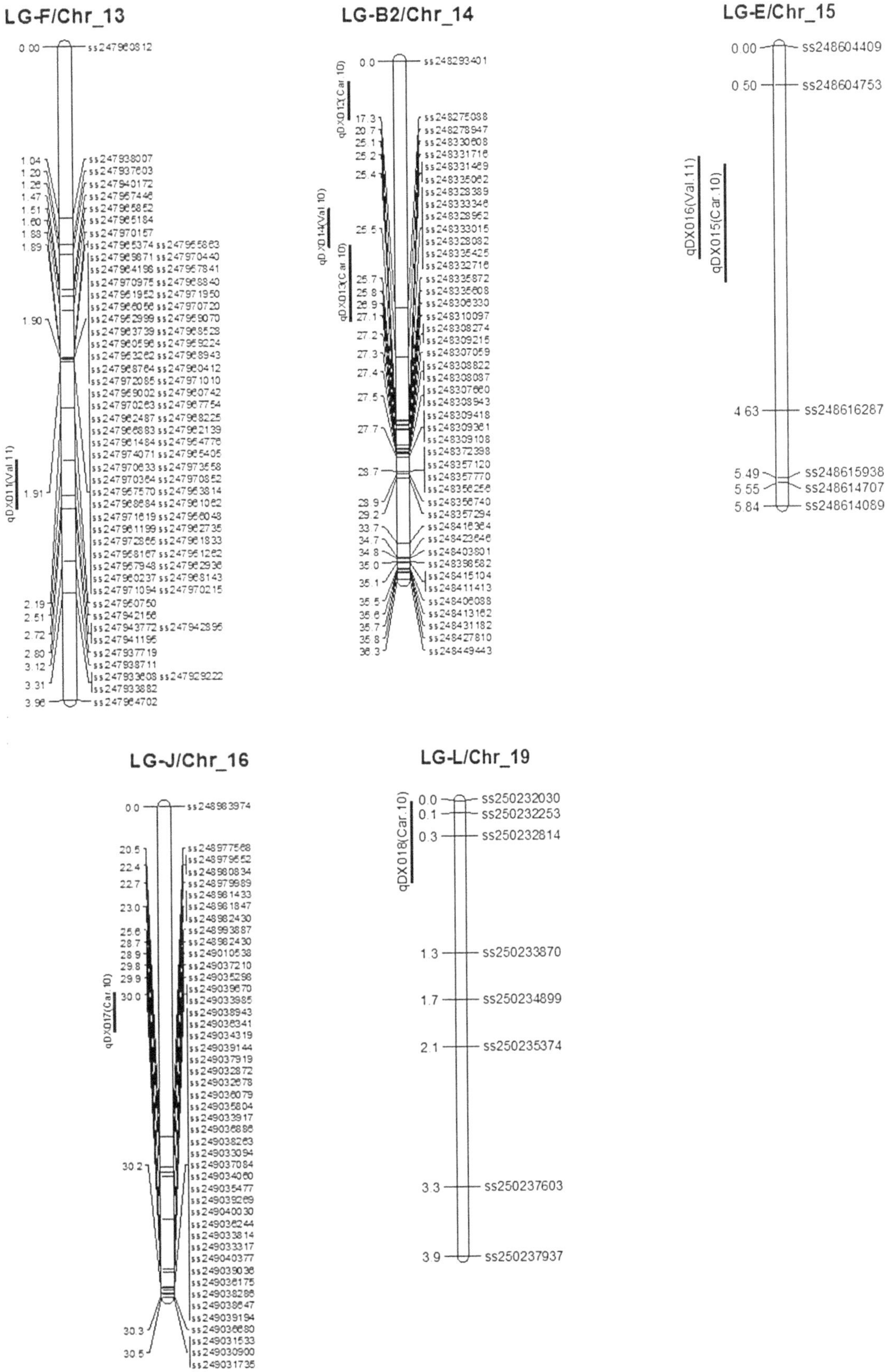

Fig. 2 continued

by 'Forrest' population, but in different positions. In this study, new QTL were identified on LG-K (one QTL), B2 (three QTL) and E (two QTL) that have not previously reported. A QTL on LG-L was identified by Kassem et al. (2012), while we are reporting three QTL in different region.

Nineteen SDS QTL identified in this report are on LG or chromosomes that are found in previous studies, but in different positions. Most of previous QTL were mapped through simple linear regression methods (SIM), not by CIM. Different types of markers and different populations that they were grown under greenhouse conditions were used in previous studies. In this study, identification of QTL was performed based SNP genotyping and phenotype data that were collected in two different seasons and locations. Also, SDS rating based on DX as proposed by Njiti et al. (1996) is traditionally used by Southern Illinois University Breeding Program and successfully has led to the development of resistant varieties (Kantartzi et al. 2012a, b). Therefore, QTL identified in this study over three different environments can be useful for soybean breeders developing new SDS resistant cultivars through marker-assisted breeding.

Acknowledgments Authors would like to thank the Department of Defense (DoD) for funding this work (Grant # W911NF-11-1-0178 awarded to M. A. Kassem and S. K. Kantartzi).

References

Akond M, Kantartzi SK, Meksem K, Song Q, Wang D, Wen Z, Lightfoot DA, Kassem MA (2013) A SNP-based genetic linkage map of soybean using the SoySNP6K Illumina infinium bead-chip genotyping array. J Plant Genome Sci 1(3):80–89

Anderson J (2012) Evaluation of soybean recombinant inbred lines for yield potential and resistance to sudden death syndrome. Southern Illinois University Carbondale, Paper 837

Aoki T, O'Donnell K, Homma Y, Lattanzi AR (2003) Sudden-death syndrome of soybean is caused by two morphologically and phylogenetically distinct species within the *Fusarium solani* species complex–*F. virguliforme* in North America and *F. tucumaniae* in South America. Mycologia 95:660–684

Barbazuk WB, Emrich SJ, Chen HD, Li L, Schnable PS (2007) SNP discovery via 454 transcriptome sequencing. Plant J 51:910–918

Barton NH, Turelli M (1989) Evolutionary quantitative genetics: how little do we know? Annu Rev Genet 23:337–370

Chang SJC, Doubler TW, Kilo V, Suttner RJ, Klein JH, Schmidt ME, Gibson PT, Lightfoot DA (1996) Two additional loci underlying durable field resistance to soybean sudden-death syndrome (SDS). Crop Sci 36:1624–1628

Choi IY, Hyten DL, Matukumalli LK, Song Q, Chaky JM, Quigley CV, Chase K, Lark KG, Reiter RS, Yoon MS, Hwang EY, Yi SI, Young ND, Shoemaker RC, Van Tassel CP, Specht JE, Cregan PB (2007) A soybean transcript map: gene distribution, haplotype and single nucleotide polymorphism analysis. Genetics 176:685–696

Coles ND, McMullen MD, Balint-Kurti PJ, Pratt RC, Holland JB (2010) Genetic control of photoperiod sensitivity in maize revealed by joint multiple population analysis. Genetics 184:799–812

Cooper RL, Martin RJ, McBlain BA, Fioritto RJ, St. Martin SK, Calip-DuBois A, Schmitthenner AF (1990) Registration of for dual 'Ripley' soybean. Crop Sci 30:963

Farias Neto A, Hartman GL, Pedersen WL, Li S, Bollero G, Diers BW (2006) Irrigation and inoculation treatments that increase the severity of soybean sudden death syndrome in the field. Crop Sci 46:2547–2554

Fan JB, Oliphant A, Shen R, Kermani BG, Garcia F, Gunderson KL, Hansen M, Steemers F, Butler SF, Deloukas P (2003) Highly parallel SNP genotyping, cold spring harbor symposia on quantitative biology. Cold Spring Harbor Laboratory Press, New York

Fehr WR, Caviness CE, Burmood DT, Pennington JS (1971) Stage of development descriptions for soybeans, *Glycine max* (L.). Crop Sci 11:929–931

Gongora-Canul CC (2010) Epidemiological studies on the infection process and symptom expression of soybean sudden death syndrome. Iowa State University. Paper 11510

Hartwig EE, Epps JM (1973) Registration of 'Forrest' soybeans. Crop Sci 13(2):287

Hartman GL, Huang YH, Nelson RL, Noel GR (1997) Germplasm evaluation of *Glycine max* for resistance to *Fusarium solani*, the causal organism of sudden death syndrome. Plant Dis 81:515–518

Hnetkovsky N, Chang SJC, Doubler TW, Gibson PT, Lightfoot DA (1996) Genetic mapping of loci underlying field resistance to soybean sudden death syndrome (SDS). Crop Sci 36:393–400

Hsu CY, An C, Saha S, Ma DP, Jenkins JN, Scheffler B, Stelly DM (2008) Molecular and SNP characterization of two genome specific transcription factor genes GhMyb8 and GhMyb10 in cotton species. Euphytica 159(1–2):259–273

Hyten DL, Cannon SB, Song Q, Weeks N, Fickus EW, Shoemaker RC, Specht JE, Farmer AD, May GD, Cregan PB (2010a) High-throughput SNP discovery through deep resequencing of a reduced representation library to anchor and orient scaffolds in the soybean whole genome sequence. BMC Genom 11:38

Hyten DL, Choi IY, Song Q, Specht JE, Carter TE, Shoemaker RC, Hwang EY, Matukumalli LK, Cregan PB (2010b) A high density integrated genetic linkage map of soybean and the development of a 1,536 universal soy linkage panel for QTL mapping. Crop Sci 50:960–968

Iqbal MJ, Meksem K, Njiti VN, Kassem My A, Lightfoot DA (2001) Microsatellite markers identify three additional quantitative trait loci for resistance to soybean sudden-death syndrome (SDS) in Essex × Forrest RILs. Theor Appl Genet 102(2/3):187–192

Kantartzi S, Klein J, Schmidt M (2012a) Registration of 'Saluki 4411' soybean with resistance to sudden death syndrome and HG Type 0 (Race 3) soybean cyst nematode. J Plant Regist 6(3):298–301

Kantartzi S, Klein J, Schmidt M (2012b) Registration of Saluki 4910 soybean (*Glycine max* (L.) Merr.) with high yield and disease resistance. J Plant Regist 7(1):31–35

Kassem MA, Shultz J, Meksem K, Cho Y, Wood AK, Iqbal MJ, Lightfoot DA (2006) An updated 'Essex' by 'Forrest' linkage map and composite interval mapping of QTL underlying six soybean traits. Theor Appl Genet 113(6):1015–1026

Kassem MA, Meksem K, Wood AJ, Lightfoot DA (2007) Loci underlying SDS and SCN resistance mapped in the 'Essex' by 'Forrest' soybean recombinant inbred lines. Rev Biol Biotech 6(1):2–10

Kassem MA, Ramos L, Leandro L, Mbofung G, Hyten DA, Kantartzi SK, Grier RL, Cianzio S, Meksem K (2012) The First 'PI438489B' by 'Hamilton' SNP-based genetic linkage map of soybean [*Glycine max* (L.) Merr.] and quantitative trait loci (QTL) that underlie SDS resistance. J Plant Genome Sci 1(2):18–30

Kazi S, Shultz J, Bashir R, Afzal J, Njiti V, Lightfoot DA (2008) Separate loci underlie resistance to soybean sudden death syndrome in 'Hartwig' by 'Flyer'. Theor Appl Genet 116:967–977

Nicod JC, Largiader CR (2003) SNPs by AFLP (SBA): a rapid SNP isolation strategy for non-model organisms. Nucl Acids Res 31(5):1–5

Njiti VN, Suttner RJ, Gray LE, Gibson PT, Lightfoot DA (1996) Rate-reducing resistance to *Fusarium solani f. sp. phaseoli* underlies to soybean sudden death syndrome field resistance in soybean because of SDS caused by the fungus *F. solani* field. Crop Sci 37:132–138

Njiti V, Gray L, Lightfoot DA (1997) Rate-reducing resistance to *Fusarium solani* f.sp. phaseoli [nee: glycines] underlies field resistance to soybean sudden-death syndrome (SDS). Crop Sci 37:1–12

Njiti VN, Doubler TW, Suttner RJ, Gray L, Gibson PT, Lightfoot DA (1998) Resistance to soybean sudden-death syndrome and *Fusarium solani* f.sp. *glycine* in near-isogenic lines. Crop Sci 38:472–477

Njiti VN, Johnson JE, Torto GA, Gray LE, Lightfoot DA (2001) Inoculum rate influences selection for field resistance to soybean sudden death syndrome in the greenhouse among recombinant inbred lines and cultivars. Crop Sci 41:1726–1731

Njiti VN, Meksem K, Iqbal MJ, Johnson JE, Kassem MA, Zobrist KF, Kilo VY, Lightfoot DA (2002) Common loci underlie field resistance to soybean sudden death syndrome in Forrest, Pyramid, Essex, and Douglas. Theor Appl Genet 104:294–300

Njiti VN, Myers O, Schroeder D, Lightfoot DA (2003) Roundup ready soybean: glyphosate effects on *Fusarium solani* root colonization and sudden death syndrome. Agron J 95:1140–1145

Ohmes A (2013) Soybean: Sudden death syndrome. Missouri Ag News. University of Missouri, St. Louis

Roff DA (1997) Evolutionary quantitative genetics. Chapman & Hall, New York

Roy KW (1997) *Fusarium solani* on soybean roots: nomenclature of the causal agent of sudden death syndrome and identity and relevance of *F. solani* form B. Plant Dis 81:259–266

Rupe JC (1989) Frequency and pathogenicity of *Fusarium solani* recovered from soybeans with sudden death syndrome. Plant Dis 73:581–584

SoyBase (2010) http://soybase.agron.iastate.edu/. Accessed June 2010

Srour A, Afzal AJ, Blahut-Beatty L, Hemmati N, Simmonds DH, Li W, Liu M, Town CD, Sharma H, Arelli P, Lightfoot DA (2012) The receptor like kinase at Rhg1-a/Rfs2 caused pleiotropic resistance to sudden death syndrome and soybean cyst nematode as a transgene by altering signaling responses. BMC Genom 13:368

Stephens PA, Nickell CD, Vodkin LO (1993) Pink flower color associated with increased protein and seed size in soybean. Crop Sci 33:1135–1137

Van Ooijen JW, Voorrips RE (2001) JoinMap®3.0: software for the calculation of genetic linkage maps. Plant Research International, Wageningen

Vuong TD, Sleper DA, Shannon JG, Nguyen HT (2010) Novel quantitative trait loci for broad-based resistance to soybean cyst nematode (*Heterodera glycines* Ichinohe) in soybean PI 567516C. Theor Appl Genet 121(7):1253–1266

Vuong TD, Wu X, Pathan MDS, Valliodan B, Nguyen HT (2007) Genomics approaches to soybean improvements. In: Varshney RK, Tuberosa R (eds) Genomics-assisted crop improvement. Genomics applications in crops, vol 2. pp 243–279

Wang S, Basten CJ, Zeng ZB (2005) Windows QTL cartographer 2.5. Department of Statistics, NCSU, Raleigh, NC. http://statgen.ncsu.edu/qtlcart/WQTLCart.htm

Westphal A, Abney TS, Xing LJ, Shaner GE (2008) Sudden death syndrome of soybean. The Plant Health Unstrictur

Wilcox JR, Roach MT, Abney TS (1989) Registration of 'Spencer' soybean. Crop Sci 29:830–831

Wrather JA, Koenning SR, Anderson TR (2003) Effect of diseases on soybean yields in the United States and Ontario (1999–2002). Plant Health Progress.

6

An extracellular thermo-alkali-stable laccase from *Bacillus tequilensis* SN4, with a potential to biobleach softwood pulp

Sonica Sondhi · Prince Sharma · Nancy George · Prakram Singh Chauhan · Neena Puri · Naveen Gupta

Abstract Degradation of residual lignin in kraft pulp by chemical bleaching is implicated in causing environmental pollution. The use of thermo- and alkali-tolerant bacterial laccases is considered to be important biological alternative to chemical processing. Laccases from *Bacillus* species have shown promise in this respect but their intracellular/spore bound presence make their industrial application economically unfeasible. We report here on a novel extracellular active thermo-alkali-stable laccase (SN4 laccase) which is active at 90 °C and pH 8.0 using 2,6-dimethoxyphenol as substrate from *Bacillus tequilensis* SN4. SN4 laccase retained 27 % activity for 5 min at 100 °C and more than 80 % activity for 24 h at 70 °C. The enzyme is also stable at a higher pH (9.0–10.0). Enzyme production was optimized by submerged fermentation. Relatively high yields (18,356 nkats ml^{-1}) of SN4 laccase was obtained in a medium containing 650 μM $MnSO_4$, 350 μM $FeSO_4$, and 3.5 % ethanol. A 764-fold increase in laccase activity was observed under optimal conditions. In addition, reduction in kappa number and increase in brightness of softwood pulp by 28 and 7.6 %, respectively, were observed after treatment with SN4 laccase without a mediator. When *N*-hydroxybenzotriazole was used as a mediator, the kappa number was decreased to 47 % and brightness was increased to 12 %.

S. Sondhi · P. Sharma · N. George · P. S. Chauhan · N. Gupta (✉)
Department of Microbiology, BMS Block, Panjab University, Chandigarh, India
e-mail: gupta27_naveen@yahoo.com

N. Puri
Department of Industrial Microbiology, Guru Nanak Khalsa College, City Centre Road, Yamunanagar 135001, Haryana, India

Keywords Extracellular laccase · *Bacillus tequilensis* SN4 · Thermo-alkali-stable laccase · Response surface methodology · Pulp biobleaching

Introduction

Laccases (benzenediol:oxygen oxidoreductases; EC 1.10.3.2) are multicopper oxidases, which catalyze the oxidation of a wide variety of organic and inorganic compounds with concomitant four-electron reduction of molecular oxygen to water. They catalyze the oxidation of both phenolic and non-phenolic substrates (Imran et al. 2012). In general, laccases oxidize phenols and aromatic amines such as methoxyphenols, phenols, polyphenols, anilines, aryl diamines, hydroxyindole, benzenethiols, and some cyanide complexes of metals such as $[Mo(CN)_8]^{4-}$, $[Fe(CN)_6]^{4-}$, $[Os(CN)_6]^{4-}$, and $[W(CN)_8]^{4-}$ (Santhanam et al. 2011). Laccases are very useful enzymes with respect to their applications in industry. They have found uses in biotechnological applications such as biobleaching, xenobiotics bioremediation, decolorization of textile dyes, biosensors, food industry, and plastic degradation (Shraddha et al. 2011).

Laccases are widely distributed in nature. They were first observed in the exudates of Japanese lacquer tree *Rhus vernicifera* (Yoshida 1883) and have henceforth been found in almost all spheres of life but have been most extensively studied in fungi including *Ascomycetes, Basidiomycetes,* and *Deuteromycetes* (Brijwani et al. 2010).

Till date, laccases from only a few bacteria (*Azospirillum lipoferum, Marinomonas mediterranea, Bacillus subtilis,* etc.) have been characterized (Sharma et al. 2007). Bacterial laccases have several properties that are not characteristics of fungal laccases such as stability at high temperature (Martins et al. 2002; Miyazaki 2005) and high pH (Singh et al. 2007), and salt tolerance (Ruijssenaars and Hartmans 2004). Also, the production of fungal laccases in large quantity is problematic due to the accumulation of fungal biomass. Still, the use of bacterial laccases could not be exploited on an industrial scale as most of the reported laccases in bacteria are intracellular or spore bound, which makes their production and purification in large quantity extremely difficult. However, analysis of several bacterial genomes showed that there are 76 % of putative laccase genes that carry the signal sequences, which could enable them to be secreted outside the cytoplasm (Ausec et al. 2011). With this viewpoint in mind, we have designed the present study for the isolation of bacteria producing extracellular laccase stable at high temperature and high pH that is suitable for industrial applications.

Moreover, for hyperproduction of this enzyme to meet the industrial standards, an attempt was made to optimize the enzyme production. In addition to the conventional stepwise approaches, recent statistics-based experimental designs including Plackett–Burman (PB) and response surface methodology (RSM) have been used (Niladevi and Prema 2008). The ability of laccase to biobleach softwood pulp was also explored.

Materials and methods

Chemicals

Guaiacol, 2,2′-Azino bis(3-ethylbenzthiazoline-6-sulfonate) (ABTS), and syringaldazine (SGZ) were purchased from Sigma (USA). Other chemicals were obtained from Himedia (India) and were of an analytical grade. Unbleached softwood sulfite pulp (Kappa number 12.71 and brightness 25) used in this study was provided by Ballarpur Industries limited, Yamunanagar, Haryana, India.

Laccase assay

The enzyme assay was performed at 90 °C for 5 min in 0.1 M phosphate buffer (pH 8.0) using 2,6-dimethoxyphenol (2 mM) (DMP) as substrate. The change in absorbance due to the oxidation of DMP was monitored at 470 nm ($\varepsilon = 14{,}800\ M^{-1}\ cm^{-1}$) in a UV–Visible spectrophotometer. The enzyme unit was expressed in nanokatals ml^{-1}. One nkat of enzyme activity was defined as nanomoles of substrate converted into product per second by one ml of enzyme.

Isolation of laccase-producing bacteria

Laccase-producing bacteria were isolated on M162 medium (Degryse et al. 1978) supplemented with 0.2 % yeast extract, 0.2 % tryptone, 100 μM $CuSO_4$, and 2 mM guaiacol. The sludge samples from effluent treatment plant of paper and textile industry and soil samples from the areas where wood was decaying were enriched and appropriate dilutions were plated. The plates were incubated at 37 °C for 48 h. The colonies showing reddish brown color were selected.

Selection of bacterial strains producing extracellular laccase

Selected bacterial cultures were screened for the presence of laccase enzyme. For this, 0.1 % of overnight-grown bacterial culture was inoculated in M162 broth containing 0.2 % yeast extract, 0.2 % tryptone, and 100 μM $CuSO_4$. After 48 h, the culture supernatant was obtained by centrifugation at 7,826×*g* for 15 min and was used as extracellular enzyme. Cells obtained were washed twice, resuspended in 5 ml of 0.1 M Tris–HCl buffer (pH 8.0), disrupted by ultrasonication, and centrifuged at 7,826×*g* for 10 min at 4 °C. The supernatant obtained from the cell extract was used as intracellular enzyme. The laccase activity was examined in both intracellular as well as extracellular enzyme preparations using DMP as substrate. The strains exhibiting extracellular laccase activities were selected. Furthermore, the oxidation of SGZ, ABTS, and tyrosine was also assayed to confirm the presence of true laccase activity.

Time course studies of isolate SN4 for growth and enzyme yield

For analyzing the growth properties as well as laccase production, M162 medium was inoculated with 0.1 % of 24-h-old culture of isolate SN4 and incubated at 37 °C, 150 rpm for 0–120 h. After specific time intervals, optical density was monitored at 600 nm and appropriate dilutions of the culture were plated on M162 agar plates for colony-forming unit (cfu) counts. The culture was centrifuged at 7,826×*g* for 10 min, and the extracellular as well as intracellular laccase activity was assayed according to the standard assay conditions.

Analysis to show the absence of laccase activity in spores

Spores of the isolate SN4 were obtained according to the method given by Schaeffer et al. (1965). The spores were harvested and suspended in 50 mM Tris–HCl buffer (pH 8.0), and laccase activity was assayed in spore suspension.

Identification of the isolate

The morphology of the isolate was studied by Gram staining. Physiological and biochemical characterization was done according to Bergey's Manual of Determinative Bacteriology (Holt et al. 1994). Furthermore, the organism was identified based on 16S rDNA sequence, wherein the genomic DNA was isolated and 16S rDNA fragment was amplified. The product was sequenced using universal primers. Sequences homologous to isolate SN4 were obtained using EzTaxon. Sequences with high query coverage and homology were selected for phylogenetic analysis (Chun et al. 2007). Multiple sequence alignment was done using MultAlin (version 5) (Corpet 1988), and phylogenetic tree was constructed using MEGA version 4.0 (Tamura et al. 2007).

Partial purification of SN4 laccase

The enzyme was partially purified by acetone precipitation. Chilled acetone (−20 °C) at a concentration of 60 % was added to crude enzyme preparation, kept for 2−3 h at −20 °C, and then centrifuged at 7,826×*g* for 15 min at 4 °C. The supernatant was discarded, and pellet was kept at room temperature till the residual acetone evaporated. It was then dissolved in 0.1 M Tris–HCl buffer (pH 8.0) (Adegoke et al. 2012).

Effect of temperature and pH on laccase activity and stability

Effect of temperature on laccase activity was determined by performing enzyme assay at different temperatures (55–100 °C). Thermostability of enzyme was measured over the temperature range of 65–100 °C by incubating the enzyme in thin-wall test tubes for 0–4 h. At different time intervals, aliquots of enzyme were taken and chilled in an ice-water bath. The residual enzyme activity was determined as per standard assay procedure.

The optimum pH for laccase activity using DMP (2 mM) as substrate was determined by performing the enzyme assay at different pH values ranging from 6.0 to 7.5 (0.1 M phosphate buffer), 8.0 to 9.0 (0.1 M Tris–HCl buffer), and 9.5 to 10.0 (0.1 M carbonate–bicarbonate buffer). pH stability of SN4 laccase was measured over a pH range of 7.0–10.0 by incubating the enzyme in buffers of various pH for 0–3 h at room temperature. At different time intervals, aliquots of enzyme were taken and residual enzyme activity was determined at 90 °C as per protocol.

Optimization of SN4 laccase production in submerged fermentation

Optimization by one variable at a time method (OVAT)

For increasing the laccase production from *Bacillus tequilensis* SN4, optimization of various nutritional and environmental parameters including incubation time, temperature, pH, different concentrations of $CuSO_4$ (0–1,000 μM), different nitrogen sources, inoculum size, agitation rates (0–250 rpm), and metal ions was carried out by classical OVAT method. All the experiments were carried out at 30 °C and 150 rpm for 96 h.

Experimental design of Plackett–Burman

In order to screen the key factors having significant influence on laccase production, a Plackett–Burman experimental design was formulated with nineteen parameters, based on the literature search and OVAT results. The parameter evaluated was as follows: *A*: $CuSO_4$, *B*: $MnSO_4$, *C*: $MgSO_4$, *D*: $FeSO_4$, *E*: $CoSO_4$, *F*: $Al_2(SO_4)_3$, *G*: yeast extract, *H*: wheat bran, *J*: 2,6-Xylidine, *K*: pyrogallol, *L*: catechol, *M*: ferulic acid, *N*: tryptone, *O*: vanillin, *P*: vanillic acid, *Q*: methanol, *R*: ethanol, *S*: phenol, and *T*: $(NH_4)_2SO_4$. Each factor was investigated at two levels, high (+) and low (−). A design of total 20 experiments was generated by using the software Design Expert 8.0.7.1. The average of enzyme activity obtained was taken as response. The effect of individual factors on enzyme activity was calculated according to the following equation:

$$E_i = \frac{(\Sigma P_{i+} - \Sigma P_{i-})}{N} \quad (1)$$

where E_i is the effect of parameter *i* under study, P_{i+} and P_{i-} are responses (enzyme activity) of trials at which the parameter was at its high and low level, respectively, and *N* is the total number of trials. From the Pareto chart, the factors showing highest positive effects were selected for optimization using central composite design (CCD) of RSM.

CCD of RSM

The three most significant variables chosen from the PB were optimized through CCD of RSM, while the other significant factors were fixed at a constant level (same as in the basal medium). For three variables, $MnSO_4$, $FeSO_4$, and ethanol, there were six central, eight factorial, and six axial points in CCD. The response was fitted by a second-order model to be correlated with the independent parameters. The correlation between the three parameters and the response (laccase activity) was described by the following predictive quadratic polynomial equation:

$$Y = \beta_0 + \Sigma \beta_i X_i + \Sigma \beta_{ii} X_{i2} + \Sigma \beta_{ij} X_i X_j \quad (2)$$

where Y is the predicted response [laccase activity (nkats ml^{-1})], β_0 is the constant term, β_i the linear coefficients, β_{ii} the squared coefficients, and β_{ij} the interaction coefficients. The quality of fitting by the polynomial model equation was expressed using coefficient of determination R^2. Equation 2 was used to construct 3D plots.

The experimental plan consisted of 20 trials, and the three variables viz. $MnSO_4$, $FeSO_4$, and ethanol were studied at three different levels, low (−1), medium (0), and high (+1). The optimal values of the three parameters were achieved by solving the obtained polynomial equation. In addition to it, three-dimensional plots were constructed for visual observation of the trend of maximum response and the interactive effects of the significant variables on the response.

Validation of the model

The RSM model was validated further for predicted versus actual responses. Each experiment was carried out in triplicate, and the results were compared with the predicted responses.

Biobleaching of softwood pulp

Softwood pulp procured from Ballarpur Industries Limited (BILT), Yamunanagar, Haryana, India, was washed with distilled water and dried at 55 °C in an oven overnight. Pulp with 5 % consistency was treated with 40 U g^{-1} oven-dried pulp of laccase enzyme at a temperature of 65 °C and pH 9.0 for 4 h (conditions standardized in our laboratory). Pulp treated under the same conditions but without enzyme was taken as control. One set of experiment was done without any mediator, and the other set was done using 2 mM *N*-hydroxybenzotriazole (HOBT) as mediator. Both the control and enzyme-treated pulp samples were filtered through muslin cloth, washed with distilled water, and then dried in oven at 55–60 °C overnight. Kappa number and brightness of pulp were determined according to the TAPPI method (Technical Association of pulp and paper industry, Atlanta, USA) T 236 and T 452, respectively.

Statistical analysis

All the experiments were carried out in triplicate, and the mean ± SD has been reported. Data were analyzed using analysis of variance (ANOVA) by Sigma Stat version 2.03 and values that were statistically significant ($p < 0.05$) were taken.

Results

Isolation and selection of extracellular laccase-producing bacteria

Ten bacterial isolates producing laccase were selected from the sludge samples of paper mill. These isolates were examined for extracellular and intracellular laccase activity. Isolate number 4 was found to produce an extracellular laccase, which was able to oxidize laccase-specific substrates, SGZ and ABTS, but not tyrosine. This isolate was selected and designated as isolate SN4.

The growth and enzyme production profile of SN4 was studied over a period of 120 h. The expression of laccase inside the cells started at the onset of stationary phase. Extracellular laccase activity was detected in the culture supernatant after 24 h, while no lysis of bacterial cells was observed up to 96 h (Fig. 1). No laccase activity was detected in isolate SN4 spore suspension.

Identification of the isolate

On morphological and biochemical analysis, isolate SN4 was found to be similar to the organism of Genus *Bacillus*. On 16S rDNA analysis, maximum similarity (99.93 %) (GenBank accession number KF150708) was obtained with *B. tequilensis* $10b^T$ (Gatson et al. 2006). However, SN4 was different from *B. tequilensis* $10b^T$ in some of the biochemical characteristics (oxidase, citrate, xylose, arginine dihydrolase, and Voges–Proskauer test). Therefore, it was concluded that the isolate SN4 could be a new strain of *B. tequilensis*. In this study, the isolate was designated as *B. tequilensis* strain SN4 (MTCC No. 11828).

Temperature and pH optima and stability

The optimum temperature at which SN4 laccase was found to be active was in the range of 80–90 °C, maximum being at 90 °C (Fig. 2a). SN4 laccase retained 59 and 27 % activity even at 95 and 100 °C, respectively, after 5 min of the incubation at these temperatures (Fig. 2b). It was found to be more than 80 % stable at 70 °C for 24 h (data not shown).

The optimum pH of SN4 laccase was found to be 8.0 (Fig. 3a). Moreover, this enzyme was highly alkali-stable as it retained 80–90 % of initial activity after 3-h incubation at pH 7.0–10.0 (Fig. 3b).

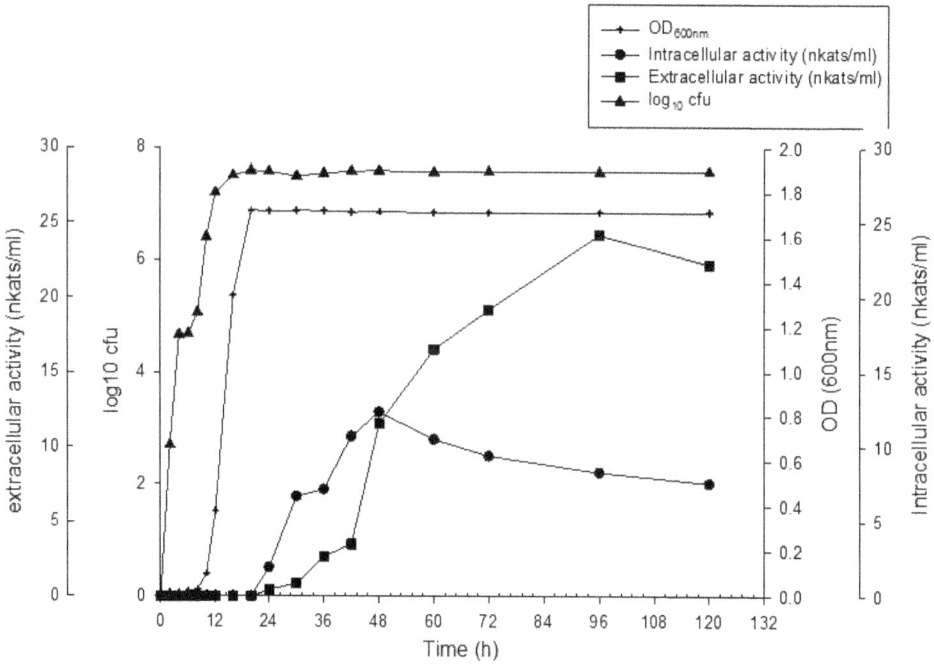

Fig. 1 Growth profile of *B. tequilensis* SN4 showing colony-forming unit (cfu), optical density (OD) at 600 nm, extracellular and intracellular enzyme production

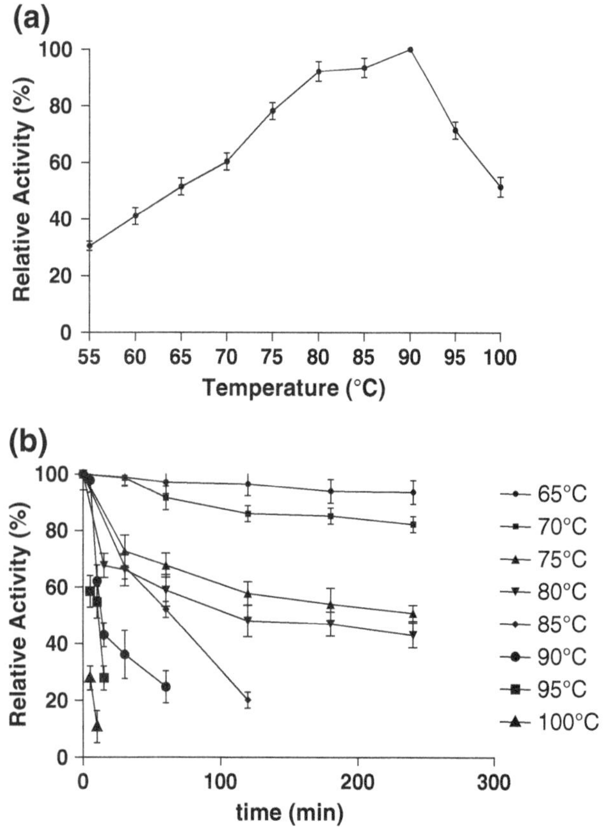

Fig. 2 Effect of temperature on laccase enzyme activity **a** optimum temperature of laccase activity **b** stability of enzyme at various temperatures

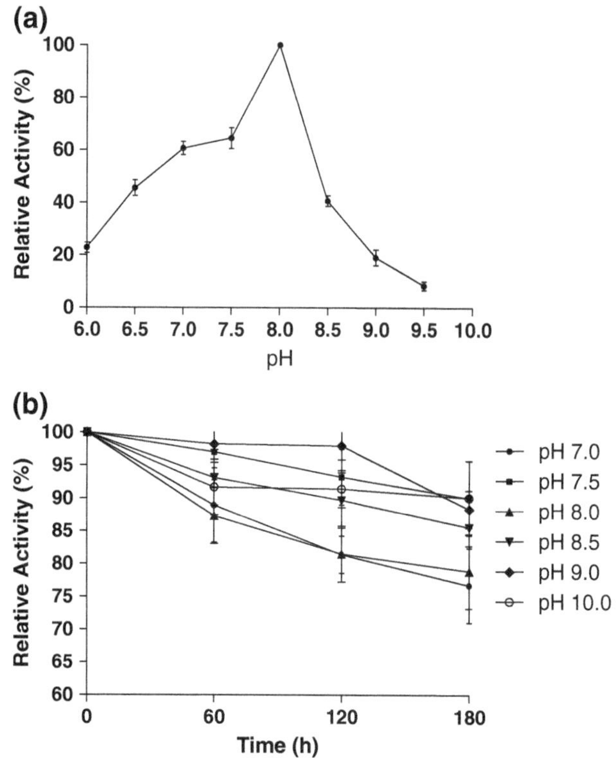

Fig. 3 Effect of pH on laccase enzyme activity **a** optimum pH **b** stability of laccase at different pH values

Optimization of enzyme production

OVAT method

In order to increase the yield of laccase, its production was optimized by classical one variable at a time method (Table 1). A remarkable increase in the enzyme activity, i.e., from 24 to 1,514 nkats ml^{-1}, was achieved under following optimized conditions viz. 96-h incubation time, 30 °C temperature, pH 8.0, 100 μM $CuSO_4$, 0.6 % yeast extract, 0.2 % tryptone, 150 rpm agitation rate, 0.3 % inoculum, and 300 μM $MnSO_4$.

Screening the significant parameters for laccase production by PB analysis

The influence of various parameters on enzyme activity was estimated and graphically represented in the form of Pareto chart (Fig. 4). Length of column represented the significance of the influence of studied parameters on enzyme activity. Out of 19 different factors, 8 factors viz. $MnSO_4$, $FeSO_4$, ethanol, tryptone, phenol, $CuSO_4$, pyrogallol, and catechol showed positive effects. Out of these, $MnSO_4$ had highest influence on enzyme activity. When the value of the effect E_i of the parameter in question was positive, the enzyme activity was more at the higher level of the parameter. However, if the value was negative, the enzyme activity was greater at the lower level of the parameter. The above stated eight factors had positive influence on enzyme activity, which indicated that the laccase production could be enhanced by these factors. Therefore, in order to obtain the maximum enzyme production, optimal concentration of $MnSO_4$, $FeSO_4$, and ethanol would be investigated with RSM.

Table 1 Optimization of laccase from *B. tequilensis* SN4 by OVAT method

Parameter	Range	Optimum range	Optimum activity (nkats ml^{-1})
Time	0–120 h	96 h	24 ± 2.07
pH	6.0–10.0	8.0	106 ± 1.38
Temperature	25–42 °C	30 °C	124 ± 1.84
$CuSO_4$	0–1,000 μM	100 μM	130 ± 2.58
Yeast extract	0.2–1.2 %	0.6 %	229 ± 0.89
Inoculum	0.05–1.0 %	0.3 %	240 ± 3.75
Agitation	0–250 rpm	150 rpm	245 ± 2.94
$MnSO_4$	0–1,000 μmol	300 μM	1,514 ± 1.25

Values represent mean ± SD ($n = 3$)

Optimization of enzyme production through RSM

Based on the PB test, $MnSO_4$, $FeSO_4$, and ethanol were identified as significant parameters, which enhanced the enzyme yield. Therefore, a CCD was formulated to investigate the optimum levels of these parameters. The experimental design and the response (enzyme activity) are presented in Table 2. By applying multiple regression analysis on the experimental data, a predictive quadratic polynomial equation was constructed to describe the correlation between enzyme activity and the three significant parameters as follows:

$$R = +16{,}696.86 + 1{,}819.23 \times A + 982.91 \times B + 1{,}546.09 \times C + 466.63 \times A \times B - 1{,}829.12 \times A \times C - 1{,}175.37 \times B \times C - 3{,}702.95 \times A^2 - 2{,}829.49 \times B^2 - 371.94 \times C^2$$

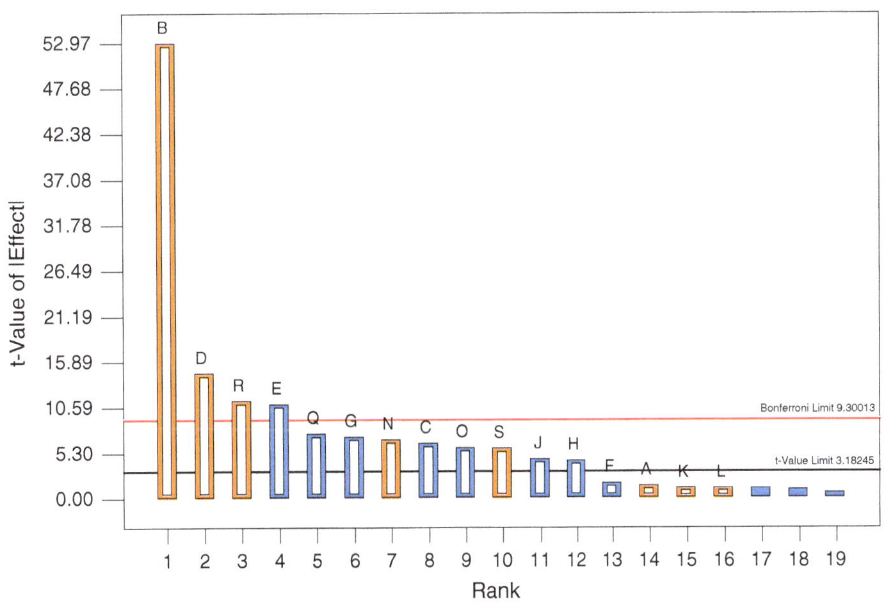

Fig. 4 Pareto chart showing significant factors (*yellow color*) and nonsignificant factors (*blue color*)

Table 2 Central composite design with predicted and actual responses

Run order	$MnSO_4$ (μM)	$FeSO_4$ (μM)	Ethanol (%)	Predicted value (nkats ml^{-1})	Actual value (nkats ml^{-1})	Residual activity
1	1,000	500	2.0	14,643.72	14,935 ± 50.97	291.28
2	300	500	5.0	10,641.03	10,845 ± 35.47	203.97
3	650	350	3.5	18,174.66	18,222 ± 48.13	47.34
4	300	200	5.0	11,812.77	11,843 ± 37.63	30.23
5	1,000	200	5.0	10,139.39	11,174 ± 30.17	34.61
6	1,000	200	2.0	9,247.46	9,365 ± 39.58	117.54
7	650	350	3.5	18,174.66	18,245 ± 27.54	70.34
8	1,000	500	5.0	11,334.16	11,509 ± 31.78	174.84
9	650	350	3.5	18,174.66	18,152 ± 25.06	−22.66
10	650	350	3.5	18,174.66	18,156 ± 37.31	−18.66
11	300	500	2.0	6,134.09	6,421 ± 38.69	286.91
12	300	200	2.0	3,104.33	3,251 ± 39.61	146.67
13	650	97.73	3.5	6,967.65	6,927 ± 25.02	−40.65
14	650	602.27	3.5	10,520.04	10,106 ± 22.47	−414.04
15	650	350	3.5	18,174.66	18,239 ± 30.92	64.34
16	1,238.63	350	3.5	9,297.63	9,085 ± 25.65	−212.63
17	650	350	0.98	12,924.89	12,579 ± 25.10	−345.89
18	61.37	350	3.5	3,549.06	3,007 ± 42.50	−242.06
19	650	350	3.5	18,174.66	18,112 ± 35.79	−62.66
20	650	350	6.02	17,464.79	18,356 ± 27.30	−108.79

Values represent mean ± SD ($n = 3$)

Table 3 Analysis of variance of response surface methodology

Source	Sum of squares	*df*	Mean square	*F* value	*p* value prob > *F*	
Model	4.171E+	9	4.635E+7	264.53	<0.0001	Significant
A	4.520E+7	1	4.520E+7	257.98	<0.0001	
B	1.319E+7	1	1.319E+7	75.31	<0.0001	
C	3.265E+7	1	3.265E+7	186.33	<0.0001	
AB	1.742E+6	1	1.742E+6	9.94	0.0103	
AC	2.677E+7	1	2.677E+7	152.77	<0.0001	
BC	1.105E+7	1	1.105E+7	63.08	<0.0001	
A^2	1.976E+8	1	1.976E+8	1,127.88	<0.0001	
B^2	1.154E+8	1	1.154E+8	658.54	<0.0001	
C^2	1.994E+6	1	1.994E+6	11.38	0.0071	
Residual	1.752E+6	10	1.752E+5			
Lack of fit	7.564E+5	5	1.513E+5	0.76	0.6148	Not significant
Pure error	9.956E+5	5	1.991E+5			
Cor total	4.189E+8	19				
Model fitting	C.V = 3.49 %	$R^2 = 0.9958$		R^2 (pred) = 0.9921	R^2 (adj) = 0.9824	

A $MnSO_4$, *B* $FeSO_4$, *C* ethanol

where *R* is the response (laccase activity); *A*, *B*, and *C* were the coded values of $MnSO_4$, $FeSO_4$, and ethanol concentration, respectively. The analysis of variance for the response surface quadratic model is summarized in Table 3. The $p < 0.0001$ indicated that the linear, interactive, and squared terms all had quite significant influence on enzyme activity. The *p* value for the lack of fit was 0.6148, indicating that this quadratic model adequately fit into the data. The determination coefficient R^2 (0.9958) indicated that the predicted and experimental values had perfect coherence with each other. The value of adjusted R^2 (0.9921) suggested that the variation of 99.21 % in the

Fig. 5 Three-dimensional response surface plots showing the effect of interaction of **a** $MnSO_4$ and $FeSO_4$ **b** $FeSO_4$ and ethanol **c** Ethanol and $MnSO_4$ on laccase production from *B. tequilensis* SN4

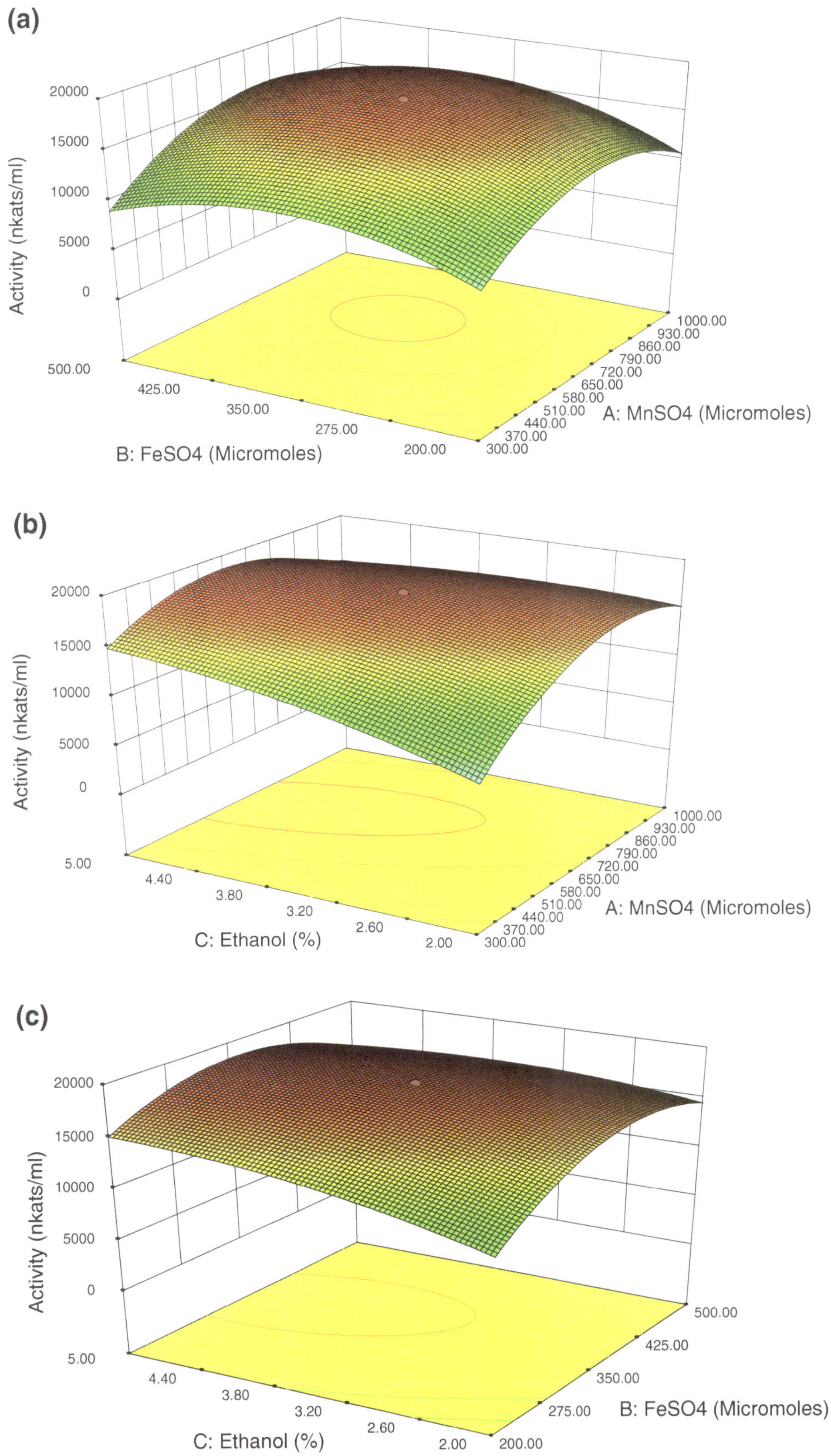

Table 4 Physicochemical properties of softwood pulp treated with laccase from *B. tequilensis* SN4

	Kappa number		Brightness	
	Kappa number	% Reduction	Brightness (% ISO)	% Increase
Control[a]	12.71 ± 0.29	–	25.00 ± 0.056	0
Laccase treated	9.147 ± 0.21	28.05	26.00 ± 0.028	7.69
LMS[#]	6.701 ± 0.28	47.29	28.00 ± 0.016	12.00

Values represent mean ± SD ($n = 3$)

[#] Laccase mediator system

[a] The control pulp was treated under same conditions with buffer but without enzyme

enzyme activity was attributed to the independent variables and only 0.79 % of the total variation could not be explained by the model. The maximum enzyme activity of 18,245 nkats ml^{-1} was obtained when 650 μM $MnSO_4$, 350 μM $FeSO_4$, and 3.5 % ethanol were added to the medium. It was observed from 3D curves (Fig. 5) that the maximum response was located inside the design boundary, which validated tested ranges of the parameters. The shape of the curve indicated whether the interactions between the factors were significant or not. The spherical curves in Fig. 5a–c indicated that the interactions between $MnSO_4$, $FeSO_4$, and ethanol were significant. According to Fig. 5, the relative significance of the impact of the parameters on enzyme activity was in the following order: $MnSO_4 > FeSO_4 >$ ethanol, which was in accordance with the results obtained from Fig. 4. Therefore, this model could be used for the prediction of laccase activity from *B. tequilensis* SN4.

Validation of the mathematic model

In order to validate the predicted results of the quadratic model, the cultivation of *B. tequilensis* SN4 was carried out under predicted optimal conditions in triplicates. The average enzyme activity of 18,356 nkats ml^{-1} ($\sim$1,101 U ml^{-1}) was obtained under optimized conditions, which was very close to the predicted value (18,174). The coherence between the experimental and estimated responses verified the validity and accuracy of the model in terms of predicting the enzyme production by *B. tequilensis* SN4. The enzyme activity obtained was 764-folds higher than the one obtained under unoptimized conditions.

Reduction in kappa number of softwood pulp with SN4 laccase

Laccase from *B. tequilensis* SN4 was applied for the biobleaching of softwood pulp. The enzyme showed 28 % reduction in kappa number and 7.6 % increase in brightness without any mediator. The use of 2 mM HOBT as mediator further improved the biobleaching by reducing the kappa number to 47 % and increasing the brightness to 12 % (Table 4).

Discussion

Most of the bacterial laccases reported so far are either intracellular or a component of bacterial spore coat protein Cot A (Sharma et al. 2007), which makes the application of bacterial laccases problematic at industrial scale. However, in a recent study on the diversity of laccase in bacterial genome, most of the potent laccase genes have been shown to carry signal sequences required for the secretion of enzyme into extracellular milieu (Ausec et al. 2011). In the present study, *B. tequilensis* SN4 was isolated from the activated sludge of paper mill effluent treatment plant, which was found to produce an extracellular laccase. In order to prove that SN4 laccase is a true extracellular enzyme and does not appear in the medium due to cell lysis, as reported in the case of γ-proteobacterium JB (Singh et al. 2007), the growth and enzyme production profile of *B. tequilensis* SN4 was studied with respect to time. It was observed that extracellular laccase production started after 24 h while no lysis of bacterial cells was observed up to 120 h. Moreover, no laccase like activity was observed in the spores as reported in other *Bacillus* spp. (Martins et al. 2002; Ruijssenaars and Hartmans 2004).

Some enzymes like tyrosinase have an overlapping substrate range with laccases. SGZ and ABTS are laccase-specific substrates. The ability of an enzyme to oxidize SGZ and ABTS, with an inability to oxidize tyrosine, is an indicator of true laccase activity (Williamson 1994). SN4 laccase was found to oxidize SGZ and ABTS but not tyrosine. This confirms that the laccase from SN4 is a true laccase. In contrast, the only other extracellular laccase from *Bacillus* sp. ADR is unable to oxidize SGZ and ABTS indicating that it is not a true laccase (Telke et al. 2011).

Besides being extracellular, some other properties of enzymes like stability at high temperature and pH values are required for it to be exploited on industrial scale. SN4 laccase was found to have optimum activity at 90 °C and was also found to be stable at high temperature (>80 % at 70 °C for 24 h). Moreover, the laccase from *B. tequilensis* SN4 is also stable at high pH. To the best of our knowledge, no other extracellular bacterial laccase has been reported with such high thermo-alkali stability. These properties also make SN4 laccase an important biomolecule for structural and functional studies.

For industrial applications, laccase production was optimized by classical OVAT and by statistical designs including PB and RSM. A 63.08-fold increase in laccase activity was observed with OVAT method. In PB, out of 19 different parameters tested, eight were found to have positive effect on laccase production. Out of these, $MnSO_4$ followed by $FeSO_4$ and ethanol were having maximum effect on enzyme production. Gene expression of laccase has been reported to be regulated mainly by Cu^{2+} but other metal ions such as Mn^{2+}, Fe^{2+}, and ethanol have also been recognized as potent inducers of laccase (Gonzalez et al. 2013). Mn^{2+} has been reported to be an effective inducer of laccase in case of *Pluerotus sajor-caju* (Soden and Dobson 2001) and *Coprinus comatus* (Lu and Ding 2010). Laccase production from *C. comatus* has been increased from 50 to 225 U (4.5-fold increase) by supplementation of 0.8 mM Mn^{2+} in the medium (Lu and Ding 2010). Different studies have shown that laccase production is regulated by metal ions such as Cu^{2+}, Fe^{2+}, and Mn^{2+} by gene expression induction or through translational or posttranslational regulation (Soden and Dobson 2001; Fonseca et al. 2010). Increase in laccase production from *B. tequilensis* SN4 by metal ions in this study can be because of similar reasons. The positive influence of organic solvents on laccase production has also been reported (Lee et al. 1999; Dhakar and Pandey 2013). Addition of ethanol to the production media increased the laccase production by 65-folds in *Pycnoporus cinnabarinus* (Lomascolo et al. 2003). These three variables were chosen to further optimize laccase production through RSM. A 764-fold increase in laccase production was observed in comparison with the unoptimized value. This enzyme yield is higher than the other reported laccases not only from bacteria but also from fungi. The use of $MnSO_4$, $FeSO_4$, and ethanol in place of known laccase inducers like veratryl alcohol and ferulic acid makes the laccase production from *B. tequilensis* SN4 both economically as well as ecologically significant.

Laccases have natural tendency to degrade lignin (Virk et al. 2012). As SN4 laccase is extracellular, thermostable, alkali-stable, and can be produced economically in large quantities, it can be a useful candidate for application in pulp biobleaching. The use of partially purified SN4 laccase was explored for biobleaching of softwood pulp. The enzyme showed significant reduction in kappa number and increase in brightness with as well as without mediator. Reduction in kappa number without using mediator further makes SN4 laccase an attractive candidate for application in pulp biobleaching.

Conclusion

It can be concluded from the present study that *B. tequilensis* SN4 produces a novel, highly thermostable and alkali-stable extracellular laccase. The production of SN4 laccase has been optimized to give higher yield at significantly lower cost. Moreover, the enzyme is capable of making significant reduction in kappa number and increase in the brightness of pulp even without the use of mediator, thus making the process industrially viable.

Acknowledgments The above work was supported by the Department of Science and Technology, New Delhi, India.

Conflict of interest The author states that there is no conflict of interest.

References

Adegoke AE, Kola OJ, Yadav A, Chandral BT (2012) Improvement of laccase production in *Pleurotus pulmonarius*-LAU 09 by mutation. J Microbiol Res 2(1):11–17

Ausec L, Zakrzewski M, Goesmann A, Schluter A, Mandic-Mulec I (2011) Bioinformatic analysis reveals high diversity of bacterial genes for laccase-like enzymes. PLoS One 6(10):e25724

Brijwani K, Rigdon A, Vadlani PV (2010) Fungal laccases: production, function, and applications in food processing. Enzyme Res 2010:149748

Chun J, Lee JH, Jung Y, Kim M, Kim S, Kim BK, Lim YW (2007) EzTaxon: a web-based tool for the identification of prokaryotes based on 16S ribosomal RNA gene sequences. Int J Syst Evol Microbiol 57:2259–2261

Corpet F (1988) Multiple sequence alignment with hierarchical clustering. Nucleic Acids Res 16(22):10881–10890

Degryse E, Glansdorff N, Pierard A (1978) A comparative analysis of extreme thermophilic bacteria belonging to the genus Thermus. Arch Microbiol 117:189–196

Dhakar K, Pandey A (2013) Laccase production from a temperature and pH tolerant fungal strain of *Trametes hirsuta* (MTCC 11397). Enzyme Res 2013:869062

Fonseca MI, Shimizu E, Zapata PD, Villalba LL (2010) Copper inducing effect on laccase production of white rot fungi native from Misiones (Argentina). Enzyme Microb Tech 46:534–539

Gatson JW, Benz BF, Chandrasekaran C, Satomi M, Venkateswaran K, Hart ME (2006) *Bacillus tequilensis* sp. nov., isolated from a 2000-year-old Mexican shaft-tomb, is closely related to *Bacillus subtilis*. Int J Syst Evol Microbiol 56:1475–1484

Gonzalez JC, Medinal SC, Rodriguez A, Osma JF, Alméciga-Díaz CJ, Sanchez OF (2013) Production of *Trametes pubescens* laccase under submerged and semi-solid culture conditions on agro-industrial wastes. PLoS One 8(9):e73721

Holt JG, Krieg NR, Sneath PHA, Staley JT, Williams ST (1994) Bergey's manual of determinative bacteriology, 9th edn. Williams and Wilkins, Waverly

Imran M, Asad MJ, Hadri SH, Sajid M (2012) Production and industrial applications of laccase enzyme. J Cell Mol Biol 10(1):1–11

Lee IY, Jung KH, Lee CH, Park YH (1999) Enhanced production of laccase in *Trametes versicolor* by the addition of ethanol. Biotechnol Lett 21:965–968

Lomascolo A, Record E, Herpoel-Gimbert I, Delattre M, Robert JL, Georis J, Dauvrin T, Sigoillot JC, Asther M (2003) Overproduction of laccase by a monokaryotic strain of *Pycnoporus cinnabarinus* using ethanol as inducer. J Appl Microbiol 94:618–624

Lu X, Ding S (2010) Effect of Cu^{2+}, Mn^{2+} and aromatic compounds on the production of laccase isoforms by *Coprinus comatus*. Mycoscience 51:68–74

Martins LO, Soares CM, Pereira MM, Teixeira M, Costa T, Jones GH, Henriques AO (2002) Molecular and biochemical characterization of a highly stable bacterial laccase that occurs as structural component of the *Bacillus subtilis* endospore coat. J Biol Chem 277(21):18849–18859

Miyazaki K (2005) A hyperthermophilic laccase from *Thermus thermophilus* HB27. Extremophiles 9:415–425

Niladevi KN, Prema P (2008) Effect of inducers and process parameters on laccase production by *Streptomyces psammoticus* and its application in dye decolourization. Bioresour Technol 99:4583–4589

Ruijssenaars HJ, Hartmans S (2004) A cloned *Bacillus halodurans* multicopper oxidase exhibiting alkaline laccase activity. Appl Microbiol Biotechnol 65:177–182

Santhanam N, Vivanco JM, Decker SR, Reardon KF (2011) Expression of industrially relevant laccases: prokaryotic style. Trends Biotechnol 29(10):480–489

Schaeffer P, Millet J, Aubert JP (1965) Catabolic repression of bacterial sporulation. Proc Natl Acad Sci USA 54:704–711

Sharma P, Goel R, Capalash N (2007) Bacterial laccases. World J Microbiol Biotechnol 23:823–832

Shraddha Shekhar R, Sehgal S, Kamthania M, Kumar A (2011) Laccase: microbial sources, production, purification, and potential biotechnological applications. Enzyme Res 2011:217861

Singh G, Capalash N, Goel R, Sharma P (2007) A pH-stable laccase from alkali-tolerant γ-proteobacterium JB: purification, characterization and indigo carmine degradation. Enzyme Microb Technol 41:794–799

Soden DM, Dobson ADW (2001) Differential regulation of laccase gene expression in *Pleurotus sajor-caju*. Microbiology 147:1755–1763

Tamura K, Dudley J, Nei M, Kumar S (2007) MEGA4: molecular evolutionary genetics analysis (MEGA) software version 4.0. Mol Biol Evol 24:1596–1599

Telke AA, Ghodake GS, Kalyani DC, Dhanve RS, Govindwar SP (2011) Biochemical characteristics of a textile dye degrading extracellular laccase from a *Bacillus* sp. ADR. Bioresour Technol 102:1752–1756

Virk A, Sharma P, Caplash N (2012) Use of laccase in pulp and paper industry. Biotechnol Prog 28:1

Williamson PR (1994) Biochemical and molecular characterization of the diphenol oxidase of *Cryptococcus neoformans*: identification as a laccase. J Bacteriol 176(3):656–664

Yoshida H (1883) Chemistry of lacquer (urushi). J Chem Soc 43:472–486

7

Strain improvement and optimization studies for enhanced production of erythromycin in bagasse based medium using *Saccharopolyspora erythraea* MTCC 1103

C. Subathra Devi · Anuj Saini · Shubham Rastogi · S. Jemimah Naine · V. Mohanasrinivasan

Abstract In the present study, *Saccharopolyspora erythraea* MTCC 1103 was used for the enhanced production of erythromycin. To enhance the yield of erythromycin, effects of various parameters such as bagasse concentration, organic nitrogen source, inorganic nitrogen source, pH and temperature were analysed. It was found that bagasse can be used as an alternate carbon source in erythromycin production medium. Erythromycin production in the new formulation of bagasse based medium was found to be 512 mg/L which was 28 % higher than glucose based medium. Strain improvement was done by random UV-mutagenesis. When compared to wild type strain, mutant strain showed 40 % higher yield in production medium. Erythromycin potency assay and HPLC analysis were performed to confirm the presence of erythromycin in the partially purified samples. These optimized conditions could be used for the commercial production of this unique antibiotic which gave significant industrial perspectives.

Keywords Erythromycin · *Saccharopolyspora erythraea* · Bagasse · UV-mutagenesis

Introduction

The discovery of novel antimicrobial compounds is currently the thirst area of research (Jemimah Naine et al. 2012). The production of complex compounds from technically convenient microorganism is an emerging route to the chemical diversity found in the surrounding environment (Jiang et al. 2013). *Saccharopolyspora erythraea* formerly identified as *Streptomyces erythraeus* is a Gram-positive filamentous bacterium. It is one of the most widely exploited strains for the industrial-scale production of erythromycin A. It is a macrolide antibiotic with broad-spectrum activity against Gram-positive pathogens (Oliynyk et al. 2007). In view of the high clinical importance of erythromycin and its derivatives, extensive efforts have been devoted to increase the erythromycin production by *S. erythraea* which has even been studied as a model system for antibiotic production (McDaniel et al. 2001; Leadlay 1997). Meanwhile, erythromycin productivity has also been enhanced by the optimization of fermentation process (El-Enshasy et al. 2008; Zou et al. 2009). Over the past 50 years, the *S. erythraea* strain improvement has been carried out mainly by multiple rounds of random mutagenesis and selection (Adrio and Demain 2006; Carata et al. 2009). Moreover, a series of derivatives of erythromycin, derived by chemical and biotechnological transformation, have been shown to have anti-parasitic, anti-neoplastic, immunosuppressant, neurotrophic, anti-inflammatory and gastro enteric therapeutic activities (Mironov et al. 2004). Erythromycin for medication purpose is generally prescribed in stearate form which is effective against *Streptococcal* and *Staphylococcal infections* (Ishii et al. 1998). Poultry farms are also dependent on this antibiotic for the treatment of bacterial and *Mycoplasma* infections in cattle. Erythromycin has its subtypes as erythromycin A (Er-A), Er-B, Er-C, Er-D, Er-G, etc., out of these types Er-A is clinically most important and specific. It is used for the synthesis of next generation azithromycin and clarithromycin. Er-B and Er-C are produced as by-products during its synthesis which have serious side effects hence not in use (Zou et al. 2010). Mutation-based

C. Subathra Devi (✉) · A. Saini · S. Rastogi · S. Jemimah Naine · V. Mohanasrinivasan
Industrial Biotechnology Division, School of Biosciences and Technology, VIT University, Vellore 632014, Tamil Nadu, India
e-mail: csubathradevi@vit.ac.in; subaresearch@rediffmail.com

enhancement in production has gained tremendous importance over the last decade. UV radiation studies have come to the forefront owing to their drastic increase in the production rate. As the antibiotic is produced by a cluster of genes, involving several operons (polycistronic), hence there are very little chances of adverse affects in this technique. Recently, double point mutation of rpsL and rsmG gene in *S erythraea* has been done which has lead to a dramatic increase in erythromycin production (Tanaka et al. 2009).

The primary aim of this work was to search for a cheap carbon source which is also readily available to suffice the basic need of a microbe without compromising the quality of the final product, erythromycin. The remarkable carbon source bagasse is a cheap and readily available substrate for the high yield of antibiotic. It laid the foundation for the future course of action which was to optimize substrate (bagasse) concentration at standard temperature and pH for maximum erythromycin production. After keeping the concentration of substrate as constant, other parameters like nitrogen source (organic and inorganic), pH, temperature were optimized to enhance the production of erythromycin, as well mainly focused on economic considerations without undermining the quality. The strain improvement studies were then carried out on the same MTCC 1103 strain so as to increase the production rate further, which has its immense value on the commercial scale. All these studies underwent had a sole purpose of reducing the cost of the primary antibiotic, erythromycin which is indispensable for the creation of future generation antibiotics like azithromycin.

Materials and methods

Microorganism and inoculum preparation

Saccharopolyspora erythraea MTCC 1103 was obtained from the Microbial Type Culture Collection, Chandigarh and maintained in ISP-2 solid medium (Enshasy et al. 2007). pH of the medium was adjusted to 7.2 and kept for sterilization. The organism was then inoculated into the sterile medium. The number of spores was counted by spread plate method. A volume of 1 mL of spore suspension (10^{10} spores/mL) was inoculated into 50 mL of seeding medium and incubated at 28 °C for 96 h on a rotary shaker at 200 rpm. 5 % seed culture was inoculated into 250 mL of fermentation medium and incubated at 28 °C for 11 days on a rotary shaker at 200 rpm (Rostamza et al. 2008). All the experiments were performed in triplicates.

Production medium and cultivation conditions

Production medium composed of (g/L): glucose-20, corn steep liquor-4; $NH_4\ (SO_4)_2$-3; NaCl-2.5 and $CaCO_3$-5 were used (Enshasy et al. 2007). The pH of the medium was adjusted to 7.2 and kept for sterilization. Bagasse was obtained from local shops and washed thoroughly with warm water and sun dried. Dried bagasse was converted into fine powder using a grinder. It was sterilized separately and added to the medium before inoculation.

Determination of biomass dry weight

After 11 days, the production broth was filtered through a Whatman No. 1 filter paper. The biomass on filter paper was carefully transferred to a sterile petri plate. The filtered broth was then centrifuged at 10,000 rpm for 20 min at 4 °C. The pellet obtained was separated from supernatant and added into the earlier petri plate containing biomass. Total biomass containing plate was kept in hot air oven at 50 °C for drying. After drying, biomass was scraped and carefully measured using an accurate weighing balance.

Standard curve of erythromycin

Erythromycin concentration was determined using UV–visible spectroscopy (Wankhade et al. 2012). Erythromycin standard was prepared by varying concentration ranging from 0.1 to 0.5 mg/mL. Subsequent amount of erythromycin was weighed and dissolved in 1 mL acetonitrile–water solution (1:1) and equal volume of concentrated sulphuric acid was added. This mixture was kept for incubation at 50 °C for 30 min. After incubation period, absorbance of various concentrations was determined at 480 nm using UV–visible spectroscopy. From the obtained values, a standard curve for erythromycin was obtained which was used for estimating unknown concentration of erythromycin.

Determination of erythromycin

After 11 days incubation, the production broth was centrifuged at 10,000 rpm for 20 min at 4 °C. Then the supernatant was transferred to sterile conical flask and equal volume of chloroform was added to it. Supernatant and chloroform formed two separate phases. The supernatant chloroform mixture was incubated overnight at room temperature on a shaker at 100 rpm. Chloroform was allowed to evaporate and erythromycin was separated. The concentration of erythromycin was then estimated using UV/Vis spectroscopy and compared to the standard curve of erythromycin.

HPLC analysis

Qualitative estimation of erythromycin was done by HPLC system equipped with a UV detector at 205 nm in C18 column. Acetonitrile:methanol:0.2 M ammonium acetate:water (45:10:10:35) was used as mobile phase at column temperature of 40 °C. Sample injection volume was 50 μL (Rostamza et al. 2008).

Erythromycin potency

Micrococcus luteus MTCC 4821 was employed for the antibacterial assay of the produced antibiotic (Holder and Boyce 1994). Samples were bioassayed against *Micrococcus luteus* MTCC 4821 using agar well diffusion assay. Crude extract from the broth as well as partially purified samples were used for the assay. Standard erythromycin concentration of 1 mg/mL was used as positive control and distilled water as negative control.

Erythromycin production and growth rate of *Saccharopolyspora erythraea*

The organisms were transferred from ISP-2 slants to 100 mL ISP-2 broth. The broth was incubated at room temperature on a rotary shaker at 200 rpm. Then 5 % (v/v) inoculum was added into 7 different sterile flasks containing production broth and incubated at room temperature on a rotary shaker at 200 rpm. Every day one flask was used for the estimation of erythromycin titre and biomass dry weight. Data were recorded and the values were plotted to obtain the growth curve of the organism.

Optimization of process parameters for erythromycin production using bagasse as a carbon source

Important process parameters such as carbon source concentration, organic nitrogen source, inorganic nitrogen source, temperature and pH have been optimized to obtain a new formulation which could ensure maximum erythromycin production. The process of optimization encompassed a total of five different experiments concerning five different parameters. Erythromycin titre and dry weight of biomass are the two important parameters which are considered for optimization.

Strain improvement studies by random UV-mutagenesis

Spore suspensions of *S. erythraea* MTCC 1103 (10^{10} spores/mL) were exposed to UV-irradiation at a distance of 6 cm for 20, 40, 60, 80, 100, 120,140 and 160 s from the UV Strata linker (wavelength 280 nm) (Khaliq et al. 2009). Experiment was carried out in dark to avoid any possible photoreaction. UV exposed spore suspensions were spread on starch casein medium plates and incubated at room temperature for 7 days.

Results

Erythromycin production and growth rate of *Saccharopolyspora erythraea*

Growth rate pattern of the wild type strain was studied in the production medium. Figure 1 shows the kinetics of dry weight biomass and erythromycin yield over a 7 days growth period. Results have suggested that the erythromycin production started from day 3 and it reached maximum on day 5 and remained constant. The maximum erythromycin concentration of 0.411 (g/L) was obtained (Fig. 1). The biomass content increased gradually and it reached the maximum on day 4. After 4 days, it entered the declined phase. The specific growth rate (μ_w) and doubling time (t_{dw}) of the wild type strain was found to be 0.735 day^{-1} and 0.94 day, respectively. The standard erythromycin curve was obtained by a linear relationship. It was used for the estimation of unknown concentration of erythromycin in samples (Fig. 2).

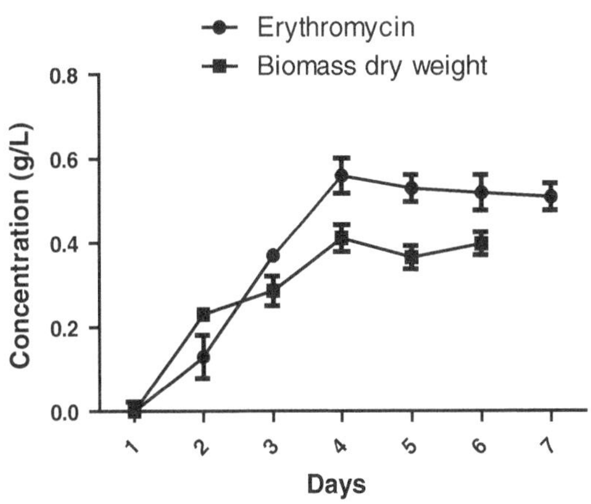

Fig. 1 Erythromycin production and growth rate of *Saccharopolyspora erythraea* MTCC 1103 (wild type strain)

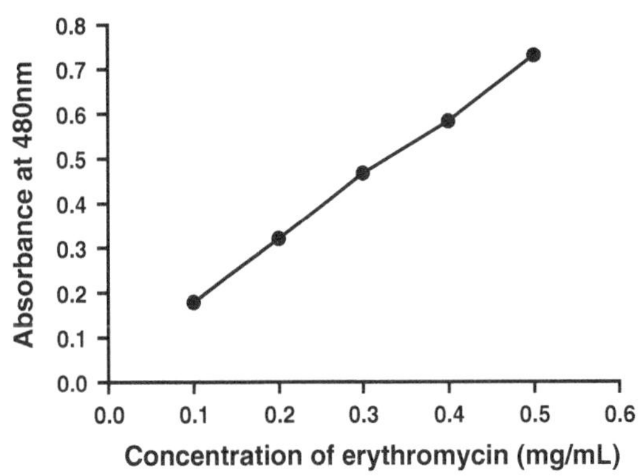

Fig. 2 Standard erythromycin curve

Optimization

Bagasse concentration

The effect of bagasse concentration on erythromycin production rate and biomass content has been represented in the Fig. 3a. The maximum concentration of erythromycin (412 mg/L) and biomass content (650 mg/L) was obtained in the production medium supplemented with 3 % bagasse. Minimum production of erythromycin (5.2 mg/mL) and biomass (100 mg/L) was obtained in 1 % bagasse supplemented production medium.

Organic nitrogen source

Nitrogen source forms a very important component of media for actinomycete growth and antibiotic production. From the results it was found out that corn steep liquor was best followed by yeast extract, malt extract, casein and peptone. The effect of organic nitrogen source on erythromycin production and biomass production has been represented in the Fig. 3b. The other media components were kept constant and 3 % bagasse was used. For corn steep liquor, concentration of erythromycin was found to be 416 mg/L and dry weight of cell biomass as 560 mg/L. In peptone added medium, erythromycin production was as low as 120 mg/L and hence it is least effective for erythromycin production. Yeast extract and malt extract had almost similar erythromycin production rate with 356 and 310 mg/L, respectively.

Inorganic nitrogen source

Inorganic nitrogen source has been found to be more effective than organic nitrogen source for actinomycetes growth. Various ammonium salts were used for optimization of inorganic nitrogen source. The effect of inorganic nitrogen source on erythromycin production and biomass production has been represented in the Fig. 3c. Corn steep liquor was used as an organic nitrogen source and rest of the parameters were kept constant. Ammonium chloride was found to be best followed by ammonium sulphate. In ammonium chloride supplemented medium, concentration of erythromycin was found to be 419 mg/L and dry weight of cell biomass as 630 mg/L whereas in ammonium sulphate it was found to be 412 and 610 mg/L, respectively. Based on the results, it was concluded that both ammonium chloride and ammonium sulphate can be used effectively as inorganic nitrogen source.

pH

pH plays an important role in growth of *S. erythraea* and erythromycin production. During the course of this experiment it was found out that the organism is very sensitive to pH. The effect of pH on erythromycin production and biomass production has been represented in the Fig. 3d. It was found that biomass was high at pH 7 and slightly higher at pH 7.5. But the erythromycin production rate was high at pH 7. At pH 7, the concentration of erythromycin was found to be 461 mg/L and dry weight of cell biomass as 500 mg/L. From the results, the optimum pH range for the growth of *S. erythraea* was found to be 7–7.5 whereas the optimum pH for erythromycin production is 7. Thus very high and low pH affected the erythromycin production rate.

Temperature

Like pH, temperature also plays an important role for the growth of *S. erythraea* and erythromycin production. Actinomycetes are sensitive to temperature. Different range of temperature has been selected ranging from 4 to 47 °C. The effect of temperature on erythromycin production and biomass production has been represented in the Fig. 3e. The optimum temperature for erythromycin production was found to be 28 °C. Highest biomass and erythromycin production was obtained at 28 °C. The betterment of growth and erythromycin production was also observed at 37 °C. Results suggested that optimum temperature for erythromycin production ranges from 25 to 35 °C. At 28 °C, the concentration of erythromycin was found to be 512 mg/L and dry weight of cell biomass as 650 mg/L. These values are obtained after optimizing all the other parameters and thus they indicate maximum erythromycin production and cell biomass dry weight which could be obtained in bagasse based medium.

Strain improvement studies by UV-mutagenesis

When the *S. erythraea* spore suspension was exposed to UV for less than a minute (10, 20, 30, 40, 50 s), many colonies are observed on the plates. This indicates the organism was not mutated and has grown luxuriously. When the culture was exposed for 1 min, very few colonies were observed on the plate which indicates possible mutation with the strain. No colonies were observed on plates, when the exposure time was more than a minute.

Erythromycin production and growth rate of mutant *Saccharopolyspora erythraea*

The growth rate pattern of the mutant strain was studied in the production broth. When compared to wild type strain, the mutated strain showed maximum growth rate with increased biomass and erythromycin concentration. Specific growth rate (μ_w) and doubling time (t_{dw}) of the mutant strain was 0.795 day^{-1} and 0.872 day, respectively. With

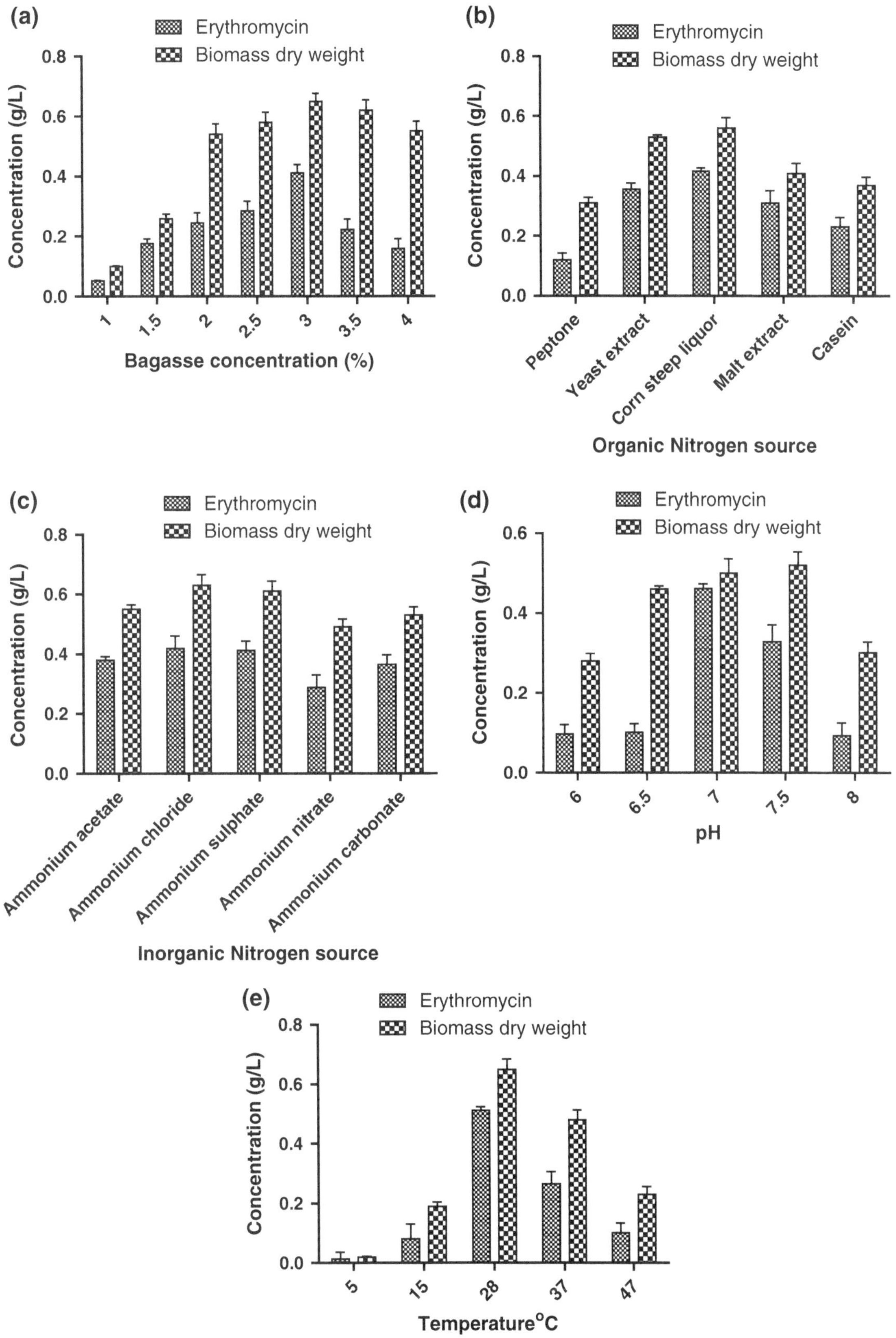

Fig. 3 Optimization of **a** bagasse concentration, **b** organic nitrogen source, **c** inorganic nitrogen source, **d** pH, **e** temperature

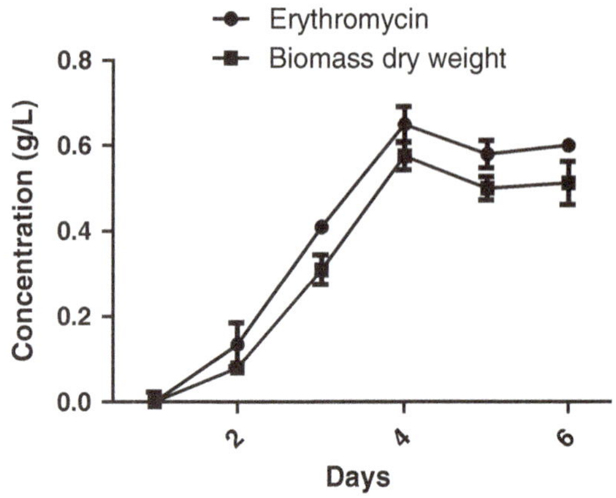

Fig. 4 Erythromycin production and growth rate of *Saccharopolyspora erythraea* MTCC 1103 (mutant strain)

Table 1 Comparison of erythromycin production in wild type and mutant strain

Type	Erythromycin (g/L)
Wild type	0.411 ± 0.024
Mutant	0.576 ± 0.012

increased biomass, erythromycin concentration was also found to be higher. Maximum erythromycin production was noticed on day 4 and remained constant. Increase in biomass was obtained from day 2 and reached maximum by day 4. Growth rate of the organisms gradually declined after day 4 (Fig. 4). Erythromycin produced by mutant and wild type strain was found to be 0.576 and 0.411 g/L (Table 1). The results clearly indicate that the productivity of erythromycin was enhanced up to 40 % than the wild strain.

Erythromycin potency assay

The clear inhibitory zone of 20 and 24 mm against *Micrococcus luteus* MTCC 4821 was observed for crude erythromycin extracted from wild type and mutated strain. The partially purified extract showed 18 and 22 mm respectively (Fig. 5a–e; Table 2). When compared to crude samples, maximum zone of inhibition was observed in partially purified samples.

HPLC analysis

Samples extracted from wild type and mutant strain of *S. erythraea* in standard production medium and wild type strain in bagasse medium were subjected to HPLC analysis. Figure 6a–c shows the HPLC profile of erythromycin. Retention time and area represent the amount of erythromycin present in crude extracts (Table 3). Hence the retention time of all the peaks were found closer to 5 min which corresponds to the standard erythromycin retention time (Abdulla et al. 2008).

Discussion

The importance of a new and cheap carbon source for erythromycin production was uncovered in the previous studies. Different production media are used in the fermentation industry. Molasses has been widely used in the production medium of erythromycin (Enshasy et al. 2007). In this research, effect of bagasse on growth of *S. erythraea* MTCC 1103 and erythromycin production was evaluated. The results suggested that bagasse can be effectively used for production of erythromycin. Bagasse concentration has been identified as

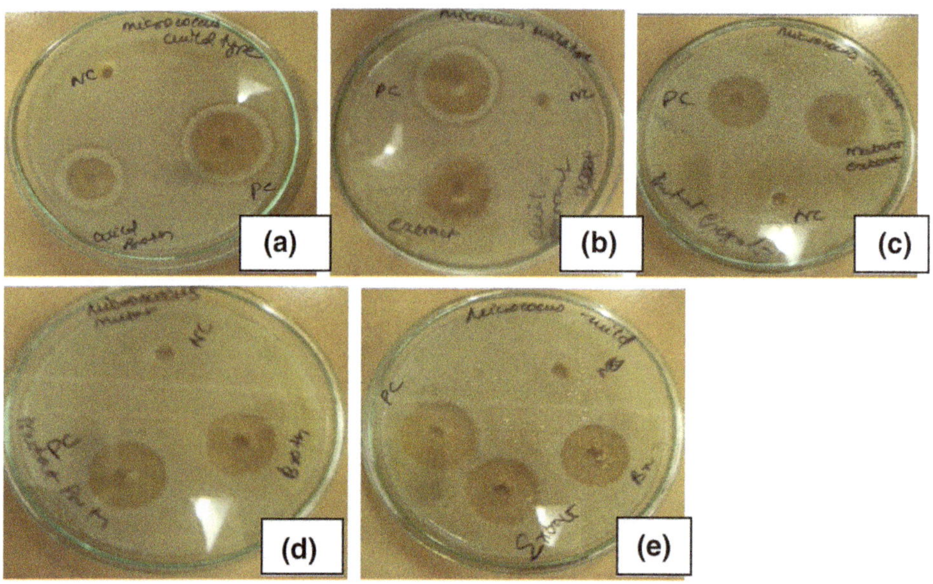

Fig. 5 Erythromycin potency against *Micrococcus luteus* MTCC 4821. **a** Crude extract of wild type strain in standard production medium, **b** partially purified extract of wild type strain in standard production medium, **c** crude extract of mutant strain in standard production medium, **d** partially purified extract of mutant strain in standard production medium, **e** partially purified and crude extract of wild type strain in bagasse medium

Table 2 Zone of inhibition of wild type and mutant in standard production medium and wild type in bagasse medium against *Micrococcus luteus* MTCC 4821

	Diameter of zone of inhibition (in cm)		
	Erythromycin produced by wild type strain in production broth	Erythromycin produced by mutant strain in production broth	Erythromycin produced by wild type strain in bagasse production broth
Positive control (standard erythromycin 1 mg/mL)	2.6 ± 0.134	2 ± 0.05	2 ± 0.052
Negative control (distilled water)	0	0	0
Crude extract	2 ± 0.067	1.8 ± 0.034	1.9 ± 0.21
Partially purified extract	2.4 ± 0.05	2.2 ± 0.09	2 ± 0.19

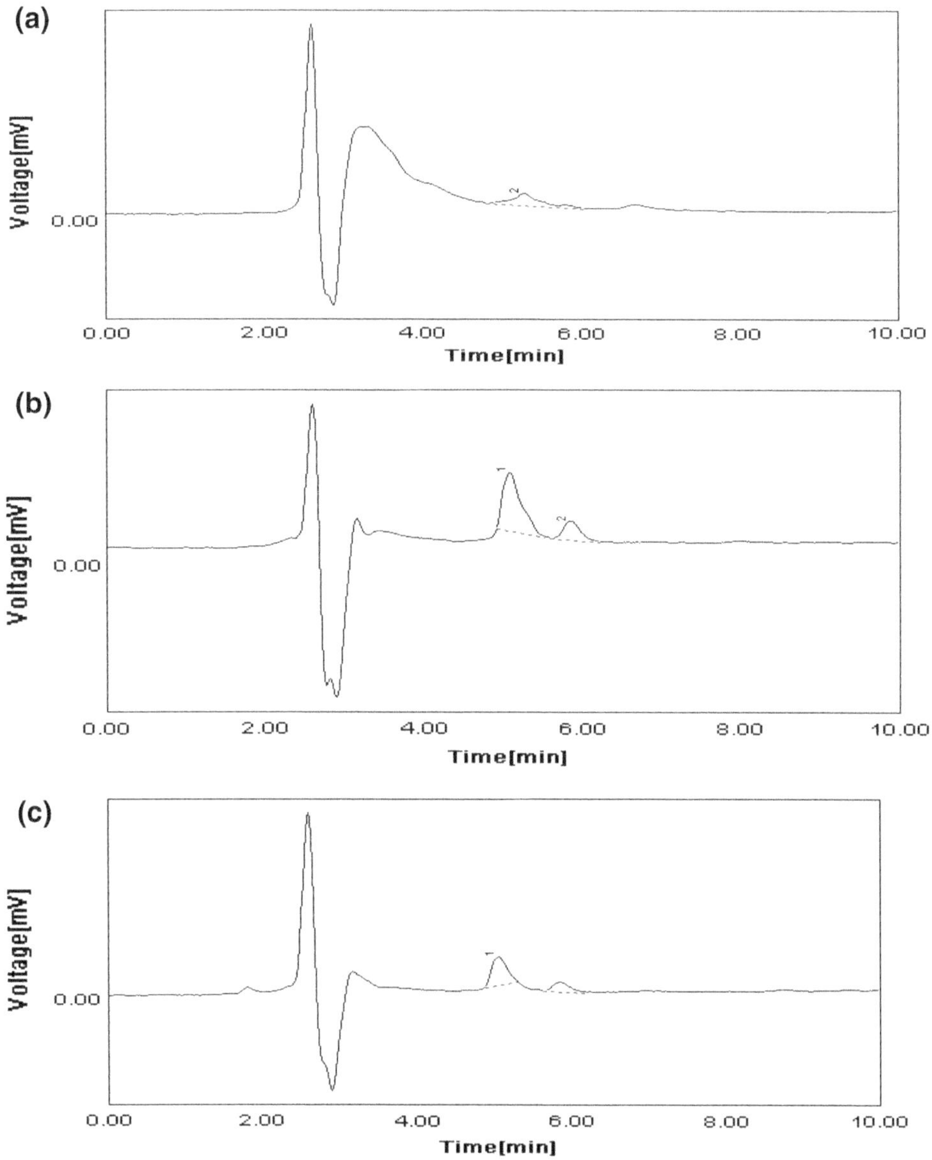

Fig. 6 HPLC chromatogram of partially purified samples. **a** Wild type strain in standard medium, **b** mutant strain in standard medium, **c** wild type strain in bagasse medium

an important parameter which affects erythromycin production as it was being used as a carbon source instead of glucose. This is the foremost report of bagasse being used for erythromycin production. Lower bagasse concentration could have resulted in low yield and higher concentration can result in lower yield because of substrate inhibition. The optimum bagasse concentration was found to be 3 % (w/v) in comparison to 2 % (w/v) conventionally used (Enshasy et al.

Table 3 Retention time and area under the peak for HPLC chromatogram

Strain and medium type	Retention time (min)	Area under the peak (mV s)
Wild type in standard medium	5.2333	133.4693
Mutant in standard medium	5.0500	289.2130
Wild type in bagasse medium	5.0167	168.2511

2007). Erythromycin production in bagasse based medium was found to be 512 mg/L which was 28 % higher compared to glucose based medium. In molasses-based medium, the antibiotic production reached up to 600 mg/L using *S. erythraea* NCIMB 8594 which is comparable to bagasse based medium. Considering the economy of sources, bagasse was better than glucose. Bagasse is cheaper than glucose and molasses. Higher antibiotic yield was obtained in glucose based medium and comparable yield to molasses-based medium.

Optimization of media parameters is essential to improve secondary metabolites production. Bagasse concentration was one of the most important parameter, as less concentration would mean compromising the production rate and higher concentration of bagasse can lead to substrate inhibition as seen in case of glucose. This lead to transient repression of antibiotic formation, maximum repression was reported at 20 mg/mL glucose concentration (Escalante et al. 1982). The optimum pH and temperature were found to be 7.0 and 28 °C. Corn steep liquor was found to be the best organic nitrogen source which corresponds to earlier findings (Enshasy et al. 2007). When compared to other salts ammonium chloride was found to be the best inorganic nitrogen source for the production of erythromycin. Earlier studies using molasses-based formulation also showed similar results with ammonium sulphate and ammonium chloride (Enshasy et al. 2007). The mutated organism showed higher growth rate and erythromycin production when compared to wild type. Earlier study involving *S. erythraea* FL2267 showed 50 % increase in antibiotic production by engineering the methylmalonyl-CoA metabolite node (Reeves et al. 2007). UV-mutagenesis was much easier to perform than metabolic engineering of specific pathways and results are comparable too.

Several studies were previously reported on *streptomyces* (Sejiny 1991). The growth kinetics of wild type and mutant *S. erythraea* was compared to the specific growth rate constant. i.e. μ_m is greater than μ_w, and doubling time of mutant was less than the wild type strain, i.e. t_{dm} is lesser than t_{dw}. This indicates that mutant strain grew faster than wild type strain. HPLC profile of the samples confirmed the presence of erythromycin. Earlier studies on HPLC chromatogram of erythromycin indicates that retention time of 5.88 min (Abdulla et al. 2008). Area under the peak for wild type strain using bagasse as carbon source is more than that of wild type using glucose as carbon source and less than the mutant strain. This suggests that erythromycin produced by wild type in bagasse is more than that produced in glucose. Also the erythromycin produced by mutant is greater than wild type in glucose and bagasse which is correlated with our earlier findings.

Conclusion

The primary aim of the study was to find a new and cheap carbon source for erythromycin production and enhance the production of erythromycin by random UV-mutagenesis of the strain *S. erythraea*. After identifying bagasse as a new carbon source it was important to optimize the process parameters for maximum production of erythromycin. Extensive literature review led to the identification of five process parameters i.e. bagasse concentration, organic nitrogen source, inorganic nitrogen source, pH and temperature. It was found out that sucrose can easily be replaced by bagasse for erythromycin production. Bagasse is cheap and easily available as waste from sugarcane industries. By random UV-mutagenesis studies, a mutant strain was obtained which showed 40 % increase in erythromycin production which is significant. The mutant strain grew faster than the wild type strain. Erythromycin is one of the most important antibiotics used in day-to-day basis and is widely manufactured and used. Thus if it is produced at a cheaper rate it would be good for the society. Further research can include production of erythromycin in bagasse medium using mutant strain and optimizing parameters for enhanced production.

Acknowledgments We are greatly indebted to Vellore Institute of Technology for the constant encouragement, help and support for extending necessary facilities.

References

Abdulla HI, Abed H, Abdulla AH, Raad M. Helmi (2008) Estimation of erythromycin concentration in saliva of healthy volunteers. MDJ 5(1):31–37

Adrio JL, Demain AL (2006) Genetic improvement of processes yielding microbial products. FEMS Microbiol Rev 14:187–214

Carata E, Peano C, Tredici SM, Ferrari F, Tala A, Corti G, Bicciato S, De Bellis G, Alifano P (2009) Phenotypes and gene expression profiles of *Saccharopolyspora erythraea* rifampicin-resistant (rif) mutants affected in erythromycin production. Microb Cell Fact 8:18

El-Enshasy HA, Mohamed NA, Farid MA, El-Diwany AI (2008) Improvement of erythromycin production by *Saccharopolyspora erythraea* in molasses based medium through cultivation medium optimization. Bioresour Technol 14:4263–4268

Enshasy HA, Mohamed NA, Farid MA, Diwang AIEI (2007) Improvement of erythromycin production by *Saccharopolyspora erythraea* in molasses based medium through cultivation medium optimization. Bioresour Technol 99(54):4263–4268

Escalante L, Lopez H, Carmen Mateos R, Lara F, Sanchez S (1982) Transient repression of erythromycin formation in *Streptomyces erythraeus*. J Gen Microbiol 128(9):2011–2015

Holder IA, Boyce ST (1994) Agar well diffusion assay testing of bacterial susceptibility to various antimicrobials in concentrations non-toxic for human cells in culture. Burns 20(5):426–429

Ishii K, Saito Y, Itai S, Nemoto M, Takayama K, Nagai T (1998) Comparative study of pharmacokinetic parameters between clarithromycin and erythromycin stearate in relation to their physicochemical properties. Drug Dev Ind Pharm 24(2):129–137

Jemimah Naine S, Nasimunislam N, Vaishnavi B, Mohanasrinivasan V, Subathra Devi C (2012) Isolation of soil actinomycetes inhabiting Amirthi forest for the potential source of bioactive compounds. Asian J Pharm Clin Res 5(4):189–192

Jiang M, Fang L, Pfeifer BA (2013) Improved heterologous erythromycin A production through expression plasmid re-design. Biotechnol Prog 29(4):862–869

Khaliq S, Akhtar K, Ghauri A, Iqbal R, Mukhtar Khalid A, Muddassar M (2009) Change in colony morphology and kinetics of tylosin production after UV and gamma irradiation mutagenesis of *Streptomyces fradiae* NRRL-2702. Microbiol Res 164(27):469–477

Leadlay PF (1997) Combinatorial approaches to polyketide biosynthesis. Curr Opin Chem Biol 14:162–168

McDaniel R, Licari P, Khosla C (2001) Process development and metabolic engineering for the overproduction of natural and unnatural polyketides. Adv Biochem Eng Biotechnol 14:31–52

Mironov VA, Sergienko OV, Nastasiak IN, Danilenko VN (2004) Biogenesis and regulation of biosynthesis of erythromycins in *Saccharopolyspora erythraea*: a review. Prikl Biokhim Mikrobiol 14:613–624

Oliynyk M, Samborskyy M, Lester JB, Mironenko T, Scott N, Dickens S, Haydock SF, Leadlay PF (2007) Complete genome sequence of the erythromycin-producing bacterium *Saccharopolyspora erythraea* NRRL23338. Nat Biotechnol 14:447–453

Reeves AR, Brikun IA, Cernota WH, Leach BI, Gonzalez MC, Weber JM (2007) Engineering of the methylmalonyl-CoA metabolite node of *Saccharopolyspora erythraea* for increased erythromycin production. Metab Eng 9:293–303

Rostamza M, Noohi A, Hamedi J (2008) Enhancement in production of erythromycin by *Saccharopolyspora erythraea* by the use of suitable industrial seeding-media. DARU 16(1):13–17

Sejiny MJ (1991) Growth phases of some antibiotics producing *Streptomyces* and their identification. J King Abdulaziz Univ 3:21–29

Tanaka Y, Komatsu M, Tokuyama S, Kaji A, Ikeda H, Ochi K (2009) Antibiotic overproduction by *rpsL* and *rsmG* mutants of various actinomycetes. Appl Environ Microbiol 75(14):4919–4922

Wankhade R, Bhalerao S, Panchory S, Pundir A, Pradhan R (2012) Analysis of erythromycin and benzoyl peroxide in combined dosage form by UV-visible spectrophotometry. Int J Pharm Pharm Sci 4(4):527–531

Zou X, Hang HF, Chu J, Zhuang YP, Zhang SL (2009) Enhancement of erythromycin A production with feeding available nitrogen sources in erythromycin biosynthesis phase. Bioresour Technol 14:3358–3365

Zou X, Zeng W, Chen C, Qi X, Qian J, Chu J, Zhuang Y, Zhang S, Li W (2010) Fermentation optimization and industrialization of recombinant *Saccharopolyspora erythraea* strains for improved erythromycin A production. Biotechnol Bioprocess E 15:959–968

8

Biodegradation of poly (ε-caprolactone) (PCL) film and foam plastic by *Pseudozyma japonica* sp. nov., a novel cutinolytic ustilaginomycetous yeast species

Fatma F. Abdel-Motaal · Magdi A. El-Sayed · Soad A. El-Zayat · Shin-ichi Ito

Abstract Aliphatic polyesters poly (ε-caprolactone) (PCL) and foam plastic have been shown to be biodegradable by microorganisms, which possess cutinolytic enzymes. *Pseudozyma japonica*-Y7-09, showed both high growth and enzyme activity on Yeast malt (YM) medium fed with PCL film than on YM medium. The hydrolytic enzyme activity of the culture on *p*-nitrophenyl butyrate indicated the occurrence of cutinase enzyme. This activity was confirmed by the degradation of PCL film which reached to the maximum (93.33 %) at 15 days and the degradation of foam plastic which reached 43.2 % at 30 days. These results suggest that the extracellular cutinase enzyme of *Pseudozyma japonica*-Y7-09 may be useful for the biological degradation of plastic wastes.

Keywords Plastic degradation · PCL degradation · Foam degradation · Yeast · *Pseudozyma*

Introduction

Plastics become the essential ingredients to provide a quality due to their versatility. These are now rival metals in breadth of use and in severity of applications because of their flexibility, toughness, excellent barrier and physical properties and ease of fabrication (Seymour 1971; Andrady et al. 1998; Meenakshi et al. 2002; Fang and Fowler 2003; Orhan et al. 2004). The growing amount of plastic waste is generating more and more environmental problems worldwide. To overcome this problem, the biodegradation of plastics has been subjected to extensive studies during the past three decades. Biodegradation of plastics is seen by many as a promising solution to this problem as it is environmentally-friendly. Some types of plastics have been shown to be biodegradable by microorganisms which produce enzymes. Those biodegradable plastics such as aliphatic polyesters (polycaprolactone, PCL) which are used mainly in thermoplastic polyurethanes, resins for surface coatings, adhesives for shoes and synthetic leather and fabrics, also serve to make stiffeners for shoes and orthopaedic splints, and fully biodegradable compostable bags, sutures and fibres. The chemical structure of a PCL trimer is similar to two cutin monomers, which are inducer for cutinase activity. This knowledge was helpful to find microorganisms which are able to degrade PCL (Premraj and Doble 2005). Although numerous studies (El-Shafei et al. 1998; Howard 2003; Nishida and Tokiwa 1993; Tokiwa et al. 2009) have been reported on the biodegradation of different types of plastics, the published literature on the biodegradation of plastic foams appears to be scarce. Foam biodegradation is carried out by the enzymes associated with some microorganisms like bacteria and fungi (Gautam et al. 2007).

Recently, a special focus has been given to the endophytic microorganisms that live inside the plant tissue without causing any immediate, overt effects (Hirsch and Braun 1992). Endophytes are known to produce bioactive natural products, which offer an enormous potential for exploitation for medicinal, agricultural and industrial uses (Tan and Zou 2001; Zhang et al. 2006). Special concerns are given to endophytic yeasts isolated from different

F. F. Abdel-Motaal · M. A. El-Sayed (✉) · S. A. El-Zayat
Department of Botany, Faculty of Science, Aswan University, Aswân 81528, Egypt
e-mail: magradi2000@yahoo.com

S. Ito
Department of Biological and Environmental Sciences, Faculty of Agriculture, Yamaguchi University, Yamaguchi 753-8515, Japan

plants (Larran et al. 2001, 2002; Cao et al. 2002; Tian et al. 2004; Nassar et al. 2005; Xin et al. 2009). Among them, smut yeast-like fungi of the genus *Pseudozyma* have attracted considerable attention as producers of enzymes (Pandey et al. 1999). Seo et al. (2007) reported that *Pseudozyma jejuensis* is able to degrade plastics waste. However, very few information is available so far about the biochemical activity of endophytic *Pseudozyma* spp.

Although PCLs have been shown to be degraded by enzymes secreted by a number of bacteria (Benedict et al. 1983), little studies have been done using yeast. A new yeast strain isolated recently from the medicinal plant *Hyoscyamus muticus* in our laboratory was identified as a novel species of the genus *Pseudozyma japonica*-Y7-09, (Abdel-Motaal et al. 2009). Evaluation of the PCL and foam plastic biodegradation by this new species was the objective of this study.

Materials and methods

Fungal strain and media

The yeast *strain Pseudozyma japonica*-Y7-09, from *Hyoscyamus muticus* (Egyptian henbane) was isolated as a new species of the genus *Pseudozyma*.

Yeast culture and PCL films, foam plastics degradation assay

The *Pseudozyma* strain was grown on the yeast malt agar (YMA) (glucose 1.0 % w/v, peptone 0.5 % w/v, yeast extract 0.3 % w/v, malt extract 0.3 % w/v, agar 2.0 % w/v), and the plates were incubated for 2 days at 25 °C. A single colony of yeast species was inoculated into 5 ml of YM broth and cultured at 30 °C. After 24 h, 1 ml of culture broth was inoculated into 1,000-ml flask contain 50 ml YM broth and incubated at 30 °C with shaking at 200 rpm over night. Sterilized dry PCL film of known weight was added to each culture in addition to positive control (YM broth culture of each *Pseudozyma* species without PCL film) and negative control (YM broth with PCL film only). Flasks were incubated at 30 °C with shaking at 200 rpm for 15 days. Three replicas from each flask have been done considering the different time intervals (3, 5, 8, 12 and 15 days) to study the degradation behaviour. Under the same conditions, foam plastics were incubated in yeast culture but for longer time (30 days). The PCL films and foams were taken out carefully in different intervals time, washed thoroughly by double distilled water to remove any media components or yeast cells if present on the their surface and were completely vacuum dried at 30 °C over night. The percentage of weight loss was recorded.

Culture turbidity

Growth rate of yeast culture incubated with and without PCL films and foams was measured and assayed according to the culture turbidity (OD_{600nm}).

pH values

Effect of variable pH values on the growth rate, protein concentrations and enzyme activity have been studied.

Preparation of polycaprolactone (PCL) film and market's foam plastic

Poly (ε-caprolactone) (PCL) with molecular weight 70,000–100,000 Da was purchased from Sigma-Aldrich Chemical Co. PCL film was prepared in small glass petri dishes (3 cm) by dissolving polymer in acetone and left for air drying. The prepared PCL films and the 1 cm quadrate cutting foams (polystyrene with molecular weight 100,000–400,000 Da) were weighed and sterilized in 80 % ethanol for 2 min, well dried and then kept under UV lamp for 30 min and fed into the pre-grown yeast cultures.

Cutinolytic enzyme assays

Spectrophotometric assay was used to determine the cutinase activity with *p*-nitrophenyl butyrate (PNB; Sigma, ST Louis, MO, USA) and *p*-nitrophenyl palmitate (PNP; Sigma, ST Louis, MO, USA) as substrates (Kim et al. 2003).

96-well microplate was used for the hydrolysis reactions at 30 °C for 4 min. Each well contained 200 μl of an enzyme/substrate solution (solution A), comprising 106.7 μl of phosphate buffer (0.1 M, pH 8.0), 13.3 μl of Triton X-100 (4 g l^{-1}), 13.3 μl of cell-free culture broth and 66.7 μl of substrate (PNB or PNP) solution. The PNB and PNP solutions were prepared with variable concentrations ranging from 0 to 1,000 mg l^{-1}. The initial velocity [i.e. the initial maximum rate of change in absorbance ($\Delta OD_{405nm}\ s^{-1}$)] was measured using a Molecular Devices EMax 96-well microplate reader (MDS Analytical Technologies, Toronto, Canada), and the enzyme-free solution A was used as a blank. As the eight wells in each of the 12 columns of the 96-well microplate contained an equal content of enzyme and substrate, the eight initial velocities measured for each column were used to calculate an average initial velocity for a specific reaction condition, resulting in an SD <5 %.

Protein assay

Protein concentrations were determined according to Bradford method using Bio-Rad reagent (Bradford 1976).

Results and discussion

Condition favouring the degradation of polycaprolactone

The rate of degradation of biodegradable plastics is controlled not only by their chemical structure but also by environmental conditions such as temperature (Kitamoto et al. 2011). The yeast *strain Pseudozyma japonica*-Y7-09 showed the best growth rate and degradation percentage of polycaprolactone (PCL) and foam plastics in yeast malt broth media at 30 °C and at pH 6 with shacking at 200 rpm. Maximal activity of the PNB and PNP was found in the same conditions.

Enzyme activity of *Pseudozyma japonica*-Y7-09

Pseudozyma japonica-Y7-09 showed higher growth on YM medium fed with PCL film than on YM medium (Fig. 1). The substrate-specific hydrolysis and high accessibility to PNB are intrinsic properties of cutinase that have never been observed for hydrolysis by lipases and esterases (Kim et al. 2003). Isolation of natural cutinase from variable microbial sources would be easy using this distinct characteristic. The hydrolytic activities of cell-free supernatant from *Pseudozyma japonica*-Y7-09 culture broth on two *p*-nitrophenyl esters (PNB and PNP) were estimated for a range of substrate concentrations on time courses. The hydrolytic activities are defined as the initial maximum rate of hydrolysis at each substrate concentration as described previously (Kim et al. 2003; Seo et al. 2007). On the time course, the hydrolytic enzyme activity of the culture on PNB rapidly increased after 3 days and the maximum activity was at day 5. The initial maximum rate of PNB hydrolysis in YM and PCL medium was higher than that in YM medium (Fig. 2), whereas the hydrolytic activity on PNP was very low with the same range of substrate concentrations, even in YM and PCL medium. Therefore, it is

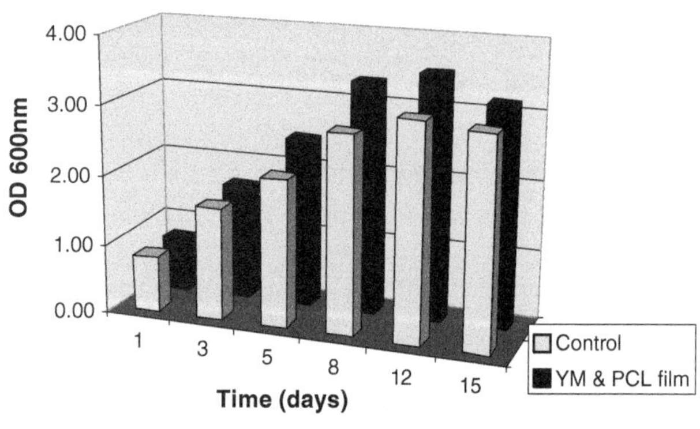

Fig. 1 Growth measured by culture turbidity (OD_{600nm}) in control (culture in YM) and culture in YM fed with PCL film (YM and PCL film)

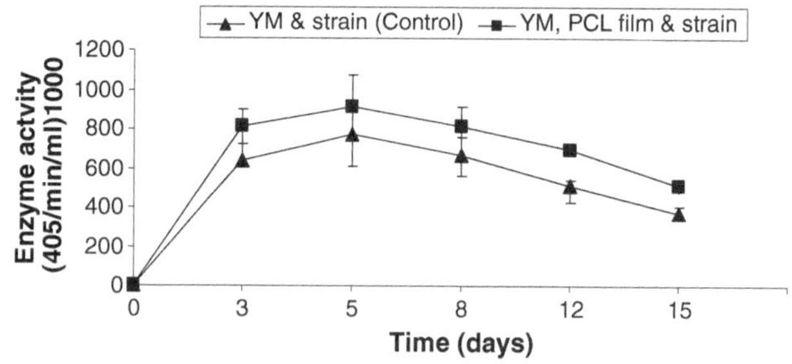

Fig. 2 Activity of enzyme secreted by *Pseudozyma japonica*-Y7-09 in yeast malt culture fed and that not fed with PCL film

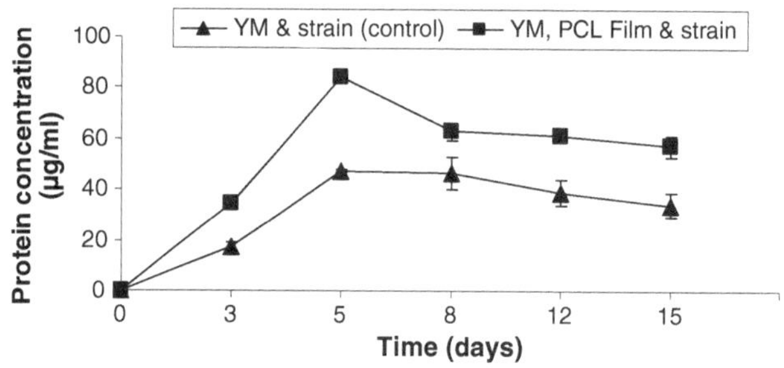

Fig. 3 Total protein concentration secreted by *Pseudozyma japonica*-Y7-09 in yeast malt culture fed and that not fed with PCL film

likely that the cutinase production by *Pseudozyma japonica*-Y7-09 was induced by PCL. These results are similar to those obtained by Kim et al. 2003, Seo et al. 2007 using *Pseudozyma jejuensis OL71*.

During growth of *Pseudozyma japonica*-Y7-09 culture broth, the protein concentration rapidly increased after 3 days with a maximum concentration at 5 days similar to enzyme activity. Protein concentration recorded higher concentration in PCL feed YM medium than that in PCL free YM medium (Fig. 3).

PCL degradation by *Pseudozyma japonica*-Y7-09

PCL dimers and trimers, products of enzymatic PCL degradation, are structurally similar to cutin degradation products that are known to be inducers of cutinase activity. This implies that oligomers of PCL may be natural inducers and substrates of cutinase (Nishida and Tokiwa 1993).

In the present study, PCL degradation was achieved in the conditions favouring the yeast strain's growth in the YM medium in Erlenmeyer flasks as described above for 15 days. The weight loss of the degraded film was recorded and compared as a function of time. Under those conditions, partial degradation of PCL film (26.88 %) was recorded at the third day of incubation. That degradation reached to the maximum (93.33 %) at the 15 days (Figs. 4, 5). Weight loss was negligible in the same medium but free yeast cells (Figs. 4, 5). In an early study (Oda et al. 1995)

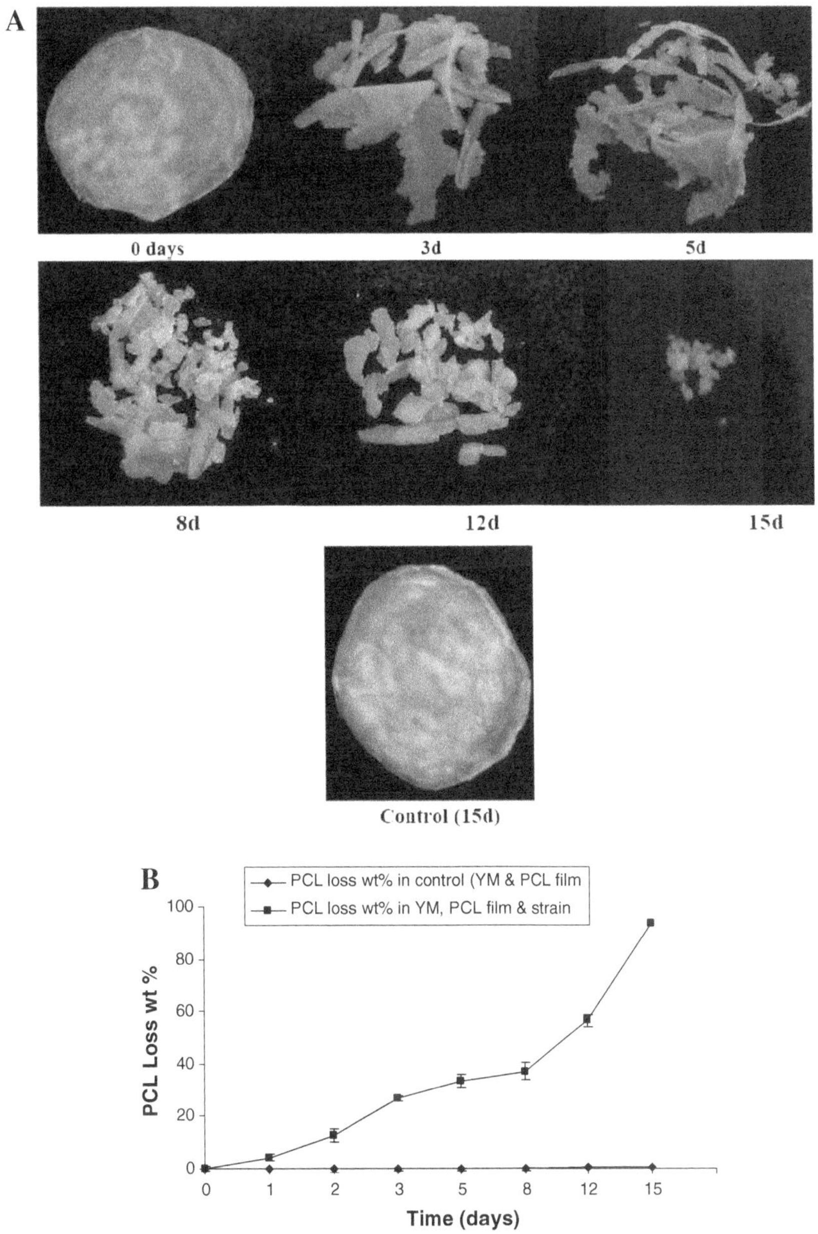

Fig. 4 **a** Degradation of PCL film after 0, 3, 5, 8, 12 and 15 days incubation time by *Pseudozyma japonica*-Y7-09 comparing with control. **b** Weight loss percentage of PCL film after 0, 3, 5, 8, 12 and 15 days incubation time by *Pseudozyma japonica*-Y7-09 comparing with control

isolated five strains of filamentous fungi of which *Paecilomyces lilacinus* was able to degrade 10 % of PCL after 10 days incubation. Xu et al. (2007) investigated the ability of several stock bacterial strains to degrade PCL. After 30 days of incubation, the weight loss of PCL film by the strain *Lysinibacillus* sp. 70,038 was 9 % while other strains showed less weight loss degradation.

Foam degradation by *Pseudozyma japonica*-Y7-09

During the growth of *Pseudozyma japonica*-Y7-09 in YM medium fed with foam disc for 30 days, the extracellular culture broth secretion exhibited an ability to degrade the foam disc. The loss weight percentage of the foam disc within 30 days incubation time was 43.2 % (Fig. 5). This ability was slower as compared to PCL degradation but it is promising for the biodegradation of foam since foams are considered to be highly resistant to biodegradation when compared with other types of synthetics plastics (Gautam et al. 2007). An increase in molecular weight results in a decline of polymer degradability by microorganisms. High molecular weights result in a sharp decrease in solubility making polymers unfavourable for microbial attack due to difficulties of the substrate to be assimilated through the cellular membrane and then further degraded by cellular enzymes. The breakdown of large polymers to carbon dioxide (mineralization) requires several different organisms, with one breaking down the polymer into its

Fig. 5 Degradation of foam plastic by *Pseudozyma japonica*-Y7-09 incubated in yeast malt media for 30 days

constituent monomers, one able to use the monomers and excreting simpler waste compounds as by-products and one able to use the excreted wastes. Our yeast strain is capable of being one in a potential series of microbial strains used for degradation of plastic foam.

Conclusions

Pseudozyma japonica-Y7-09 as a new yeast species exhibited high ability to degrade PCL and foam plastic. The hydrolytic activities of cell-free supernatant of *Pseudozyma japonica*-Y7-09 culture broth on two *p*-nitrophenyl esters (PNB and PNP) and its high accessibility to PNB are intrinsic properties of cutinase production. Further detailed studies of the enzymes produced by *Pseudozyma japonica*-Y7-09 to degrade other types of plastics are currently being undertaken, and the results of these will be important for the development of new technological processes for the biodegradation of plastic pollutants.

Acknowledgments We thank Dr. Kazunobu Matsushita (Yamaguchi University) for technical assistance.

Conflict of interest The authors declare that they have no conflict of interest in the publication.

References

Abdel-Motaal FF, El-Zayat SA, Kosaka Y, El-Sayed MA, Nassar MS, Ito S (2009) Four novel ustilaginomycetous anamorphic yeast species isolated as endophytes from the medicinal plant *Hyoscyamus muticus*. Asian J Plant Sci 8(8):526–535

Andrady AL, Hamid SH, Hu X, Torikai A (1998) Effects of increased solar ultraviolet radiation on materials. J Photochem Photobiol B Biol 46:96–103

Benedict CV, Cameron JA, Huang SJ (1983) Polycaprolactone degradation by mixed and pure cultures of bacteria and a yeast. J Appl Polym Sci 28:335–342

Bradford MM (1976) A rapid and sensitive method for the quantitation of microgram quantities of protein utilizing the principle of protein–dye binding. Anal Biochem 72:248–254

Cao LX, You JL, Zhou SN (2002) Endophytic fungi from *Musa acuminate* leaves and roots in South China. World J Microbiol Biotechnol 18:169–171

EI-Shafei H, EI-Nasser NHA, Kansoh AL, Ali AM (1998) Biodegradation of disposable polyethylene by fungi Streptomyces species. Polym Degrad Stab 62:361–365

Fang J, Fowler P (2003) The use of starch and its derivatives as biopolymer sources of packaging materials. Food Agric Environ 1:82–84

Gautam R, Bassi AS, Yanful EK (2007) A review of biodegradation of synthetic plastic and foams. Appl Biochem Biotech 1:85–108

Hirsch GU, Braun U (1992) Communities of parasitic microfungi. In: Winterhoff W (ed) Handbook of vegetation science: Fungi in vegetation science, vol 19. Kluwer Academic, Dordrecht, Netherlands, pp 225–250

Howard GT (2003) Biodegradation of polyurethane: a review. Int Biodeter Biodegrad 40:245–252

Kim YH, Lee J, Moon SH (2003) Uniqueness of microbial cutinases in hydrolysis of p-nitrophenyl esters. J Microbiol Biotechnol 13:57–63

Kitamoto HK, Shinozaki Y, Cao X, Morita T, Konishi M, Tago K, Kajiwara H, Koitabashi M, Yoshida S, Watanabe T, Sameshima-Yamashita Y, Nakajima-Kambe T, Tsushima S (2011) Phyllosphere yeasts rapidly break down biodegradable plastics. AMB Express 29:1–44

Larran S, Monaco C, Alippi HE (2001) Endophytic fungi in leaves of *Lycopersicon esculentum* mill. World J Microbiol Biotechnol 17:181–184

Larran S, Perello A, Simon MR, Moreno V (2002) Isolation and analysis of endophytic microorganisms in wheat (*Triticum aestivum* L.) leaves. World J Microbiol Biotechnol 18:683–686

Meenakshi P, Noorjahan SE, Rajini R, Venkateswarlu U, Rose C, Sastry TP (2002) Mechanical and microstructure studies on the modification of CA film by blending with PS. Bull Mater Sci 25:25–29

Nassar AH, El-Tarabily KA, Sivasithamparam K (2005) Promotion of plant growth by an auxin-producing isolate of the yeast *Williopsis saturnus* endophytic in maize (*Zea mays* L.) roots. Biol Fertil Soil 42:97–108

Nishida H, Tokiwa Y (1993) Distribution of poly(-hydroxybutyrate) and poly(-caprolactone) aerobic degrading microorganisms in different environments. J Environ Polym Degrad 1:227–233

Oda Y, Asari H, Urakami T, Tonomura K (1995) Microbial degradation of poly(3-hydroxybutyrate) and polycaprolactone by filamentous fungi. J Ferment Bioeng 80(3):265–269

Orhan Y, Hrenovic J, Buyukgungor H (2004) Biodegradation of plastic compost bags under controlled soil conditions. Acta Chim Slov 51:579–588

Pandey A, Benjamin S, Soccol CR, Nigam P, Krieger N, Soccol VT (1999) The realm of microbial lipases in biotechnology. Biotechnol Appl Biochem 29:119–131

Premraj R, Doble M (2005) Biodegradation of polymers. Ind J Biotechnol 4:186–193

Seo H, Um H, Min J, Rhee S, Cho T, Kim Y, Lee J (2007) *Pseudozyma jejuensis* sp. nov., a novel cutinolytic ustilaginomycetous yeast species that is able to degrade plastic waste. FEMS Yeast Res 7:1035–1045

Seymour RB (1971) Additives for polymers, introduction to polymer chemistry. McGraw-Hill Book Company, New York, pp 268–271

Tan RX, Zou WX (2001) Endophytes: a rich source of functional metabolites. Nat Prod Rep 18:448–459

Tian XL, Cao LX, Tan MH, Zeng GQ, Jia YY, Han WQ, Zhou SN (2004) Study on the communities of endophytic fungi and endophytic Actinomycetes from rice and their antipathogenic activities in vitro. World J Microbiol Biotechnol 20:303–309

Tokiwa Y, Calabia BP, Ugwu CU, Aiba S (2009) Biodegradability of plastics. Int J Mol Sci 10:3722–3742

Xin G, Glawe D, Doty S (2009) Characterization of three endophytic, indole-3-acetic acid-producing yeasts occurring in *Populus* trees. Mycol Res 113:973–980

Xu S, Yamaguchi T, Osawa S, Suye S (2007) Biodegradation of poly(6-caprolactone) film in the presence of *Lysinibacillus* sp. 70038 and characterization of the degraded film. Biocontrol Sci 12(3):119–122

Zhang HW, Song YC, Tan RX (2006) Biology and chemistry of endophytes. Nat Pro Rep 23:753–771

9

Improved saccharification of steam-exploded *Pinus radiata* on supplementing crude extract of *Penicillium* sp.

Hamish Cameron · Sylke H. Campion · Tripti Singh · Alankar A. Vaidya

Abstract Commercially available enzymes do not contain all the necessary softwood-specific accessory enzymes to obtain high saccharification efficiency. In this work, six saprophytic fungi obtained from *Pinus radiata* plantation site were screened for the putative softwood-specific accessory enzyme, β-mannanase. A *Penicillium* sp. was found to produce β-mannanase in both solid (31.6 units/g of dry biomass) and liquid (117 units/g of dry biomass) cultures using locust bean gum as an inducer after 2 weeks of incubation. The saccharification of steam-exploded *Pinus radiata* was 7.8 % w/w improved when the crude extract of *Penicillium* sp. was added to a mixture of commercial enzymes.

Keywords Accessory enzyme · Fungi · Hemicellulase · β-Mannanase · Saccharification · Softwood

Introduction

Woody biomass can be used as a feed-stock for the production of sugars which can be converted to biofuels and bio-chemicals (Zhu and Pan 2010). Enzymes can be used to break down wood into sugars but this is an expensive step due to the cost of enzymes (Berlin et al. 2007). An alternative strategy is to decrease the cost of enzymes by producing a crude enzyme extract on-site, rather than using a commercial enzyme. Moreover, commercially available enzymes are limited in number and composition and have generally been optimized for non-woody herbaceous biomass (Banerjee et al. 2010; Berlin et al. 2007). In the case of herbaceous and hardwood plant species, the hemicellulose is xylan rich whereas for softwood it is mannan rich (Juturu and Wu 2013). In softwood, the presence of hemicellulose may limit the hydrolysis of cellulose in the absence of accessory enzymes (Várnai et al. 2011).

β-Mannanase (E.C. 3.2.1.78) is a key softwood-specific accessory enzyme as it cleaves the main chain of both glucomannans and galactoglucomannans (GGMs) into oligosaccharides (Várnai et al. 2011; Yang et al. 2009). Another strategy for obtaining a complete enzyme mixture involves blending crude enzyme extracts having softwood-specific accessory enzyme with commercial enzymes. This approach has great potential since no activities are lost and it can lead to enzyme synergy and better saccharification efficiency (Yang et al. 2009).

In the screening of fungi for specific enzymes capable of degrading softwood biomass, it is crucial to search in the ecological niche where the desired biomass is growing. Hence, in this work, saprophytic fungi that grow in *Pinus radiata* plantations were selected (Visser et al. 2013; Singh et al. 2008; Zhang and Lynd 2004). Two fungi, each from the categories of brown rot, soft rot and mould were selected. The effect of solid versus liquid medium and incubation time on the extracellular production of β-mannanase from these six species was studied. For the first time, improvement in saccharification of steam-exploded *Pinus radiata* substrate is demonstrated after supplementing commercial enzyme with crude enzyme preparation from New Zealand native *Penicillium* sp.

H. Cameron
Faculty of Science and Engineering, University of Waikato, Private Bag 3105, Hamilton, New Zealand

S. H. Campion · T. Singh · A. A. Vaidya (✉)
Scion, Te Papa Tipu Innovation Park, Private Bag 3020, Rotorua 3046, New Zealand
e-mail: Alankar.vaidya@scionresearch.com

Materials and methods

All fungi were obtained from Scion's mycological depository. The commercial enzymes Celluclast 1.5L and Novo-188 were obtained from Novozymes A/S. All other chemicals were purchased from Sigma-Aldrich and were used as received. An individual culture of each fungus was transferred from their glycerol stocks (40 % w/v glycerol) aseptically onto 1.5 % w/v malt agar nutrient medium and incubated in the dark at 75 % relative humidity and 25 °C. Each fungus was grown for 2 weeks till mycelium and fungal colonies are clearly seen. These plates are used as the stock cultures in sub-culturing onto the fresh solid or liquid media.

Growth on solid medium

Malt agar (0.5 % w/v) was prepared by dissolving 10 g of malt and 12 g of agar in 2 L deionized water with heating. While still liquid, it was divided equally into two 1 L flasks. One flask contained 5 g locust bean gum (LBG; 0.5 % w/v) and the other flask was kept as a control. The pH was adjusted to 5.0 using 0.4 % HCl or 0.5 M NaOH and both flasks were autoclaved at 121 °C for 20 min. The warm media were poured into pre-sterilized Petri-plates (40 mL per plate). The plates were allowed to cool to room temperature and then each plate was inoculated with 5-mm diameter fungal plugs punched from stock culture plates. Duplicate plates were prepared for each fungus.

Growth in liquid media

Broth was prepared by dissolving, KH_2PO_4 1 g, $MgSO_4$ 0.5 g, ammonium tartrate 0.2 g, NaH_2PO_4 0.2 g, $CaCl_2$ 0.05 g, $FeSO_4$ 0.05 g, $CuSO_4$ 0.01 g, $ZnSO_4$ 0.005 g, $MnSO_4$ 0.005 g per litre of deionized water. Two litres of liquid medium was prepared and divided equally into two 1 L flasks. One flask contained 5 g locust bean gum (LBG; 0.5 % w/v) and the other flask was kept as control. In 100 mL capacity conical flasks, 40 mL of either control or locust bean gum (LBG) liquid medium was poured and flasks autoclaved at 121 °C for 20 min and then cooled to room temperature. Each flask was inoculated at room temperature using 5-mm diameter plugs punched from stock culture plates. Duplicate flasks were prepared for each fungus.

Fungal growth and crude enzyme recovery

The inoculated samples were incubated in the dark at 25 °C and at 75 % relative humidity without shaking. At defined time intervals (7, 14, 21 and 28 days), duplicate samples of each fungus were removed and crude β-mannanase was recovered from solid medium as follows—at each time point, culture grown on solid medium was cut into 1 × 1 × 0.5 cm cubes and transferred into separate 50 mL capacity centrifuge (Falcon) tubes and crushed to fine particles. A 30 mL of 0.05 M Na-citrate buffer (pH 5) was added and the mixture was incubated in a rotatory shaker at 15 °C for 18 h at 220 rpm. Then, each sample was centrifuged at 10,000×*g* for 20 min. The supernatant was decanted and the precipitate was washed once with 10 mL of 0.05 M Na-citrate buffer pH 5 and centrifuged at 10,000×*g* for 20 min. The washing was mixed with the supernatant and kept in a refrigerator at 8 °C for further use. The solid mass precipitated was freeze dried and weighed. The weight of the dry biomass was calculated from the difference in dry weight of fungal culture grown on locust bean gum—solid medium and the control. Crude β-mannanase recovery from liquid medium—the contents from each flask were transferred to a 50 mL capacity centrifuge (Falcon) tube and centrifuged at 10,000×*g* for 20 min. The supernatant was decanted and kept in a refrigerator at 8 °C for further use. The precipitated biomass was freeze dried and weighed.

Enzyme assays

β-Mannanase activity in the supernatant was measured using a partially depolymerized Carob-galactomannan conjugated with Remazol brilliant blue R (Megazyme, Ireland) as the substrate. The substrate stock solution was prepared at 1 % w/v in 0.2 M sodium acetate buffer pH 5.0. The substrate stock and crude enzyme supernatant were incubated separately at 40 °C for 3 h before the assay was performed. In 5 mL capacity glass tubes, 0.5 mL of stock substrate solution was mixed with 0.5 mL of crude enzyme solution and incubated at 40 °C for 2.5 h. The reaction was terminated by the addition of 2.5 mL of ethanol (95 % v/v). After brief vortexing, the absorbance of supernatant was measured at 590 nm. Enzyme activity was determined from a standard calibration curve ($R^2 = 0.99$) of Remazol brilliant blue R solution (0.008–0.00016 μM) prepared in 0.2 M sodium acetate buffer pH 5. One unit of β-mannanase activity was defined as the amount of enzyme which released 0.001 μM of dye under assay conditions. Each assay was performed in quadruplicate and the average value is presented in Fig. 1. β-Glucosidase (E.C. 3.2.1.21) and β-mannosidase (E.C. 3.2.1.25) activity were each measured using 10 mM of standard chromogenic substrates, i.e. *p*-nitrophenyl-β-D-glucoside and *p*-nitrophenyl-β-D-mannoside, respectively (Bailey and Nevalainen 1981). One unit of activity was defined as the amount of enzyme which released 1 μM of *p*-nitrophenol from their respective standard substrates under assay conditions.

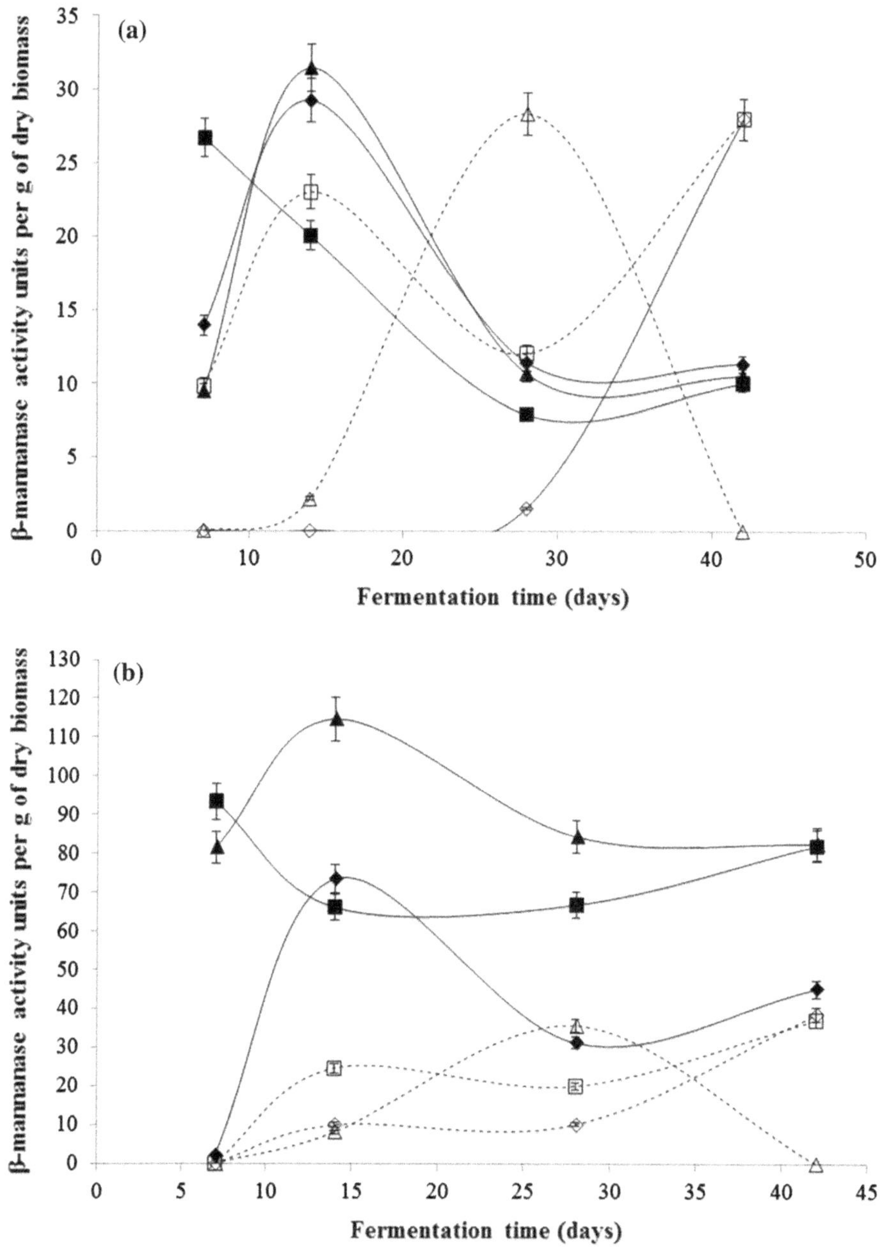

Fig. 1 **a** Crude β-mannanase recovery from solid medium. **b** Crude β-mannanase recovery from liquid medium. Data shown here are average of quadruplicate measurements with corresponding relative standard deviations. *Filled triangles Penicillium* sp., *empty triangles Cladosporum herbarum*, *filled squares Oligoporus placenta*, *empty squares Lenzites trabea*, *filled diamonds Cephalosporium* sp. and *empty diamonds Chaetomium globosum*

Enzymatic hydrolysis of steam exploded (SEW) *Pinus radiata*

SEW was produced in the steam explosion apparatus (Newman et al. 2013). 0.75 kg chips were taken with a moisture content of 60 %, impregnated with SO_2 (3 % w/w) and heated with steam at 215 °C for 3 min. The SEW pulp was washed four times with water and oven dried at 110 °C to obtain a 54 % yield. This substrate was hydrolysed using commercial enzyme in the form of Celluclast 1.5L and Novo-188 at 10 FPU:12.5 CBU per g of substrate in a 5-mL saccharification mixture with and without addition of the freeze dried crude extract of *Penicillium* sp. at 100 mg/mL. A control experiment was run in parallel in which hydrolysis of SEW was performed using only crude extract of *Penicillium* sp. at 100 mg/mL in 5-mL mixture. All experiments were incubated at 180 rpm agitation speed in an inclined vibratory shaker for 72 h. Experiments were carried using a substrate concentration of 1.5 % on a dry basis. The sugar composition (monomeric and oligomeric) of the saccharified filtrate was determined using ion chromatography as described elsewhere (Vaidya and Singh 2012).

Results and discussion

Cell growth for all six test fungi was normal when grown on control medium (without locust bean gum) but no β-mannanase activity was detected. This indicates that the presence of locust bean gum did not affect cell growth but did lead to extracellular secretion of enzymes required for the degradation of glucomannan. All fungi (except *Cladosporum herbarum* and *Lenzites trabea*) showed almost 200–400 % higher β-mannanase activity units per g of dry biomass when grown in liquid medium compared to solid medium (Fig. 1a, b). The improved enzyme production in liquid medium can be due to better accessibility of growing hyphae to locust bean gum.

Generally, the time profile of enzyme production is similar for a given fungi in solid and in liquid medium. An early induction period (7 days) was seen in the case of the brown rot *Oligoporus placenta* whereas *Lenzites trabea* only produced β-mannanase after a 14-day growth period and maximum activity occurred only after 42 days. Soft rot fungi had different induction responses to locust bean gum. *Cephalosporium* sp. produced the most activity at 14 days but then had an inconsistent response to locust bean gum, whereas *Chaetomium globosum* reached maximum production after 42 days. The two mould fungi *Penicillium* sp. and *Cladosporum herbarum* showed the highest β-mannanase activity at 14 and 28 days incubation period, respectively, and activity reduced thereafter.

Among six fungi tested, *Penicillium* sp. showed highest β-mannanase activity per g of dry biomass in both liquid and solid media. Moreover, it also showed low β-mannosidase activity (12.9 units) in accordance with the rare appearance of fungal β-mannosidase in the literature (Maijala et al. 2012; Valášková and Baldrian 2006). Both β-mannanase and β-mannosidase activities are relevant to softwood glucomannan hydrolysis. Interestingly, none of the fungi expressed β-glucosidase activity either in liquid or solid medium.

Hydrolysis of SEW using Celluclast 1.5L:Novo-188 at 10 FPU:12.5 CBU per g of substrate gave 2.10 ± 0.1 g of total glycosyl units (monomeric + oligomeric) per litre, which correspond to 26.1 ± 1.3 % w/w of polysaccharide conversion (Table 1). No mannosyl or galactosyl units were released from SEW using a mixture of commercial enzymes only. This is to be expected because the commercial enzyme preparation Celluclast 1.5L contains α-L-arabinofuranosidase, endo-xylanase and β-xylosidase suitable for degrading hardwood hemicellulose but lacks mannan-degrading enzymes, such as β-mannanase and/or β-mannosidase, which act as accessory enzymes in the hydrolysis of glucomannan and GGMs in softwood hemicellulose (Juturu and Wu 2013). When the commercial enzyme was supplemented with the crude enzyme preparation from *Penicillium* sp. the production of monomeric glucosyl units increased from 1.85 ± 0.1 to 2.48 ± 0.1 g/L. This corresponds to a 7.8 ± 0.4 % w/w improvement in monomeric glucosyl released. This improvement in release of glucosyl units (g/L) during saccharification is due to the synergistic action of following enzymes—cellulase available from Celluclast 1.5L and crude extract of *Penicillium* sp. (for details refer to supplementary information), β-glucosidase available from Novo-188 and the presence of β-mannanase and β-mannosidase in the crude extract of *Penicillium* sp. Furthermore, from control experiment in which only crude extract of *Penicillium* sp. is used for SEW hydrolysis, a small amount of monomeric glucosyl units (0.10 g/L) released because of lack of β-glucosidase activity in the crude extract. This in turn resulted in an accumulation of oligomeric glucosyl units (0.41 g/L; 5 % w/w cellulose conversion). Only in presence of crude extract of *Penicillium* sp., the conversion of soluble polymeric mannan to monomeric and oligomeric mannosyl units is achieved albeit the original hemicellulose content in SEW is small (0.3 % w/w) (Table 1). To corroborate this result further a comparison of crude extract of *Penicillium* sp. against commercial enzymes (both were taken on equal protein basis of 10 mg/mL) in hydrolysis of locust bean gum (1 % w/w) showed that using only crude extract twofold enhanced activity was obtained compared to commercial enzyme.

Thus, the crude preparation of *Penicillium* sp. has a boosting effect in the hydrolysis of pretreated softwood substrate and emphasizes the importance of a feedstock specific mixture of hemicellulase and cellulase enzymes (Gao et al. 2011). Várnai et al. (2011) reported a 4 % w/w increase in the hydrolysis of steam pretreated spruce when

Table 1 Saccharification of SEW

	Monomeric sugars g/L (% w/w conversion)			Oligomeric sugars g/L (% w/w conversion)		
	Glucose	Mannose	Galactose	Glucosyl units	Mannosyl units	Galactosyl units
Commercial enzyme	1.85 (23.0)[a]	0.00	0.00	0.25 (3.1)[a]	0.00	0.00
Penicillium sp. crude extract (control)	0.10 (1.24)[a]	0.01 (33.3)[a]	0.00	0.41 (5.08)[a]	0.02 (66.7)[a]	0.01 (66.7)[a]
Commercial enzyme + *Penicillium* sp. extract	2.48 (30.8)[a]	0.01 (33.3)[a]	0.00	0.00	0.02 (66.7)[a]	0.01 (66.7)[a]

All values in the table are the mean of triplicate experiments and have a standard deviation of ±5 %. The sugar composition (% w/w) of original SEW is—glucosyl units = 53.6; mannosyl units = 0.3; galactosyl units = 0.1

[a] The numbers shown in the parentheses are % w/w conversion of sugars from the original SEW sample taken in 5 mL saccharification

Trichoderma reesei β-mannanase supplemented at 1:4 w/w on protein basis with cellulases mixture (Várnai et al. 2011). This work indicates that the effectiveness of a mixture of commercial cellulases can be improved by adding biomass-specific accessory enzymes and, therefore, has commercial implications in the production of biofuel enzymes. Additional work on enzyme purification and improving production of β-mannanase by whole genome mutagenesis of *Penicillium* sp. or heterologous expression of the β-mannanase gene into *Pichia pastoris* are a subject matter of future communication.

Acknowledgments This work was financially supported by the Scion core funding in Bioenergy. The authors wish to thank Daniel van de Pas for SEW preparation and Katrina Martin for analysis of the substrate using ion chromatography. The help offered by Dr. Ruth Falshaw and Dr. Ian Suckling in proof reading this manuscript is also acknowledged.

Conflict of interest The authors declare that they have no conflict of interests in this research paper.

References

Bailey MJ, Nevalainen KMH (1981) Induction, isolation and testing of stable *Trichoderma reesei* mutants with improved production of solubilizing cellulase. Enzyme Microb Technol 3:153–157

Banerjee G, Scott-Craig JS, Walton JD (2010) Improving enzymes for biomass conversion: a basic research perspective. Bioenergy Res 3:82–92

Berlin A, Maximenko V, Gilkes N, Saddler J (2007) Optimization of enzyme complexes for lignocellulose hydrolysis. Biotechnol Bioeng 97:287–296

Gao D, Uppugundla N, Chundawat S, Yu X, Hermanson S, Gowda K, Brumm P, Mead D, Balan V, Dale BE (2011) Hemicellulases and auxiliary enzymes for improved conversion of lignocellulosic biomass to monosaccharides. Biotechnol Biofuels 4:5.

Juturu V, Wu JC (2013) Insight into microbial hemicellulases other than xylanases: a review. J Chem Technol Biotechnol 88:353–363

Maijala P, Kango N, Szijarto N, Viikari L (2012) Characterization of hemicellulases from thermophilic fungi. Antonie Van Leeuwenhoek 101:905–917

Newman RH, Vaidya AA, Campion SH (2013) A mathematical model for the inhibitory effects of lignin in enzymatic hydrolysis of lignocellulosics. Bioresour Technol 130:757–762

Singh T, Vesentini D, Singh AP, Daniel G (2008) Effect of chitosan on physiological, morphological and ultrastructural characteristics of wood degrading fungi. Int Biodeterior Biodegrad 62:116–124

Vaidya A, Singh T (2012) Pre-treatment of *Pinus radiata* substrates by basidiomycetes fungi to enhance enzymatic hydrolysis. Biotechnol Lett 34:1263–1267

Valášková V, Baldrian P (2006) Degradation of cellulose and hemicelluloses by the brown rot fungus *Piptoporus betulinus*—production of extracellular enzymes and characterization of the major cellulases. Microbiology 152:3613–3622

Várnai A, Huikko L, Pere J, Siika-aho M, Viikari L (2011) Synergistic action of xylanase and mannanase improves the total hydrolysis of softwood. Bioresour Technol 102:9096–9104

Visser EM, Falkoski DL, de Almeida MN, Maitan-Alfenas GP, Rezende ST, Guimarães VM (2013) Production and application of an enzyme blend from *Chrysoporthe cubensis* and *Penicillium pinophilum* with potential for hydrolysis of sugarcane bagasse. Bioresour Technol 144:587–594

Yang X, Jiang Z, Song H, Jiang S, Madzak C, Ma L (2009) Cell-surface display of the active mannanase in *Yarrowia lipolytica* with a novel surface-display system. Biotechnol Appl Biochem 54:171–176

Zhang YHP, Lynd LR (2004) Toward an aggregated understanding of enzymatic hydrolysis of cellulose: noncomplexed cellulase systems. Biotechnol Bioeng 88:797–824

Zhu JY, Pan XJ (2010) Woody biomass pretreatment for cellulosic ethanol production: technology and energy consumption evaluation. Bioresour Technol 101:4992–5002

10

Biogenic silver nanoparticles production and characterization from native stain of *Corynebacterium* species and its antimicrobial activity

B. Gowramma · U. Keerthi · Mokula Rafi · D. Muralidhara Rao

Abstract In the present study, synthesis, characterization, and the antibacterial activity of silver nanoparticles from native isolate of *Corynebacterium glutamicum* has been reported. Silver nanoparticles were synthesized by challenging the dried biomass of *C. glutamicum* with aqueous diamine silver ([Ag $(NH_3)^2]^+$) containing 1 mM $AgNO_3$. Synthesized silver nanoparticles (AgNPs) were characterized by ultraviolet–visible spectroscopy and energy-dispersive X-ray (EDX) spectroscopy analysis. Morphological study of silver nanoparticles was carried out using transmission electron microscopy (TEM) and scanning electron microscope (SEM). The spherical morphology of silver nanoparticles was confirmed from SEM image. The TEM image showed the average particle size of silver nanoparticles was about 15 nm. Silver nanoparticles synthesized from *C. glutamicum* were found to have enhanced antimicrobial activity against selected pathogenic strains. Silver nanoparticles from pure strains of *Corynebacterium* species was done by many investigators, but as per the present literature, this is the first report on the production of silver nanoparticles using a native strain of *Corynebacterium*.

Keywords Diamine silver · Silver nanoparticles · Native strain · EDX · TEM · SEM · Antimicrobial activity

B. Gowramma · U. Keerthi · M. Rafi · D. Muralidhara Rao (✉)
Department of Biotechnology, Sri Krishnadevaraya University, Anantapuram 515003, Andhra Pradesh, India
e-mail: rao.muralidhara@gmail.com

Introduction

Nanotechnology is referring to the ability for designing, characterization, production and application of structures, devices and systems by controlling shape and size at the nano scale (Mansoori et al. 2007) and it is a promising emerging industry, which is bringing us exiting new products. Currently, there is a growing need to use environmental friendly nanoparticles that do not produce toxic wastes in their synthesis protocol (Vahabi et al. 2011). Many methods have been designed to synthesize nanoparticles and the most important aspects of nanotechnology rely on the synthesis of nanoparticles with well-defined sizes, shapes and controlled monodispersity (Pugazhenthiran et al. 2009). Biotechnological route has emerged as a safe and alternative process in the synthesis of nanoparticles by employing ambient biological resources (Baker et al. 2013). Studies have reported that the biological methods depict an inexpensive and eco-friendly route for synthesis of nanoparticles. Synthesis of nanoparticles have been demonstrated from some microbes, among them the bacterial species used have included *Escherichia coli* (Natarajan et al. 2010), *Bacillus cereus* (Sunkar and Nachiyar 2012), *Corynebacterium* sp. (Arun et al. 2013), while yeast species have included MKY3 yeast strain (Kowshik and Ashataputre, 2003), *Saccharomyces cerevisiae* BU-MBT CY-1 (Selvakumar et al. 2011), fungi included *T. asperellum* (Mukherjee et al. 2008), *Aspergillus clavatus* (Verma et al. 2010) *Trichoderma Reesei* (Vahabi et al. 2011), *Aspergillus terreus* (Li et al. 2012), algae *Cyanobacteria* (Sudha et al. 2013) and lichen *Parmotrema praesorediosum* (Mie et al. 2013) are able to absorb and accumulate metal and can be used in the reduction of environmental pollution and also for the recovery of metals from waste. The adaptation to heavy metal-rich environments is resulting in microorganisms that express activities, such as biosorption, bioprecipitation, extracellular sequestration, transport mechanisms, and chelation and such resistance mechanism forms the basis for the use of microorganisms in production of nanoparticles (Baker et al. 2013).

Among the noble metals, silver (Ag) is the metal of choice in the field of biological system, living organisms and medicine (Parashar et al. 2009). Many studies have proved that microorganisms can produce nanoparticles either by enzymatic or non-enzymatic reduction mechanism. Ahmad et al. (2003) had shown that NADH-dependent enzymes are responsible for the biosynthesis of nanoparticles. Few researchers reported that nanoparticles were produced without the involvement of biological enzymes. Liu et al. (2000) produced $Au3+$ nanoparticles from dried cells of *Bacillus megaterium.* Studies on the absorption of $Ag+$ by some microorganisms have been reported (Sneha et al. 2010). In these cases, no involvement of enzymes was observed; this non-enzymatic reduction mechanism suggested that some organic functional groups of microbial cell walls could be responsible for the production process under certain conditions (Lin et al. 2001). Dried biomass of some microorganisms, such as *Lactobacillus* A09, *Bacillus megaterium* D01 also has the ability to reduce $Ag+$ ions through the interaction between $Ag+$ and some groups on the microbial cell walls (Fu et al. 2000).

When compared with all the other types of nanomaterials silver nanoparticles have proved to be the most effective antimicrobial agents also they have shown great promise in terms of biomedical applications, not only due to their large surface area to volume ratio (Bhattacharya and Mukherjee 2008; Hirst et al. 2009), but also different biomedical activities (Hussain and Ferguson 2006). In particular, because of the recent advances in research on metal nanoparticles, Ag-NPs have received special attention as a possible antimicrobial agent (Baker et al. 2005; Firdhouse et al. 2013). A recent study showed that yeast and *E. coli* was inhibited at a low concentration of AgNPs, the study of mechanisms revealed that free radicals and oxidative stress was responsible for the antibacterial activities (Kim et al. 2007). Disease causing microbes that have become resistant to drug therapy are an increasing public health problem. Therefore, there is a vital need to develop new bactericides.

Current work was focused on the synthesis and characterization of silver nanoparticles from native isolate of *Corenebacterium glutamicum* by non-enzymatic method and the assessment of antibacterial activity against pathogenic bacteria.

Materials and methods

Chemicals

Peptone, beef extract, yeast extract, bacto tryptone, agar agar, potato dextrose, silver nitrate ($AgNO_3$), $NH_3{\cdot}H_2O$ (25 % w/w, AR), NaOH, NaCl, HNO_3, etc.

Bacterial culture for silver nanoparticles production

The bacterial strain *C. glutamicum* was isolated from native soil and characterized performing biochemical tests. The strain was maintained at 4 °C on nutrient agar slants as well as sub cultured from time to time to regulate its viability. *C. glutamicum* is a small, non-motile, gram-positive soil bacterium. It is non-pathogenic, non-spore forming, grows quickly, has relatively few growth requirements, has no extracellular protease secretion, and used to produce many amino acids.

Microbial cultures to test antimicrobial sensitivity

Bacterial strains *Staphylococcus aureus* MTCC3160, *E. coli* MTCC40, *Salmonella enterica* MTCC3917, *Pseudomonas aeruginosa* MTCC424, *Klebsiella pneumoniae* MTCC3384, and *Shigella flexneri* MTCC1457 were procured from Institute of Microbial Type Culture Collection (MTCC), Chandigarh, India and *Bacillus subtilis, Bacillus flexus* were isolated from the native soil of Sri Krishnadevaraya University, Anantapuram, AP India and the cultures were maintained at 4 °C on nutrient agar slants.

Preparation of diamine silver

Diamine silver complex ($[Ag(NH_3)^2]^+$) was prepared by adding dilute ammonia solution ($NH_3{\cdot}H_2O$, 25 % w/w, AR) into aqueous solution of silver oxide (Ag_2O) until the precipitate of Ag_2O was transformed into soluble $[Ag(NH_3)_2]^+$ (Ag when treated with alkali $AgNO_3$ forms silver oxide, which in case of NH_4OH dissolves to form complex ion) (Vogel 1956).

Production of biomass

C. glutamicum cultures were maintained by subculturing at monthly intervals and growth conditions were optimized. Luria Broth (LB) (1 % bactotryptone, 0.5 % yeast extract, 1 % NaCl, pH 7.0 ± 0.2) was used for growing the organism. 250 ml of LB was prepared using Milli-Q water, autoclaved at 121 ± 1 °C for 15 min and inoculated with a fresh batch of the bacteria, *C. glutamicum.* The culture flasks were incubated for 72 h at 30 °C with shaking at 120 rpm. After 72 h of growth, the biomass were harvested by centrifugation at 5,000 rpm for 10 min and the collected cell pellet was washed three times with deionized water to remove the culture medium and dried overnight in an oven at 60 °C. Silver nitrate will form precipitation with the salts present in the water, to avoid that Milli-Q water is used throughout the experiment.

Synthesis of Ag nanoparticles from *C. glutamicum* biomass

C. glutamicum at the concentration of 0.5 g/l dried biomass was resuspended with deionized water and added to aqueous solutions of freshly prepared 15 ml of a diamond silver complex containing 1 mM silver nitrate and made final volume to 25 ml with deionized water then incubated for 24 h at room temperature in a orbital shaker at 150 rpm. Control (without the diamine silver, only biomass) was also run along with the experimental conditions. After 24 h of incubation, 1 ml sample was with drown and centrifuged. Then, the supernatant was assayed for of silver nanoparticle formation. The amount of silver nanoparticles produced by *C. glutamicum* was determined by adopting the method recommended by Zhang et al. (2005). The concentration of Ag in the supernatant was (Q, mg g^{-1}) calculated as:

$$Q = [(Ci - Cb)V]/M$$

where V was the volume of sample solution (L), Ci and Cr were the initial and equilibrium Ag concentration (mg/l) in solution, respectively, and M was the weight of dried biomass (g).

UV–Vis spectrophotometer analysis

UV–Vis analysis of the reaction solutions was carried out at room temperature using Thermo scientific-EVOLUTION201 spectrophotometer equipped with matching quartz cells at a resolution of 1 nm from 200 to 800 nm. The supernatant was diluted with deionized water (1:1) and the absorption was measured. It is generally recognized that UV–Vis spectroscopy could be used to examine the size and shape controlled nanoparticles in aqueous suspensions (Wiley et al. 2006).

Transmission electron microscopy (TEM)

TEM grids were prepared by sonicating the AgNPs sample solution for 5 min and placing a few drops on the 300 mesh carbon-coated copper grid and dried for the complete evaporation of water under a lamp and operated at an accelerating voltage of 200 kV using EM2000Fx-II, transmission electron microscope, which is a 200 kV HRTEM from JEOL, Japan, to characterize the sample after usual alignment procedures. In situ LCD camera is used to record the pictures.

Scanning electron microscope (SEM)

Sample for scanning electron microscopic (SEM) analysis was prepared as thin films on a carbon-coated copper grid by just dropping a very small amount of the sample on the grid, extra solution was removed using a blotting paper and then the film on the SEM grid were allowed to dry by putting it under a mercury lamp for 5 min. The instrument used is a JEOL 840 with resolution at 20 kV: 10 nm.

Energy-dispersive X-ray (EDX) analysis

Energy-dispersive X-ray spectrometers take advantage of the photon nature of light. EDX analysis was performed by measuring the energy and intensity distribution of X-ray signals generated by a focused electron beam on a specimen operating at 120 kV using ESEM Quanta 200, FEI EDX instrument. The data are used to obtain the elemental composition of the material.

Antimicrobial assay

Pure cultures of bacteria to be tested were grown in nutrient broth for 24 h at room temperature. In this method, sterile nutrient agar plate was prepared. Bacterial pathogens used in the present experiment were spread over the agar plate using sterile spreader. The plates were allowed to dry and using a sterile well-cutter with the diameter of 6.0 mm two wells were made in each agar plate, one for control (labeled as C) and the other for test (labeled as C). Subsequently, 30 µl (5 µg) of nanoparticles suspension was introduced into the wells labeled as 'T' another well left empty without adding silver nanoparticles maintained as a control of the inoculated nutrient agar plates and then incubated at 25 °C for 24 h, measured the diameter of inhibitory zones in mm after incubation (Lee et al. 2010). The assays were performed in triplicates and the mean values were recorded.

Results and discussion

Silver nanoparticles were synthesized successfully from dried cells of native *C. glutamicum* with aqueous diamine silver solution. The presence of nanoparticles in the medium was confirmed by the change in color from colorless to brown or deep yellow shown in Fig. 1b, where control (without diamine silver) showed no color formation in the culture when incubated for the same period and conditions (Fig. 1a). The color intensity increased with a period of incubation due to the reduction in silver nanoparticles. The

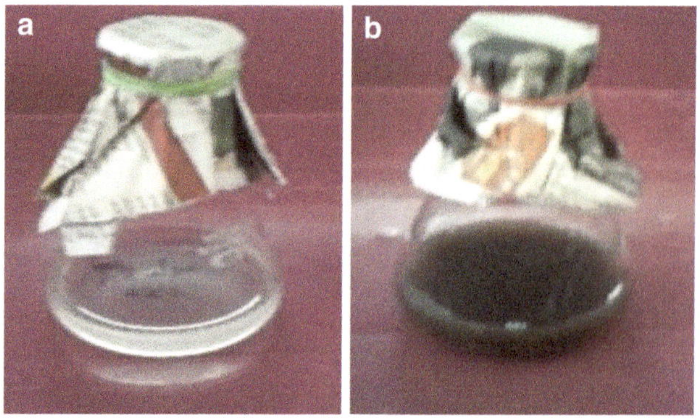

Fig. 1 *C. glutamicum* biomass containing diamine silver before (**a**) and after (**b**) production

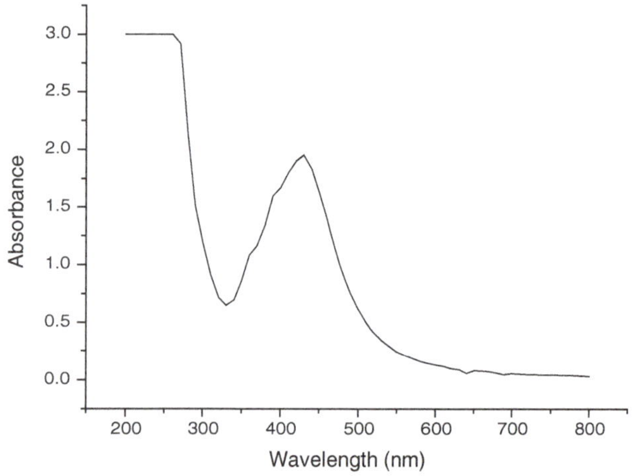

Fig. 2 UV–Vis spectra of silver nanoparticles showing maximum absorbance at 450 nm

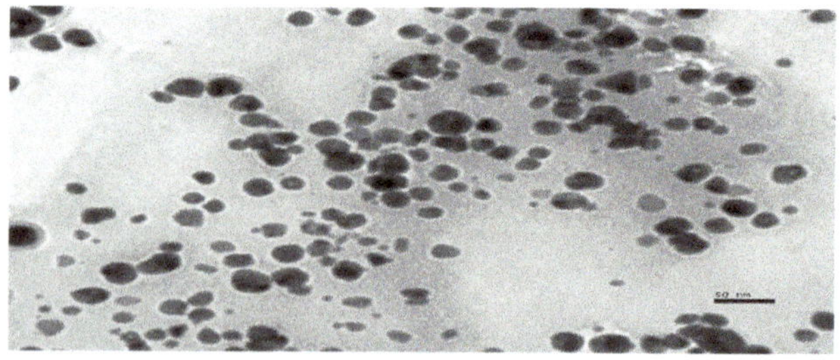

Fig. 3 TEM image of synthesized AgNPs

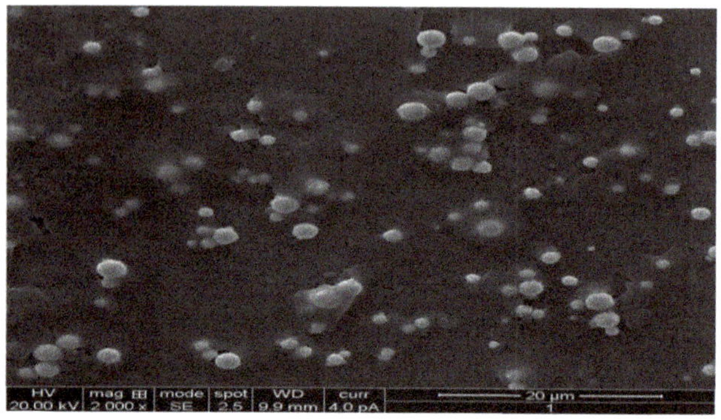

Fig. 4 SEM image of silver nanoparticles

characteristic brown color of colloidal silver solution is due to the excitation of surface plasmon vibrations in the nanoparticle and provides a convenient spectroscopic signature of their formation (Saifuddin et al. 2009). The yield of silver nanoparticles from native strain of *C. glutamicum* was found good and the concentration of silver nanoparticles present in the supernatant was estimated 280 mg/gram dried biomass.

UV–Vis spectrophotometer analysis

The silver nanoparticles were characterized by UV–Vis spectroscopy, one of the most widely used techniques for structural characterization of silver nanoparticles (Sun et al. 2001). Biological method of silver nanoparticles synthesis exhibit strong absorption of electromagnetic waves in the visible range due to their optical resonant property, called occurs due to its collective oscillation of conduction electrons, combined with the incident light (Kreibig and Vollmer 1995). Figure 2 shows the UV–Vis absorption spectra of the 96 h old AgNPs sample at 440 nm. The spectra recorded from the *C. glutamicum* reaction vessel at different reaction times and observed increased intensity in absorption spectra of silver solution with time, indicating the formation of increased number of silver nanoparticles in the solution (the graph is not included); this is similar to the findings of Esumi et al. (2000), Zhang et al. (2005). The prepared aqueous solution of AgNPs showed a strong absorption band between 410 and 440 nm, which is a typical absorption band of spherical Ag nanoparticles due to their surface plasmon.

TEM analysis of silver nanoparticles

The well-known technique for imaging solid materials at atomic resolution is TEM. The technique was employed to visualize the size and shape of Ag nanoparticles. From the results, it is observed that most of the Ag nanoparticles were spherical in shape. Figure 3 shows the TEM image of individual silver nanoparticles with the mean particle size estimated as 15 nm.

SEM analysis

The scanning electron microscopy has been employed to characterize the size, shape and morphologies of forming silver nanoparticles. The SEM image of the sample was shown in Fig. 4. From the results morphology of AgNPs was more clearly seen, the particle shapes were circular and poly-dispersed and the size ranged between 15 and

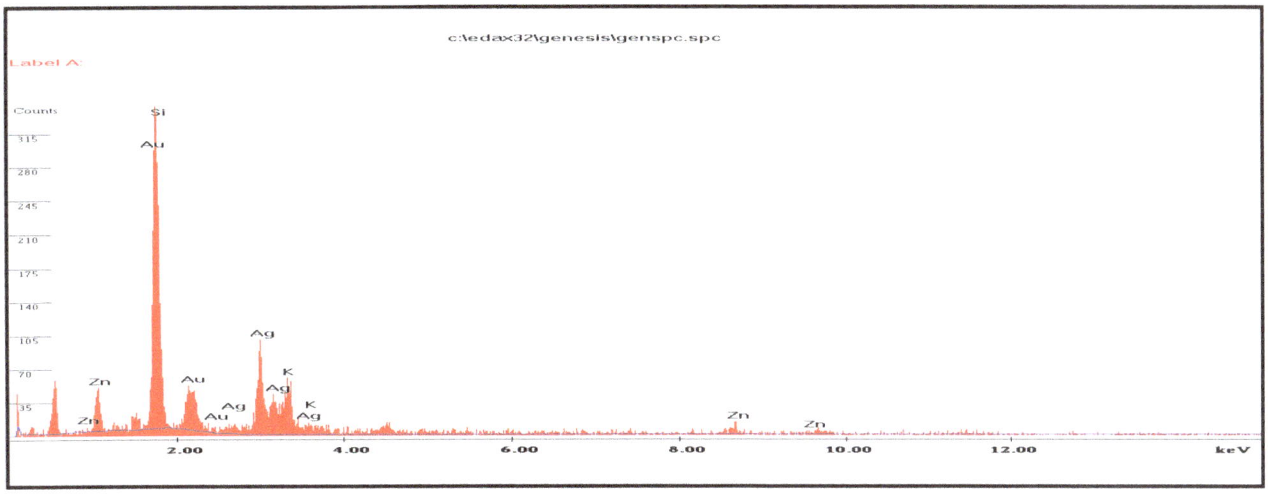

Fig. 5 Energy-dispersive X-ray spectrum of nanoparticles

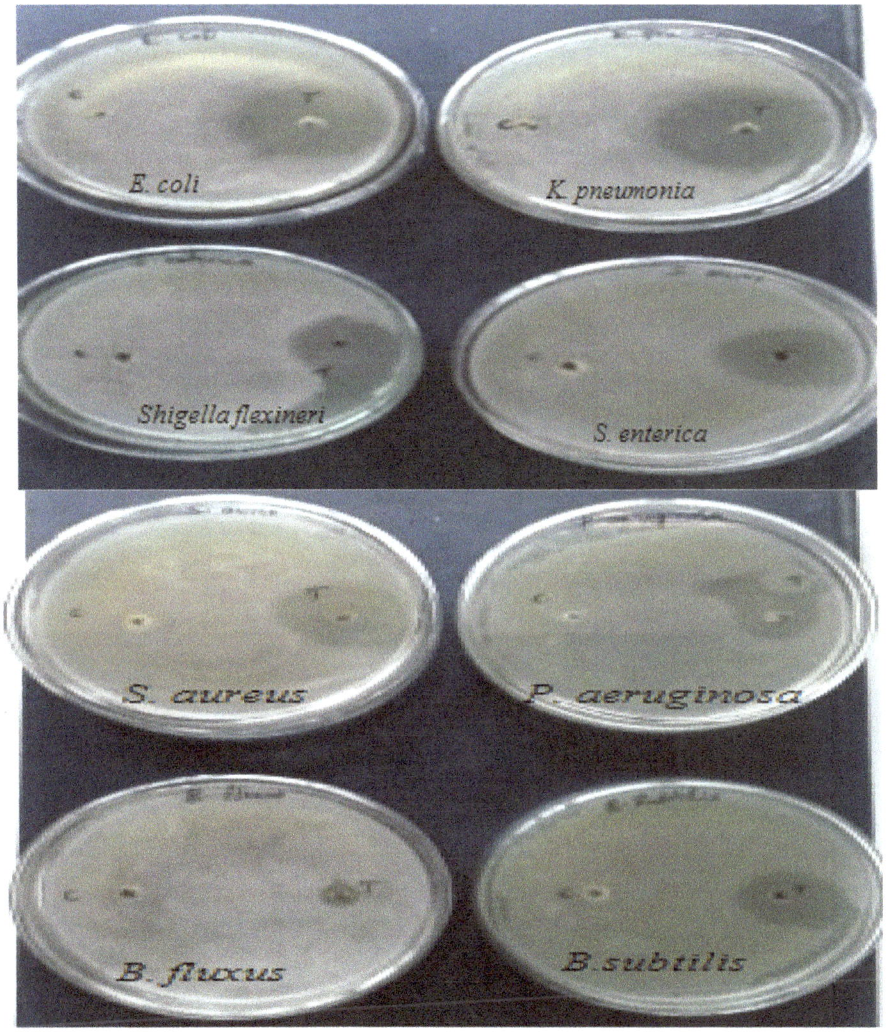

Fig. 6 Antimicrobial activity of silver nanoparticles against bacterial pathogens

20 nm. The size of obtaining nanoparticles was less when compared with the nanoparticle size of Arun et al. (2013), where they got diameter ranged from 24.55 to 32.02 nm, but similar with the results (Zhang et al. 2005). A report by Gurunathan et al. (2009a) showed that by controlling the environment of nanoparticle synthesis, silver nanoparticles

of various sizes and shapes could be synthesized. At acidic pH, the size of the nanoparticles ranged 45 nm, whereas at pH 10 the size is just 15 nm. It is also due to increase in size of the particles with growth time (Saha et al. 2011).

EDX studies

EDX spectrum of silver nanoparticles was checked and was found to contain a great deal of silver. Figure 5 shows strong peaks from the silver atoms in the nanoparticles are observed at 2.6, 3.0, 3.2, and 3.6 keV. The presence of the optical absorbance band at ~3 eV reveals the presence of pure metallic silver nanoparticles (Magudapathy et al. 2001). The graph also shows the presence of zinc (Zn), potassium (K)and aurum (Au) in the EDX picture of silver nanoparticles. This is probably due to the presence of substrate over which the NP sample was held during SEM microscopy.

Antimicrobial activity

Microbial contamination of air, water and soil due to different types of microorganisms creates problems in living conditions, in the public health and industrial fields. As a result, an increased occurrence of antibiotic resistant genes in many bacterial species is found in human beings and animals (Raffi et al. 2010). Emergence of new bacterial strains resistant to current antibiotics has become a serious public health issue, which raised the need to develop new bactericidal materials. Utilization of silver as a disinfecting agent is not new, and silver compounds were shown to be effective against both aerobic and anaerobic bacteria by precipitating bacterial cellular proteins and by blocking the microbial respiratory chain system (Gravante et al. 2009; Monteiro et al. 2009). In the present study, 5 µg of the nanoparticles was taken as final products for antimicrobial assay. The antimicrobial activity of the synthesized silver nanoparticles was studied against *E. coli, Klebsiella pneumonia, Shigella flexneri, Salmonella enterica, Staphylococcus aureus, Pseudomonas aueroginosa, and Bacillus flexus Bacillus subtilis* using well diffusion technique (Fig. 6) and the assay was performed in triplicates. Test and control were maintained in all the plates. In the control zone of inhibition was not observed. In the test zone, inhibitory zone was clearly observed and the mean diameter of inhibitory zones around each well with AgNPs is measured. From the mean values, highest antimicrobial activity was observed against *K. pneumonia*, which was 18 mm in diameter. Although the cultures of *E. coli, S. aureus, S. enterica, S. flexneri, P. aueroginosa, B. subtilis* and *B. flexus* were also showed clear zones of inhibition, which were about 16, 15, 15, 14, 12, 8 and 4 mm in diameter, respectively. Many other studies (Durán et al. 2007; Shirley et al. 2010; Tollamadugu et al. 2011; Arun et al. 2013; Firdhouse and Lalitha 2013; Madhavaraj et al. 2013) proved the antimicrobial activity of AgNPs and the nanoparticles from native strain have shown comparatively good inhibitory zone.

Conclusion

Silver nanoparticles from novel *C. glutamicum* strain were successfully synthesized, and the characterization of the nanoparticles was carried out using UV–Vis spectroscopy, TEM, SEM, and energy-dispersive X-ray analysis. Native *C. glutamicum* strain is greatly reduced diamine silver into silver nanoparticles at the concentration of 280 mg/g dried biomass. Synthesized nanoparticles were showing surface plasmon resonance (SPR) between 410 and 450 nm. The TEM analysis indicates the presence of 15 nm size of silver nanoparticles. SEM images clearly evident that the synthesized silver nanoparticles were spherical in shape. Chemical analysis performed with the use of EDX revealed that the powder contains of silver. The antimicrobial potential of synthesized Ag nanoparticles was tested against infectious bacteria by well diffusion assay. The AgNPs was showing strong antimicrobial potential against the tested bacteria. The present study is eco-friendly need cheaper cultivation requirements, higher growth rates on laboratory scales. Present findings not only confirm the nanoparticle formation, but also implicate the efficient antimicrobial property of silver nanoparticles.

Acknowledgments The authors are sincerely thankful to Indian Institute of Sciences, Bangalore, for providing SEM, TEM and EDX facilities. Dr. DMR is thankful to CSIR, Govt. of India.

Conflict of interest The authors declare that they have no conflict of interest in the publication.

References

Ahmad A, Mukherjee P, Senapati S, Mandal D, Khan MSI, Kumar R, Sastry M (2003) Extracellular biosynthesis of silver nanoparticles using the fungus Fusarium oxysporum. Colloids Surf B 28:313–318

Arun P, Shanmugaraju V, Renga Ramanujam J, Senthil Prabhu S, Kumaran E (2013) Biosynthesis of silver nanoparticles from *Corynebacterium* sp. and its antimicrobial activity. Int J Cur Microbiol App Sci 2:57–64

Baker C, Pradhan A, Pakstis L, Pochan DJ, Shah SI (2005) Synthesis and antibacterial properties of silver nanoparticles. J Nanosci Nanotechnol 5:244–249

Baker S, Harini BP, Rakshith D, Satish S (2013) Marine microbes: invisible nanofactories. J Pharm Res 6:383–388

Bhattacharya R, Mukherjee P (2008) Biological properties of naked metal nanoparticles. Adv Drug Delivery Rev 60:1289–1306

Durán N, Marcato PD, De S, Gabriel IH, Alves OL, Esposito E (2007) Antibacterial effect of silver nanoparticles produced by fungal process on textile fabrics and their effluent treatment. J Biomed Nanotechnol 3:203–208

Esumi K, Hosoya T, Suzuki A (2000) Formation of gold and silver nanoparticles in aqueous solution of sugarpersubstituted poly (amidoamine) dendrimers. J Colloid Interf Sci 226:346–352

Firdhouse MJ, Lalitha P (2013) Fabrication of antimicrobial perspiration pads and cotton cloth using *Amaranthus dubius* mediated silver nanoparticles. J Chem 2013:1–5

Fu JK, Liu YY, Gu PY, Tang DL, Lin ZY, Yao BX, Weng SZ (2000) Spectroscopic characterization on the biosorption and bioreduction of Ag(I) by *Lactobacillus* sp A09. Acta Phys-Chim Sin 16:779–782

Gravante G, Caruso R, Sorge R (2009) Nanocrystalline silver: a systematic review of randomized trials conducted on burned patients and an evidence-based assessment of potential advantages over older silver formulations. Ann Plast Surg 63:201–205

Gurunathan S, Lee KJ, Kalishwaralal K, Sheikpranbabu S, Vaidyanathan R, Eom SH (2009) Antiangiogenic properties of silver nanoparticles. Biomaterials 30:6341–6350

Hirst SM, Karakoti AS, Tyler RD, Sriranganathan N, Seal N, Reilly CM (2009) Small 5:2848–2856

Hussain S, Ferguson C (2006) Best evidence topic report. Silver sulphadiazine cream in burns. Emerg Med J 23:929–932

Kim JS, Kuk E, Yu KN, Kim JH (2007) Antimicrobial effects of silver nanoparticles. Nanomed Nanotechnol Biol Med 3:95–101

Kowshik M, Ashataputre S (2003) Extracellular synthesis of silver nanoparticles by a silver-tolerant yeast strain MKY3. Nanotechnology 14:95–100

Kreibig U, Vollmer M (1995) Optical properties of metal clusters. Springer series in material science, vol 25. Springer-Verlag, New York, pp 32–35

Lee SM, Song KC, Lee BS (2010) Antibacterial activity of silver nanoparticles prepared by a chemical reduction method. Korean J Chem Eng 27:688–692

Li G, He D, Qian Y, Guan B, Gao S, Cui Y, Yokoyama K, Wang L (2012) Fungus-mediated green synthesis of silver nanoparticles using *Aspergillus terreus*. Int J Mol Sci 13:466–476

Lin ZY, Fu JK, Wu JM, Liu YY, Cheng Hu (2001) Preliminary study on the mechanism of non enzymatic bioreduction of precious metal ions. Acta Phys-Chim Sin 17:477–480

Liu YY, Fu JK, Chen P, Yu XS, Yang PC (2000) Studies on biosorption of Au3+ by *Bacillus megaterium*. Acta Microbiol Sin 40:425–429

Madhavaraj L, Sethumadhavan VV, Gim GH, Kathiravan MN, Kim SW (2013) Synthesis, characterization and evaluation of antimicrobial efficacy of silver nanoparticles using *Paederia foetida* L. leaf extract. Int Res J Biol Sci 2:28–34

Magudapathy P, Gangopadhyay P, Panigrahi BK, Nair KGM, Dhara S (2001) Phy B 299:142–146

Mansoori GA, George TF, Zhang G, Assoufid L (2007) Molecular building blocks for nanotechnology. Springer, New York

Mie R, Samsudin MW, Din LB, Ahmad A, Ibrahim N, Adnan SNA (2013) Synthesis of silver nanoparticles with antibacterial activity using the lichen Parmotrema praesorediosum. Int J Nanomed 9:121–127

Monteiro DR, Gorup LF, Takamiya AS (2009) The growing importance of materials that prevent microbial adhesion: antimicrobial effect of medical devices containing silver. Int J Antimicrob Agents 34:103–110

Morones JR, Elechiguerra JL, Camacho A, Holt K, Kouri JB, Ramirez JT, Yacaman MJ (2005) The bactericidal effect of silver nanoparticles. Nanotechnology 16:2346

Mukherjee P, Roy M, Mandal BP, Dey GK, Mukherjee PK, Ghatak J, Tyagi AK, Kale SP (2008) Green synthesis of highly stabilized nanocrystalline silver particles by a non-pathogenic and agriculturally important fungus *T. asperellum*. Nanotechnology 19:103–110

Natarajan K, Selvaraj S, Ramachandra Murty V (2010) Microbial production of silver nanoparticles. Dig J Nanomater Biostructures 5:135–140

Parashar V, Parashar R, Sharma B, Pandey AC (2009) Dig J Nanomat Biostructures 4:45–50

Pugazhenthiran N, Anandan S, Kathiravan G, Udaya Prakash NK, Simon Crawford S, Ashok kumar M (2009) Microbial synthesis of silver nanoparticles by Bacillus sp. J Nanoparticle 11:1811–1815

Raffi M, Mehrwan S, Bhatti TM, Akhter JI, Hameed A, Yawar W, Masood ul Hasan M (2010) Investigations into the antibacterial behavior of copper nanoparticles against *Escherichia coli*. Ann Microbiol 60:75–80

Saha S, Bera K, Jana PC (2011) Growth time dependence of size of nanoparticle of ZnS. Int J Soft Comput Eng 5:23–26

Saifuddin N, Wong CW, Nur yasumira AA (2009) Rapid biosynthesis of silver nanoparticles using culture supernatant of bacteria with microwave irradiation. E-J Chem 6:61–70

Selvakumar R, Arul Jothi N, Jayavignesh V, Karthikaiselvi K, Antony Geny Immanual, Sharmila PR, Kavitha S, Swaminathan K (2011) As(V) removal using carbonized yeast cells containing silver nanoparticles. Water Res 45:583–592

Shirley, Dayanand A, Sreedhar B, Dastager Syed G (2010) Antimicrobial activity of silver nanoparticles synthesized from novel *Streptomyces* species. Dig J Nanomat Biostructures 5:447–451

Sneha K, Sathishkumar M, Mao J, Kwak IS, Yun YS (2010) *Corynebacterium glutamicum*-mediated crystallization of silver ions through sorption and reduction processes. Chem Eng J 162:989–996

Sudha SS, Karthic R, Rengaramanujam J (2013) Microalgae mediated synthesis of silver nanoparticles and their antibacterial activity against pathogenic bacteria. Indian J Exp Biol 51:393–399

Sun YP, Atorngitjawat P, Meziani MJ (2001) Preparation of silver nanoparticles via rapid expansion of water in carbon dioxide microemulsion into reductant solution. Langmuir 17:5707–5710

Sunkar S, Nachiyar CV (2012) Biogenesis of antibacterial silver nanoparticles using the endophytic bacterium *Bacillus cereus* isolated from *Garcinia xanthochymu*. Asian Pac J Trop Biomed 12:953–959

Tollamadugu NVKV, Prasad T, Subba Rao Kambala V, Venkata Naidu R (2011) A critical review on biogenic silver nanoparticles and their antimicrobial activity. Curr Nanosci 7:531–544

Vahabi K, Mansoori GA, Karimi S (2011) Biosynthesis of silver nanoparticles by fungus *Trichoderma Reesei*. Insci J 1:65–79

Verma VC, Kharwar RN, Gange AC (2010) Biosynthesis of antimicrobial silver nanoparticles by the endophytic fungus *Aspergillus clavatus*. Nanomedicine 5:33–40

Vogel AI (1956) A text book of practical organic chemistry including qualitative organic analysis. In: Longman group limited, 3rd edn, p 330

Wiley BJ, Im SH, McLellan J, Siekkkinen A, Xia Y (2006) Maneuvering the surface plasmon resonance of silver nanostructures through shape-controlled synthesis. J Phy Chem B 110:15666–15675

Zhang HR, Li QB, Lu YH, Sun DH, Ling XP, He N, Zheng SZ (2005) Biosorption and bioreduction of diamine silver complex by *Corynebacterium*. J Chem Technol Biot 80:285–290

11

Gordonia sp.: a salt tolerant bacterial inoculant for growth promotion of pearl millet under saline soil conditions

Monika Kayasth · Varun Kumar · Rajesh Gera

Abstract In the resent study, a diazotrophic bacterial isolate JPA2 having the ability to tolerate salinity (6 % NaCl) and plant growth-promoting features was isolated from rhizospheric soil of weed *Chenopodium murale* growing in saline soil of Pindara (EC 11.47 dS m^{-1}), district Jind (Haryana). The nitrogen fixing ability of the isolate was confirmed by *nifH* gene amplification and acetylene reduction assay (38.9 nmol ethylene h^{-1} mg^{-1} protein). The potential of strain JPA2 to promote growth of pearl millet was investigated by inoculation experiment which showed significant increase in plant height (51.1, 39.9 and 28.8 %) and dry weight (55.9, 36.4 and 35.5 %) over uninoculated control plants at EC 0, 6, 8 dS m^{-1}, respectively. The strain JPA2 was Gram +ve and identified as *Gordonia* sp. on the basis of partial 16S rRNA gene sequencing and biochemical characterization. It is concluded that salt tolerant diazotrophic *Gordonia* sp. can be considered as a beneficial microbe for agriculture in saline soils.

Keywords Salt tolerant · Plant growth-promoting rhizobacterium · *nifH* · Pearl millet · *Gordonia* sp.

M. Kayasth · R. Gera (✉)
Department of Microbiology, CCS Haryana Agricultural University, Hisar 125004, India
e-mail: rajeshgera1967@gmail.com

V. Kumar
Department of Biotechnology and Bioinformatics, Jaypee University of Information Technology, Waknaghat 173234, Himachal Pradesh, India

Introduction

Saline soils are extensive in the plains of Haryana state (Choubey et al. 2009) particularly in arid and semi-arid regions with ~2,32,985 hectares of area is considered as salt affected. Soils containing excess amount of water soluble salts are called saline soils. The concentration of salt in the root-zone of soil goes so high that it badly affects growth of the most crop species. But some cereal species are well adapted to semi-arid and arid tropics, where salinity and drought are the major problems due to limited water supply. Among the cereals, pearl millet (*Pennisetum glaucum*) is the only major crop that has high level of tolerance to saline soils. It can be cultivated even in the most sandy infertile soils and saline environments where no other cereal crop can survive (Kumar et al. 2010). However, to survive in saline soils, the increased demand of nitrogen in pearl millet could be met by the exploitation of diazotrophic bacteria (Latake et al. 2009). Saline rhizosphere habitats are blessed with some free-living nitrogen fixing bacteria which siphon out appreciable amount of nitrogen from reservoirs and enrich the soil. So, it is important to exploit rhizobacteria which often play crucial role in increasing crop productivity of the plants they colonize due to their close proximity with the roots they inhabit. Plant growth-promoting rhizobacteria (PGPR) stimulate the plant growth by producing Indole-3-acetic acid (IAA) and synthesizing siderophores that chelate iron and make it available to plant root (Kumar and Gera 2013). The application of PGPR as crop inoculants for biofertilization would be an attractive alternative to decrease the use of chemical fertilizers which effect environmental pollution.

The genus *Gordonia*, a coryneform bacterial member, is widespread in nature and comprises over 33 species, with broader characteristics such as the ability to degrade hydrocarbons (Xue et al. 2003), toxic environmental pollutants, xenobiotics (Hernandez et al. 2001), and natural compounds that are not readily biodegradable. *Gordonia* sp. may play an important role during waste water treatment (Bendinger et al. 1992) and also used as PGPR for Zea mays (Hong et al. 2010). Even though the higher number of *Gordonia* species described, to the best of our knowledge, there are no in-depth studies on the occurrence of nitrogen fixing species within the genus in saline soils and their possible role in growth promotion of pearl millet. Therefore, in the present investigation, diazotrophic *Gordonia* sp. was isolated from rhizospheric soil of weed growing in saline soils and was further characterized for its PGP features. In addition, the inoculation effects of *Gordonia* sp. on the growth of pearl millet were evaluated in artificially created saline conditions under pot house experiment.

Materials and methods

A salt affected barren field of Pindara village (latitude: 29°30′ N, longitude: 76°34′ E), district Jind, Haryana was chosen for sampling which harbors only few weeds such as *C. murale*. The weed rhizospheric soil sample was collected carefully by uprooting the root system and soil was placed in the sterile plastic bags for transport to the laboratory. Physiochemical analysis of soil properties, including pH, EC, organic carbon (Kalembasa and Jenkinson 1973) and total N (Bruel et al. 1946) was estimated. Three morphotypes were obtained on Jensen's medium (Jensen 1951) and maintained on respective medium slants at 4 °C. To investigate their nitrogen fixing ability, the genomic DNA of cultures was subjected to *nifH* gene amplification using primers: *nifH* for (5′-TAY GGN AAR GGN GGHATY GGY ATC-3′) and *nifH* rev (5′-ATR TTR TTN GCN GCR TAV ABB GCC ATC AT-3′) (Sarita et al. 2007). The PCR reaction mixture (25 μl) contained 0.5 μl of dNTP mixture (10 mM), 0.5 μl of Taq DNA polymerase (3 U μl^{-1}), 10× PCR assay buffer with 25 mM $MgCl_2$ (2.5 μl), 2 μl of each primer (10 pmol each), 2 μl template DNA (0.1–0.14 μg/μl) and 15.5 μl sterile water. PCR protocol conditions were: initial denaturation at 95 °C for 4 min, 35 cycles of denaturation at 94 °C for 30 s, annealing at 54 °C for 1 min and extension at 72 °C for 1 min followed by a final extension at 72 °C for 5 min. The PCR was then set on hold at 4 °C. Amplified PCR product was separated by electrophoresis on 2 % agarose gel stained with ethidium bromide and photographed under UV illumination using Gel Documentation system (DNR Bio-Imaging Systems). The isolate having *nifH* gene designated as JPA2 was further examined for acetylene reduction according to the method of Bridage et al. (1982).

For the identification of the strain JPA2, its 16S rRNA region was amplified and partially sequenced (Bioserve, Hyderabad, India) by primer walking using five different internal primers (16SEQ2R, 16SEQ3F, INS16SREV, 16SEQ4R and 16SEQ4F). The resulting 16S rRNA gene sequence was compared with sequences in GenBank database using the BLAST-N algorithm to identify sequences with a high degree of similarity. Comparable reference sequences were downloaded in FASTA format from NCBI database (http://www.ncbi.nlm.nih.gov). All the sequences were aligned with ClustalW program and phylogenetic tree was constructed using MEGA4 software through neighbor-joining method. Bootstrap analysis was used to evaluate the tree topology of the neighbor-joining data with 1,000 resamplings. The partial 16S rRNA gene sequence determined in this study was deposited in the GenBank database (http://www.ncbi.nlm.nih.gov/Genbank/index.html) under the accession number JQ437540. Bacterial culture JPA2 was further characterized morphologically and subjected to a battery of biochemical tests such as nitrate reduction, hydrolysis of Aesculin and starch (Cowan and Steel 1965). Sugar utilization pattern was assayed according to Collins and Lyne (1984).

The ability of the strain JPA2 to produce IAA was detected using the rapid assay (Bric et al. 1991), ammonia production examined using Nessler's reagent (Cappuccino and Sherman 1992) while siderophore production tested by CAS assay (Schwyn and Neilands 1987). Jensen's medium modified to the saline concentration of 2–10 % NaCl was used to examine the capability of strain JPA2 to tolerate different concentrations of salinity.

In vivo pot test for plant growth promotion was carried out at CCS Haryana Agricultural University, Hisar to evaluate the effect of salt tolerant PGPR isolate using pearl millet as test crop. Seeds of pearl millet (HHB 146: a salt tolerant variety) obtained from Bajra section, Agriculture college, CCS HAU, Hisar, were surface sterilized by exposing to 95 % ethanol and immersing in 0.2 % of $HgCl_2$ solution for 3 min. The seeds were then subjected to five washings with sterile distilled water. Earthen pots with drainage hole were filled with 5 kg of soil and two salinity levels (EC 6, 8 dS m^{-1}) were maintained through irrigation water by addition of different concentration of salts such as NaCl, $CaCl_2 \cdot 2H_2O$, $MgCl_2$ and $MgSO_4 \cdot 7H_2O$ while pots of EC 0 dS m^{-1} were irrigated with normal distilled water. Recommended dose of commercial fertilizer at the rate of 80 N and 40 P kg ha^{-1} was supplied to each pot through urea and single super phosphate, respectively. Nitrogen was applied in two splits, while full dose of phosphorus was applied as basal dose. Experiment was laid out in randomized block design with three replications per treatment. Four seeds were planted per pot at a uniform depth of 1.5 cm. For preparation of bacterial formulation, the strain JPA2 was cultured in LB medium and incubated on an orbital shaker at 150 rpm at 30 °C. 1 ml of this overnight grown bacterial suspension (10^6–10^7 cells ml^{-1}, 0.5 O.D. at 540 nm) was applied to each pearl millet seed. However, in control, seeds were treated with sterile non-inoculated LB medium. The pots were kept in sunlight, irrigated and observed everyday for 45 days. On 46th day, plants were harvested carefully at vegetative stage. Shoot length was measured after harvesting and samples were dried in oven at 75 °C. After 4 days, shoot dry weight (g/pot) was calculated and statistical significance of the data was determined by ANOVA (Analysis of Variance) using MINITAB version 15.0 and Tukey's tests ($P = 0.05$).

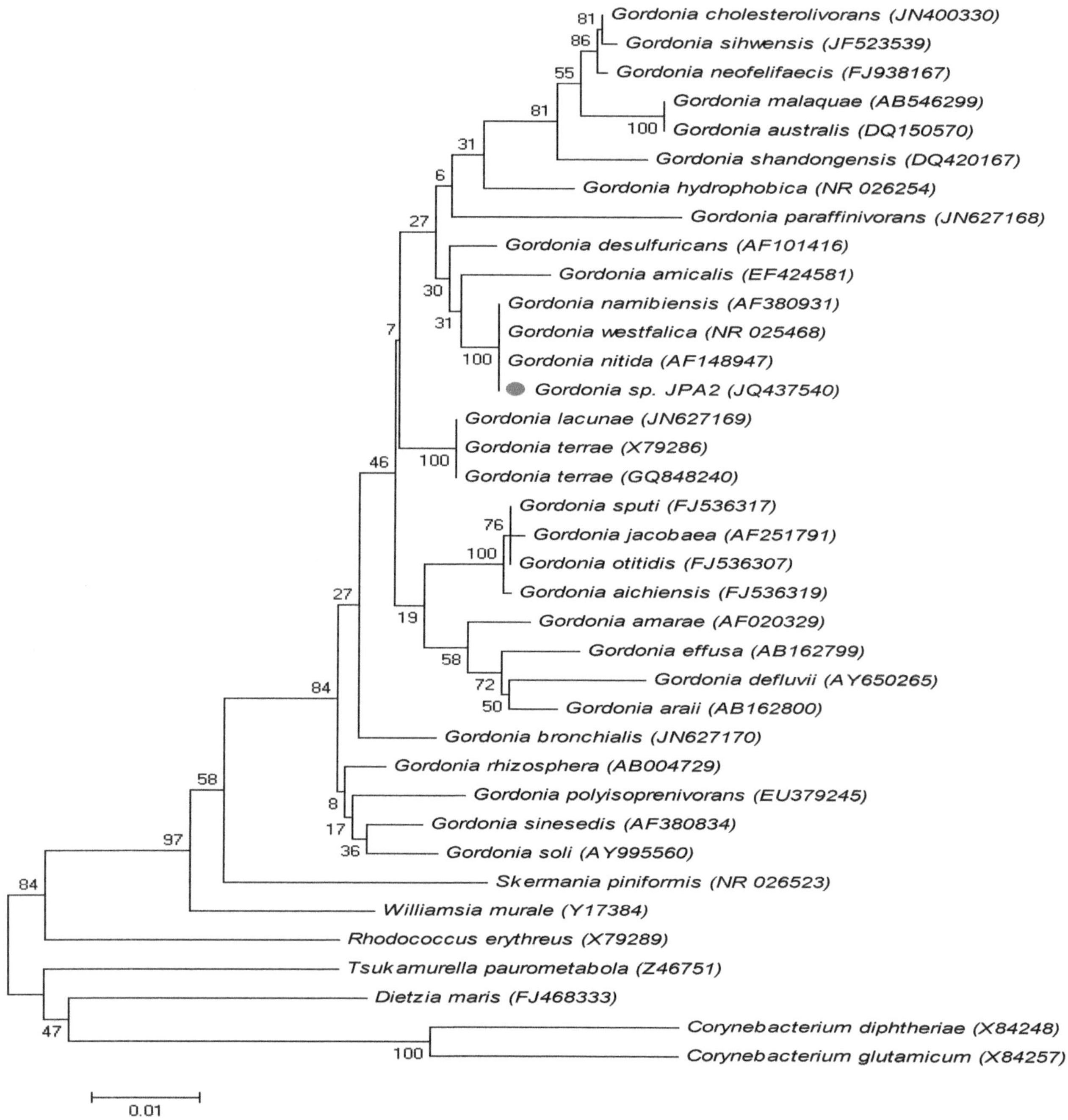

Fig. 1 Phylogenetic tree showing the position of strain JPA2 *Gordonia* species and some other related taxa based on 16S rRNA gene sequences. Distances were calculated by neighbor-joining method. Bootstrap values (expressed as percentages of 1,000 replications) are shown at branch points. Genbank accession numbers are given in parentheses. The *scale bar* indicates 0.01 substitutions per nucleotide position

Results and discussion

In the present investigation, a salt tolerant PGPR strain was isolated from rhizosphere of weed *Chenopodium murale* grown in saline fields of Pindara village. The rhizospheric soil sample has the following physiochemical properties: pH 6.24; 11.47 dS m^{-1} EC; 0.13 % organic carbon and 0.12 % total N. Three different morphotypes were selected on the basis of their colony color, shape and size on Jensen's nitrogen free media. Amplification of *nifH* gene segment yielded the product of expected size (420 bp) from DNA template of only one isolate JPA2. Further, considerable acetylene reduction of 38.9 nmol ethylene h^{-1} mg^{-1} protein confirmed the presence of nitrogenase activity.

Identification of the strain JPA2 was done on the basis of partial 16S rRNA gene sequencing from Bioserve, Hyderabad (India). Figure 1 represents the phylogenetic position of the detected partial 16S rRNA sequence of JPA2 compared to sequences published in GenBank database. Phylogenetic analyses of the strains based on the NJ method with 1,000 bootstrap sampling resulted in two major clusters and strain JPA2 fell within the cluster comprising the *Gordonia* sp. The bootstrap resampling value for the major cluster comprising distantly related *Gordonia* sp. was 84 % and *Rhodococcus erythreus* (X79289) was used as an outgroup. The identification was further confirmed by morphological and biochemical characterization of the isolate. Colony color of JPA2 was yellowish and cells were Gram +ve cocci as observed in other *Gordonia* sp. Exhibition of biochemical traits such as nitrate reduction, aesculin hydrolysis and starch hydrolysis, assimilation of sugars like glycerol, D- cellobiose and *N*-Acetyl D-glucosamine confirmed it as *Gordonia* sp.

The isolate was able to produce significant amount of IAA (193.27 μg ml^{-1}), ammonia (0.085 μg ml^{-1}) and positive siderophore production was also detected. Strain JPA2 was found to be salt tolerant as it could tolerate up to 6 % NaCl.

To study the effect of bacterial inoculation, a pot house experiment was conducted and JPA2 was used as an inoculant for pearl millet. JPA2 had more stimulating effect on plant growth and showed significant increase in plant height (51.1, 39.9 and 28.8 %) and plant dry weight (55.9, 36.4 and 35.5 %) over uninoculated control plants at EC 0, 6, 8 dS m^{-1}, respectively (Table 1). The higher plant height and shoot dry weight response to inoculant compared to control under different salinity levels clearly showed the beneficial role of this rhizobacterium. Such an improvement might be attributed to N_2-fixing capacity as well as the ability of this PGPR strain to produce growth-promoting substances. The enhancing effect of seeds treated with bioinoculant on growth and yield of pearl millet has been reported by many researchers. (Latake et al. 2009). Tanawy (2009) isolated bacteria from highly saline soil (EC 23 dS m^{-1}) that showed a high ability for promoting plants establishment under adverse saline conditions by producing plant growth promoter substances and fixing a high amount of nitrogen in plant ecosystem. Several studies have also shown that application of PGPR *Azospirillum*, *Bacillus* and *Enterobacter* strains as bioinoculants improves plant growth (Tahir et al. 2013). PGPR *Gordonia* sp. S2RP-17 showed plant growth promotion in Zea mays (Hong et al. 2010). Similar to these reports, our findings also demonstrated that strain JPA2 may be used as a bioinoculant for plant growth promotion.

In conclusion, all data in the current work emphasize that the inspected strain JPA2 fixed the nitrogen biologically producing substances of plant growth promoters (IAA, ammonia and siderophores) that enhanced growth and productivity of pearl millet plants when inoculated with this strain, under normal and saline conditions. Also this work proved that the isolated strain JPA2 is saline tolerant (6 % NaCl tolerance) and had the ability to tolerate high soil salinity. Therefore, this strain proved to be potential candidate for the development of bioinoculant for crop plants growing in saline soils.

Table 1 Effect of JPA2 on plant height and dry weight of the pearl millet plants

Treatment	EC-0 (dS m^{-1})		EC-6 (dS m^{-1})		EC-8 (dS m^{-1})	
	Plant height (cm)	Dry weight (g/plant)	Plant height (cm)	Dry weight (g/plant)	Plant height (cm)	Dry weight (g/plant)
Control	67.83 ± 2.061	3.70 ± 0.080	59.85 ± 1.507	2.85 ± 0.063	53.88 ± 1.882	2.14 ± 0.069
100 % RDF + JPA2	102.5 ± 2.425***	5.77 ± 0.051***	83.75 ± 3.152**	3.89 ± 0.040***	69.42 ± 0.842**	2.90 ± 0.046***

Data represent the mean ± SEM in each group. Group 1: plant height; Group 2: dry weight

** $P < 0.01$

*** $P < 0.001$

Acknowledgments This research work is a part of the AMAAS project funded by National Bureau of Agriculturally Important Microorganisms (NBAIM) and Mau Nath Bhanjan. The authors are grateful to these funding agencies.

Conflict of interest The authors declare that they have no conflict of interest.

References

Bendinger B, Kroppenstedt RM, Klatte S, Altendorf K (1992) Chemotaxonomic differentiation of coryneform bacteria isolated from biofilters. Int J Syst Bacteriol 42:474–486

Bric JM, Bostock RM, Silvestone SE (1991) Rapid in situ assay for indole acetic acid production by bacteria immobilized on a nitrocellulose membrane. Appl Environ Microbiol 57:535–538

Bridage GL, Davidson MS, Sparling GP (1982) Critical evaluation of the acetylene reduction test for estimating the nitrogenase activity of nitrogen fixing bacteria associated with the roots of wheat and barley. Soil Biol Biochem 14:27–35

Bruel D, Holter H, Linderstrom-Lang K (1946) C R Lab Carlsberg. Scr Chim 25:289

Cappuccino JG, Sherman N (1992) Microbiology: a laboratory manual. Addison-Wesley, New York

Choubey VK, Singh O, Srivastava SL (2009) Study of hydrological soil properties of salt affected areas around Gohana, Sonipat district, Haryana. Earth Sci India 2(3):211–223

Collins CH, Lyne PM (1984) Microbiological methods, 5th edn. Butterworth, London

Cowan ST, Steel KJ (1965) Manual for the identification of medical bacteria. Cambridge University Press, London

Hernandez P, Fayolle GF, Vandecasteele JP (2001) Biodegradation of ethyl *t*-butyl ether (ETBE), methyl *t*-butyl ether (MTBE) and *t*-amyl methyl ether (TAME) by *Gordonia terrae*. Appl Microbiol Biotechnol 55:117–121

Hong SH, Ryu HW, Kim J, Cho KS (2010) Rhizoremediation of diesel-contaminated soil using the plant growth-promoting rhizobacterium *Gordonia* sp. S2RP-17. Biodegradation.

Jensen V (1951) Notes on biology of *Azotobacter*. Proc Soc Appl Bacteriol 74:89–93

Kalembasa SJ, Jenkinson DS (1973) A comparative study of titrimetric and gravimetric methods for the determination of organic carbon in soil. J Sci Food Agric 24:1089–1090

Kumar V, Gera R (2013) Isolation of a multi-trait plant growth promoting *Brevundimonas* sp. and its effect on the growth of Bt-cotton. 3Biotech.

Kumar A, Kumar R, Yadav VPS, Kumar R (2010) Impact assessment of frontline demonstrations of Bajra in Haryana state. Indian Res J Ext Education 10(1):105–108

Latake SB, Shinde DB, Bhosale DM (2009) Effect of inoculation of beneficial microorganisms on growth and yield of Pearl millet. Indian J Agric Res 43(1):61–64

Sarita S, Priefer U, Prell J, Sharma PK (2007) Diversity of *nifH* gene amplified from rhizosphere soil DNA. Curr Sci 94:109–114

Schwyn B, Neilands JB (1987) Universal chemical assay for the detection and determination of siderophores. Anal Biochem 160:47–56

Tahir M, Mirza MS, Zaheer A, Dimitrov MR, Smidt H, Hameed S (2013) Isolation and identification of phosphate solubilizer *Azospirillum*, *Bacillus* and *Enterobacter* strains by 16S rRNA sequence analysis and their effect on growth of wheat (*Triticum aestivum* L.). Aust J Crop Sci 7(9):1284–1292

Tanawy EA (2009) Acquainting with salt tolerant endophytic bacteria isolated from rice plant grown in highly saline soil in Egypt. Int J Acad Res 1(2):72–79

Xue Y, Xuesong X, Zhou R, Liu R, Liang F, Ma Y (2003) *Gordonia paraffinivorans* sp. nov., a hydrocarbon-degrading actinomycete isolated from an oil-producing well. Int J Syst Evol Micriobiol 53:1643–1646

12

Structure-based functional annotation of putative conserved proteins having lyase activity from *Haemophilus influenzae*

Mohd. Shahbaaz · Faizan Ahmad · Md. Imtaiyaz Hassan

Abstract *Haemophilus influenzae* is a small pleomorphic Gram-negative bacteria which causes several chronic diseases, including bacteremia, meningitis, cellulitis, epiglottitis, septic arthritis, pneumonia, and empyema. Here we extensively analyzed the sequenced genome of *H. influenzae* strain Rd KW20 using protein family databases, protein structure prediction, pathways and genome context methods to assign a precise function to proteins whose functions are unknown. These proteins are termed as hypothetical proteins (HPs), for which no experimental information is available. Function prediction of these proteins would surely be supportive to precisely understand the biochemical pathways and mechanism of pathogenesis of *Haemophilus influenzae*. During the extensive analysis of *H. influenzae* genome, we found the presence of eight HPs showing lyase activity. Subsequently, we modeled and analyzed three-dimensional structure of all these HPs to determine their functions more precisely. We found these HPs possess cystathionine-β-synthase, cyclase, carboxymuconolactone decarboxylase, pseudouridine synthase A and C, D-tagatose-1,6-bisphosphate aldolase and aminodeoxychorismate lyase-like features, indicating their corresponding functions in the *H. influenzae*. Lyases are actively involved in the regulation of biosynthesis of various hormones, metabolic pathways, signal transduction, and DNA repair. Lyases are also considered as a key player for various biological processes. These enzymes are critically essential for the survival and pathogenesis of *H. influenzae* and, therefore, these enzymes may be considered as a potential target for structure-based rational drug design. Our structure–function relationship analysis will be useful to search and design potential lead molecules based on the structure of these lyases, for drug design and discovery.

Keywords *Haemophilus influenzae* · Hypothetical protein · Sequence analysis · Structure analysis · Function prediction · Gene annotation · Structure–function relationship

Introduction

Haemophilus influenzae strain Rd KW20 was the first organism whose genome was successfully sequenced in 1995 by Craig Venter's group (Fleischmann et al. 1995). The genome of *H. influenzae* contains 1,740 protein-coding genes, 2 transfer RNA genes, and 18 other RNA genes in a single circular chromosome of 1,830,140 base pairs (Fleischmann et al. 1995). This organism belongs to the family *Pasteurellaceae* (Christensen 2008; Kuhnert 2008), a non-typeable (NTHi) causing bacteremia and acute bacterial meningitis in infants, children and adults (Murphy and Sethi 1992; Sethi and Murphy 2001). *H. influenzae* is part of the normal nasopharyngeal flora in humans and is often coupled with otitis media, chronic bronchitis and community-acquired pneumonia (Apisarnthanarak and Mundy 2005; Eldika and Sethi 2006). *H. influenzae* is obligatory to human and requires β-nicotinamide adenine dinucleotide

Mohd. Shahbaaz
Department of Computer Science, Jamia Millia Islamia, New Delhi 110025, India

F. Ahmad · Md. Imtaiyaz Hassan (✉)
Center for Interdisciplinary Research in Basic Sciences, Jamia Millia Islamia, Jamia Nagar, New Delhi 110025, India
e-mail: mihassan@jmi.ac.in

and heme for growth (Markel et al. 2007; Morton et al. 2004a). Due to its inability to produce iron-containing heme for the cytoplasmic enzymes, it uses multiple mechanisms to obtain heme (Stojiljkovic and Perkins-Balding 2002) that include various heme acquisition proteins like the heme utilization protein, Hup (Morton et al. 2004b) and heme-binding lipoprotein Hbp A (Morton et al. 2005). Mechanism for heme acquisition indicates a strict regulation of iron homeostasis in *H. influenzae,* and it is important for its survival and virulence (Morton et al. 2004a).

The sequenced genome of various organisms on comparative genomics analysis shows that a significant portion of the genes encodes "hypothetical proteins (HPs)", i.e. functionally uncharacterized proteins but found in organisms (Galperin and Koonin 2004). HPs are predicted to be expressed from an open reading frame, but no experimental evidences are available for their function. The majority of HPs is believed to be a product of pseudogenes in human as well as of other organisms and constitute an extensive fraction of their proteomes (Desler et al. 2012; Galperin 2001). HPs have unique sequences and are essential determinants of species-specific phenotypic properties, such as pathogenicity in a given organism (Desler et al. 2012; Galperin 2001). These determinants are considered as a potent drug targets in pathogenic organisms (Kumar et al. 2014; Tsoka and Ouzounis 2000). Experimental characterization of these HPs in a model organism, such as *Escherichia coli, Saccharomyces cerevisiae, H. influenzae* etc., will be helpful in complete understanding of their biological systems at the molecular level (Nimrod et al. 2008; Park et al. 2012; Pidugu et al. 2009). The recent functional annotation of previously uncharacterized tRNA modification enzymes (Alexandrov et al. 2002; Jackman et al. 2003; Soma et al. 2003) of the deoxyxylulose pathway (Eisenreich et al. 2001, 2004), and of the central role of cyclic diguanylate in bacterial signaling (Galperin 2004; Jenal 2004) emphasizes the importance HP functional characterization.

A precise prediction of protein functions depends on accurate protein folding, which is revealed by their three-dimensional structure (Dobson 2003). Similarity in protein structure leads to similarity in biological function that cannot be predicted by sequence analysis alone, for the protein sequence is less conserved than the tertiary structure of a protein (Illergard et al. 2009). Available information on protein structure is useful for characterization of binding motifs and catalytic cores present in proteins (Shapiro and Harris 2000). Knowledge of protein structure is important for rational drug design that mainly relies on the structure of the target protein to narrow the searching of lead compound and for refinement of the compound (Capdeville et al. 2002; Klebe 2000). Hence, structure prediction of HPs may deliver some evidences about their biological function and help in the drug discovery of better therapeutics for treating diseases caused by *H. influenzae.*

Table 1 List of hypothetical protein with lyase activity in *H. influenzae* strain Rd KW20

S. no.	Accession no	Gene ID	Protein product	Uniprot ID	Protein name
1	NC_000907.1	949660	NP_438613.1	P44717	HP HI0452
2	NC_000907.1	950684	NP_438775.1	P44782	HP HI0617
3	NC_000907.1	950339	NP_439586.2	P44197	HP HI1435
4	NC_000907.1	950454	NP_439740.1	P45267	HP HI1598
5	NC_000907.1	950030	NP_439212.1	Q57498	HP HI1053
6	NC_000907.1	949991	NP_439177.1	P44095	HP HI1016
7	NC_000907.1	950004	NP_439172.1	P44093	HP HI1011
8	NC_000907.1	950653	NP_438618.1	P44720	HP HI0457

Recently, we have annotated the function of HPs from *H. influenzae* (Shahbaaz et al. 2013). During extensive sequence analysis of HPs from *H. influenzae,* we predicted that eight HPs possess lyase activity as listed in Table 1. Since working on the structure-based drug design therefore we are looking for the novel therapeutic targets (Hassan et al. 2007a, b; Thakur et al. 2013; Thakur and Hassan 2011). Lyase enzymes are important for virulence and survival of pathogens in the host. These enzymes provide essential nutrients and are involved in modifying the local environment for the favorable growth (Bjornson 1984). Two examples of such enzymes are cystathionine β-lyase and isocitrate lyase. The formation and degradation of cystathionine cause the mobilization of sulfur from Cys, which is required for the biosynthesis of methionine in bacteria and is important for the virulence (Ejim et al. 2004). Similarly, the isocitrate lyase is an enzyme of the glyoxylate pathway (an anaplerotic pathway of the TCA cycle), which cleaves isocitrate to glyoxylate and succinate, which are potential drug targets for control of many human diseases caused by various pathogens (Britton et al. 2001). Thus, lyase enzymes are used for sequence and structure analysis for further study in this paper.

Materials and methods

Sequence retrieval

The genome analysis of *H. influenzae* shows that 1,657 proteins are present in its proteome (http://www.ncbi.nlm.nih.gov/genome/). We have selected all 429 proteins (Table S1 and S2) with a name HP separately and perform rigorous sequence analyses using fasta sequences retrieved from UniProt (http://www.uniprot.org/) (Shahbaaz et al.

2013). Sequences of proteins having lyase-like activity were used for further structure prediction and analysis.

Sub-cellular localization

To characterize a protein as a drug and/or vaccine target, sub-cellular localization is essentially important. We used sub-cellular localization tools, PSORTb (Yu et al. 2010b), PSLpred (Bhasin et al. 2005) and CELLO (Yu et al. 2006). SignalP 4.1 (Emanuelsson et al. 2007) was used for signal peptide prediction and SecretomeP (Bendtsen et al. 2005) for identifying protein involvement in non-classical secretory pathway. TMHMM (Krogh et al. 2001) and HMMTOP (Tusnady and Simon 2001) were used to predict the propensity of a protein to be a membrane protein.

Sequence comparisons

We predicted the function of HPs using conserved sequence patterns in protein families (Chen and Jeong 2000). BLASTp (Altschul et al. 1990) and HHpred (Soding et al. 2005) based on hidden Markov models were used for remote protein homology detection against various protein databases such as PDB (Bernstein et al. 1978), SCOP (Hubbard et al. 1999), CATH, etc.

Function prediction

The domain analysis allows more precise function prediction (Reid et al. 2007). We used various publicly available databases for functional annotation such as Pfam (Punta et al. 2012), PANTHER (Mi et al. 2005), SMART (Letunic et al. Letunic et al. 2012), SUPERFAMILY (Gough et al. 2001), CATH (Sillitoe et al. 2013), CDART (Geer et al. 2002), SYSTERS (Meinel et al. 2005), ProtoNet (Rappoport et al. 2012) and SVMProt (Cai et al. 2003). CDART and SMART were used for similarity search based on domain architecture and profiles rather than by direct sequence similarity. MOTIF (Kanehisa 1997) and InterProScan (Quevillon et al. 2005) help to identify signature sequence in a particular protein. The MEME suite (Bailey et al. 2009) has been used to perform motif-sequence database searching.

Virulence factor analysis

Virulence factors are considered as potential drug/vaccine targets (Baron and Coombes 2007). We used VICMpred (Saha and Raghava 2006) and Virulentpred (Garg and Gupta 2008) for identifying potential virulence factors. Both are SVM-based methods used to predict bacterial virulence factors from protein sequences with a significant accuracy.

Functional protein association networks

To understand the role of a protein in certain biological pathways, protein–protein interaction analysis is very useful. Here, we used STRING (version-9.05) (Szklarczyk et al. 2011) to predict protein interaction partners of HPs. The interactions include direct (physical) and indirect (functional) associations, experimental or co-expression.

Structure prediction

Three-dimensional structure of proteins P44197 and P44720 was predicted using the homology modeling (Marti-Renom et al. 2000) on MODELLER module present in the Discovery Studio 3.5 (Accelrys 2013). We have identified templates using sequence similarity search methods PSI-BLAST (Altschul et al. 1997) and found pseudouridine synthases RluC (Mizutani et al. 2004) and aminodeoxychorismate lyase from *Escherichia coli* (PDB id: 2RIF) as a suitable template for P44197 and P44720, respectively. The template and query sequences were aligned and finally used for modeling protein structure in MODELLER (Eswar et al. 2007).

Since we do not find any suitable template for other HPs with lyase activity, MODELLER has no role in structure prediction of such proteins. We used threading or fold recognition and ab initio modeling protocol for predicting the structure of such proteins. The threading or fold recognition (Xu et al. 2008) methods were applied for aligning the primary sequence with those proteins present in the Protein Data Bank (PDB) with similar folds. It uses sequence–structure alignment approaches for modeling. We used Sparksx server (Yang et al. 2011) for predicting the structure of P44782, P45267 and Q57498, which uses the SPINE X server for predicting secondary structure, torsion angles and solvent accessibility with higher accuracy. For protein P44093, we used RaptorX (Peng and Xu 2011), as sparksx models are not found suitable for further studies which use NEFF to measure the amount of information in the sequence profile of a protein. However, sparksx along with PSI-BLAST and HHpred is considered to be the best server for predicting the hard targets in Critical Assessment of protein Structure Prediction (CASP) proceedings (Kryshtafovych et al. 2011).

ROBETTA server (Kim et al. 2004) was used for structure prediction of HP P44717, which uses ab initio or de novo methods to predict the structure of proteins whose structural analogs do not exist in the PDB or could not be successfully identified by threading. ROBETTA server generates structure from scratch, since it uses a new alignment method, called K*Sync, to align the query sequence onto the parent structure. Then it models variable regions by allowing them to explore conformational space with fragments in a fashion similar to the de novo protocol in context of the template. When no structural homolog is available, domains are modeled with the Rosetta de novo protocol (Misura et al. 2006), which allows the full length of the domain to explore conformational space via fragment insertion, producing a large decoy ensemble from which the final models are selected.

We used I-TASSER (Roy et al. 2010) for predicting the structure of HP P44095 that first generates three-dimensional (3D) atomic models from multiple threading alignments and iterative structural assembly simulations. Using the structural matching of the 3D models with other known proteins it inferred function of the protein and the output contains full-length secondary and tertiary structure predictions, and functional annotations on ligand-binding sites, Enzyme Commission numbers and Gene Ontology terms in the results (Roy et al. 2010). Since the unavailability of proper prediction, we used the COACH (Yang et al. 2013) for protein–ligand-binding site prediction and COFACTOR (Roy et al. 2012) for structure-based biological function annotation of HP, for proper function analysis of P44095.

Predicted models were further refined using a side-chain refinement protocol of Discovery Studio 3.5, which uses force fields such as CHARMM (Brooks et al. 2009) and SCWRL4 (Krivov et al. 2009), and predicts positions of the side chains which are used for refinement of predicted protein structures. The final models were further validated by PROCHECK suite (Laskowski et al. 1996), a module of SAVES (Structural Analysis and Verification Server).

Structure analysis

Three-dimensional structure of homologous proteins often remains more conserved than their sequence (Chothia and Lesk 1986). Structural similarities are more reliable than that of sequences for grouping together distant homologues (Taylor and Orengo 1989). Three-dimensional structure of all models was analyzed by various servers for function prediction. Identifying the functional region of protein is an important step towards characterizing its molecular function. We used POCASA (Yu et al. 2010a), Pocket-Finder (Laurie and Jackson 2005) and firestar server (Lopez et al. 2007) for prediction of functionally important residues in HP and the PPM server for calculating rotational as well as translational positions of transmembrane and peripheral proteins in cell membranes (Lomize et al. 2012). It can be applied to newly determine experimental protein structures or theoretical models. Since structural motifs are associated with catalytic functions, motif finding is also essential for precise function prediction (Singh and Saha 2003). We used ProFunc (Laskowski et al. 2005) web server for predicting the function of proteins from its atomic coordinates based on identification of functional motifs. DALI server (Holm and Rosenstrom 2010) was used for structure comparison and fold similarity search. PyMOL (DeLano 2002), a molecular graphics system, is used for visualization of protein structure.

Results and discussion

After the extensive sequence and structure analysis, we found eight HPs with characteristic function, namely, cystathionine-β-synthase, cyclase family protein, carboxymuconolactone decarboxylase, pseudouridine synthase A and C, tagatose-1,6-bisphosphate aldolase and aminodeoxychorismate lyase proteins (Table 2). We successfully predicted and analyzed the three-dimensional structure of HPs with lyase activity, using various computational tools. Final models were further validated to check their stereo-chemical parameters such as bond angle and bond length. All eight models P44782, P44717, P45267, Q57498, P44197, P44095, P44720 and P44093 show significant validation score on SAVES server (http://nihServer.mbi.ucla.edu/SAVES/). Structure analysis helped us to predict the function of each HP precisely (Table 3). Detailed structural analysis outcomes for each protein are described here, separately.

HP P44717

Sub-cellular localization prediction of HP P44717 indicates that this HP is localized in the cytoplasmic membrane. Signal peptide prediction suggests the presence of a signal peptide, and indicates that its translocation occurs through non-classical secretory pathway. HMMTOP (Tusnady and Simon 2001) and TMHMM (Krogh et al. 2001) analyses suggest that HP P44717 consists of four transmembrane helix, and may work as a transporter protein (Table S3). Similarity search revealed its high closeness to the cystathionine-β-synthase (CBS) domain protein. HHpred (Soding et al. 2005) also suggests high similarity with magnesium and cobalt efflux protein CorC (PDB ID: 4HG0) and CBS domain protein (PDB ID: 3LHH) (Table S4). Sequence analyses of P44717 suggest that P44717 is comprised of two domains, CorC/HlyC and CBS domain. By applying the clustering algorithms, we analyzed that HP P44717 is a member of cluster 143846 (CBS domain protein) of SYSTERS (Meinel et al. 2005) and cluster 4144744 (Transporter-associated region) of ProtoNet (Rappoport et al. 2012), and a member of transmembrane family protein. Sequence-based motifs are discovered using InterProScan (Quevillon et al. 2005) and MOTIF tools (Kanehisa 1997). We further observed that its sequence possesses cystathionine-β-synthase motif and

Table 2 List of sequence-based predicted function of HPs with lyase activity and motif discovered using MEME of *H. influenzae* strain Rd KW20

S. no.	Cluster[a]	Uniprot ID	MEME results							Consensus[b] function
			Motif 1		Motif 2		Motif 3		MAST function prediction	
			Start	Site	Start	Site	Start	Site		
1	Cluster 118	P44717	262	GYIESH	199	TQEHYL	405	YGKYKF	UPF0053 protein	Cystathionine-beta-synthase
2	Cluster 187	P44782	152	VKLTPVTGRSHQLRL HMLALGHPILGDKFY	56	FCEPAHRLDMATSGIIVFALSKAADRELK RQFREREPKKHYQAIVWGH	18	YQDNHLCVVNKPSG	Ribosomal large subunit pseudouridine synthase A	RNA pseudouridylate synthase rluA (lyase)
3	Cluster 187	P44197	152	VKLIPHTGRKHXLRXHMK HVFHPIXGDTQY	46	HVFPIHRLDRPTSGVLLFALSSEIANL MCEQFEQKYVQKSYLAVVRGY	6	YQDGFLVAVNKPAG	tRNA pseudouridine synthase C	Ribosomal large subunit pseudouridine synthase C rluC
4	Cluster 114	P45267	335	RVYFERM	14	AISPQI	253	FSQDFM	No result	CYTH-like adenylate cyclase
5	Cluster 196	Q57498	1	MFTDWK	59	TRCESC	21	KQYPKM	No result	Carboxymuconolactone decarboxylase
6	Cluster 131	P44095	128	GIWFPCTW	21	PNHCGTHM	11	TPFSSF	No result	Cyclase family protein
7	Cluster 45	P44093	68	WLKENGCTQFYFKYCST	281	IHNENYIE	151	NLMRLM	No result	D-tagatose-1,6-bisphosphate aldolase
8	Cluster 89	P44720	315	DGSGGH	123	WRKDLENAPH	338	RWYRSQ	UPF0755 protein	Aminodeoxychorismate lyase

[a] CLUSS predicted clusters

[b] Consensus result form on the basis of values present in supplementary Table S2 & S3

Table 3 List of structure-based predicted function and validation of HP with lyase activity in *H. influenzae* strain Rd KW20

S. no.	Uniprot ID	Template	Identity (%)	RMSD	Ramachandran	Proposed function
1	P44717 (robetta model)	Magnesium and cobalt efflux protein CorC, 4HG0 CorC/HlyC transporter-associated domain of a CBS domain protein, 2PLS	23 60	3.400 0.598	99 % (91.7 % core 6.5 % allow 0.8 % gener 1.0 % disall)	CBS domain associated CorC/HlyC transporter
2	P44782 (sparksx model)	Template-pseudouridine synthase RluA, 2I82	60	0.174	99.5 % (91.1 % core 8.4 % allow 0.0 % gener 0.5 % disall)	Pseudouridine synthase RluA
3	P44197 (Modeller structure)	Pseudouridine synthase RluC, 1XPI Pseudouridine synthase RluC, 1V9K	34 33	0.379 0.334	99.5 % (88.2 % core 8.8 % allow 2.5 % gener 0.5 % disall)	Pseudouridine synthase RluC
4	P45267 (Sparksx model)	Putative adenylate cyclase, 2GFG	39	0.918	98.8 % (88.9 % core 9.6 % allow 0.3 % gener 1.2 % disall)	Adenylate cyclase
5	Q57498 (Sparksx model)	Template—Carboxymuconolactone decarboxylase family protein, 1VKE	31	0.300	99 % (94.2 % core 2.9 % allow 1.9 % gener 1.0 % disall)	Carboxymuconolactone decarboxylase
6	P44095 (I-TASSERmodel)	Manganese dependent isatin hydrolase, 4J0N (A) Metal-dependent hydrolase with cyclase activity, 1R61	54 27	2.752 1.139	98.3 % (88.4 % core 9.1 % allow 0.8 % gener 1.7 % disall)	Metal-dependent hydrolase with cyclase activity
7	P44093 (Raptorx model)	Putative tRNA synthase, 3DQQ	56	0.215	99.7 % (93.1 % core 6.6 % allow 0.0 % gener 0.3 % disall)	Putative tRNA synthase
8	P44720 (Modeller)	Predicted aminodeoxychorismate lyase protein, 2R1F	49	0.378	99.7 % (89.8 % core 8.3 % allow 1.6 % gener 0.3 % disall)	Aminodeoxychorismate lyase protein

CBS domain motif. Another motif discovery search tool, MEME suite (Bailey et al. 2009) has detected three motifs, namely, 262′-GYIESH, 199′-TQEHYL and 405′-YGKYKF (Table 2). Based on all these observations, we suggest that HP P44717 may function as cystathionine-β-synthase and present in the inner plasma membrane to work as a transporter for various inorganic salts.

The virulence factor analysis shows that HP P44717 is a non-virulent protein. We further analyzed the functional protein association networks of HP P44717 using an online server STRING. This protein is showing close interaction with thymidylate kinase, 16S ribosomal RNA methyltransferase, virulence-associated protein D, hemolysin, DNA repair protein and GTP-binding protein (Figure S1).

We predicted three-dimensional structure of P44717 using online server Robetta (Kim et al. 2004). The predicted structure was energy minimized, and validated using online server SAVES showing 99 % of residues in the allowed region of the Ramachandran plot (Hooft et al. 1997; Ramachandran et al. 1963). The root-mean-square deviation (rmsd) of the model with respect to the templates 4HG0 and 2PLS was 3.400 and 0.598 Å^2, respectively (Table 3), indicating close functionality. Structure comparison and analysis further revealed that HP P44717 shows the presence of three functional domains, namely, CBS domain, cytochrome *c* oxidase polypeptide and CorC_HlyC domain (Fig. 1a). The overall structure of P44717 is comprised of 18 α-helices and nine β-strands including a CBSX subunit that contains two conserved CBS domains which form a pair facing each other, and are connected by two fundamental β-strands (β1–β2 and β3–β5) and bordered by six α-helices. The presence of helix α14 and helix α16 is the main feature of identifying dimeric CBS domain proteins, and both helices are critical for the dimer assembly of CBSX subunit (Fig. 1b). The CorC_HlyC domain contains four β-strands (β6–β9) arranged as antiparallel β-strand and two α-helices (α17 and α18) that contain binding sites for metal ion binding to amino acids Ile413, Asp414, Thr415 (β8) and Asp421 (β9) (Figure S2 A and B). We also detected an AMP-binding site in P44717 structure at Val227 using Firestar server analysis (Figure S2 C). The active site cavity was predicted by Pocket-Finder (Laurie and Jackson 2005) showing residues Val227, Tyr228, Leu229, Asp230, Phe233, Glu237, Val238, Thr241, Leu242 and Ile250 which are presumably responsible for the lyase activity (Fig. 1c).

In order to find structurally similar proteins in the PDB related to the P44717, we used DALI server (Holm and

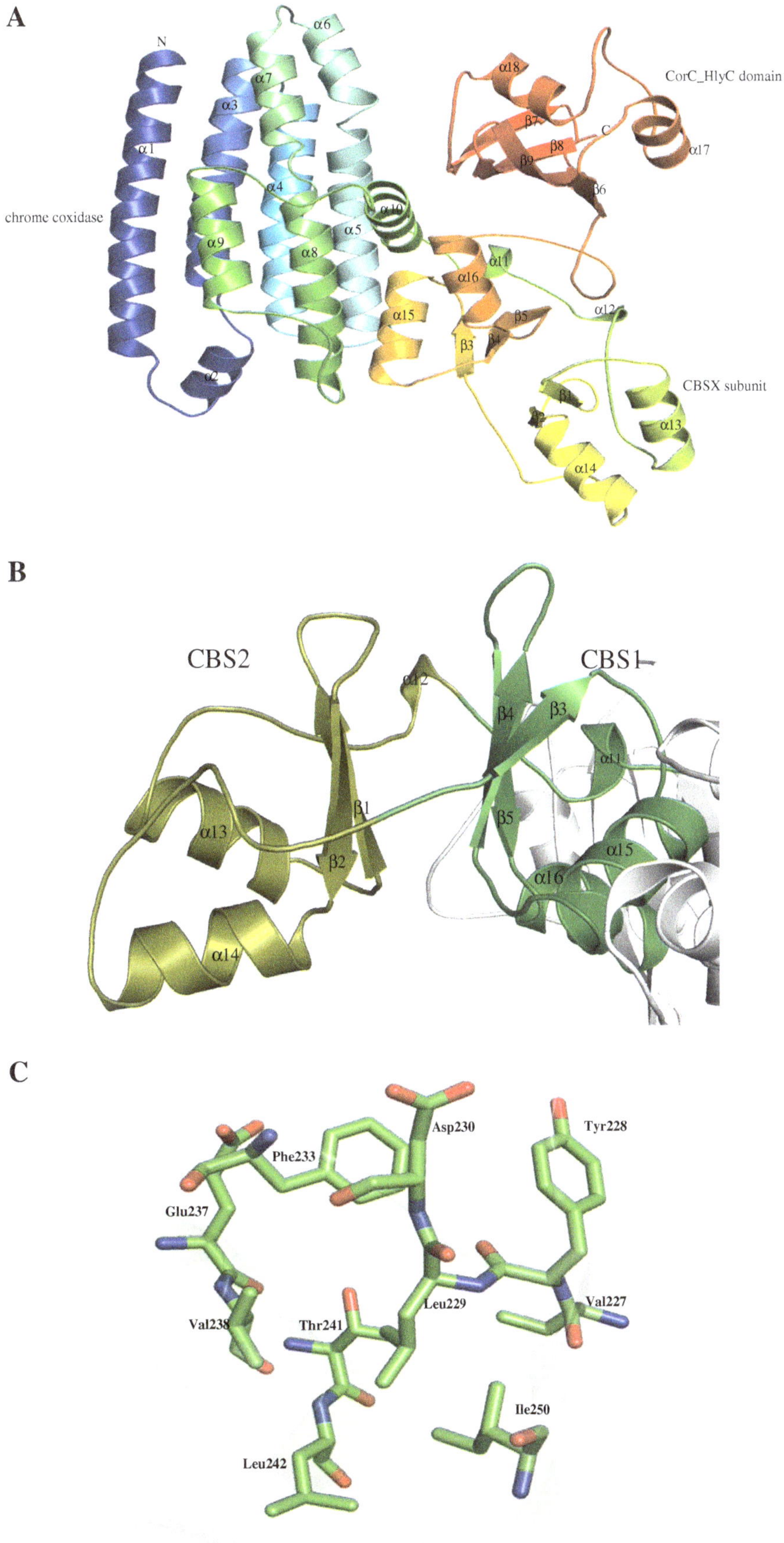

Fig. 1 Representation of model structure of HP P44717. **a** Showing CBS, CorC_HlyC and cytochrome c oxidase domains. **b** Diagram showing the monomer of CBSX with CBS1 and CBS2 domain. **c** Residues present in the active site pocket are illustrated in stick

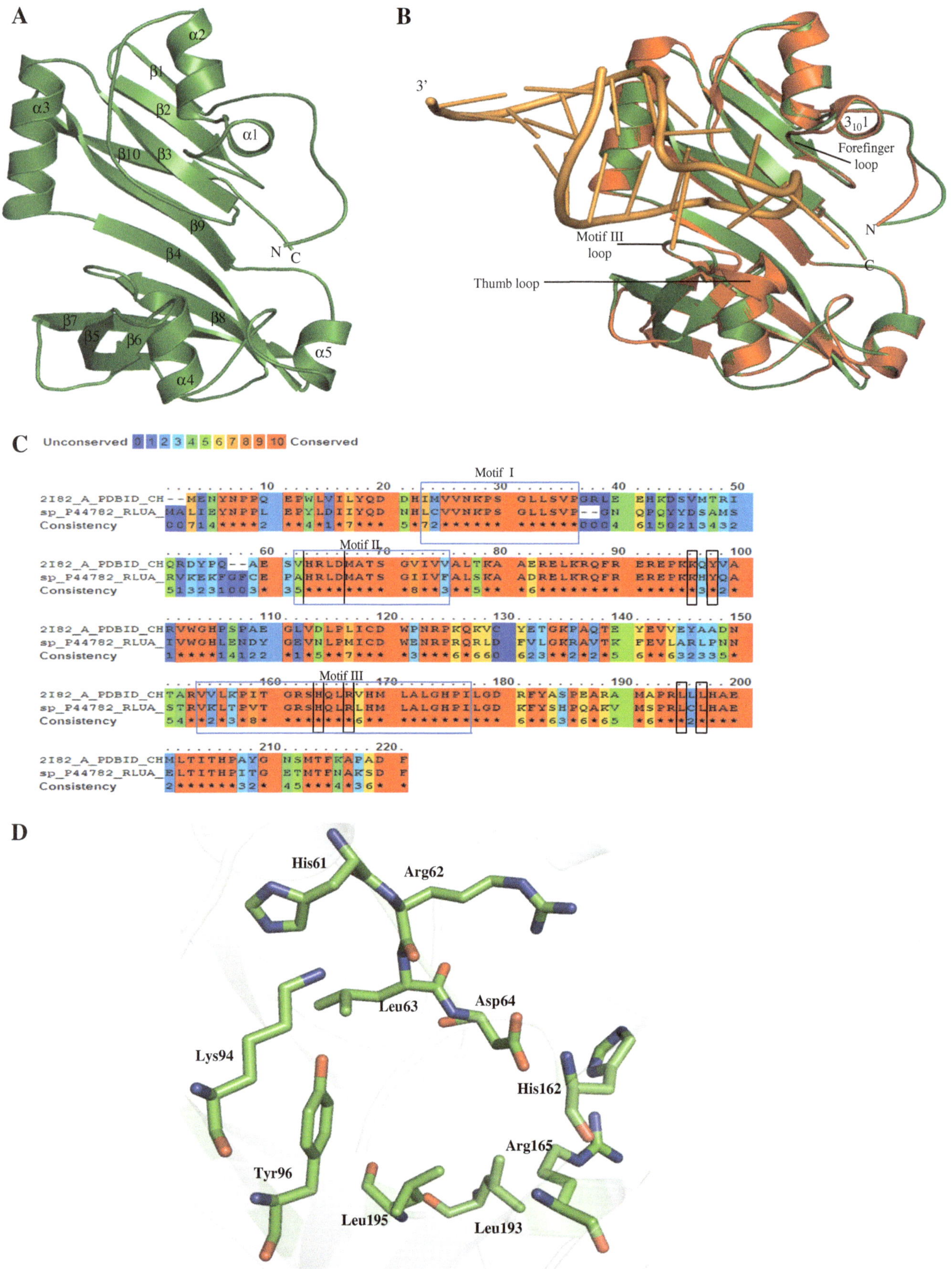
A
α2
β1
β2
α3
α1
β10
β3
β9
β4
N
C
β7
β5
β6
β8
α5
α4
B
3'
3101
Forefinger loop
N
Motif III loop
Thumb loop
C
C
Unconserved 0 1 2 3 4 5 6 7 8 9 10 Conserved
Motif I
Motif II
Motif III
2I82_A_PDBID_CH
sp_P44782_RLUA_
Consistency
D
His61
Arg62
Leu63
Asp64
Lys94
His162
Arg165
Tyr96
Leu195
Leu193

◀ **Fig. 2** Representation of model structure of HP P44782. **a** Overall structure is represented in cartoon. **b** Superposition of the RluA–RNA complex (PDB ID: 2I82; *red*) with free RluA (P44782; *dark green*). The forefinger and thumb loops that "grasp" the RNA are indicated with motif III loop, and contrast the absence of the thumb loop in P44782 structure. **c** Multiple sequence alignment of P44782 and Rlu A (PDB ID: 2I82) with *red* color indicates residues that are conserved in two proteins. Motifs I, II, and III are as defined by Koonin (1996) highlighted in *blue boxes* and conserved active site residues in *black boxes*. **d** Predicted active site residues shown in stick

Rosenstrom 2010). The results clearly indicate a significant similarity of the CBS domain containing proteins such as magnesium and cobalt efflux protein (Z score = 13.8), hemolysin-like protein (Z score = 13.4), etc., which are among the top hits. In each case, residues at positions 80–150 in the sequence show a close resemblance with rmsd of 0.8–3.2 Å^2, despite a sequence similarity of 34 % in the aligned region. We also found a close structural similarity to the magnesium and cobalt efflux protein. Furthermore, ProFunc server (Laskowski et al. 2005) was used to predict the function of the HP on the basis of sequence and structure comparisons. We found 11 motifs in the InterPro database (Hunter et al. 2011) with Cor-C_HlyC, CBS and hemolysin related sequence motifs. There are two significant ligand-binding templates also present in the P44717 structure. All these analyses strongly suggest that HP P44717 contains CBS domain and it has CorC/HlyC transporter function (Table 3).

The CBS domain, also known as 'Bateman domain', is a conserved domain present in prokaryotes and eukaryotes (Ignoul and Eggermont 2005). Structure analyses of bacterial CBS domains show that two CBS domains form an intramolecular dimeric structure (CBS pair). Human hereditary diseases, such as hypertrophic cardiomyopathy, homocystinuria, myotonia congenital, retinitis pigmentosa, etc., can be caused by mutations in CBS domains of cystathionine-β-synthase, inosine 5′-monophosphate dehydrogenase, AMP kinase, and chloride channels, respectively, which also affect multimerization and sorting of proteins, channel gating, and ligand binding (Bateman 1997). Recent experiments show that CBS domains can bind adenosine-containing ligands such as ATP, AMP, or *S*-adenosylmethionine, which indicate that CBS domains may function as sensors of intracellular metabolites (Bateman 1997).

HP P44782

HP P44782 is localized in the cytoplasm and lacks signal peptide, therefore, it is not secreted (Table S3). The sequence analysis showed a significant similarity of HP P44782 with 23S rRNA/tRNA pseudouridine synthase A (Table S1 and S2), which was further confirmed by domain annotation and cluster analysis using online tool SYSTERS and ProtoNet. Furthermore, a pseudouridine synthase, RsuA/RluB/C/D/E/F motif, and pseudouridine synthase, the RluA motif, are present in the sequence of HP P44782. MEME suite has further confirmed three distinct motifs, namely, 152′-VKLTPVTGRSHQLRLHMLALGHPILGDKFY, 56′-FCEPAHRLDMATSGIIVFALSKAADRELKRQFREREPKKHYQAIVWGH, 18′-YQDNHLCVVNKPSG (Table 2), an indication of ribosomal large subunit pseudouridine synthase A. Interacting partners of HP P44782 are ATP-dependent helicase, tRNA pseudouridine synthase B, tRNA pseudouridine synthase A, tRNA-dihydrouridine synthase A, glp protein, lipoprotein signal peptidase, glycerol-3-phosphate regulon repressor, 16S pseudouridylate 516 synthase, *S*-adenosyl-L-methionine-dependent methyltransferase (Figure S1) indicating its role in the pseudouridine synthesis.

Sparksx predicted the structure of P44782 using ribosomal large subunit pseudouridine synthase A (PDBID: 2I82), a lyase, as a template. The final model was validated on PROCHECK (Laskowski et al. 1996) which shows that 99.5 % residues are in the allowed region of Ramachandran plot. Model structure showed rmsd of 0.174 Å^2 from the template. The overall structure of P44782 is elongated and is comprised of mixed α/β fold (Fig. 2a, b). The structure is a conserved Ψ synthase fold, with an eight-stranded β sheet core, and additional strand extends the core β sheet which is also encircled by helices and loops on one face. ASL-RNA binds with its helical axis nearly parallel to the plane of the β sheet core of the predicted P44782 model (Fig. 2b). We found three motifs in sequence of HP P44782 (Fig. 2c) (Koonin 1996). Motif II is the part of the active site while motif I plays an architectural role by supporting motif II (Hoang and Ferre-D'Amare 2001, 2004). The superimposition of structures of P44782 and pseudouridine synthase RluA (PDB code: 2I82) indicates that C-terminal side of the motif I forms a projection that packs against minor groove face of the anticodon loop. N-terminal half of motif III (conserved among Ψ synthases of the RluA and RsuA families) interacts with the RNA backbone on the 5′- side of the ASL (Fig. 2b). The "thumb loop" and "forefinger loop" are found in pseudouridine synthase RluA (Hoang et al. 2006). Structure alignment showed the presence of forefinger loop but not the thumb loop in the HP P44782. The forefinger loop makes RluA enzyme bind with the minor groove of the substrate RNA. Previously determined crystal structure shows that families of pseudouridine synthases have universal conservation in the active sites as an aspartate, a basic residue, and a tyrosine (Del Campo et al. 2004; Koonin 1996; Ramamurthy et al. 1999). Amino acids corresponding to Asp64, Tyr96, and Arg165 are conserved in all ψ synthases, and the active site of RluA (Hoang et al. 2006) is comprised of residues His 61, Arg 62, Leu 63, Asp64, Lys94, Tyr96, His162, Arg165, Leu193 and Leu195 (Fig. 2c, d).

Structure similarity search using DALI server provides many significant hits for P44782 having ribosomal large subunit pseudouridine synthase activity, with ribosomal large subunit pseudouridine synthase C (Z score = 27.3), ribosomal large subunit pseudouridine synthase F (Z score = 12.8), ribosomal large subunit pseudouridine synthase E (Z score = 12.7), etc., among the top hits. We observe an rmsd of 0.4–3.0 Å^2 for residues 180–250 in each match with 60 % sequence identity. Structure of P44782 was further analyzed using ProFunc server showing five sequence pseudouridine synthase motif and six significant ligand-binding templates. The ProFunc and Dali server searches clearly indicate the presence of pseudouridylate synthase activity in the HP P44782 (Table S4).

The isomerization of uridine to pseudouridine (Ψ) within RNA, discovered often in highly conserved locations in ribosomal and transfer RNA, is a ubiquitous process reported to be present in early stages in the evolution of life (Ofengand et al. 2001). Pseudouridine (Ψ) synthases are enzymes that catalyze site-specific isomerization of uridine residues in cellular RNAs. Ψ synthases can be classified into five families, namely, RluA, RsuA, TruA, TruB, and TruD on the basis of amino acid sequence (Kaya and Ofengand 2003), and structure analyses revealed that in spite of minimal sequence similarity, the cores of all Ψ synthases adopt the same fold (Ferre-D'Amare 2003; Foster et al. 2000; Sivaraman et al. 2002) and enzymatic core includes an aspartate residue that is conserved among all five families of Ψ synthases. On site-directed mutagenesis this Asp was found to be essential for catalytic activity of such enzyme (Conrad et al. 1999; Ramamurthy et al. 1999).

HP P44197

HP P44197 is localized in the cytoplasm, and lacks signal peptide and transmembrane helix (Table S3). This protein showed high sequence similarity to tRNA pseudouridine synthase C. The domain annotation shows that P44197 is a member of pseudouridine synthase RsuA/RluD family, showing pseudouridylate synthase activity (Table S2). Inter ProScan and MOTIF tools further confirmed a pseudouridine synthase motif in the HP P44197 (Table 2). Predicted functional partners of HP P44197 are exodeoxyribonuclease VII small subunit, exodeoxyribonuclease VII small subunit, 1-deoxy-D-xylulose-5-phosphate synthase, aspartate-semialdehyde dehydrogenase, tRNA pseudouridine synthase B, and tRNA pseudouridine synthase A which provide an insight of pseudouridine synthase activity for this HP (Figure S1).

Ribosomal large subunit pseudouridine synthase C (PDB ID: 1V9K) is a lyase used as template for predicting the structure of HP. Ramachandran plots show 99.5 % of residues in allowed region. An rmsd of 0.379 Å^2 with template was observed (Table 3). The predicted structure of P44197 is also an elongated, mixed α/β fold and possesses six α-helices and 10 β-strands (Fig. 3a), with an eight-stranded β sheet core. This structure is also a conserved Ψ synthase fold. Two sequence motifs (motif I and II) are present in the sequence of P44197 (Fig. 3b) (Koonin 1996), where motif II is the part of the active site, and architectural role was played by motif I (Koonin 1996). The active site of P44197 includes amino acid residues His51, Arg52, Leu53, Asp54, Lys84, Tyr86, His162, Arg164, Leu199, and Leu201 (Fig. 2c), in which Asp54 and Arg164 are conserved in the active site and mainly responsible for the catalytic activity (Del Campo et al. 2004).

The structure of HP P44197 was further analyzed using ProFunc and Dali servers. This analysis revealed a close similarity to the ribosomal large subunit pseudouridine synthase C (Z score = 34.5), ribosomal large subunit pseudouridine synthase D, (Z score = 26.7), etc. The best match showed 35 % sequence identity over aligned 217 residues. ProFunc server finds five pseudouridine synthase signature sequence, six ligand-binding templates and one DNA-binding template in the HP P44197. All these observations strongly suggest pseudouridylate synthase C activity in this HP (Table 3).

Similarly, RluC and RluD are homologous enzymes which convert three specific uridine bases to pseudouridine in *E. coli* ribosomal 23S RNA to pseudouridine, namely, bases 955, 2504, and 2580 in the case of RluC and 1911, 1915, and 1917 in the case of RluD and both contain N-terminal S4 RNA-binding domain (Mizutani et al. 2004). A loss of RluD which acts as an RNA chaperone is important for ribosome assembly. The loss of RluC results in reduced growth rate, and has no significant effect of impairing growth (Mizutani et al. 2004).

HP P45267

The sequence analysis of HP P45267 showed that it is localized in the cytoplasm and is a non-secretary protein (Table S3). This protein showed a significant sequence similarity to the adenylate cyclase, which was further confirmed by domain analysis which suggested the presence of adenylate cyclase-like domain. The cluster analysis also suggests that HP P45267 belongs to the cluster of protein involved in the cAMP biosynthetic process. Analyzing SVMProt suggests that the HP P45267 is a member of zinc-binding protein and belongs to the DNA-binding protein family. Motif search showed the presence of CYTH-like phosphatases and adenylate cyclase motifs (Table 2). This is a virulent protein, which may be involved in the cellular process. The interacting partners of HP P45267 are Hsf-like protein, phosphatase, glutamate-ammonia-ligase, adenylyl transferase, DNA repair protein,

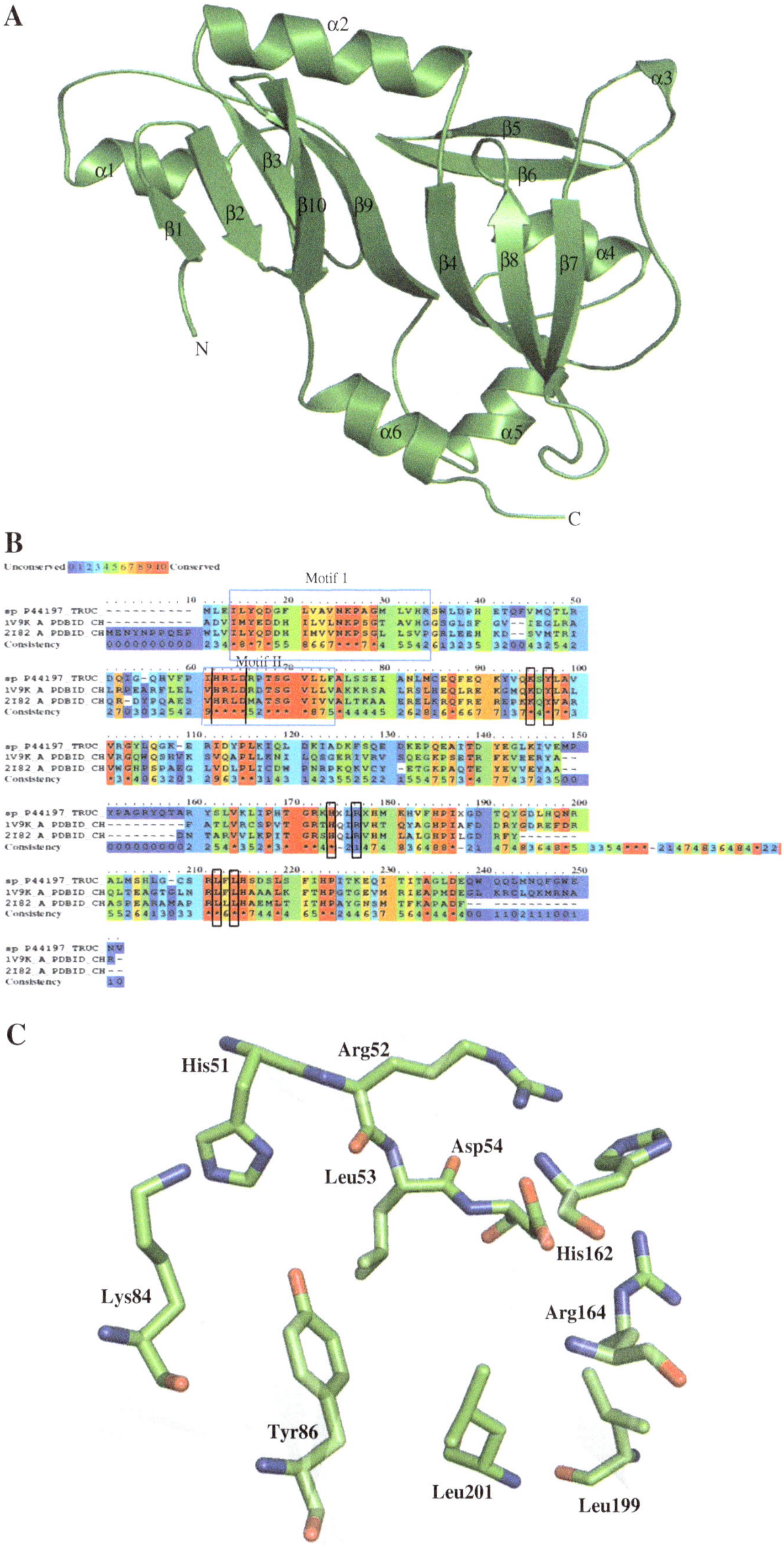

Fig. 3 Representation of model structure of HP P44197. **a** Cartoon model showing overall structure. **b** Multiple sequence alignment showing conserved residues in *red color*, with a catalytic domain of P44197, Rlu C and *E. coli* Rlu A (PDB ID: 2I82) distinguished residues in *black box* and motifs I and II in *blue box*. **c** A detailed description of P44197 active site in which Asp54 is surrounded by arginine residues from both the sides namely Arg52 and Arg164

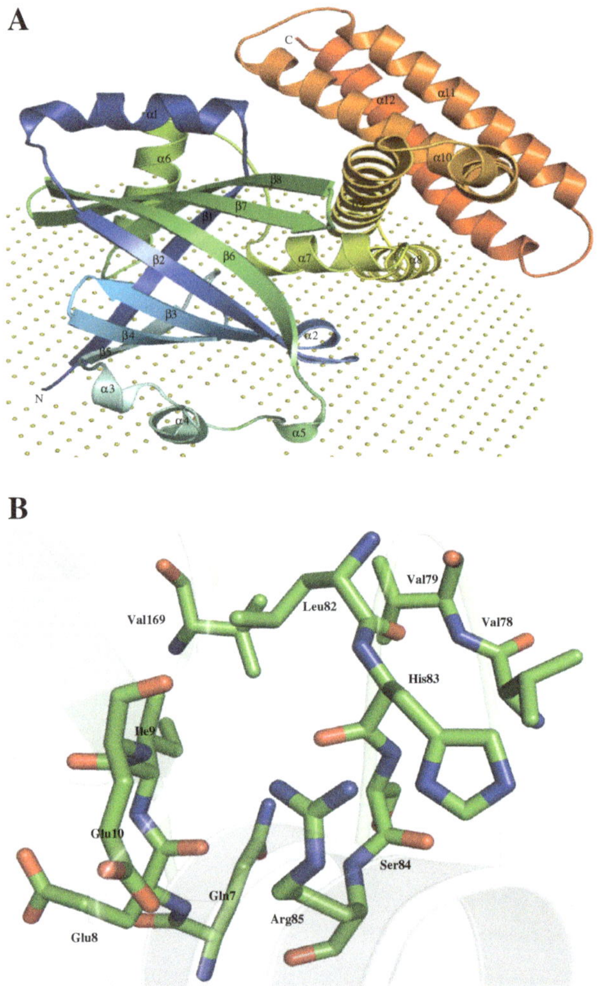

Fig. 4 Representation of model structure of HP P45267. **a** Cartoon diagram showing N-terminal part is shown in *blue* while C-terminal in *red*. Collection of non-bonded spheres represents membrane. **b** Stick representation of P45267 active site with Glu10 residue is proposed to be involve in cAMP binding

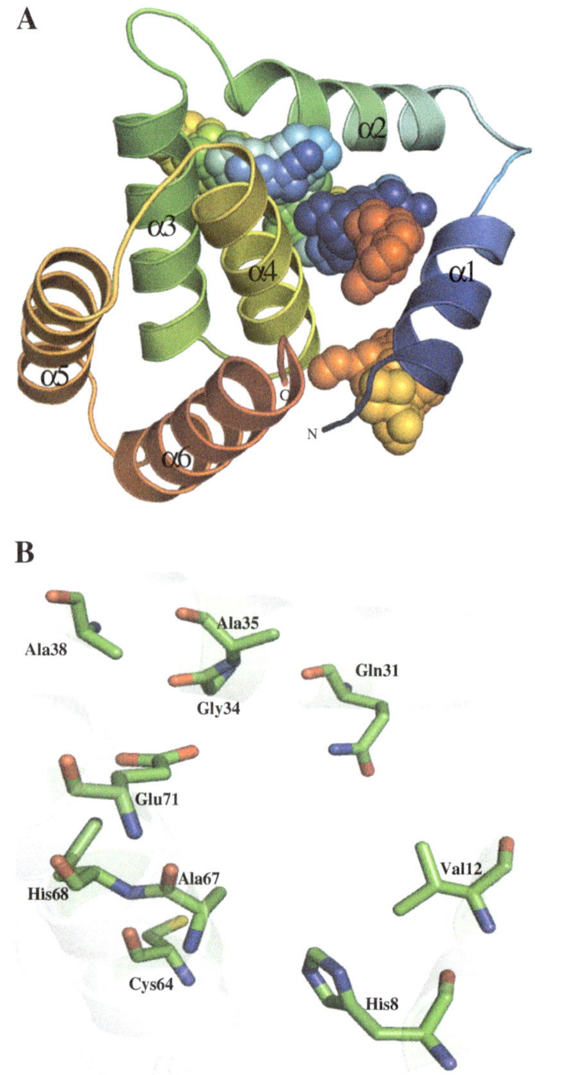

Fig. 5 Representation of model structure of HP Q57498. **a** Cartoon model showing overall structure with active site residues. **b** A description of active site with His68 residue represents the citrate-binding site

outer membrane protein P2 (Figure S1). All these findings strongly suggest that the HP P45267 possesses CYTH-like adenylate cyclase activity and play significant roles in the survival of *H. influenzae,* and may be considered as a potential drug target (Carbonetti 2010).

Structure of P45267 was predicted by fold recognition algorithm on the sparksx server using adenylate cyclase (PDB ID: 2GFG) as template. PROCHECK reports that 98.8 % of residues are present in the allowed region of the Ramachandran plot. Overall structure contains 12 α-helices and 8 β-strands. Interestingly, eight stranded anti-parallel barrels (β1–β8) surrounded by nine long α-helices and three small helical regions fold were adopted by this protein (Fig. 4a) which looks like a cup or bottle with helical handles on both sides. The structure analysis shows that residues Met1, Val79, Pro108 and Phe109 are attached to the membrane. We found three AMP-binding residues Glu10, Arg62 and Arg127 in the P45267 (Figure S3). The active site may contain residues Gln7, Glu8, Ile9, Glu10, Val78, Val79, Leu82, His83, Ser84, Arg85, and Val169 (Fig. 4b).

The structure analysis results showed the presence of adenylate activity in this HP due to the high structural resemblance with the CYTH-like phosphatase (*Z* score = 34.5), adenylate cyclase 2 (*Z* score = 13.5), etc. We also found a similar search in ProFunc analysis, which shows the presence of 3 CYTH-like phosphatase motifs in this HP. These observations suggest that the HP P45267 may be a CYTH-like phosphatase and adenylate cyclase enzyme, and it is essential for pathogenesis (Table 3).

Adenylyl cyclase (AC, EC 4.6.1.1) catalyzes the cyclization of ATP to form cyclic AMP (cAMP), and is an important signaling molecule present in most of cells (Taussig and Gilman 1995). The liver cells produce cAMP, which is responsible for phosphorylase activation in response to extracellular adrenaline. It is found that adenylyl cyclase is involved in a wide variety of signal transduction mechanisms in different cell types (Sutherland 1972). In prokaryotes, expression of metabolic pathways such as the lactose operon, which are otherwise inactive when glucose is abundant, are triggered by cAMP (Black et al. 1980), and in eukaryotes, cAMP is involved in hormonal signal transduction or transmits stimulus from extracellular receptors to activate cell-specific mechanisms (Sutherland 1972). Danchin (1993) identified six distinct non-homologous classes of AC.

HP Q57498

We obtained variable results for sub-cellular localization of HP Q57498. PSLpred (Bhasin et al. 2005) analysis revealed that this protein is localized in the periplasmic region. However, CELLO (Yu et al. 2006) tool suggests that it is a cytoplasmic protein. This is a non-secreted protein and shows the absence of transmembrane helix (Table S3). Similarity search shows a significant similarity to the alkyl hydroperoxidase (AhpD) core protein and carboxymuconolactone decarboxylase (CMD). Domain annotation and cluster analysis further confirmed that this protein belongs to a CMD family protein. Motif discovery shows that this HP contains a typical CMD motif, which was further confirmed by a motif search on MEME suite (Table 2). This HP is found to be a virulent protein on the basis VirulentPred. This structural information of this protein may be utilized for drug design and discovery. This HP interacts with transcriptional regulator, autonomous glycyl radical cofactor GrcA, dihydrolipoamide dehydrogenase, and type III restriction–modification system endonuclease-like protein. This indicates the functional significance of HP Q57498 (Figure S1).

Sparksx server (Yang et al. 2011) modeled the sequence of HP Q57498 where CMD proteins (PDB ID: 1VKE and 2QEU) were used as a template and 99 % residues in allowed region of the Ramachandran plot. The Q57498 predicted structure is showing all α-helix topology (six α-helices), with active cavities present between α1, α2 and α4 (Fig. 5a). We found citrate-binding sites in Q57498 which include residues corresponding Tyr30 and His68 (Figure S4). His68 is also present in the predicted active cavity along with His8, Val12, Gln31, Gly34, Ala35, Ala38, Cys64, Ala67, His68, and Glu71 (Fig. 5b). The ProFunc and Dali server searches, further used to validate the predicted function for Q57498, indicate the CMD activity in the HP. The Dali search mainly includes gamma-CMD (*Z* score = 11.6), CMD family protein (*Z* score = 10.7), etc., which are in top hits, showing the overall rmsd range of 0.7–3.1 Å^2 over 97–133 residues. ProFunc validated the Dali results and showed the presence of five CMD and AhpD-like sequence motifs. ProFunc and Dali servers have also confirmed the structural similarity of HP Q57498 with that of CMD (Table 3).

The 3-oxoadipate pathway branches, namely catechol and protocatechuate, cause degradation of aromatic compounds. This pathway is important for the bacteria, unites at the common intermediate 3-oxoadipate enol-lactone, and the enzyme, CMD is involved in protocatechuate catabolism and gene fusion event leads to expression of CMD in bacteria (Eulberg et al. 1998).

HP P44095

The HP P44095 is localized in the cytoplasm with no signal peptide and transmembrane helix (Table S3). Similarity search and sequence analysis show that HP P44095 is highly similar to the protein of cyclase family and metal-dependent hydrolase with cyclase activity (Table S2). Furthermore, cluster analysis shows that this protein belongs to the cyclase activity proteins containing clusters. Motif analysis tools have further predicted a putative cyclase motif and glucose transporter type 3 (GLUT3) signatures, respectively, in the HP P44095. All these analyses led to a conclusion that HP P44095 presumably works as a protein with cyclase activity. String analysis suggests various proteins such as gluconate permease, 3-hydroxyisobutyrate dehydrogenase, putative aldolase, and glycerol-3-phosphate regulon as a functional networking partner of the HP P44095 (Figure S1).

We used I-TASSER structure prediction server to compute the structure of P44095 using metal-dependent hydrolase with cyclase activity (PDB ID: 1R61) and manganese-dependent isatin hydrolase (PDB ID: 4J0N) as templates. A final model was validated with PROCHECK which shows that 98.3 % residues are in the allowed region of the Ramachandran plot. P44095 adopts mixed α/β fold with five β-strands and six α-helices (Fig. 6a). Residues Glu57, Pro60, Pro74, Tyr75, Ala76, Ala77, and Ile78 are responsible for membrane attachment in P44095. The structure and firestar server, COACH and COFACTOR analyses show the presence of three zinc-binding sites present at His23, His27, and Asp29 (Figure S5). Pocket-Finder predicts the active site residues as Met1, His23, Cys24 and Gly25 (Fig. 6b). We further analyzed this HP using ProFunc and Dali servers. Results showed a close structural similarity to the hydrolase (*Z* score = 13.9) and alpha amylase (*Z* score = 3.2). Furthermore, ProFunc found three cyclase motifs. ProFunc and Dali servers

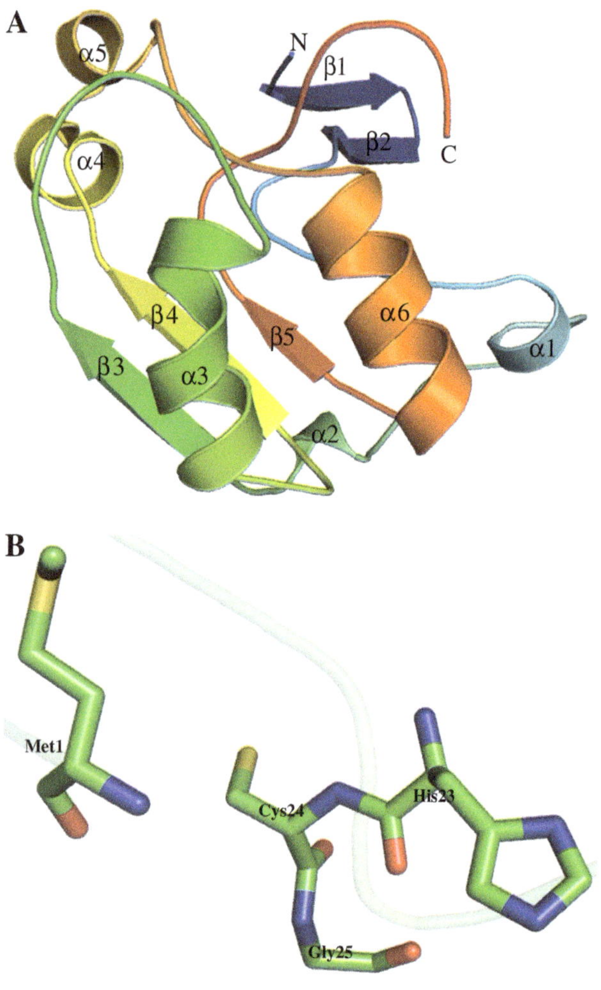

Fig. 6 Representation of model structure of HP P44095. **a** Overall structure of P44095 shown in cartoon model with N-terminal is shown in *blue* and C-terminal in *red*. **b** Representation of the active site residues of P44095 in stick model

suggest that HP P44095 belongs to the cyclase family and possesses metal-dependent hydrolase activity (Table S4).

HP P44093

HP P44093 shows cytoplasmic localization devoid of any transmembrane helix as well as the signal peptide, hence it may not be secreted (Table S3). This protein shows high sequence similarity to the 4-hydroxy-3-methyl but-2-enyl diphosphatereductase. We are unable to find any significant hit in most of the searches performed in various databases. However, SUPERFAMILY (Gough et al. 2001) shows that HP P44093 is a member of YgbK-like family, and PANTHER (Mi et al. 2005) predicts that this HP belongs to D-tagatose-1,6-bisphosphate aldolase family. This protein belongs to the cluster of proteins with ygbK-like activity. The classification system SVMProt (Cai et al. 2003) search shows that this HP belongs to the carbon–carbon lyases family. Motif finding tools suggest the presence of D-tagatose-1,6-bisphosphate aldolase motif and glycosyl hydrolases family 1 active site motifs in the HP P44093. Interaction partners of HP P44093 are aldolase, 3-hydroxyisobutyrate dehydrogenase, glycerol-3-phosphate regulon repressor, gluconate permease, L-fuculose phosphate aldolase, and D-xylose transporter subunit XylF indicating this protein plays a significant role in the carbohydrate metabolism (Figure S1).

We used RaptorX sever for predicting the structure of P44093 using putative tRNA synthase (PDB ID: 3DQQ) as a template (Table 3) and found that 99.7 % residues in allowed region of the Ramachandran plot. The structure of P44093 contains 17 β-strands and 18 α-helices, and attains α/β fold with N-terminus covering one end and the C-terminus the other in a mixed parallel and anti-parallel fold (Fig. 7a). The residues Pro54, Asn56, Lys90, Ile281, Leu315, Ile318, Gln319, His320, Gln321, and Phe322 are membrane-attaching residues in P44093 (Fig. 7b). The firestar server was unable to detect any ligand-binding sites in P44093, while COACH and COFACTOR consensus show Ile7, Asn8 and Ile10 may be magnesium- and beta-D-glucose-binding sites. The P44093 also shows a conserved dihydroxyacetone phosphate-binding site (Figure S6), and the amino acid residues Pro243, Thr244, Thr248, Asn349, Trp382, Ala408, Gln409 and Phe412 may form the active cavity of the protein as suggested by the Pocket-Finder. Results of ProFunc and Dali servers show a close resemblance to the formimidoylglutamase (Z score = 7.2) and arginase-1 (Z score = 7.2) with overall rmsd of 1.5–4.0 Å^2. Furthermore, three motifs with D-tagatose-1,6-bisphosphate aldolase were predicted in the sequence of HP P44093. Function assignment by ProFunc and Dali servers confirms the previously annotated results that the given HP is a D-tagatose-1,6-bisphosphate aldolase and tRNA synthase, respectively (Table S4).

It is known that tagatose-1,6-bisphosphate aldolase (TBPA), a tetrameric class II aldolase, catalyzes the reversible condensation of dihydroxyacetone phosphate with glyceraldehyde 3-phosphate to produce tagatose 1,6-bisphosphate. A comparison of TBPA with related fructose-1,6-bisphosphate aldolase (FBPA) shows common features associated with the mechanism. Major products of these enzymes catalyzed reactions differ in the chirality at a single position (Hall et al. 2002).

HP P44720

Sub-cellular localization prediction with Psortb and PSLpred shows that HP P44720 is localized in cytoplasm, while CELLO suggests a periplasmic localization.

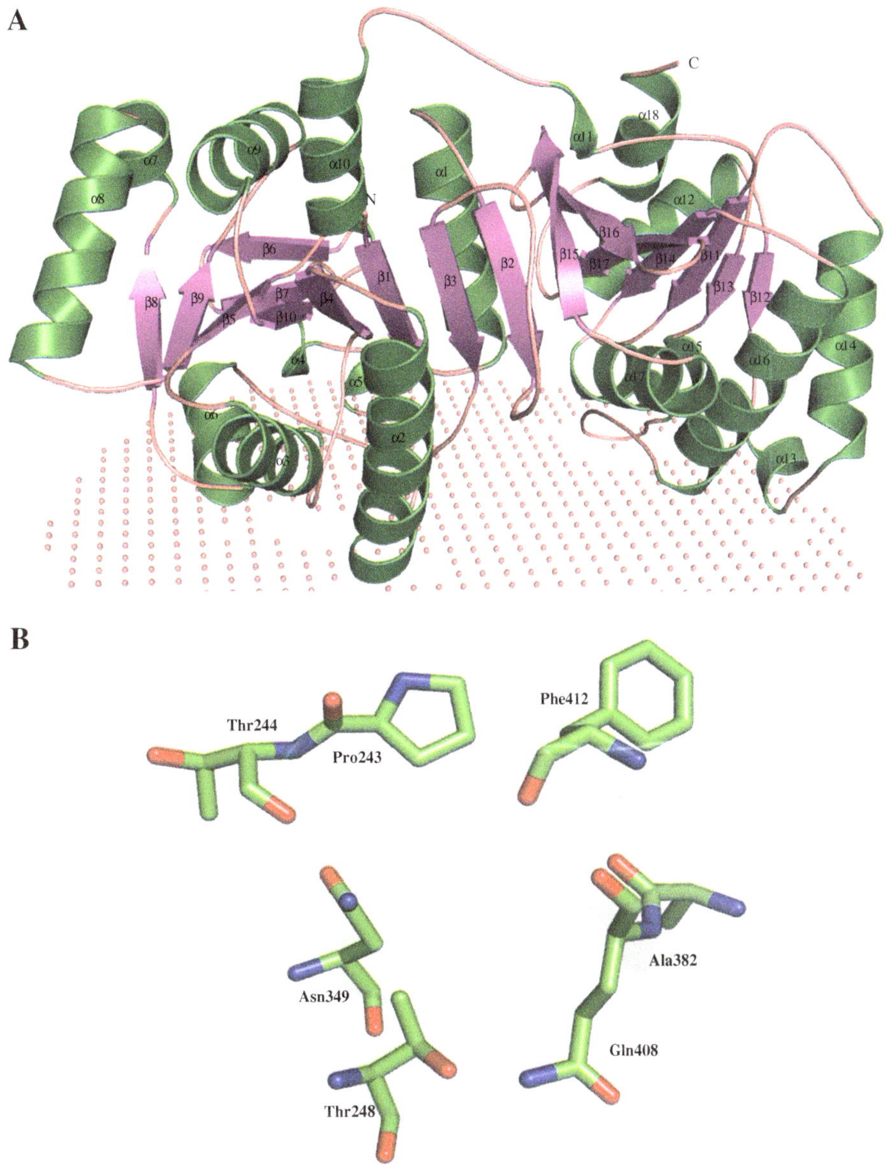

Fig. 7 Representation of model structure of HP P44093.
a Three-dimensional structure represented in cartoon model with membrane represented as non-bonded spheres.
b Representation of the active site residues of P44093 in stick model

Similarity search for HP P44720 shows high sequence similarity to aminodeoxychorismate lyase (Table S1 and S2). SUPERFAMILY shows that this HP is a member of zinc finger protein 425-like domain family. The cluster analysis using SYSTERS and ProtoNet shows that the protein belongs to the cluster of proteins with aminodeoxychorismate lyase activity. The motif discovery suggested the presence of zinc finger C2H2-type/integrase DNA-binding domain motif and aminodeoxychorismate lyase active site motif, respectively. The interaction partners of HP P44720 are DNA polymerase III subunit delta, thymidylate kinase, β-hexosaminidase, protease, Holliday junction resolvase-like protein, 3-oxoacyl-(acyl carrier protein) synthase I, acyl carrier protein indicating this protein plays a significant role in the pathogen metabolic system (Figure S1).

Structure of P44720 was predicted using amino deoxychorismate lyase protein (PDB ID: 2R1F) as a template (Table 3). The energy minimized model shows 99.7 % residues in the allowed region of the Ramachandran plot. The structure of P44720 shows six β-strands and 12 α-helices, and it assumes α/β fold with N-terminus covering one end and the C-terminus the other in a parallel fold (Fig. 8a). The membrane-attached residues in P44720 is Lys2, Phe4, Leu5, Ile6, Ala7 and Ile8 (Fig. 8b). Since Lys and Glu residues are important for catalytic activity of aminodeoxychorismate lyase (Nakai et al. 2000), we assume that residues corresponding to Tyr168, Pro169, Met191, Lys197, Ala198, Pro293, Ser294, Glu295, Leu298, Gln299, and Ala302 may form the active site cavity as suggested by the Pocket-Finder. Both ProFunc and Dali server searches suggested the presence of amino

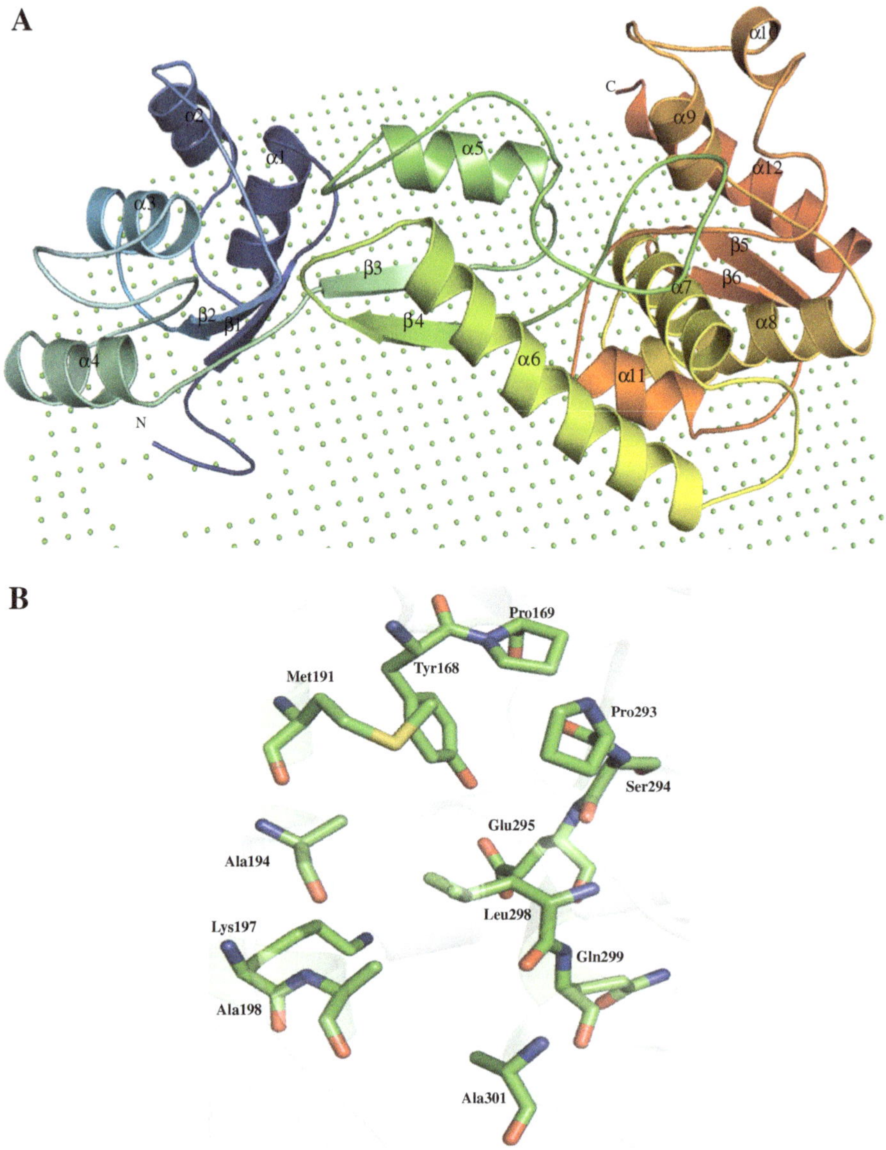

Fig. 8 Representation of model structure of HP P44720.
a Cartoon model representation of overall structure in which membrane is represented by non-bonded atoms.
b Representation of the active site residues of P44720 in stick model

deoxychorismate lyase in this HP. Moreover, ProFunc shows seven motifs in the HP P44720. Function allocation by ProFunc and Dali servers further confirms that given HP is an amino deoxychorismate lyase (Table S4).

Chorismate is important in the biosynthesis of many important aromatic products such as anthranilate, prephenate, *p*-aminobenzoate and *p*-hydroxybenzoate, since it acts as a branch point precursor for the intermediary metabolites in bacteria (Green and Nichols 1991). 4-amino-4-deoxychorismate (ADC), formed from chorismate by the action of the enzyme p-aminobenzoate synthase, encoded by pabA and pabB, the product of pabC, is amino deoxychorismate lyase (ADCL) with pyridoxal 5′-phosphate (PLP) as a cofactor, converts ADC to *p*-aminobenzoate and pyruvate (Nichols et al. 1989; Ye et al. 1990), and along with *p*-aminobenzoate synthase is a key enzyme in the biosynthesis of *p*-aminobenzoate.

Conclusions

Our study combines a number of bioinformatics tools for function predictions of previously not assigned proteins in the *H. influenzae* genome. We stressed that besides sequence analysis, structure can be used as a framework to explain known functional properties. Here we have combined the latest versions of several protein databases, protein motifs, features from the amino acid sequence, structure prediction, structure analysis, structure–function relationship, as well as a pathway and genome context methods to assign a precise function to hypothetical proteins for which no any experimental information is available. We used the sequence and structure-based methods for functional annotation of HPs from *H. influenzae* with lyase activity. The structure of these enzymes was predicted by most of the available methods, but we chose those models with best quality. HPs P44717, P44782, P44197, P45267, Q57498 and P44720 are categorized as lyase enzyme which was further confirmed by structure prediction and analysis. We also observed the reliability of different methods for predicting the three-dimensional structure of HPs and found that homology and fold recognition methods were only applicable to highly evolutionary conserved proteins and ab initio servers, especially I-TASSER gave good result for P44095 which is not predicted by other methods very precisely. Our in silico approach may be combined with experimental methods and make a leading contribution in functional elucidation of the protein structures.

Acknowledgments Authors sincerely thank Indian Council of Medical Research for financial assistance (Project No. BIC/12(04)/2012).

Conflict of interest Authors declare no conflict of interest.

References

Accelrys (2013) Discovery studio modeling environment, Release 3.5. Accelrys Software Inc., San Diego

Alexandrov A, Martzen MR, Phizicky EM (2002) Two proteins that form a complex are required for 7-methylguanosine modification of yeast tRNA. RNA 8:1253–1266

Altschul SF, Gish W, Miller W, Myers EW, Lipman DJ (1990) Basic local alignment search tool. J Mol Biol 215:403–410

Altschul SF, Madden TL, Schaffer AA, Zhang J, Zhang Z, Miller W, Lipman DJ (1997) Gapped BLAST and PSI-BLAST: a new generation of protein database search programs. Nucleic Acids Res 25:3389–3402

Apisarnthanarak A, Mundy LM (2005) Etiology of community-acquired pneumonia. Clin Chest Med 26:47–55

Bailey TL, Boden M, Buske FA, Frith M, Grant CE, Clementi L, Ren J, Li WW, Noble WS (2009) MEME SUITE: tools for motif discovery and searching. Nucleic Acids Res 37:W202–W208

Baron C, Coombes B (2007) Targeting bacterial secretion systems: benefits of disarmament in the microcosm. Infect Disord Drug Targets 7:19–27

Bateman A (1997) The structure of a domain common to archaebacteria and the homocystinuria disease protein. Trends Biochem Sci 22:12–13

Bendtsen JD, Kiemer L, Fausboll A, Brunak S (2005) Non-classical protein secretion in bacteria. BMC Microbiol 5:58

Bernstein FC, Koetzle TF, Williams GJ, Meyer EF Jr, Brice MD, Rodgers JR, Kennard O, Shimanouchi T, Tasumi M (1978) The Protein Data Bank: a computer-based archival file for macromolecular structures. Arch Biochem Biophys 185:584–591

Bhasin M, Garg A, Raghava GP (2005) PSLpred: prediction of subcellular localization of bacterial proteins. Bioinformatics 21:2522–2524

Bjornson HS (1984) Enzymes associated with the survival and virulence of gram-negative anaerobes. Rev Infect Dis 6(Suppl 1):S21–S24

Black RA, Hobson AC, Adler J (1980) Involvement of cyclic GMP in intracellular signaling in the chemotactic response of *Escherichia coli*. Proc Natl Acad Sci USA 77:3879–3883

Britton KL, Abeysinghe IS, Baker PJ, Barynin V, Diehl P, Langridge SJ, McFadden BA, Sedelnikova SE, Stillman TJ, Weeradechapon K et al (2001) The structure and domain organization of *Escherichia coli* isocitrate lyase. Acta Crystallogr D Biol Crystallogr 57:1209–1218

Brooks BR, Brooks CL 3rd, Mackerell AD Jr, Nilsson L, Petrella RJ, Roux B, Won Y, Archontis G, Bartels C, Boresch S et al (2009) CHARMM: the biomolecular simulation program. J Comput Chem 30:1545–1614

Cai CZ, Han LY, Ji ZL, Chen X, Chen YZ (2003) SVM-Prot: web-based support vector machine software for functional classification of a protein from its primary sequence. Nucleic Acids Res 31:3692–3697

Capdeville R, Buchdunger E, Zimmermann J, Matter A (2002) Glivec (STI571, imatinib), a rationally developed, targeted anticancer drug. Nat Rev Drug Discov 1:493–502

Carbonetti NH (2010) Pertussis toxin and adenylate cyclase toxin: key virulence factors of *Bordetella pertussis* and cell biology tools. Future Microbiol 5:455–469

Chen R, Jeong SS (2000) Functional prediction: identification of protein orthologs and paralogs. Protein Sci 9:2344–2353

Chothia C, Lesk AM (1986) The relation between the divergence of sequence and structure in proteins. EMBO J 5:823–826

Christensen H (2008) Taxonomy and biodiversity of members of Pasteurellaceae. Academic Press, Caister

Conrad J, Niu L, Rudd K, Lane BG, Ofengand J (1999) 16S ribosomal RNA pseudouridine synthase RsuA of *Escherichia coli*: deletion, mutation of the conserved Asp102 residue, and sequence comparison among all other pseudouridine synthases. RNA 5:751–763

Danchin A (1993) Phylogeny of adenylyl cyclases. Adv Second Messenger Phosphoprotein Res 27:109–162

Del Campo M, Ofengand J, Malhotra A (2004) Crystal structure of the catalytic domain of RluD, the only rRNA pseudouridine synthase required for normal growth of *Escherichia coli*. RNA 10:231–239

DeLano WL (2002) The PyMOL molecular graphics system. In: Schrödinger L (ed) DeLano Scientific, San Carlos, CA, USA.

Desler C, Durhuus JA, Rasmussen LJ (2012) Genome-wide screens for expressed hypothetical proteins. Methods Mol Biol 815:25–38

Dobson CM (2003) Protein folding and misfolding. Nature 426:884–890

Eisenreich W, Rohdich F, Bacher A (2001) Deoxyxylulose phosphate pathway to terpenoids. Trends Plant Sci 6:78–84

Eisenreich W, Bacher A, Arigoni D, Rohdich F (2004) Biosynthesis of isoprenoids via the non-mevalonate pathway. Cell Mol Life Sci 61:1401–1426

Ejim LJ, D'Costa VM, Elowe NH, Loredo-Osti JC, Malo D, Wright GD (2004) Cystathionine beta-lyase is important for virulence of *Salmonella enterica serovar typhimurium*. Infect Immun 72:3310–3314

Eldika N, Sethi S (2006) Role of nontypeable *Haemophilus influenzae* in exacerbations and progression of chronic obstructive pulmonary disease. Curr Opin Pulm Med 12:118–124

Emanuelsson O, Brunak S, von Heijne G, Nielsen H (2007) Locating proteins in the cell using TargetP, SignalP and related tools. Nat Protoc 2:953–971

Eswar N, Webb B, Marti-Renom MA, Madhusudhan MS, Eramian D, Shen MY, Pieper U, Sali A (2007) Comparative protein structure modeling using MODELLER. Curr Protoc Protein Sci Chapter 2: Unit 2.9

Eulberg D, Lakner S, Golovleva LA, Schlomann M (1998) Characterization of a protocatechuate catabolic gene cluster from *Rhodococcus opacus* 1CP: evidence for a merged enzyme with 4-carboxymuconolactone-decarboxylating and 3-oxoadipate enol-lactone-hydrolyzing activity. J Bacteriol 180:1072–1081

Ferre-D'Amare AR (2003) RNA-modifying enzymes. Curr Opin Struct Biol 13:49–55

Fleischmann RD, Adams MD, White O, Clayton RA, Kirkness EF, Kerlavage AR, Bult CJ, Tomb JF, Dougherty BA, Merrick JM et al (1995) Whole-genome random sequencing and assembly of *Haemophilus influenzae* Rd. Science 269:496–512

Foster PG, Huang L, Santi DV, Stroud RM (2000) The structural basis for tRNA recognition and pseudouridine formation by pseudouridine synthase I. Nat Struct Biol 7:23–27

Galperin MY (2001) Conserved 'hypothetical' proteins: new hints and new puzzles. Comp Funct Genomics 2:14–18

Galperin MY (2004) Bacterial signal transduction network in a genomic perspective. Environ Microbiol 6:552–567

Galperin MY, Koonin EV (2004) 'Conserved hypothetical' proteins: prioritization of targets for experimental study. Nucleic Acids Res 32:5452–5463

Garg A, Gupta D (2008) VirulentPred: a SVM based prediction method for virulent proteins in bacterial pathogens. BMC Bioinform 9:62

Geer LY, Domrachev M, Lipman DJ, Bryant SH (2002) CDART: protein homology by domain architecture. Genome Res 12:1619–1623

Gough J, Karplus K, Hughey R, Chothia C (2001) Assignment of homology to genome sequences using a library of hidden Markov models that represent all proteins of known structure. J Mol Biol 313:903–919

Green JM, Nichols BP (1991) p-Aminobenzoate biosynthesis in *Escherichia coli*. Purification of aminodeoxychorismate lyase and cloning of pabC. J Biol Chem 266:12971–12975

Hall DR, Bond CS, Leonard GA, Watt CI, Berry A, Hunter WN (2002) Structure of tagatose-1,6-bisphosphate aldolase. Insight into chiral discrimination, mechanism, and specificity of class II aldolases. J Biol Chem 277:22018–22024

Hassan M, Kumar V, Somvanshi RK, Dey S, Singh TP, Yadav S (2007a) Structure-guided design of peptidic ligand for human prostate specific antigen. J Pept Sci 13:849–855

Hassan MI, Kumar V, Singh TP, Yadav S (2007b) Structural model of human PSA: a target for prostate cancer therapy. Chem Biol Drug Des 70:261–267

Hoang C, Ferre-D'Amare AR (2001) Cocrystal structure of a tRNA Psi55 pseudouridine synthase: nucleotide flipping by an RNA-modifying enzyme. Cell 107:929–939

Hoang C, Ferre-D'Amare AR (2004) Crystal structure of the highly divergent pseudouridine synthase TruD reveals a circular permutation of a conserved fold. RNA 10:1026–1033

Hoang C, Chen J, Vizthum CA, Kandel JM, Hamilton CS, Mueller EG, Ferre-D'Amare AR (2006) Crystal structure of pseudouridine synthase RluA: indirect sequence readout through protein-induced RNA structure. Mol Cell 24:535–545

Holm L, Rosenstrom P (2010) Dali server: conservation mapping in 3D. Nucleic Acids Res 38:W545–W549

Hooft RW, Sander C, Vriend G (1997) Objectively judging the quality of a protein structure from a Ramachandran plot. Comput Appl Biosci 13:425–430

Hubbard TJ, Ailey B, Brenner SE, Murzin AG, Chothia C (1999) SCOP: a structural classification of proteins database. Nucleic Acids Res 27:254–256

Hunter S, Jones P, Mitchell A, Apweiler R, Attwood TK, Bateman A, Bernard T, Binns D, Bork P, Burge S et al (2011) InterPro in 2011: new developments in the family and domain prediction database. Nucleic Acids Res 40:D306–D312

Ignoul S, Eggermont J (2005) CBS domains: structure, function, and pathology in human proteins. Am J Physiol Cell Physiol 289:C1369–C1378

Illergard K, Ardell DH, Elofsson A (2009) Structure is three to ten times more conserved than sequence—a study of structural response in protein cores. Proteins 77:499–508

Jackman JE, Montange RK, Malik HS, Phizicky EM (2003) Identification of the yeast gene encoding the tRNA m1G methyltransferase responsible for modification at position 9. RNA 9:574–585

Jenal U (2004) Cyclic di-guanosine-monophosphate comes of age: a novel secondary messenger involved in modulating cell surface structures in bacteria? Curr Opin Microbiol 7:185–191

Kanehisa M (1997) Linking databases and organisms: GenomeNet resources in Japan. Trends Biochem Sci 22:442–444

Kaya Y, Ofengand J (2003) A novel unanticipated type of pseudouridine synthase with homologs in bacteria, archaea, and eukarya. RNA 9:711–721

Kim DE, Chivian D, Baker D (2004) Protein structure prediction and analysis using the Robetta server. Nucleic Acids Res 32:W526–W531

Klebe G (2000) Recent developments in structure-based drug design. J Mol Med (Berl) 78:269–281

Koonin EV (1996) Pseudouridine synthases: four families of enzymes containing a putative uridine-binding motif also conserved in dUTPases and dCTP deaminases. Nucleic Acids Res 24:2411–2415

Krivov GG, Shapovalov MV, Dunbrack RL Jr (2009) Improved prediction of protein side-chain conformations with SCWRL4. Proteins 77:778–795

Krogh A, Larsson B, von Heijne G, Sonnhammer EL (2001) Predicting transmembrane protein topology with a hidden Markov model: application to complete genomes. J Mol Biol 305:567–580

Kryshtafovych A, Moult J, Bartual SG, Bazan JF, Berman H, Casteel DE, Christodoulou E, Everett JK, Hausmann J, Heidebrecht T et al (2011) Target highlights in CASP9: experimental target structures for the critical assessment of techniques for protein structure prediction. Proteins 79(Suppl 10):6–20

Kuhnert P (2008) Pasteurellaceae: biology, genomics and molecular aspects. Caister Academic Press, Norwich

Kumar K, Prakash A, Tasleem M, Islam A, Ahmad F, Hassan MI (2014) Functional annotation of putative hypothetical proteins from *Candida dubliniensis*. Gene 543:93–100

Laskowski RA, Rullmannn JA, MacArthur MW, Kaptein R, Thornton JM (1996) AQUA and PROCHECK-NMR: programs for checking the quality of protein structures solved by NMR. J Biomol NMR 8:477–486

Laskowski RA, Watson JD, Thornton JM (2005) ProFunc: a server for predicting protein function from 3D structure. Nucleic Acids Res 33:W89–W93

Laurie AT, Jackson RM (2005) Q-SiteFinder: an energy-based method for the prediction of protein–ligand binding sites. Bioinformatics 21:1908–1916

Letunic I, Doerks T, Bork P (2012) SMART 7: recent updates to the protein domain annotation resource. Nucleic Acids Res 40:D302–D305

Lomize MA, Pogozheva ID, Joo H, Mosberg HI, Lomize AL (2012) OPM database and PPM web server: resources for positioning of proteins in membranes. Nucleic Acids Res 40:D370–D376

Lopez G, Valencia A, Tress ML (2007) Firestar—prediction of functionally important residues using structural templates and alignment reliability. Nucleic Acids Res 35:W573–W577

Markel TA, Crisostomo PR, Wang M, Herring CM, Meldrum KK, Lillemoe KD, Meldrum DR (2007) The struggle for iron:

gastrointestinal microbes modulate the host immune response during infection. J Leukoc Biol 81:393–400

Marti-Renom MA, Stuart AC, Fiser A, Sanchez R, Melo F, Sali A (2000) Comparative protein structure modeling of genes and genomes. Annu Rev Biophys Biomol Struct 29:291–325

Meinel T, Krause A, Luz H, Vingron M, Staub E (2005) The SYSTERS protein family database in 2005. Nucleic Acids Res 33:D226–D229

Mi H, Lazareva-Ulitsky B, Loo R, Kejariwal A, Vandergriff J, Rabkin S, Guo N, Muruganujan A, Doremieux O, Campbell MJ et al (2005) The PANTHER database of protein families, subfamilies, functions and pathways. Nucleic Acids Res 33:D284–D288

Misura KM, Chivian D, Rohl CA, Kim DE, Baker D (2006) Physically realistic homology models built with ROSETTA can be more accurate than their templates. Proc Natl Acad Sci USA 103:5361–5366

Mizutani K, Machida Y, Unzai S, Park SY, Tame JR (2004) Crystal structures of the catalytic domains of pseudouridine synthases RluC and RluD from *Escherichia coli*. Biochemistry 43:4454–4463

Morton DJ, Bakaletz LO, Jurcisek JA, VanWagoner TM, Seale TW, Whitby PW, Stull TL (2004a) Reduced severity of middle ear infection caused by nontypeable *Haemophilus influenzae* lacking the hemoglobin/hemoglobin–haptoglobin binding proteins (Hgp) in a chinchilla model of otitis media. Microb Pathog 36:25–33

Morton DJ, Smith A, Ren Z, Madore LL, VanWagoner TM, Seale TW, Whitby PW, Stull TL (2004b) Identification of a haem-utilization protein (Hup) in *Haemophilus influenzae*. Microbiology 150:3923–3933

Morton DJ, Madore LL, Smith A, Vanwagoner TM, Seale TW, Whitby PW, Stull TL (2005) The heme-binding lipoprotein (HbpA) of *Haemophilus influenzae*: role in heme utilization. FEMS Microbiol Lett 253:193–199

Murphy TF, Sethi S (1992) Bacterial infection in chronic obstructive pulmonary disease. Am Rev Respir Dis 146:1067–1083

Nakai T, Mizutani H, Miyahara I, Hirotsu K, Takeda S, Jhee KH, Yoshimura T, Esaki N (2000) Three-dimensional structure of 4-amino-4-deoxychorismate lyase from *Escherichia coli*. J Biochem 128:29–38

Nichols BP, Seibold AM, Doktor SZ (1989) Para-aminobenzoate synthesis from chorismate occurs in two steps. J Biol Chem 264:8597–8601

Nimrod G, Schushan M, Steinberg DM, Ben-Tal N (2008) Detection of functionally important regions in "hypothetical proteins" of known structure. Structure 16:1755–1763

Ofengand J, Malhotra A, Remme J, Gutgsell NS, Del Campo M, Jean-Charles S, Peil L, Kaya Y (2001) Pseudouridines and pseudouridine synthases of the ribosome. Cold Spring Harb Symp Quant Biol 66:147–159

Park SJ, Son WS, Lee BJ (2012) Structural analysis of hypothetical proteins from *Helicobacter pylori*: an approach to estimate functions of unknown or hypothetical proteins. Int J Mol Sci 13:7109–7137

Peng J, Xu J (2011) RaptorX: exploiting structure information for protein alignment by statistical inference. Proteins 79(Suppl 10):161–171

Pidugu LS, Maity K, Ramaswamy K, Surolia N, Suguna K (2009) Analysis of proteins with the 'hot dog' fold: prediction of function and identification of catalytic residues of hypothetical proteins. BMC Struct Biol 9:37

Punta M, Coggill PC, Eberhardt RY, Mistry J, Tate J, Boursnell C, Pang N, Forslund K, Ceric G, Clements J et al (2012) The Pfam protein families database. Nucleic Acids Res 40:D290–D301

Quevillon E, Silventoinen V, Pillai S, Harte N, Mulder N, Apweiler R, Lopez R (2005) InterProScan: protein domains identifier. Nucleic Acids Res 33:W116–W120

Ramachandran GN, Ramakrishnan C, Sasisekharan V (1963) Stereochemistry of polypeptide chain configurations. J Mol Biol 7:95–99

Ramamurthy V, Swann SL, Paulson JL, Spedaliere CJ, Mueller EG (1999) Critical aspartic acid residues in pseudouridine synthases. J Biol Chem 274:22225–22230

Rappoport N, Karsenty S, Stern A, Linial N, Linial M (2012) ProtoNet 6.0: organizing 10 million protein sequences in a compact hierarchical family tree. Nucleic Acids Res 40:D313–D320

Reid AJ, Yeats C, Orengo CA (2007) Methods of remote homology detection can be combined to increase coverage by 10% in the midnight zone. Bioinformatics 23:2353–2360

Roy A, Kucukural A, Zhang Y (2010) I-TASSER: a unified platform for automated protein structure and function prediction. Nat Protoc 5:725–738

Roy A, Yang J, Zhang Y (2012) COFACTOR: an accurate comparative algorithm for structure-based protein function annotation. Nucleic Acids Res 40:W471–W477

Saha S, Raghava GP (2006) VICMpred: an SVM-based method for the prediction of functional proteins of Gram-negative bacteria using amino acid patterns and composition. Genomics Proteomics Bioinform 4:42–47

Sethi S, Murphy TF (2001) Bacterial infection in chronic obstructive pulmonary disease in 2000: a state-of-the-art review. Clin Microbiol Rev 14:336–363

Shahbaaz M, Hassan MI, Ahmad F (2013) Functional Annotation of conserved hypothetical proteins from *Haemophilus influenzae* Rd KW20. PLoS One 8:e84263

Shapiro L, Harris T (2000) Finding function through structural genomics. Curr Opin Biotechnol 11:31–35

Sillitoe I, Cuff AL, Dessailly BH, Dawson NL, Furnham N, Lee D, Lees JG, Lewis TE, Studer RA, Rentzsch R et al (2013) New functional families (FunFams) in CATH to improve the mapping of conserved functional sites to 3D structures. Nucleic Acids Res 41:D490–D498

Singh R, Saha M (2003) Identifying structural motifs in proteins. Pac Symp Biocomput 8:228–239

Sivaraman J, Sauve V, Larocque R, Stura EA, Schrag JD, Cygler M, Matte A (2002) Structure of the 16S rRNA pseudouridine synthase RsuA bound to uracil and UMP. Nat Struct Biol 9:353–358

Soding J, Biegert A, Lupas AN (2005) The HHpred interactive server for protein homology detection and structure prediction. Nucleic Acids Res 33:W244–W248

Soma A, Ikeuchi Y, Kanemasa S, Kobayashi K, Ogasawara N, Ote T, Kato J, Watanabe K, Sekine Y, Suzuki T (2003) An RNA-modifying enzyme that governs both the codon and amino acid specificities of isoleucine tRNA. Mol Cell 12:689–698

Stojiljkovic I, Perkins-Balding D (2002) Processing of heme and heme-containing proteins by bacteria. DNA Cell Biol 21:281–295

Sutherland EW (1972) Studies on the mechanism of hormone action. Science 177:401–408

Szklarczyk D, Franceschini A, Kuhn M, Simonovic M, Roth A, Minguez P, Doerks T, Stark M, Muller J, Bork P et al (2011) The STRING database in 2011: functional interaction networks of proteins, globally integrated and scored. Nucleic Acids Res 39:D561–D568

Taussig R, Gilman AG (1995) Mammalian membrane-bound adenylyl cyclases. J Biol Chem 270:1–4

Taylor WR, Orengo CA (1989) Protein structure alignment. J Mol Biol 208:1–22

Thakur PK, Hassan I (2011) Discovering a potent small molecule inhibitor for gankyrin using de novo drug design approach. Int J Comput Biol Drug Des 4:373–386

Thakur P, Kumar J, Ray D, Anjum F, Hassan MI (2013) Search of potential inhibitor against New Delhi metallo-beta-lactamase 1 from a series of antibacterial natural compounds using docking approach. J Nat Sci Biol Med 4:51–56

Tsoka S, Ouzounis CA (2000) Recent developments and future directions in computational genomics. FEBS Lett 480:42–48

Tusnady GE, Simon I (2001) The HMMTOP transmembrane topology prediction server. Bioinformatics 17:849–850

Xu J, Jiao F, Yu L (2008) Protein structure prediction using threading. Methods Mol Biol 413:91–121

Yang Y, Faraggi E, Zhao H, Zhou Y (2011) Improving protein fold recognition and template-based modeling by employing probabilistic-based matching between predicted one-dimensional structural properties of query and corresponding native properties of templates. Bioinformatics 27:2076–2082

Yang J, Roy A, Zhang Y (2013) Protein–ligand binding site recognition using complementary binding-specific substructure comparison and sequence profile alignment. Bioinformatics 29:2588–2595

Ye QZ, Liu J, Walsh CT (1990) p-Aminobenzoate synthesis in *Escherichia coli*: purification and characterization of PabB as aminodeoxychorismate synthase and enzyme X as aminodeoxychorismate lyase. Proc Natl Acad Sci USA 87:9391–9395

Yu CS, Chen YC, Lu CH, Hwang JK (2006) Prediction of protein subcellular localization. Proteins 64:643–651

Yu J, Zhou Y, Tanaka I, Yao M (2010a) Roll: a new algorithm for the detection of protein pockets and cavities with a rolling probe sphere. Bioinformatics 26:46–52

Yu NY, Wagner JR, Laird MR, Melli G, Rey S, Lo R, Dao P, Sahinalp SC, Ester M, Foster LJ et al (2010b) PSORTb 3.0: improved protein subcellular localization prediction with refined localization subcategories and predictive capabilities for all prokaryotes. Bioinformatics 26:1608–1615

13

Antiproliferative activity of ferulic acid-encapsulated electrospun PLGA/PEO nanofibers against MCF-7 human breast carcinoma cells

Priya Vashisth · Mohit Sharma · Kumar Nikhil · Harmeet Singh · Richa Panwar · Parul A. Pruthi · Vikas Pruthi

Abstract Ferulic acid (FA) is a polyphenolic phytonutrient which possesses strong antiproliferative effect; however, it has limited therapeutic applications due to its physiochemical instability and low bioavailability at the tumor site. In present study, these shortcomings associated with FA were overcome by fabricating FA-encapsulated poly(D,L-lactide-*co*-glycolide)/polyethylene oxide (PLGA/PEO) blend nanofibers using electrospinning technique. FESEM and fluorescence microscopic analysis imitates the smooth morphology and even distribution of FA within the polymeric nanofibers at optimum 2 wt% concentration of FA. The average diameters were recorded to be 150 ± 47.4 and 200 ± 79 nm for PLGA/PEO and FA-encapsulated PLGA/PEO nanofibers, respectively. The encapsulation, compatibility, and physical state of FA within the nanofibers were further confirmed by FTIR, TGA and XRD analysis. In vitro drug delivery studies demonstrated initial burst liberation of FA within 24 h followed by a sustained release for the subsequent time. MTT assay revealed the effectiveness of FA-encapsulated nanofibers against human breast carcinoma cells (MCF-7) cells as compared to control. FESEM and fluorescence microscopic analysis further confirmed the apoptotic effect of FA-encapsulated PLGA/PEO nanofibers against MCF-7. These fabricated nanofibers hold enormous potential to be used as a therapeutic agent for various biomedical applications.

P. Vashisth · M. Sharma · K. Nikhil · H. Singh · R. Panwar · P. A. Pruthi · V. Pruthi (✉)
Department of Biotechnology, Indian Institute of Technology Roorkee, Roorkee 247667, India
e-mail: vikasfbs@gmail.com; vikasfbs@iitr.ernet.in

Keywords Antiproliferative activity · Electrospinning · Ferulic acid · Nanofiber · PEO · PLGA · Drug delivery

Introduction

Phenolic nucleus and unsaturated side chain of FA (4-hydroxy-3-methoxycinnamic acid) readily form a resonance stabilized phenoxy radicals which is responsible for its potent antioxidant capacity. This ubiquitous polyphenolic pharmaceutically active constituent arises from the metabolism of phenylalanine and tyrosine, occurs mainly in seeds, leaves and cell walls of plants such as wheat, rice and oats (Srinivasan et al. 2007). Studies have shown that FA can act as an ingredient of anti-aging supplements, can reduce the level of cholesterol and triglycerides as well as can reduce the risk of cardiovascular dysfunction and inflammatory disease (Ou and Kwok 2004; Murakami et al. 2002). As a multifunctional drug FA exhibits major pharmacological activities such as antioxidant, metal chelation and modulation of enzymatic activities that are responsible for its action to combat deadly infections (Davis and Milner 2010; Ren et al. 2003; Kandaswami et al. 2005). However, despite the extensive potential and beneficiary properties, the therapeutic applications of FA are mostly hampered by its poor solubility in aqueous solution, physiochemical instability, less bioavailability and short half shelf life in the human body fluids (Ouimet et al. 2013; Soobrattee et al. 2005). It has been shown that incorporation of such pharmaceutical active components into polymeric matrix enhanced its bioavailability, hydrophilicity as well as stability (Merlin et al. 2012; Gu et al. 2012). In recent years, biodegradable and biocompatible polymeric nanofibers have been emerged as potential drug carriers. In the same context, Xie and Wang (2006) designed the PLGA-based electrospun fibers for the sustained release of paclitaxel to treat C6 glioma in vitro. Xu et al. (2008) developed doxorubicin hydrochloride containing core–sheath PEG-PLA electrospun nanofibers and studied its release profiles. Electrospinning is a promising method for fabricating ultrafine nanofibers containing anticancer drugs as an effective drug carrier for postoperative local chemotherapy. In a typical electrospinning process, a strong electrostatic field is applied to a polymer solution placed in a syringe with a capillary orifice. When the surface tension of the polymer solution is overcome by the applied electric force, a fiber is extruded from the syringe tip (Huang et al. 2003). This unique technique can produce the fibers with diameters ranging from nanometers to several microns (Vashisth et al. 2013). In this investigation, two biodegradable polymers, PLGA and PEO were used as copolymer blend to fabricate electrospun nanofibers. PLGA is FDA-approved biopolymer and widely used for different therapeutic applications (Vashisth et al. 2013). However, due to its high cost of PLGA polymer, in this study, another biopolymer (PEO) was used along with it, to fabricate cost effective electrospun nanofibers. These fabricated nanofibers were further explored as a drug-carrier for FA. The encapsulation, distribution, physical state and compatibility parameter of these nanofibers were examined. Subsequently, the in vitro release profile of FA from the nanofibers and antitumor activity against MCF-7 breast carcinoma cell lines were also evaluated for confirming their potential therapeutic applications.

Materials and methods

Materials

PLGA (average Mw ~45,000) and PEO (Mw ~900,000) were purchased from Sigma-Aldrich (St. Louis, MO). Michigan Cancer Foundation-7 (MCF-7) and Human Embryonic Kidney (HEK-293) cell lines were procured from National Center for Cell Science (NCCS), Pune, India. FA, dichloromethane (DCM), *N,N*-dimethylformamide (DMF), 3-(4,5-dimethylthiazol-2-yl)-2,5-diphenyltetrazolium bromide (MTT), cell culture-grade dimethyl sulfoxide (DMSO), phosphate buffer saline (PBS), acridine orange (AO), ethidium bromide (EtBr), Dulbecco's modified Eagle medium (DMEM) and all analytical grade chemicals were from Himedia (India).

Nanofibers fabrication

PLGA and PEO in a blend ratio of 1:1 were prepared in a solvent mixture of DCM/DMF (4:1, v/v) to prepare a blank electrospinning solution at a final concentration of 2 wt%. FA-encapsulated PLGA/PEO solutions were obtained by dissolving FA at different concentrations (1, 2, 4, 6 and 8 wt%) w.r.t the total polymer concentration in the PLGA/PEO polymeric solution. The FA-free and FA-encapsulated PLGA/PEO solutions were then carefully placed into a 5 mL syringe, attached with a metallic needle (21 G). Electrospinning was carried out under a fixed electric field (15 kV), feeding rate (0.3 mL h^{-1}) by means of a single syringe pump (Harvard apparatus 11 plus syringe pumps, US). A piece of aluminum foil was used to collect the nanofibers with the horizontal distance of 12 cm from the needle tip. All electrospinning processes were carried out under ambient conditions (temperature 25 ± 2 °C, relative humidity 60 ± 1 %). The resultant fibers were further dried for 24 h in desiccators to remove the residual organic solvent and moisture.

Nanofibers characterization

Surface morphology of the nanofibers was observed using field emission scanning electron microscopy (FESEM; Quanta 200F Model, FEI, Netherland) at an accelerated voltage of 15 kV. For sample preparation the nanofibers were cut into rectangular pieces (1 × 1 cm) and sputter coated (sputter coater: Biotech SC005, Switzerland) with gold for 1 min. The average diameters of the nanofibers were measured at over 50 different points of FESEM images (by using image J analyzer software). The morphology of FA-encapsulated PLGA/PEO nanofibers was also confirmed by fluorescent microscopy (Evos fl, AMG groups, USA). The physical state of FA in the electrospun nanofibers was further examined by X-ray diffraction (XRD) analysis. The XRD patterns were recorded with Cu Kα radiation over the 2θ range from 5° to 100° with the scanning rate of 2° min^{-1}. Thermogravimetric analysis (TGA) of the native FA and nanofibrous samples was performed by using a TGA instrument (EXSTAR, TG/DTA 6300). Each sample (8–10 mg) was kept under vacuum for 24 h prior to testing and then the precisely weighed samples were heated from 23 to 500 °C at a scanning rate of 10 °C min^{-1} under a nitrogen atmosphere. Infrared absorptions of the native FA, PLGA/PEO and FA-encapsulated PLGA/PEO nanofibers were recorded by Fourier transform infrared spectrometer (FTIR; Thermo Nicolet Nexus 6700, US). Scans (16) were recorded in the scanning range of 4,000–500 cm^{-1} with the resolution of 4 cm^{-1}.

Drug encapsulation efficiency

The encapsulation efficiency (EE) of FA in the nanofibers was quantified by thoroughly dissolving the FA-encapsulated nanofibers (containing 2 wt% FA) in a mixture of DCM/DMF (4:1) and the amount of released FA was measured by using UV–Vis spectrophotometer (Lasany double beam LI-2800) at 319 nm. The amount of FA in the fibers was calculated from the obtained data against a predetermined calibration curve for the drug. The encapsulation efficiency of the FA was determined as follows:

EE (%) = Actual FA content in nanofibers (mg)/Theoretical FA content in nanofibers (mg) × 100.

In vitro drug release

In vitro release studies of FA from the 2 wt% FA-encapsulated PLGA/PEO nanofibers were performed by incubating 10 mg nanofibrous mat in 30 mL PBS (pH 7.4). The incubated samples were maintained in a thermostat (37 °C) at 50 rpm. At predetermined time intervals, 3 ml of samples was withdrawn from dissolution medium and the OD was measured at the 319 nm using UV/Vis spectrophotometry.

In vitro cytotoxicity

In vitro cytotoxicity of the PLGA/PEO and 2 wt% FA-encapsulated PLGA/PEO nanofibers toward MCF-7 cells was evaluated using MTT assay (Mosmann 1983). Briefly, the UV sterilized nanofibrous samples were first placed in 24-well microtiter plates individually followed by the addition of MCF-7 cells (5×10^5 cells mL^{-1}) along with DMEM medium supplemented with 10 % fetal bovine serum (FBS). The plates were then incubated in humidified atmosphere containing 5 % CO_2 at 37 °C for 24 h. MCF-7 cells grown in presence of culture medium only were used as negative control, whereas the MCF-7 cells cultured in presence of native FA only were used as a positive control. Subsequently, the MTT solution (60 µL of 5 mg mL^{-1} stock) was added to each well containing sample, and incubated at 37 °C for 4 h. Then, the solution in the wells was removed carefully and 400 µL of DMSO was added to dissolve the MTT formazan crystals. After that, 100 µL of the dissolved formazan solution of each test sample was transferred to individual wells of 96-well plate to determine the absorbance at 570 nm (A_{570nm}) using micro-plate reader (Fluostar optima, BMG labtech, Germany). The cell viability was calculated as follows:

$$\text{Cell viability } (\%) = OD_{570nm}(\text{test samples}) / OD_{570nm}(\text{control}) \times 100$$

Analysis of change in morphology of MCF-7 tumor cells in presence of native FA and FA-encapsulated nanofibers.

The UV-sterilized nanofibrous samples were placed in 24-well plates, seeded with MCF-7 cells (5×10^5 cells mL^{-1}) and incubated at 37 °C for 24 h in a CO_2 incubator. The samples were then washed thrice with PBS (pH 7.4) immediately after the incubation period to remove any unattached MCF-7 cells. The cell morphologies for each sample were first observed using an inverted phase contrast microscope (Zeiss, Axiovert 25, Germany) at 100× magnification. For FESEM examination, the MCF-7 tumor cells adhered to nanofibers were fixed by immersing the samples into 2.5 % glutaraldehyde PBS solution at 4 °C for 4 h. Subsequently, the samples were dehydrated stepwise with graded concentrations of ethanol (25, 50, 75, 90 and 100 %).

Analysis of cell death using fluorescent microscopy

MCF-7 cells (5×10^5 cells mL^{-1}) cultured on the different formulations (control, free FA, PLGA/PEO nanofibers and FA-encapsulated PLGA/PEO nanofibers), as described above were also visualized by double staining with acridine orange (AO) and ethidium bromide (EtBr). Briefly, after 24 h of incubation, the MCF-7 tumor cells seeded samples were rinsed thrice with PBS and instantly stained with equal volumes of AO and EtBr (100 µg mL^{-1} stock solution) for 10 min. The samples were then visualized under the fluorescence microscope (Evos fl, AMG groups, USA).

Cytocompatibility evaluation

The cytocompatibility of different formulation (native FA, PLGA/PEO and FA-encapsulated PLGA/PEO nanofibers) was evaluated using HEK-293 cells. Briefly, the sterilized formulations were first placed individually in 24-well microtiter plates followed by addition of HEK-293 cells (5×10^4 cells mL^{-1}), seeded in DMEM medium supplemented with 10 % FBS, 100 U mL^{-1} penicillin, and 100 µg mL^{-1} streptomycin. The plates were then incubated in a humidified atmosphere with 5 % CO_2 at 37 °C for 24 h. Coverslips without nanofibers were used as controls. After 24 h of cell seeding, MTT assay (A_{570nm}) and FESEM studies were performed as described above.

Statistical analysis

Each measurement was performed at least in three independent experiments ($n = 3$) and the data are expressed as the mean values with ±standard deviation (SD). Statistical analysis of the differences between mean values obtained from experimental groups was performed using one-way ANOVA test. Values of $p < 0.05$ or less were considered as denoting statistical significance.

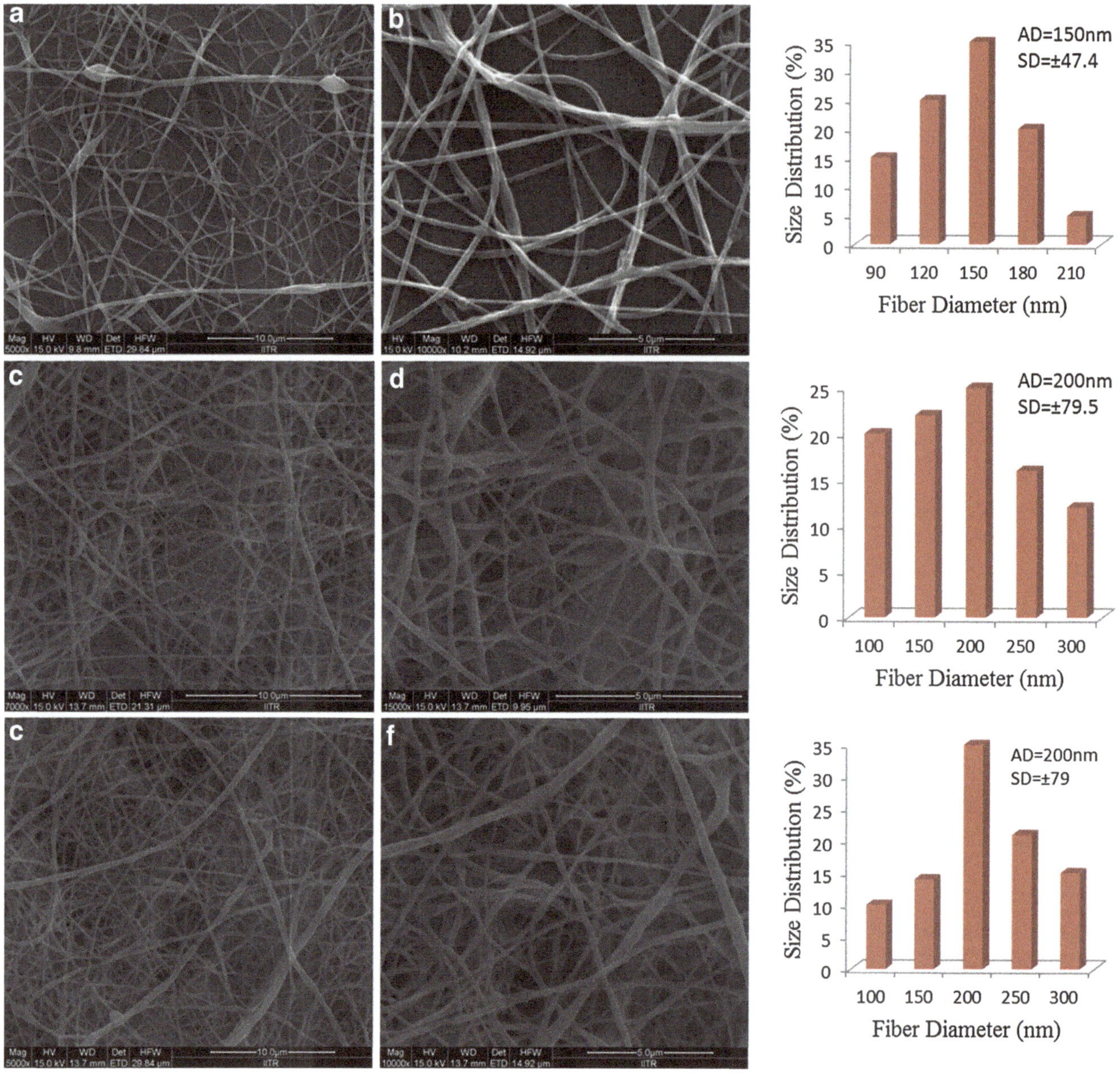

Fig. 1 FESEM micrographs and corresponding diameter distribution histogram of 2 wt% PLGA/PEO nanofibers (**a**, **b**) 1 wt% FA-encapsulated PLGA/PEO nanofibers (**c**, **d**) and 2 wt% FA-encapsulated PLGA/PEO nanofibers (**e**, **f**); *Scale bar* (**a**, **c**, **e**) = 10 μm and (**b**, **d**, **f**) = 5 μm. *SD* standard deviation, *FESEM* field emission scanning electron microscope, *AD* average diameter; $n = 5$

Results and discussion

Morphology of FA-free and FA-encapsulated PLGA/PEO nanofibers

The key processing parameters that influence the formation as well as the morphology of electrospun nanofibers include concentration of electrospinning solution, surface tension, nature of the solvent used, applied voltage, solution flow rate, and the distance between tip and collector (Zhang et al. 2012). In this study, all the above described parameters were optimized for obtaining the nanofibers with desired morphology and FA distribution. The FESEM micrographs of the PLGA/PEO nanofibers and their average diameter have been presented graphically in Fig. 1. Data obtained from FESEM characterization demonstrated the beads free uniform morphology of PLGA/PEO nanofibers. The surface morphology of FA-encapsulated PLGA/PEO nanofibers at 1 and 2 wt% concentration of FA was found to be cylindrical, smooth, and free of any bead defect but the presence of FA caused entanglement in the fabricated PLGA/PEO nanofibers. However, no FA crystals were noticed on encapsulated nanofibrous surface which suggests the homogenous dispersion of FA within the PLGA/PEO

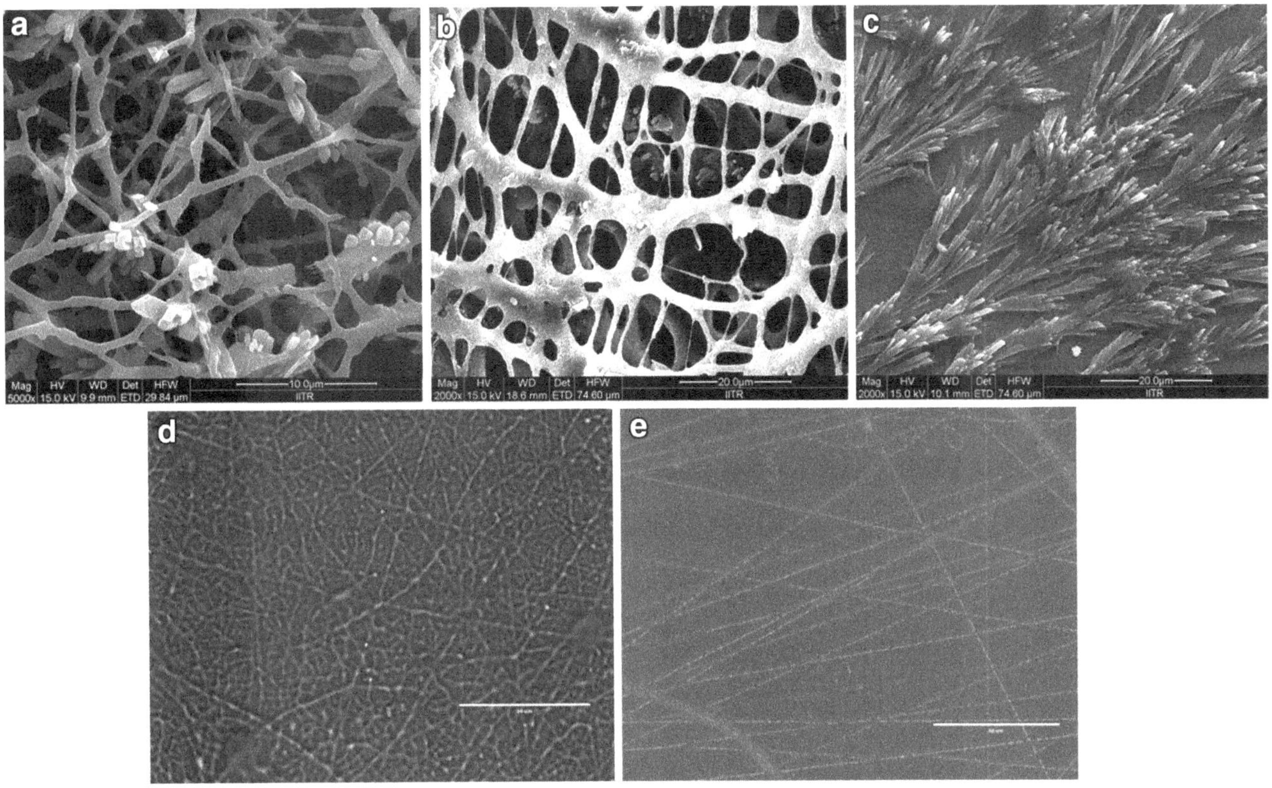

Fig. 2 FESEM micrographs of 4, 6 and 8 wt% FA-encapsulated PLGA/PEO nanofibers (**a**, **b**, **c**), respectively and overlay fluorescent micrographs of PLGA/PEO and 2 wt% FA-encapsulated PLGA/PEO nanofibers (**d**, **e**), respectively (*bar* 50 μm)

nanofibers (Fig. 1c–f). Data also depicted the increase in average fiber diameter after FA-encapsulation and recorded as 150 ± 47.4 to 200 ± 79 nm for PLGA/PEO and FA-encapsulated PLGA/PEO nanofibers, respectively (as presented in corresponding histograms in Fig. 1).

At higher FA concentration (4 wt%) the surface morphology of nanofibers changed from smooth, cylindrical, defect free to rough and damaged surfaces (Fig. 2a). Some drug crystals and several agglomerates were also observed at the exterior surface of the nanofiber. On further increasing concentrations of FA (6 and 8 wt%) instead of fiber formation, electrospraying of drug-polymeric solution was observed (Fig. 2b, c) (Almería et al. 2010). These observations suggest that the higher concentration of FA does not support electrospinning process and disrupt the morphology as well as uniformity of the electrospun nanofibers. Further, the autofluorescence properties of FA were exploited for visualizing the distribution of it within the PLGA/PEO nanofibers. The overlay fluorescent images of FA-encapsulated (containing 2 wt% FA) PLGA/PEO nanofibers as shown in Fig. 2d and e, represent the core-sheath arrangement of FA within the PLGA/PEO fibers. These observations suggested that the 2 wt% FA is the optimum/maximum amount of drug that can be encapsulated in the above fabricated nanofibers. Furthermore, it seemed that incorporation of the FA at lower concentration in the PLGA/PEO nanofibers did not affect the morphology of the resulting fibers.

Analysis of physical state of FA within PLGA/PEO nanofibers

Native FA, PLGA/PEO nanofibers and 2 wt% FA-encapsulated PLGA/PEO nanofibers were examined through XRD technique in order to reveal the physical state and distribution of the drug (FA) in the electrospun nanofibers. As shown in Fig. 3, FA exhibits several characteristic peaks at $2\theta = 9.02$, 10.4, 12.5, 15.6, 17.2, 24.4, 29.4, 35.7, 42.8, 45.8, and 50.7 (Yu et al. 2013). The XRD patterns of FA-free PLGA/PEO nanofibers exhibit typical crystalline peaks at $2\theta = 44.6$, 65.1 and 78.2. The composite FA-encapsulated PLGA/PEO nanofibers possessed the same peaks at $2\theta = 44.6$, 65.1 and 78.2 as that of PLGA/PEO nanofibers. Observations revealed that loading of the FA within the nanofibers did not change the typical crystalline nature of the PLGA/PEO nanofibers. The peaks that attributed to crystalline FA disappeared in the XRD spectrum of composite PLGA/PEO-FA nanofibers confirmed the amorphous state of FA in the composite nanofibers.

Thermal properties of the native FA, PLGA/PEO, and FA-encapsulated PLGA/PEO nanofibers were analyzed by TGA as shown in Fig. 4a. FA exhibits a single stage

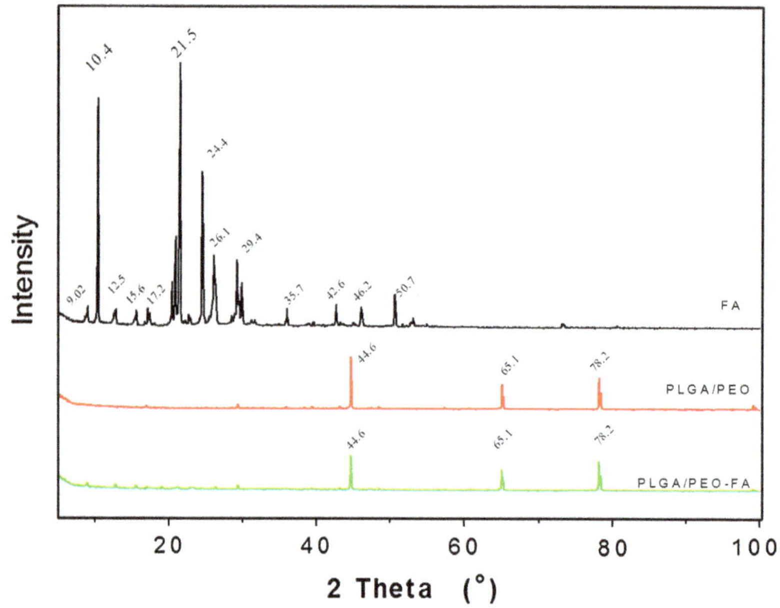

Fig. 3 X-ray diffraction patterns of the native FA and nanofibrous formulations

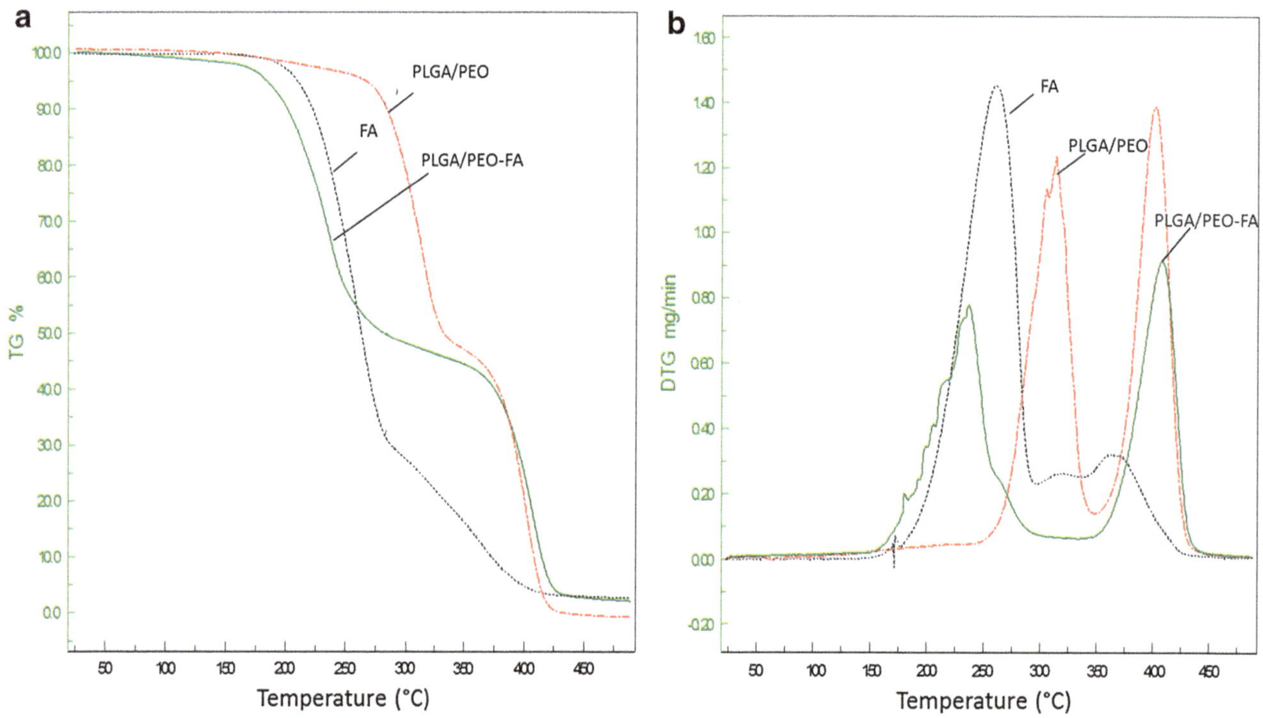

Fig. 4 TGA (**a**) and DTG (**b**) thermograms of the native FA and nanofibrous formulations

thermal degradation with decomposition starts around 173 °C which is similar to the studies of Mathew and Abraham (2008). However, the decomposition of PLGA/PEO and FA-encapsulated PLGA/PEO nanofibers was recorded in two stages. The initial weight loss in the range of 50–200 °C was due to the evaporation of physically weak and chemically strong H_2O bonding. The second stage of decomposition started at around 260 °C and completed at around 400 °C. This zone was found to be the highest thermal degradation zone which corresponds to a complex process including depolymerization as well as decomposition of bonds and units of polymers (Fouad and Elsarnagawy 2013; Ibrahim and Johan 2012). The result obtained suggested that there was no significant difference in the thermal stability of PLGA/PEO and FA-encapsulated PLGA/PEO nanofibers.

The derivative TGA curves of weight loss for native FA, PLGA/PEO and FA-encapsulated PLAG/PEO nanofibers are shown in Fig. 4b. Each temperature peak in the thermograms represents the T_{max}, corresponds to the maximum degradation rate. FA exhibits a peak of maximum degradation at 263 °C while PLGA/PEO nanofibers exhibited two peaks at 315 and 402 °C. However, the peaks of maximum decomposition for FA-encapsulated PLGA/PEO nanofibers were observed at 240 and 408 °C. The slight shifting of temperature peak toward the lower temperature was perceived in case of FA-encapsulated PLGA/PEO nanofibers as compared to native FA. This shifting indicated the presence of weak interactions between polymer components (PLGA/PEO) and FA as well as reflected the early onset of FA decomposition in nanofibers than that of native FA. This may be due to the even distributions and presence of amorphous state of FA in PLGA/PEO nanofibers which make FA molecules respond to the heat more sensitively than native FA crystalline particles.

Analysis of secondary interactions among the components

Secondary interactions in terms of hydrogen bonding, hydrophobic interactions, and electrostatic forces are often considered to be true indication of compatibility between the polymer and drug component for producing high quality, stable nanofiber. As shown in Fig. 5, FA molecule has both –OH and –C=O active groups and exhibited characteristic absorption bonds at 3,437 cm^{-1} (–OH group), 3,015, 2,922, 2,838, and 2,594 cm^{-1} (C–H bond stretching), 1,690, 1,665, and 1,619 cm^{-1} (–C=O stretching), 1,272–1,034 cm^{-1} (–C–O stretching) as also reported earlier by Yang et al. (2013) and Yu et al. (2010). The PLGA/PEO nanofibers showed the absorption peaks at 3,436 cm^{-1} (–OH stretching), 2,922 and 2,856 cm^{-1} (–CH stretching), 2,360 cm^{-1} (C≡C stretching), 1,760 and 1,630 cm^{-1} (–C=O stretching), 1,271 and 1,100 cm^{-1} (C–O stretching). All the entire sharp peaks for FA can be observed in the IR spectrum of FA-encapsulated PLGA/PEO-FA nanofibers, indicating that FA in the composite nanofibers more likely tends to form FA dimers that are necessary for constructing a crystal lattice. The shift of 1,640–1,820 cm^{-1} absorption bands to lower wave numbers in the spectrum of FA-encapsulated PLGA/PEO nanofibers illustrated the presence of weak interactions between the carbonyl groups of PLGA/PEO blend and the hydroxyl group of FA (Fig. 6). These weak interactions between the FA and PLGA/PEO nanofibers imitated the compatibility between the FA and nanofibers.

In vitro drug-release assay

The amount of drug-loading and the drug-release characteristic can be controlled by investigating the actual drug

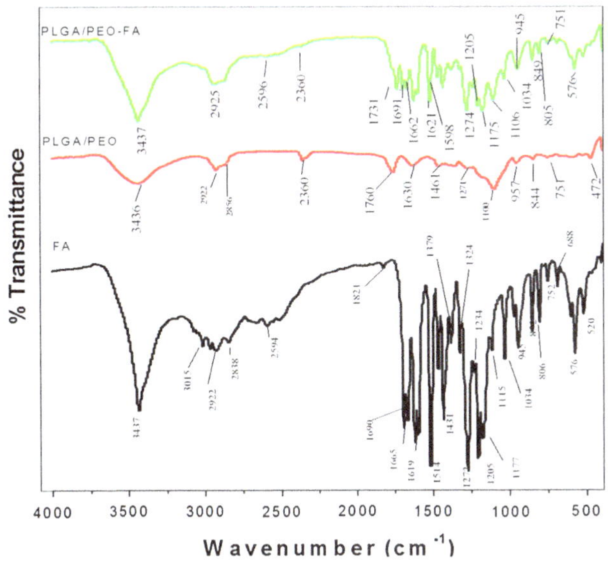

Fig. 5 FTIR analysis of the native FA and nanofibrous formulations

content in the resulted fibers (Han et al. 2009). The EE of FA in the 2 wt% FA-encapsulated PLGA/PEO nanofibers was recorded as 66 ± 1.34 %. The sustainability of FA molecules in the PLGA/PEO nanofibers was investigated by in vitro drug-release profiles in physiological condition at different time points. The release rate of encapsulated FA revealed an initial burst release within the 24 h, followed by a continuous and sustained release during the subsequent time (Fig. 7). More than half (approx. 278 μg, 53.51 %) of the encapsulated FA was released from the nanofibers within 24 h of incubation. At the end of the 240 h time interval, 89 % (approx. 459.13 μg) of FA had been found to be released from the PLGA/PEO nanofibrous matrix. The initial burst release of FA is possibly due to the dissolution/ or diffusion of FA molecules that are entrapped/ or adsorbed at the surface of polymeric nanofibers. However, the sustained release of FA in subsequent stage may be attributed either to the diffusion of encapsulated-FA from the core region ofnanofibers or due to degradation of polymeric matrix. The results obtained are also supported by the findings of Merlin et al. (2012) and indicated that these nanofibers can be potentially useful for sustained delivery of drug for cancer treatment (Cui et al. 2010).

In vitro cytotoxicity

Globally, breast cancer has become one of the most frequently diagnosed cancer and second leading causes of female malignancy in India (Indap and Radhika 2006). In spite of huge progress in cancer treatments, the assimilated or inherent resistance of breast cancer cells toward the anticancer agents is the prime hurdle for effective

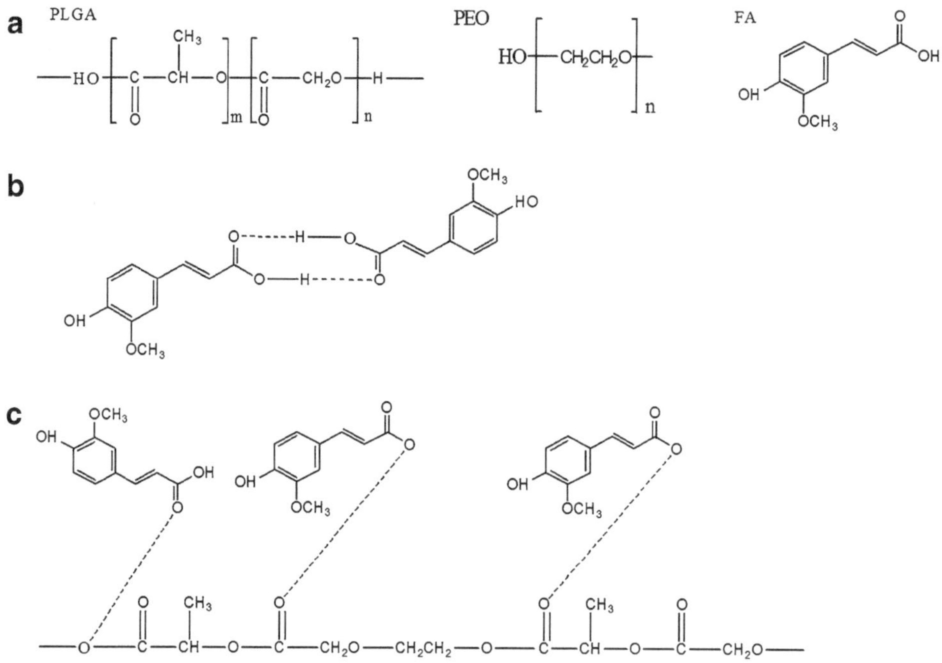

Fig. 6 **a** Molecular structures of PLGA, PEO and FA. **b** Hydrogen bonding between FA molecules. **c** Interaction between FA and PLGA-PEO molecules

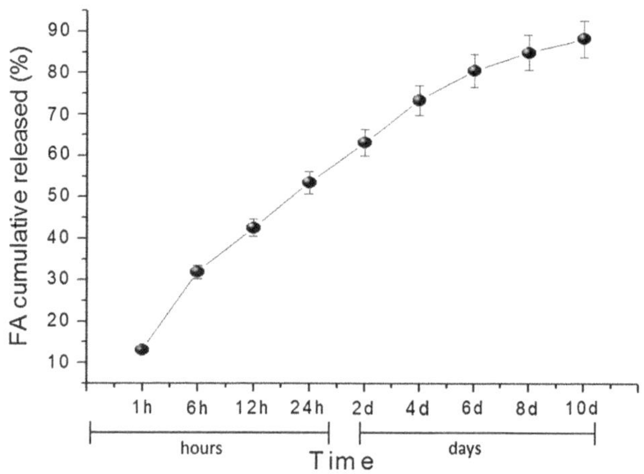

Fig. 7 Invitro drug-release profile of FA from the FA-encapsulated PLGA/PEO nanofibers at different time intervals

chemotherapeutic treatments. Hence, the search for new and effective chemotherapeutic agents is the requisite for treatment of drug-resistant breast cancers (Ignatova et al. 2011). FA is a therapeutically renowned phytochemical which has been reported to exhibit antitumor activities against colon cancer (Kawabata et al. 2000), skin tumor (Asanoma et al. 1994), and pulmonary cancer (Lesca 1983). The antiproliferative effect of FA is mainly attributed to its inhibitory effect toward formation of nitrosamine in the cells (Kuenzig et al. 1984). Till now, no report regarding to the antitumor activity of nanofibers containing FA against MCF-7 has been reported. Therefore, in present investigation, in vitro cytotoxic effects of the FA-encapsulated nanofibers toward MCF-7 cells have been evaluated. As shown in Fig. 8, the 2 wt% FA-encapsulated PLGA/PEO nanofibers possess a strong antiproliferative activity toward MCF-7 cells as compared to control. No inhibition in cell growth was recorded in control and PLGA/PEO nanofibrous sample. However, in case of native FA and FA-encapsulated PLGA/PEO nanofibers, the cell inhibition rate of 51.4 % and 67 % ($p < 0.01$) was achieved, respectively. Moreover, it could be interpret from the results that FA retains its activity even after encapsulation into nanofibrous matrix and is not affected by the electrospinning process. The polymeric nanofibers performed as preservative agent for FA which improved its physiochemical stability as well as protected it from the premature degradation (Supplementary Fig. 1). Therefore, the encapsulation of FA in PLGA/PEO nanofibers resulted into enhanced activity of encapsulated FA in comparison to native FA in solution form. Similar studies have also been reported by Ouimet et al. (2013) on FA-containing polyanhydride esters.

Morphological analysis of MCF-7 cells in presence of FA-encapsulated nanofibers

The change in cell morphologies of the MCF-7 cell in presence of native FA and FA-encapsulated PLGA/PEO nanofibers is illustrated in Fig. 9. Tumor cells cultured on PLGA/PEO nanofibers did not comprise any changes in their morphology as well as in the cell growth (Fig. 9a, c). In contrast, the MCF-7 tumor cells possessed good adhesion and proliferation properties onto the PLGA/PEO nanofibers. A great number of tumor cells that adhered to PLGA/PEO nanofibers were found to acquire normal bilateral symmetric morphology and also displayed numerous microvilli on their surface, similar to control.

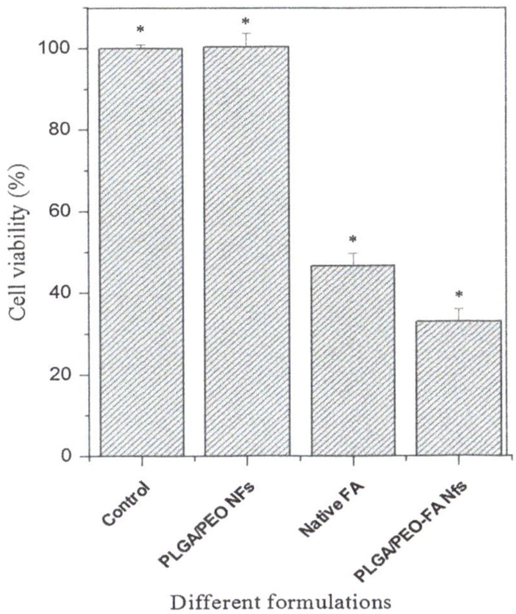

Fig. 8 Colorimetric MTT viability assay for MCF-7 cells treated with different formulations coverslips (control), PLGA/PEO nanofibers, native FA (in solution form) and FA-encapsulated PLGA/PEO nanofibers after 24 h at 570 nm. *Error bars* represent mean ± standard deviation for three independent experiments ($n = 3$). $^*p < 0.05$ indicates statistical significant difference as compared to control. (*Nfs* nanofibers)

The FESEM images depicted the significant reduction in tumor cell number/growth when cultured on FA-encapsulated nanofibers. The cell blebbings on the tumor cells surface and disappearance/reduction of the microvilli were also observed in presence of native FA solution and FA-encapsulated nanofibers (Fig. 9d). Our observations are found to be in agreement with the findings of Ignatova et al. (2011).

Analysis of cell death using fluorescent microscopy

To determine the cell cytotoxicity, double staining of MCF-7 tumor cells cultured in the presence of different formulations (control, native FA, PLGA/PEO nanofibers and FA-encapsulated PLGA/PEO nanofibers) was executed using equivalent mixture of AO and EtBr (1:1 w/w). The stained cells on different formulations were then analyzed by fluorescent microscopy. AO and EtBr stain DNA, which allows visualization and differentiation of dead and viable cells. Cells with intact membranes stained green due to passage of AO whereas EtBr stained the cells with damaged membranes, as a result, the DNA intercalation of both gives orange fluorescence (Fig. 10). No cell death was observed in control MCF-7 cells (Fig. 10a–c). Similarly, no growth inhibition was recorded in case of FA-free PLGA/PEO nanofibers (Fig. 10d–f) whereas a considerable amount of damaged MCF-7 cells were get into notice when exposed to native FA (94.3 %, $p < 0.05$) and FA-encapsulated nanofibers (99.8 %, $p < 0.05$). The cells cultured

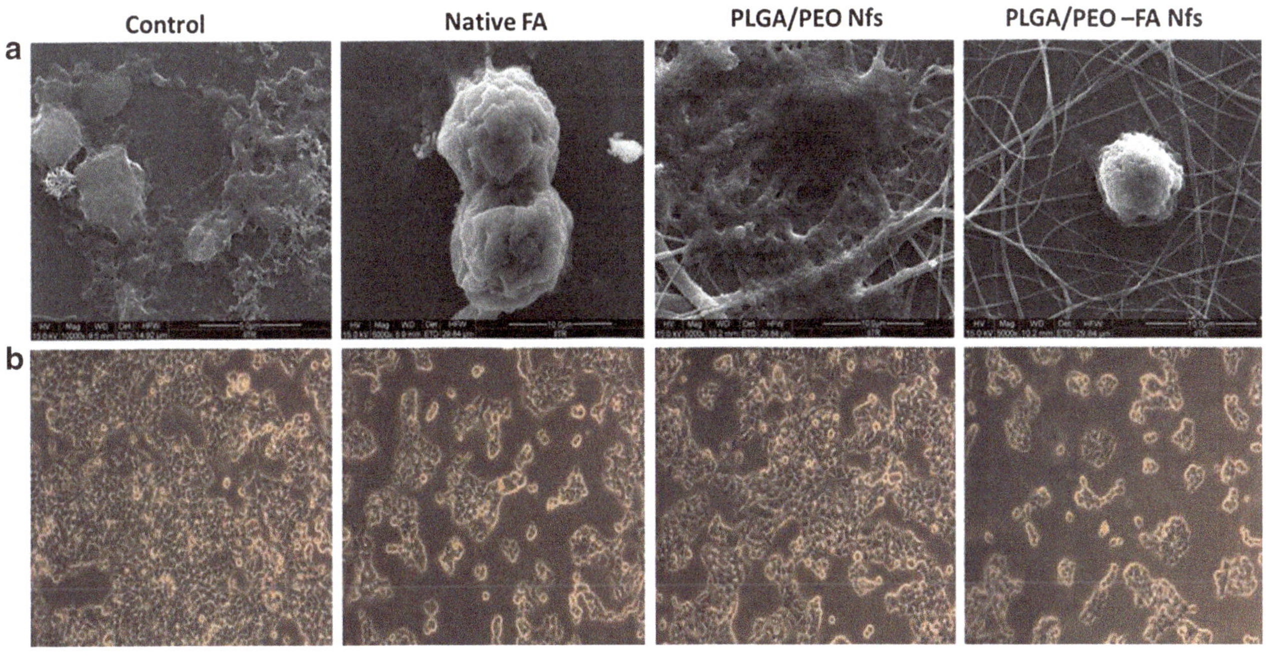

Fig. 9 The effect of different formulations (control, native FA, PLGA/PEO nanofibers and FA-encapsulated PLGA/PEO nanofibers) on cell morphology of MCF-7 cells **a** FESEM and **b** phase contrast micrographs of the MCF-7 cells after 24 h of treatment with different formulations

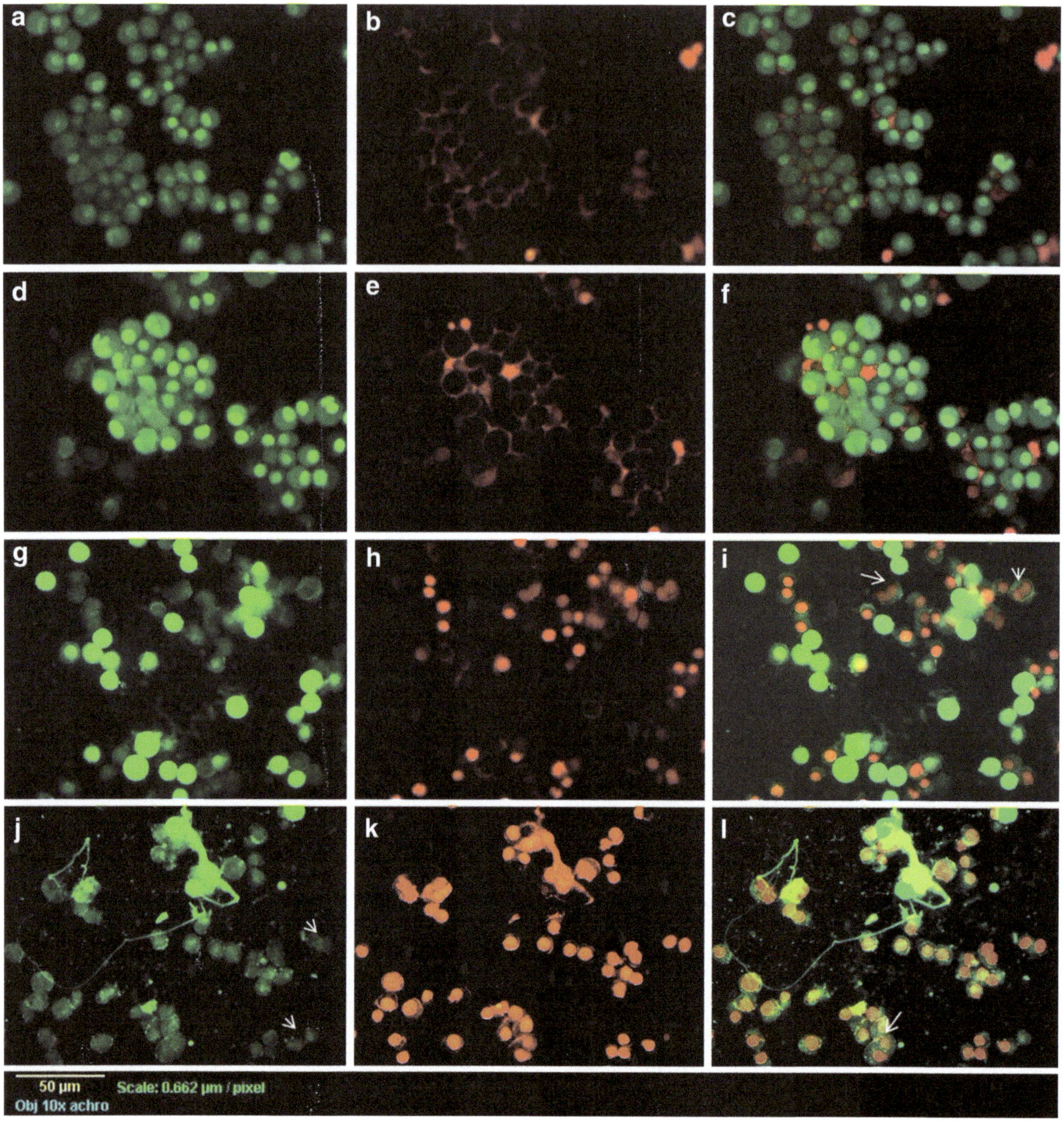

Fig. 10 Fluorescent micrographs of AO:EtBr stained MCF-7 cells incubated for 24 h **a–c** untreated cells, **d–f** cells treated with PLGA/PEO nanofibers, **g–i** treated with native FA and **j–l** cells treated with FA-encapsulated PLGA/PEO nanofibers. The significance of *arrows* is elucidated in text

on native FA (Fig. 10g–i) and the FA-encapsulated nanofibers (Fig. 10j–l) displayed the morphological signs for apoptosis such as cytoplasmic remnants, chromatin condensation and damaged wrinkled cells with orange fluorescence (as specified by arrows), indicated that induction of apoptosis is one of the major mechanism through which FA exerts its action to destroy tumor cells. The significant amount of apoptotic cells after treatment of FA-encapsulated PLGA/PEO nanofibers indicated that the encapsulated FA would be able to induce apoptosis more efficiently and possessed a substantial strong antitumor capacity as compared to native FA. These observations are in line with the work done by earlier researchers, who described the anti-apoptotic role of FA on different cancer cell lines (Khanduja et al. 2006; Chen et al. 2007; Indap and Radhika 2006).

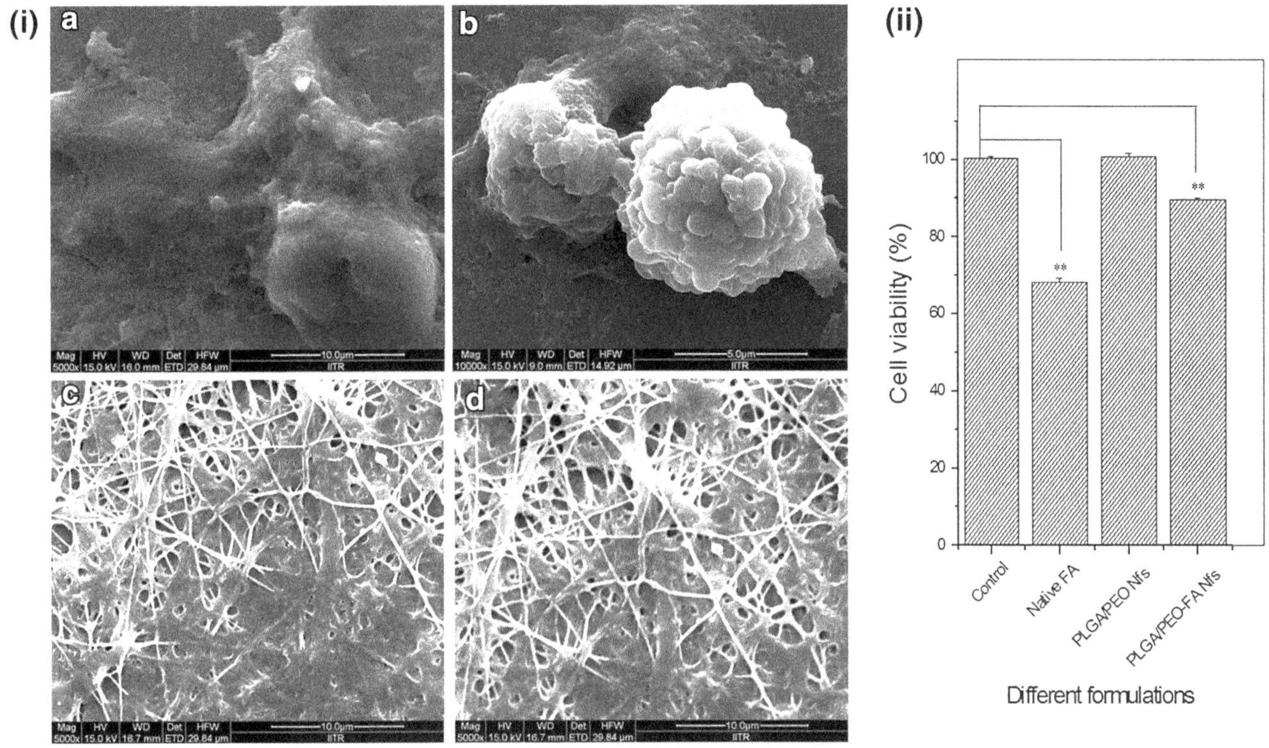

Fig. 11 **i** FESEM micrographs of the HEK-293 cells incubated for 24 h on *a* control, *b* in presence of native FA, *c* PLGA/PEO nanofibers and *d* FA-encapsulated PLGA/PEO nanofibers. **ii** MTT assay for HEK-293 cells treated with different formulations for 24 h (A_{570nm}). *Error bars* represent mean ± standard deviation for three independent experiments ($n = 3$). $^{**}p < 0.01$ indicated statistical significant difference and (*) *unmarked bar* indicated statistically insignificant difference as compared to control

Cytocompatibility assay

To evaluate the prospective biomedical applications of these FA-encapsulated PLGA/PEO nanofibers, cytocompatibility assay was carried out against HEK-293 cells using MTT assay. The viability of HEK-293 cells seeded on different formulation is shown in Fig. 11. Based on the results, it was observed that the viability of cells on PLGA/PEO and FA-encapsulated PLGA/PEO nanofibers was much higher than in presence of native FA, indicating the better cytocompatibility of FA-encapsulated nanofibers in comparison with native FA ($p < 0.01$). Data obtained, imply that the incorporation of FA within the nanofibers did not compromise the cytocompatibility of PLGA/PEO nanofibers. The cytocompatibility evaluation of FA-encapsulated PLGA/PEO nanofibers as compared to the native FA, was also performed via cell surface morphology observations (Fig. 11i). The cell adherence, proliferation and migration of HEK-293 cells cultured on both type of nanofibers was found to be similar as on native extracellular matrix. However, the blebbings on cell surface were observed in case of cells cultured in presence of native FA. The cell viability was found to be lower on FA-encapsulated nanofibers as compare to PLGA/PEO nanofibers which may be attributed to the antioxidant property of FA (Srinivasan et al. 2007).

Conclusion

FA, a potent phytonutrient, was successfully encapsulated in the blend PLGA/PEO nanofibers using electrospinning technique to improve both stability as well as efficiency of FA. Microscopic studies revealed the homogenous distribution of FA as a core-sheath structure encapsulated in PLGA/PEO polymeric matrix. MTT assay revealed the strong cytotoxic activity of FA-encapsulated PLGA/PEO nanofibers against MCF-7 cell line, which was primarily due to the initiation of apoptosis in tumor cell. Furthermore, these FA-encapsulated PLGA/PEO nanofibers demonstrated cytocompatibility when tested on embryonic kidney (HEK-293) cells. The findings suggested that the incorporation of FA in nanofibers may reduce the chemotherapeutic side effects and can be useful in providing a high local drug concentration to destroy the tumor cells. Therefore, these fabricated electrospun nanofibers are viewed as future candidates for antitumor drug delivery as well as potential local chemotherapeutic agent for breast tumor formations.

Acknowledgments This study was financially supported by Council of Scientific and Industrial Research (CSIR), Government of India. Authors are thankful to the Assistant Professor Narayan C. Mishra, Department of polymer & process engineering and Institute instrumentation Center, IITR for providing research facilities for this work.

Conflict of interest The authors declare that they have no conflict of interest in the publication.

References

Almería B, Deng W, Fahmy TM, Gomez A (2010) Controlling the morphology of electrospray-generated PLGA microparticles for drug delivery. J Colloid Interface Sci 343:125–133.

Asanoma M, Takahashi K, Miyabe M et al (1994) Inhibitory effect of topical application of polymerized ferulic acid, a synthetic lignin, on tumor promotion in mouse skin two-stage tumorigenesis. Carcinogenesis 15:2069–2071.

Chen C-J, Hsu M-H, Kuo S-C et al (2007) (2E)-*N*, *N*-dibutyl-3-(4-hydroxy-3-methoxyphenyl) acrylamide induces apoptosis and cell cycle arrest in HL-60 cells. Anticancer Res 27:343–349

Cui W, Zhou Y, Chang J (2010) Electrospun nanofibrous materials for tissue engineering and drug delivery. J Sci Technol Adv Mat 11:014108.

Davis CD, Milner JA (2010) Gastrointestinal microflora, food component and colon cancer prevention. J Nutr Biochem 20:743–752.

Fouad H, Elsarnagawy T (2013) Preparation and in vitro thermo-mechanical characterization of electrospun PLGA nanofibers for soft and hard tissue replacement. Int J Electrochem Sci 8:2293–2304

Gu Q, Xing JZ, Huang M et al (2012) SN-38 loaded polymeric micelles to enhance cancer therapy. Nanotechnology 23:205101.

Han J, Chen T-X, Branford-White CJ, Zhu L-M (2009) Electrospun shikonin-loaded PCL/PTMC composite fiber mats with potential biomedical applications. Int J Pharm 382:215–221.

Huang Z-M, Zhang Y-Z, Kotaki M, Ramakrishna S (2003) A review on polymer nanofibers by electrospinning and their applications in nanocomposites. Compos Sci Technol 63:2223–2253.

Ibrahim S, Johan M (2012) Thermolysis and conductivity studies of poly (ethylene oxide)(PEO) based polymer electrolytes doped with carbon nanotube. Int J Electrochem Sci 7:2596–2615

Ignatova M, Yossifova L, Gardeva E et al (2011) Antiproliferative activity of nanofibers containing quaternized chitosan and/or doxorubicin against MCF-7 human breast carcinoma cell line by apoptosis. J Bioact Compat Polym 26:539–551.

Indap M, Radhika S (2006) Anticancer activity of phenolic antioxidants against breast cancer cells and a spontaneous mammary tumor. Indian J Pharm Sci 68:470–474.

Kandaswami C, Lee L-T, Lee P-PH et al (2005) The antitumor activities of flavonoids. In vivo 19:895–910

Kawabata K, Yamamoto T, Hara A et al (2000) Modifying effects of ferulic acid on azoxymethane-induced colon carcinogenesis in F344 rats. Cancer Lett 157:15–21

Khanduja KL, Avti PK, Kumar S et al (2006) Anti-apoptotic activity of caffeic acid, ellagic acid and ferulic acid in normal human peripheral blood mononuclear cells: a Bcl-2 independent mechanism. Biochim Biophys Acta 1760:283–289.

Kuenzig W, Chau J, Norkus E et al (1984) Caffeic and ferulic acid as blockers of nitrosamine formation. Carcinogenesis 5:309–313.

Lesca P (1983) Protective effects of ellagic acid and other plant phenols on benzo[a]pyrene-induced neoplasia in mice. Carcinogenesis 4:1651–1653.

Mathew S, Abraham TE (2008) Characterization of ferulic acid incorporated starch–chitosan blend films. Food Hydrocolloids 22:826–835.

Merlin JPJ, Rajendra Prasad N, Shibli SM et al (2012) Ferulic acid loaded poly-d, l-lactide-co-glycolide nanoparticles: systematic study of particle size, drug encapsulation efficiency and anticancer effect in non-small cell lung carcinoma cell line in vitro. Biomed Prev Nutr 2:69–76.

Mosmann T (1983) Rapid colorimetric assay for cellular growth and survival: application to proliferation and cytotoxicity assays. J Immunol Methods 65:55–63.

Murakami A, Nakamura Y, Koshimizu K et al (2002) FA15, a hydrophobic derivative of ferulic acid, suppresses inflammatory responses and skin tumor promotion: comparison with ferulic acid. Cancer Lett 180:121–129.

Ou S, Kwok K-C (2004) Ferulic acid: pharmaceutical functions, preparation and applications in foods. J Sci Food Agric 84:1261–1269.

Ouimet MA, Griffin J, Carbone-Howell AL et al (2013) Biodegradable ferulic acid-containing poly(anhydride-ester): degradation products with controlled release and sustained antioxidant activity. Biomacromolecules 14:854–861.

Ren W, Qiao Z, Wang H et al (2003) Flavonoids: promising anticancer agents. Med Res Rev 23:519–534.

Soobrattee MA, Neergheen VS, Luximon-Ramma A et al (2005) Phenolics as potential antioxidant therapeutic agents: mechanism and actions. Mutat Res 579:200–213.

Srinivasan M, Sudheer A, Menon V (2007) Ferulic acid: therapeutic potential through its antioxidant property. J Clin Biochem Nutr 40:92–100.

Vashisth P, Nikhil K, Pemmaraju SC et al (2013) Antibiofilm activity of quercetin-encapsulated cytocompatible nanofibers against *Candida albicans*. J Bioact Compat Polym 28:652–665.

Xie J, Wang C-H (2006) Electrospun micro- and nanofibers for sustained delivery of paclitaxel to treat C6 glioma in vitro. Pharm Res 23:1817–1826.

Xu X, Chen X, Ma P et al (2008) The release behavior of doxorubicin hydrochloride from medicated fibers prepared by emulsion-electrospinning. Eur J Pharm Biopharm 70:165–170.

Yang J-M, Zha L, Yu D-G, Liu J (2013) Coaxial electrospinning with acetic acid for preparing ferulic acid/zein composite fibers with improved drug release profiles. J Colloids Surf B 102:737–743.

Yu D-G, Yang J-M, Branford-White C et al (2010) Third generation solid dispersions of ferulic acid in electrospun composite nanofibers. Int J Pharm 400:158–164

Yu D, Qian W, Wang X et al (2013) Ferulic acid-loaded shellac microparticles prepared using electro hydrodynamic atomization. Adv Mater Res 675:326–330. http://www.scientific.net/AMR.675.326

Zhang H, Lou S, Williams GR et al (2012) A systematic study of captopril-loaded polyester fiber mats prepared by electrospinning. Int J Pharm 439:100–108.

14

Aflatoxin B1 degradation during co-cultivation of *Aspergillus flavus* and *Pleurotus ostreatus* strains on rice straw

Arijit Das · Sourav Bhattacharya · Muthusamy Palaniswamy · Jayaraman Angayarkanni

Abstract Aflatoxin B1 (AFB1) produced by *Aspergillus flavus* is known to have carcinogenic and teratogenic effects on animal health. Accidental feeding of AFB1-contaminated rice straw may be detrimental to dairy cattle. White-rot basidiomycetous fungus *Pleurotus ostreatus* can grow on different agronomic wastes by synthesizing different ligninolytic enzymes. These extracellular enzymes are capable of degrading many environmentally hazardous compounds including AFB1. The present study examines the ability of different strains of *P. ostreatus* to degrade AFB1 in contaminated rice straw. Different strains of *A. flavus* were inoculated on rice straw for AFB1 production. The moldy straw was then subjected to co-cultivation by different strains of *P. ostreatus*. The extent of AFB1 degradation was determined by high performance liquid chromatography. Results indicated the presence of AFB1 in the moldy straw samples at levels of 27.95 ± 0.23 and 21.26 ± 0.55 μg/g of dry substrate for *A. flavus* MTCC 2798 and *A. flavus* GHBF09, respectively. Co-cultivation of *P. ostreatus* strains on AFB1-contaminated rice straw revealed their ability to rapidly colonize the substrate by profuse hyphal ramification. Highest degradation of AFB1 (89.41 %) was recorded in the straw containing co-cultures of *A. flavus* MTCC 2798 and *P. ostreatus* GHBBF10. Natural isolate *P. ostreatus* GHBBF10 demonstrated higher AFB1-degradation potential than *P. ostreatus* MTCC 142. This basidiomycete strain can be further exploited to effectively degrade moderate concentrations of AFB1 in contaminated moldy rice straw.

Keywords Aflatoxin B1 · *Pleurotus ostreatus* · Co-cultivation · Rice straw

A. Das (✉) · S. Bhattacharya · M. Palaniswamy
Department of Microbiology, Karpagam University, Coimbatore 641021, Tamil Nadu, India
e-mail: jit2007das@gmail.com

A. Das · S. Bhattacharya
Department of Microbiology, Centre for Advanced Studies in Biosciences, Jain University, Bangalore 560019, Karnataka, India

J. Angayarkanni
Department of Microbial Biotechnology, Bharathiar University, Coimbatore 641046, Tamil Nadu, India

Introduction

Mycotoxins are highly toxic fungal metabolites which have adverse effects on human and animal health. Among the mycotoxins, aflatoxin B1 (AFB1) is extremely hazardous for being mutagenic, carcinogenic and teratogenic (Khoshpey et al. 2011). AFB1 is produced by strains of *Aspergillus flavus* and *A. parasiticus* on a variety of agricultural commodities. Animal feeds and forages may become contaminated with AFB1 in the fields and/or during storage under conditions conducive for fungal growth (Streit et al. 2012). In the tropical countries like India, rice (paddy) straw is often used as cattle feed. However, consumption of AFB1-contaminated straw can lead to aflatoxicosis in cattle and eventually lead to transmission of these toxins or their metabolites into the milk of dairy animals (Bennett and Klich 2003).

White-rot basidiomycetous fungus such as *Pleurotus ostreatus* is able to transform AFB1 into less toxic metabolites (Guan et al. 2008). Under natural condition, *P. ostreatus* grows on dead wood but can be easily cultivated on different agro-industrial wastes (Sánchez 2010). It produces different oxidative [laccase and manganese peroxidase (MnP)] and hydrolytic (cellulase, xylanase and tannase) enzymes which are involved in utilization of lignocellulosic substrates (da Luz et al. 2012).

Degradation of AFB1 by white-rot fungi through the production of extracellular lignin modifying enzymes was previously investigated (Cohen et al. 2002). Due to lack of substrate specificity, laccase can degrade many xenobiotic compounds. A correlation between laccase activity and AFB1 degradation was previously reported by Alberts et al. (2009). Though some studies on AFB1 degradation have been conducted, very few scientific literatures are available on AFB1 degradation by *P. ostreatus* in *A. flavus*-contaminated moldy rice straw. Among *P. ostreatus*, different strains may vary in their ability to degrade AFB1. Selection of a potent basidiomycete strain may facilitate effective degradation of this hazardous toxin. Therefore, the present study was conducted to assess the ability of different strains of *P. ostreatus* to degrade AFB1 in contaminated rice straw.

Materials and methods

Chemicals and reagents

AFB1 standard was procured from Sigma (USA). All other analytical grade laboratory chemicals were purchased from S D Fine-Chem Ltd. (Mumbai, India) and Nice Chemicals (Kochi, India).

Source of fungal strains

A. flavus MTCC 2798 and *P. ostreatus* MTCC 142 were obtained from Microbial Type Culture Collection and Gene Bank, Chandigarh, India. An aflatoxigenic fungal strain identified as *A. flavus* GHBF09 (GenBank accession number KC987360) was isolated from rhizosphere soil (Das et al. 2013a). A basidiomycetous fungal strain identified as *P. ostreatus* GHBBF10 (GenBank accession number KC987361) was isolated from a decomposing tree trunk (Das et al. 2013b). Cultures of *A. flavus* and *P. ostreatus* were maintained on potato dextrose agar and glucose yeast extract agar plates, respectively.

Substrate preparation

Rice (paddy) straw was obtained from the local market in Bangalore. The straw was washed several times under running tap water, dried in hot air oven at 80 °C and cut into approximately 1 cm pieces. This was used as the substrate for co-cultivation. Dry straw (1 g) was filled in a glass culture tube measuring 2.5 cm in diameter and 15 cm long, moistened with 5 ml supplemented mineral salt solution containing (g/l) lactose, 10; yeast extract, 5; KH_2PO_4, 0.2; $MgSO_4{\cdot}7H_2O$, 0.1; NH_4Cl, 0.3; $CaCO_3$, 1.0 and distilled water, at initial pH 5.8 and autoclaved (Das et al. 2013b).

Co-cultivation of *A. flavus* and *P. ostreatus* strains

Degradation of AFB1 during co-cultivation of *A. flavus* and *P. ostreatus* strains on rice straw was investigated. Duplicate sets of four culture tubes containing the substrate were labeled. Two tubes were inoculated with spore suspension of *A. flavus* MTCC 2798 and two other tubes with *A. flavus* GHBF09. The inoculated sets of tubes were incubated at 28 ± 2 °C for 15 days for AFB1 production. Post incubation, one set of tubes was subjected to AFB1 analysis using high performance liquid chromatography (HPLC). In the second set, two alternate tubes containing moldy straw with live *A. flavus* cultures were aseptically inoculated with two mycelial plugs of *P. ostreatus* MTCC 142, while the other two tubes were inoculated with *P. ostreatus* GHBBF10. This set of inoculated tubes was incubated at 30 ± 2 °C for 15 days in dark. Mycelial growth in these tubes was carefully monitored daily. After 15 days of incubation, the level of residual AFB1 present in the tubes was determined by HPLC. A sterile uninoculated tube containing the substrate was maintained separately throughout the study for sterility check.

Extraction of AFB1

The extraction of AFB1 was performed using AOAC protocols (AOAC International 2000) with few modifications. Following incubation, the substrate colonized with fungal mycelia was homogenized in a blender with 20 ml of 90 % (v/v) methanol and filtered through coarse filter paper. To the filtrate equal volume of methylene chloride was added and AFB1 was partitioned by shaking vigorously for 15 min. The extraction procedure was repeated with the aqueous fraction. The methylene chloride fractions were pooled and concentrated in hot air oven at 40 °C overnight.

Analysis of AFB1 using HPLC

The concentrated residue was reconstituted in 1 ml of acetonitrile and filtered through 0.25 μ nylon membrane filter (Axiva Sichem Biotech, Delhi, India). The working standard solution of AFB1 (concentration of 10 μg/ml) was prepared using acetonitrile. Working standard (20 μl) containing 0.2 μg of standard AFB1 was injected into the HPLC system (Waters, USA, model number-2487, with Dual λ absorbance UV detector and binary pump system, model number-1525). A reverse phase, YMCA Triart C18 column (150 × 4.6 mm, 3 μm) was used. The mobile phase consisted of acetonitrile-methanol-water (1:1:2, v/v/v). A total run time of 14 min was maintained at a flow rate of 0.8 ml/min under ambient temperature. The absorbance of AFB1 was determined at 360 nm. Area under AFB1 absorbance peak was used to estimate its concentration and also the percentage of degradation using the following formula (Guan et al. 2010):

$$\text{Percentage of AFB1 degradation} = [(C_i - C_f)/C_i] \times 100$$

where, C_i is the initial concentration of AFB1 and C_f is the final concentration of AFB1.

Statistical analysis

The experiment was conducted in triplicate. The data were analyzed using single factor analysis of variance (ANOVA) and graphically presented as mean ± SD ($n = 3$). ANOVA was performed using Microsoft Excel 2007. P values <0.05 were considered significant with a confidence limit of 95 %.

Results and discussion

In the present study, co-cultivation of *A. flavus* and *P. ostreatus* strains on rice straw was carried out to understand the practical applicability of the basidiomycetous fungus in AFB1 degradation in rice straw which may be contaminated with AFB1 in agricultural fields.

Growth of *A. flavus* strains on rice straw and AFB1 production

After 15 days of incubation, rice straw inoculated with the tested strains of *A. flavus* turned completely moldy due to profuse mycelial growth and sporulation of the fungi (Fig. 1). HPLC analysis of these moldy straw samples revealed the presence of AFB1 at levels of 27.95 ± 0.23 and 21.26 ± 0.55 μg/g of dry substrate for *A. flavus* MTCC 2798 and *A. flavus* GHBF09, respectively. Comparatively, *A. flavus* MTCC 2798 demonstrated higher production of AFB1. Zain (2011) suggested that among various factors, the ability to produce AFB1 in foods and feeds is also dependent on fungal strain specificity and strain variation. Moreover, the presence of elevated moisture content and high C/N ratio in rice straw would have facilitated increased AFB1 production from the tested strains of *A. flavus* (Abdulla 2007).

Fig. 1 Mycelial growth of *A. flavus* in tubes containing moist rice straw as substrate. *a, b* Tubes inoculated with *A. flavus* MTCC 2798. *c, d* Tubes inoculated with *A. flavus* GHBF09. A sterile uninoculated tube (control) is shown on the *left*

Growth of *P. ostreatus* strains on moldy rice straw

The growth of oyster mushroom generally depends on the type of substrate used for its cultivation. Generally, rice straw supports higher yields of oyster mushrooms compared to wheat straw under identical cultivation conditions (Zhang et al. 2002). The presence of high initial moisture content (83 %) within the straw fibers together with utilizable lignin and cellulose would have stimulated abundant mycelial growth of *P. ostreatus* through the synthesis of lignocellulolytic enzymes like laccase and cellulase. An earlier study reported larger fruiting bodies in *P. eryngii* when cultivated on rice straw (Moonmoon et al. 2010).

Co-cultivation of *P. ostreatus* strains on rice straw already containing *A. flavus* culture revealed the ability of *P. ostreatus* strains to rapidly colonize the substrate. Following incubation, the tubes inoculated with *P. ostreatus* MTCC 142 showed moderate hyphal proliferation, whereas, the tubes inoculated with *P. ostreatus* GHBBF10 demonstrated profuse mycelial growth (Fig. 2). This basidiomycete strain might have synthesized higher titres of extracellular enzymes, which would have resulted in better utilization of the lignocellulosic substrate.

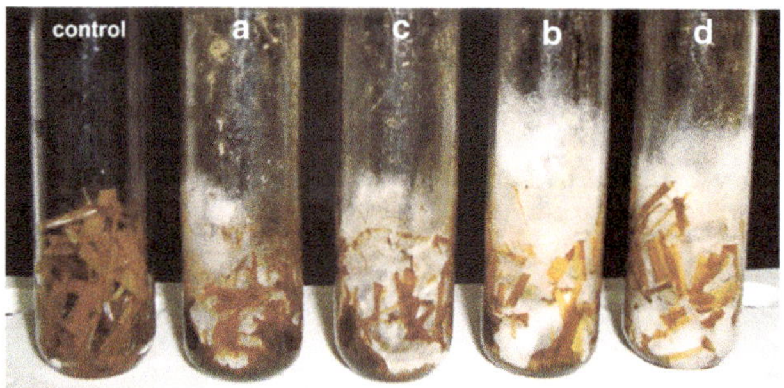

Fig. 2 Tubes containing moist rice straw with co-cultures of *A. flavus* and *P. ostreatus* strains. *a* Tube inoculated with *A. flavus* MTCC 2798 + *P. ostreatus* MTCC 142. *b* Tube inoculated with *A. flavus* MTCC 2798 + *P. ostreatus* GHBBF10. *c* Tube inoculated with *A. flavus* GHBF09 + *P. ostreatus* MTCC 142. *d* Tube inoculated with *A. flavus* GHBF09 + *P. ostreatus* GHBBF10. A sterile uninoculated tube (control) is shown on the *left*

Fig. 3 HPLC chromatograms showing AFB1 peaks **a** peak for AFB1 standard, **b** AFB1 peak after degradation by *P. ostreatus* MTCC 142, **c** AFB1 peak after degradation by *P. ostreatus* GHBBF10. AFB1 peaks have been indicated by *arrows*

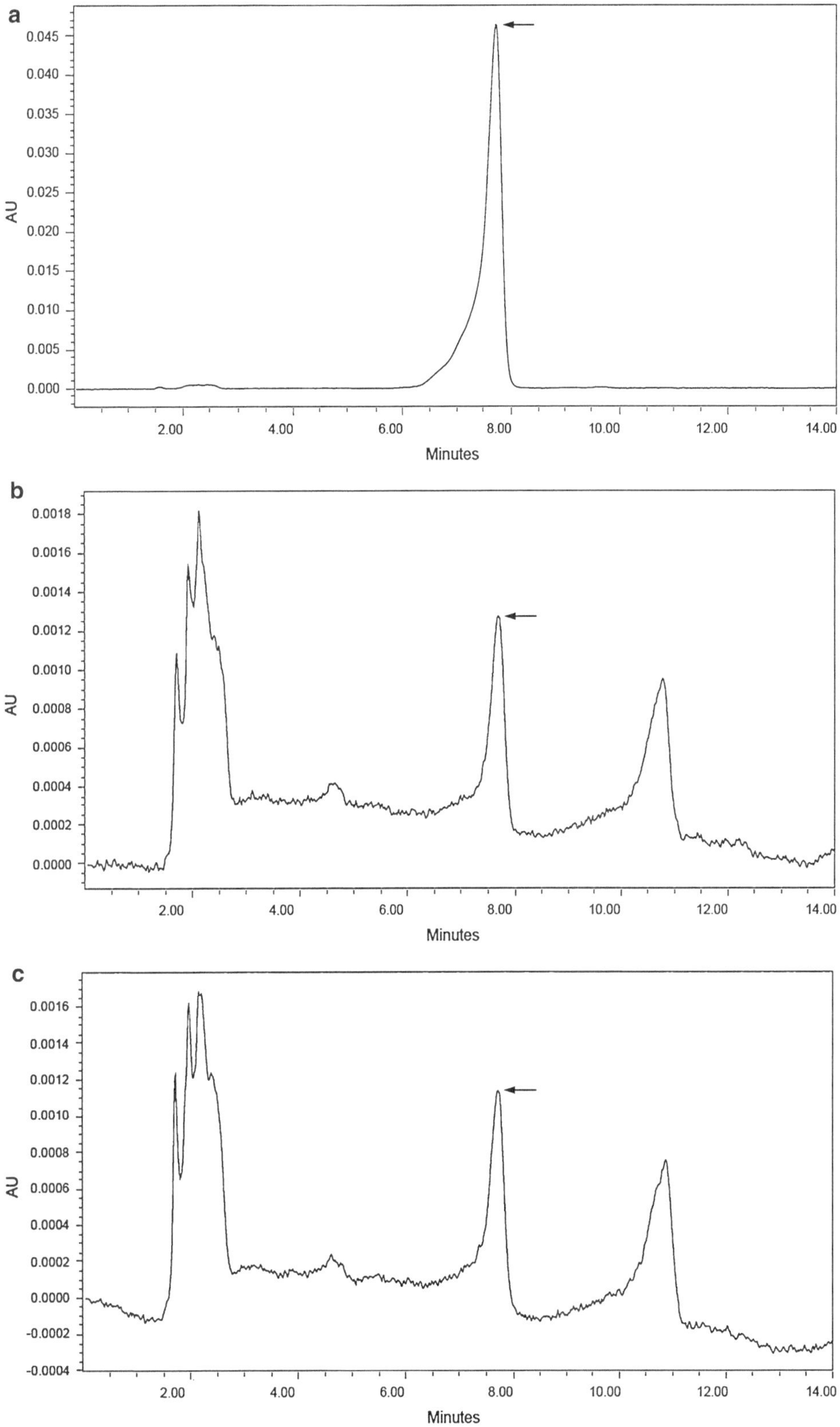

AFB1 degradation during co-cultivation of fungal strains

HPLC chromatogram revealed the peak for AFB1 standard at retention time of 7.722 min (Fig. 3a). The chromatograms for residual AFB1 after degradation by *P. ostreatus* MTCC 142 and *P. ostreatus* GHBBF10 have been illustrated in Fig. 3b and c, respectively. From the chromatograms it became evident that AFB1 was still present in the samples after co-cultivation, but at reduced concentrations. The presence of several peaks other than that for AFB1 may explain the possible formation of AFB1 adducts and various intermediate compounds during AFB1 degradation (Das et al. 2014). The highest degradation of AFB1 (89.41 %) was recorded in the tube containing co-cultures of *A. flavus* MTCC 2798 and *P. ostreatus* GHBBF10, with residual AFB1 content of 2.95 ± 0.03 μg/g (Fig. 4). The lowest degradation (81.73 %) was noted in the tube containing co-cultures of *A. flavus* GHBF09 and *P. ostreatus* MTCC 142, with residual AFB1 content of 3.88 ± 0.01 μg/g. Comparatively, higher percentages of AFB1 degradation were recorded in substrate inoculated with the natural isolate *P. ostreatus* GHBBF10. This could be attributed to its actively functional degradative enzymes' synthetic machinery owing to its natural isolation from decomposing tree trunk containing lignin. In a previous study on AFB1 degradation in contaminated rice straw by *P. ostreatus* strains, we reported that the natural isolate *P. ostreatus* GHBBF10 secreted higher titres of laccase (1.95 U/gds) and MnP (2.84 U/gds), compared to *P. ostreatus* MTCC 142 (Das et al. 2014). Degradation of recalcitrant compounds by *P. ostreatus* is mediated by secretion of degradative enzymes like laccase and MnP (Bhattacharya et al. 2014). Furthermore, analysis of AFB1 degradation by HPLC and liquid chromatography mass spectrometry revealed reduction in initial concentration of AFB1 and formation of decarbonylated and *O*-dealkylated degradation compounds, respectively (Das et al. 2014).

Due to profuse ramification of the basidiomycete mycelia, the straw containing *A. flavus* MTCC 2798 was effectively colonized and the synthesized AFB1, though present at highest concentration (27.95 ± 0.23 μg/g), was degraded to a greater extent. This observation may be further explained by the presence of yeast extract as nitrogen source in the supplemented mineral salt solution which probably enhanced the laccase production. Gomes et al. (2009) had reported that supplementation of nitrogen in rice straw resulted in 3–4 times increased production of laccase in lignocellulolytic fungi.

In a previous study, co-cultivation of *A. flavus* and *P. ostreatus* was carried out on various substrates. On wheat straw, corn cobs and millet, *A. flavus* produced aflatoxin after 3 weeks of cultivation. A subsequent cultivation of *P. ostreatus* on *A. flavus*-contaminated straw led to detoxification of the straw and corn cobs. It was found that *P. ostreatus* could liquidate colonies of *A. flavus*. However, cultivation of *P. ostreatus* in the presence of 40–100 μg of AFB1/g of substrate did not result in complete detoxification of the material even after 34 days of co-cultivation, but AFB1 concentration decreased to about one-fourth of the added amount (Ginterová et al. 1980).

The present findings are significant as biodegradation of AFB1 with microorganisms or their enzymes may be considered as the best strategy for detoxification of contaminated feedstuffs. This approach is considered as environment friendly in contrast to physico-chemical techniques of detoxification (Upadhaya et al. 2010).

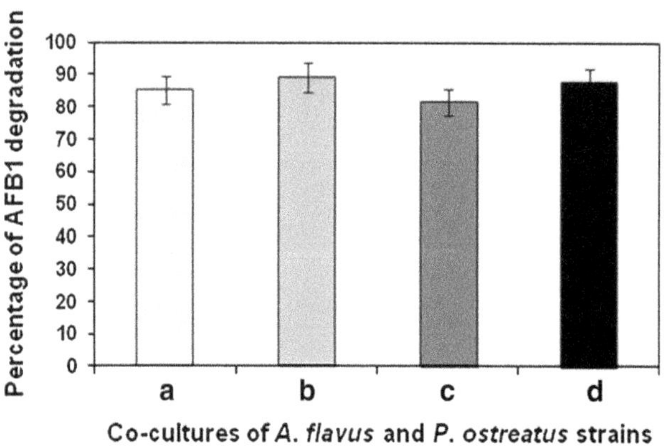

Fig. 4 Extent of AFB1 degradation during co-cultivation of fungal strains on rice straw. *a* Straw containing *A. flavus* MTCC 2798 + *P. ostreatus* MTCC 142, *b* straw containing *A. flavus* MTCC 2798 + *P. ostreatus* GHBBF10, *c* straw containing *A. flavus* GHBF09 + *P. ostreatus* MTCC 142, *d* straw containing *A. flavus* GHBF09 + *P. ostreatus* GHBBF10. Data represent mean ± SD ($n = 3$); $P < 0.05$

Conclusions

From this co-cultivation study, it might be deduced that both standard and naturally isolated strains of *P. ostreatus* could effectively degrade moderate concentrations of AFB1 in *A. flavus*-contaminated moldy rice straw. Further studies would be required to enhance the AFB1-degradation potential of the basidiomycete strains through process optimization.

Acknowledgments We wish to extend our sincere gratitude to the managements of Karpagam University, Jain University and Bharathiar University for their encouraging support. We express our heartfelt thanks to Mrs. K. Prashanthi, Jain University, Bangalore, for her assistance in HPLC analysis.

Conflict of interest The authors declare that they have no conflict of interest.

References

Abdulla HM (2007) Enhancement of rice straw composting by lignocellulolytic actinomycete strains. Int J Agric Biol 9:106–109

Alberts JF, Gelderblom WC, Botha A, van Zyl WH (2009) Degradation of aflatoxin B(1) by fungal laccase enzymes. Int J Food Microbiol 135:47–52

AOAC International (2000) Official methods of analysis of AOAC International, 17th edn. Association of Analytical Communities, Gaithersburg

Bennett JW, Klich M (2003) Mycotoxins. Clin Microbiol Rev 16:497–516

Bhattacharya S, Das A, Prashanthi K, Palaniswamy M, Angayarkanni J (2014) Mycoremediation of benzo[*a*]pyrene by *Pleurotus ostreatus* in the presence of heavy metals and mediators. 3 Biotech 4:205–211

Cohen R, Persky L, Hadar Y (2002) Biotechnological applications and potential of wood-degrading mushrooms of the genus *Pleurotus*. Appl Microbiol Biotechnol 58:582–594

da Luz JM, Nunes MD, Paes SA, Torres DP, de Cássia Soares da Silva M, Kasuya MC (2012) Lignocellulolytic enzyme production of *Pleurotus ostreatus* growth in agroindustrial wastes. Braz J Microbiol 43:1508–1515

Das A, Bhattacharya S, Palaniswamy M, Angayarkanni J (2013a) Molecular identification of *Aspergillus flavus* GHBF09 involved in aflatoxin B1 production in rice straw. Ann Biol Res 4:102–110

Das A, Bhattacharya S, Palaniswamy M, Angayarkanni J (2013b) Assessment of parameters influencing rice straw associated mycelia growth of *Pleurotus ostreatus* MTCC 142 and a wild isolate of *Pleurotus ostreatus*. Int Res J Biological Sci 2:15–21

Das A, Bhattacharya S, Palaniswamy M, Angayarkanni J (2014) Biodegradation of aflatoxin B1 in contaminated rice straw by *Pleurotus ostreatus* MTCC 142 and *Pleurotus ostreatus* GHBBF10 in the presence of metal salts and surfactants. World J Microbiol Biotechnol.

Ginterová A, Polster M, Janotková O (1980) The relationship between *Pleurotus ostreatus* and *Aspergillus flavus* and the production of aflatoxin. Folia Microbiol (Praha) 25:332–336

Gomes E, Aguiar AP, Carvalho CC, Bonfá MRB, da Silva R, Boscolo M (2009) Ligninases production by basidiomycetes strains on lignocellulosic agricultural residues and their application in the decolorization of synthetic dyes. Braz J Microbiol 40:31–39

Guan S, Ji C, Zhou T, Li J, Ma Q, Niu T (2008) Aflatoxin B_1 degradation by *Stenotrophomonas maltophilia* and other microbes selected using coumarin medium. Int J Mol Sci 9:1489–1503

Guan S, Zhao L, Ma Q, Zhou T, Wang N, Hu X, Ji C (2010) In vitro efficacy of *Myxococcus fulvus* ANSM068 to biotransform aflatoxin B_1. Int J Mol Sci 11:4063–4079

Khoshpey B, Farhud DD, Zaini F (2011) Aflatoxins in Iran: nature, hazards and carcinogenicity. Iran J Public Health 40:1–30

Moonmoon M, Uddin MN, Ahmed S, Shelly NJ, Khan MA (2010) Cultivation of different strains of king oyster mushroom (*Pleurotus eryngii*) on saw dust and rice straw in Bangladesh. Saudi J Biol Sci 17:341–345

Sánchez C (2010) Cultivation of *Pleurotus ostreatus* and other edible mushrooms. Appl Microbiol Biotechnol 85:1321–1337

Streit E, Schatzmayr G, Tassis P, Tzika E, Marin D, Taranu I, Tabuc C, Nicolau A, Aprodu I, Puel O, Oswald IP (2012) Current situation of mycotoxin contamination and co-occurrence in animal feed-focus on Europe. Toxins (Basel) 4:788–809

Upadhaya SD, Park MA, Ha JK (2010) Mycotoxins and their biotransformation in the rumen: a review. Asian Aust J Anim Sci 23:1250–1260

Zain ME (2011) Impact of mycotoxins on humans and animals. J Saudi Chem Soc 15:129–144

Zhang R, Li X, Fadel JG (2002) Oyster mushroom cultivation with rice and wheat straw. Bioresour Technol 82:277–284

15

Biochemical characterization and pharmacognostic evaluation of purified catechins in green tea (*Camellia sinensis*) cultivars of India

Jigisha Anand · Bhagwati Upadhyaya · Pramod Rawat · Nishant Rai

Abstract Green tea (GT) is derived from the leaves of *Camellia sinensis* implicated in a wide range of health attributes. In the present comprehensive study, methanolic, acetone and aqueous extract of leaves of *C. sinensis* var. *sinensis* [Kashmir (KW), Uttarakhand (IP & PN)] and *C. sinensis* var. *assamica* (Assam, AT) were explored for their phytoconstituents. Solvent extracts of GT cultivars showed rich presence of phytoconstituents in comparison with aqueous extracts. The methanolic extract of AT and acetone extract of KW showed highest total phenol content (18.32 ± 0.357 mg of GAE equivalent/g of sample) and total flavonoid content (29.25 ± 0.015 mg of catechin equivalent/g of sample), respectively. All the cultivars revealed higher free radical scavenging activity in the range of 73.80 ± 0.152 to 82.40 ± 0.004 % confirming antioxidant potentials. The HPLC analysis of purified residue procured from solvent partitioning depicted AT with highest concentration of epigallocatechin gallate (EGCg) i.e., 154.7 ± 4.949 mg/g followed by Kashmir and Uttarakhand GT cultivars. The present study revealed that Assam GT could be a potent herbal candidate with multiple nutraceutical applications. However, significant investigation of the cultivars is to be done to further explore the EGCg-dependent activity of GT for herbal drug development.

J. Anand · P. Rawat · N. Rai (✉)
Department of Biotechnology, Graphic Era University, Dehradun, India
e-mail: nishantrai1@gmail.com

B. Upadhyaya
Department of Microbiology, Guru Nanak Dev University, Amritsar, Punjab, India

Keywords Phytoconstituents · GT · *Camellia sinensis* · Flavonoid · HPLC

Abbreviations

EGCg	Epigallocatechin gallate
GT	Green tea
AT	Assam green tea
KW	Kashmir green tea
IP, PN	Uttarakhand green tea
GAE	Gallic acid equivalent
DPPH	1,1-Diphenyl-2-picrylhydrazyl

Introduction

In recent years, nutritional therapy and phytotherapy have emerged as new concepts of health aid. Plant-derived nutraceutical or functional foods have received considerable attention because of their presumed safety and potential nutritional and therapeutic effects. Nutraceuticals serve essentially as therapeutics on patients suffering with several debilitating diseases, and are useful as health giving food supplements for general population (Pandey et al. 2011).

Green tea in its purest and most unadulterated form has always influenced human health from generations, and day-by-day scientific evidences throughout the world are making people aware of health benefits associated with this herbal drink. It is obtained from dried leaves of *Camellia sinensis* (L.) Kuntze, which is an angiosperm dicot plant (Anand et al. 2012). Commercial tea cultivars are recognized under three different taxa, namely, *C. sinensis*, *C. assamica*, and *C. assamica* ssp. *lasiocalyx* (Baruah 1965). However, tea is highly heterogeneous (Gulati et al. 2009),

Fig. 1 Structures of different catechins present in green tea

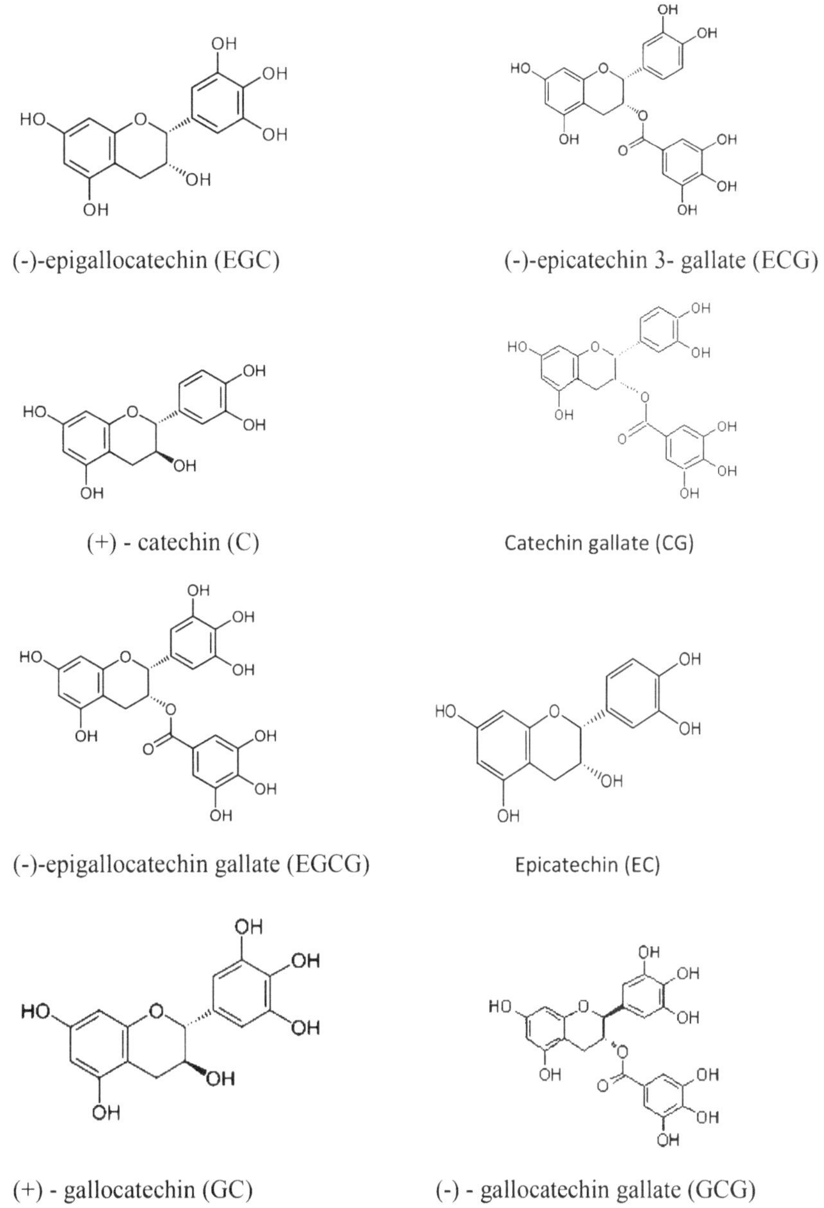

and all the above taxa freely inter-breed, resulting in a cline extending from extreme China types to those of Assam origin (Wight 1962).

There are two main varietals of *C. sinensis*, *C. sinensis* var. *sinensis*, better known as China bush, and *C. sinensis* var. *assamica*, also known as Assam bush. However, altogether, there are more than a thousand sub varieties of *C. sinensis*.

Phytochemicals are the bioactive compounds that occur naturally in plants. They include secondary metabolites, many of which are synthesized for plant defences and adaptation to environmental stress (Mcclanahan 2012). Tea is reported to contain nearly 4,000 bioactive compounds of which one-third is contributed by polyphenols (Tariq et al. 2010). The bioactive compounds present in GT leaves are alkaloids, flavonoids, steroids, phenols and terpenoids, which serve as valuable starting material for the medicine development (Lister and Wilson 2001). Among the natural polyphenols present actively in GT are the flavonoids commonly known as catechins. The flavonoids (and their fraction, catechins) are the basic phenolic compounds in green tea responsible for antioxidant activities such as neutralization of free radicals that are formed in the process of metabolism (Horzic et al. 2009). Some major catechins are (−)-epigallocatechin gallate (EGCg), (−)-epigallocatechin (EGC), (−)-epicatechin 3-gallate (ECG) and epicatechin (EC). Other minor catechins like catechin gallate (CG), (+)-gallocatechin (GC), (−)-gallocatechin gallate (GCG) and (+)-catechin (C) (Fig. 1.) are also present in tea (Yamamoto et al. 1997).

The EGCg, are the major catechin in GT which is believed to be the primary source of green tea's beneficial effects (Fujimura et al. 2002). EGCg have shown to be promising in promotion of cardio-vascular health, cancer prevention, and skin protection, to fight high cholesterol levels, infection, impaired immune function, diarrhoea, and fatigue. The most notable health benefit of GT is its powerful antioxidant potential which, at the molecular level, help prevent cellular damage from certain oxidation reactions in the body (Anand et al. 2012). The higher antioxidant activity of green tea makes it more beneficial in protecting the body from oxidative damage due to free radicals. It appeared that these antioxidants slow or halt the initiation of cancer, heart disease, suppress immune function, and accelerated aging (Hamilton-Miller 2001).

Although the catechins in tea leaves were identified long ago, the regulatory mechanism governing catechin biosynthesis remains unclear (Xiong et al. 2013). The genetic differences between the hybrids are well reflected in biochemical composition of leaves. However, biochemical composition, as varied between varieties, is yet to be fully utilized in tea taxonomy (Sanderson and Kanapathipillai 1964). It has been studied that the cultivars differ in their antimicrobial potentials, which is correlated with their inherent catechin composition. Therefore, we can say that there is a differential gene expression in GT cultivars with different morphology and catechin content (Yang et al. 2012).

The regional variation of quality within the tea growing region can be attributed to genetic diversity and its interaction with the environment (Sabhapondit et al. 2012). Environmental factors such as pedoclimatic (soil type, sun exposure, rainfall) or agronomic (culture in greenhouses or fields, biological culture, hydroponic culture, fruit yield per tree, etc.) have a crucial role in the composition of catechins (Manach et al. 2004). To ascertain diversity, careful study of secondary metabolites, especially those which are major contributors to quality, is essential. Total catechin content could be used to indicate the quality potential of tea, with high content being related to high quality (Obanda et al. 1997).

In our study, the finding suggests vast regional variation among the catechin composition of purified residues of leaves of GT cultivars collected from different regions of Indian subcontinent. Also, much variation on the quantitative analysis of different solvent extracts has been observed.

Materials and methods

Collection of the samples

Green tea leaves from Assam (AT), Kashmir (KW) and Uttarakhand region (IP and PN) of North India were collected and identified by Dr. S. K. Srivastava, Scientist 'D' from Botanical Survey of India, Dehradun. The collected leaves were washed thoroughly under fresh water and left for drying under shed for 2 weeks. The dried leaves were then powdered in a blender and kept in sealed packets in refrigerator at 4 °C until further use. The methodology followed in the present study is shown as flow diagram (Fig. 2).

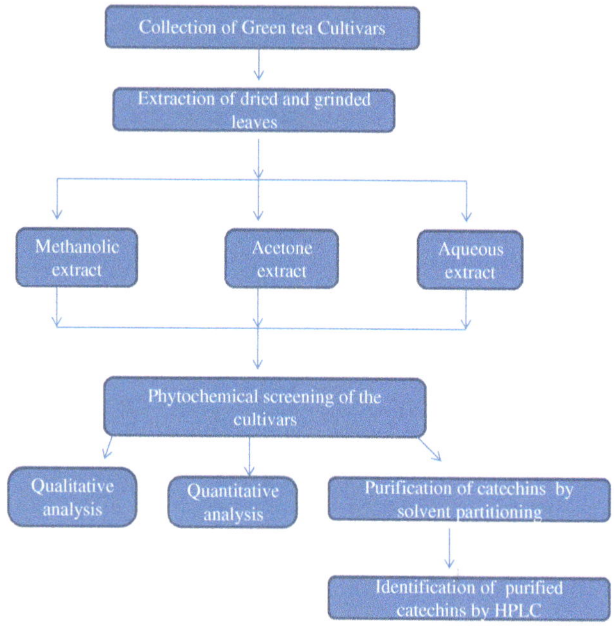

Fig. 2 Flow diagram showing phytochemical and pharmacognostic study of GT cultivars

Preparation of tea extract

For the preparation of solvent extracts, dried powdered leaves were separately soaked in 70 % methanol and 70 % acetone in ratio plant: solvent (1:10) and allowed to macerate for consecutive 2 days. After maceration, the extract was filtered using Whatman No. 1 paper and the solvents were completely evaporated at appropriate temperature till it gave a viscous mass. The crude extract was weighed and its percentage yield was recorded (Table 2). The crude extract was stored at 4 °C until further use (Archana and Abraham 2011).

Aqueous extraction of the leaves samples was done by soaking the leaves in sterilized distilled water at approximately 1:10 w/v ratio for 1 h. After soaking, the decoction was brought to boil for 60 min. The liquid extract was separated from the solids by filtration with double-layered muslin cloth and allowed to precipitate. The crude extract was weighed and its percentage yield was recorded (Table 1). The crude extract was stored at 4 °C until further use (Koh et al. 2009).

Table 1 Qualitative estimation of phytochemicals in aqueous and solvent extracts of different green tea cultivars

Phytochemicals	Methanolic extract				Acetone extract				Aqueous extract			
	AT	KW	IP	PN	AT	KW	IP	PN	AT	KW	IP	PN
Phytosterol	++	++	++	++	+++	+++	++	++	+	+	+	+
Carbohydrates	+++	+++	+++	+++	++	+++	+++	++	+	+	–	–
Saponin	–	–	–	–	–	–	–	–	–	–	–	–
Catecholic Tannins	+++	+++	+++	+++	+++	+++	+++	++	–	–	–	–
Alkaloids	+	+	+	+	+	+	+	+	+	+	+	+
Flavonoids	+++	++	++	++	+++	+++	+++	+	+	+	+	+
Anthocyanins	–	–	–	–	–	–	–	–	–	–	–	–
Proteins	+	+	+	+	+++	+++	+	+	+	+	+	+
Terpenoids	–	–	–	–	+++	++	++	–	+++	++	+++	+++
Glycosides	++	++	++	++	+++	+++	+++	+++	+	+	+	+

AT Assam green tea, *KW* Kashmir green tea, *IP* and *PN* Uttarakhand green tea

+++>++>+ indicates the intensity of color formation (difference in specific component conc. and – indicates no color (absence of specific component)

Qualitative phytochemical screening

For the qualitative phytochemical screening of all four GT cultivars, standard procedures were followed to trace out the presence of the active principles, i.e., flavonoids, phenols, alkaloids, terpenoids, glycosides, tannins, anthocyanin, saponins, carbohydrates, and amino acid (Tariq and Reyaz 2012).

Quantitative phytochemical screening

Determination of total phenolic content

Determination of total phenolic content was carried out using Folin–Ciocalteu reagent (FCR) assay by taking 20 μl of stock solution (1 mg/ml) of the extract dissolved in 80 μl of water and 500 μl of Folin–Ciocalteu reagent. The solutions were mixed and incubated in dark at room temperature for 5 min. After 5 min, 400 μl of 7.5 % sodium carbonate (Na_2CO_3) solution was added and the mixture was further incubated in dark for 30 min at room temperature. The absorbance of all samples was measured at 765 nm using colorimeter. Gallic acid was used as standard for the calibration curve that is plotted at 0.0156, 00312, 0.0625, 0.125, 0.25, 0.5, and 1.0 mg/ml of concentration, respectively. Triplicate measurements were carried out and total phenolic content was expressed as milligram of gallic acid equivalents (GAE) per gram of samples (Khan et al. 2011).

Determination of total flavonoid content

The flavonoid content was estimated by taking 100 μl of the sample in a test tube containing 400 μl of distilled water and subsequently 30 μl of 5 % sodium nitrite solution was added. After 5 min, 30 μl of 10 % aluminium chloride was added and allowed to stand for 5 min, then 20 μl of 4 % sodium hydroxide was added and the volume was adjusted up to 1 ml with distilled water. The absorbance of the mixture at 510 nm was measured immediately. Hydrated catechins were used as standard for the calibration curve that is plotted at 0.0156, 00312, 0.0625, 0.125, 0.25, 0.5, and 1.0 mg/ml of concentration, respectively. Triplicate measurements were carried out and total phenolic content was expressed as milligram of catechin equivalents per gram of samples (Subhashini et al. 2010).

Determination of the free-radical scavenging activity by the 1,1-Diphenyl-2-picrylhydrazyl (DPPH) free-radical scavenging assay

DPPH assay stock solution was prepared by dissolving 24 mg DPPH with 100 ml methanol and then stored at −20 °C until needed. The working solution was obtained by mixing 10 ml stock solution with 45 ml methanol to obtain an absorbance of 1.10 ± 0.02 units at 515 nm using spectrophotometer. 100 μl of leaf extract solution was allowed to react with 1,900 μl of the DPPH solution for 2 h in the dark. Then the absorbance was measured at 515 nm. Ascorbic acid was used as standard for the preparation of calibration curve that is plotted at 0.0156, 00312, 0.0625, 0.125, 0.25, 0.5, and 1.0 mg/ml of concentration respectively. Triplicate measurements were carried out and the percentage scavenging effect was calculated as:scavenging rate (%) = $[(A_0 - A_1)/A_0] \times 100$, where A_0 was the absorbance of the control (without extract) and A_1 was the absorbance in the presence of the extract (Azzahra et al. 2012).

Purification of GT polyphenols

The GT extracts namely AT, IP, KW, and PN were initially partitioned with water/chloroform (1:1). Then the water phase was collected and the impurities associated with the chloroform phase were discarded. As a second partition, water/ethyl acetate (1:1) was used. Polyphenol compounds such as catechins, epigallocatechin, epicatechin gallate, and epicatechin expected to have moved into the ethyl acetate layer were collected and concentrated by rota vapour for their analysis (Kyung and Yinzhe 2006).

Analysis of GT phytoconstituents by HPLC

Reagents and standard preparation HPLC grade water and acetonitrile were used for the analysis. The standard compound (−)-epigallocatechin gallate (EGCg) was obtained from Sigma Aldrich (Catalogue No. RM10179). The standard with varying concentrations (12, 10, 8, 4, and 2 mg/ml) in acetonitrile was used for the plotting of standard calibration curve.

Sample preparation The ethyl acetate residue was used to detect the EGCg concentration among the *C. sinensis* cultivars. The purified GT residue of different cultivars (AT, KW, PN and IP) was dissolved in acetonitrile for the HPLC analysis. The analytical determination of GT phytochemicals was performed using Dionex HPLC using varian, Microsorb-MV 100-5 C18 (250 × 4.6 mm) reverse phase column fitted in thermostatic column Compartment TCC-100 oven. The detector used for the analysis was UVD340U detector. Millipore syringe filter (0.2 μm) was used for the filtration of purified samples and the standard. Mobile phase contained acetonitrile:water in the ratio of 70:30. Ambient temperature was maintained at 28 °C; while the flow rate of the sample was adjusted to 0.5 ml/min. Detection wavelength was set at 280 nm. As a standard, solution of EGCg in acetonitrile was used. Volume of standard and a purified sample to be injected was 5 μl with a run time of 12 min (Yuegang et al. 2000). Identification of the specific polyphenol was carried out on the basis of their retention time. Quantification of polyphenol was directly performed by HPLC UVD340U detector using regression curve obtained by plotting the absorbance versus concentration of series of dilutions of standard EGCg.

Results

The qualitative estimation of phytoconstituents in GT cultivars indicated the presence of pytosterols, catechin tannins, alkaloids, flavonoids, terpenoids, glycosides, carbohydrates and amino acids while the analysis revealed the absence of anthocyanin and saponins (Table 1). Based on the color consistency, the methanolic and acetone extracts have comparatively shown higher presence of the glycosides, catecholic tannins, and flavonoids while the presence of terpenoids was estimated higher in acetone and aqueous extracts.

Overall, the acetone extract of *Camellia sinensis* var. *sinsensis* and *assamica* cultivars had showed higher qualitative and quantitative presence of phytoconstituents than methanolic and aqueous extract. The percentage yields for all prepared extracts from GT had been evaluated that ranges from 3.07 to 26.8 % (Table 2) with the maximum yield depicted in acetone extract of Assam GT.

Antioxidant activity, total phenol content and total flavonoid content were analyzed and their results were expressed as mean ± standard deviation. The flavonoids content in different solvent extracts of GT cultivars has been reported as catechin equivalents by reference to standard curve. It was estimated highest in acetone extract of GT cultivars with a concentration of 29.25 ± 0.015 mg of catechin equivalent/g of sample in Kashmir GT followed by aqueous extract with a concentration of 23.6 ± 0.017 mg of catechin equivalent/g of sample in PN GT extract ($P \geq 0.05$) (Fig. 3a). The least flavonoids content 8.5 ± 0.003 mg of catechin equivalent/g of sample was calculated in methanolic extract of Kashmir GT (Table 3; Figs. 3a, 4a).

Table 2 The nature and color of crude extracts of green tea cultivars and their yield

Type of extract	Solvent	Green tea cultivars	Nature and color of extract	% yield
Crude	Methanolic	AT	Powdered, dark brown	4.00
		KW	Coal tar form, black	7.33
		PN	Lustrous thick, black	12
		IP	Lustrous thick, dark green	14.6
Crude	Acetone	AT	Thick, viscous, dark brown	26.8
		KW	Particulate, dark green	5.06
		PN	Particulate, dark green	9.00
		IP	Oily particulate, dark green	3.83
Crude extract	Aqueous	AT	Liquid, dark brown	3.07
		KW	Liquid, dark brown	7.33
		PN	Liquid, dark brown	12.00
		IP	Liquid, dark brown	14.6

Percentage yield = Weight of the sample extract obtained (g) × 100/Weight of the powdered sampled used (g)

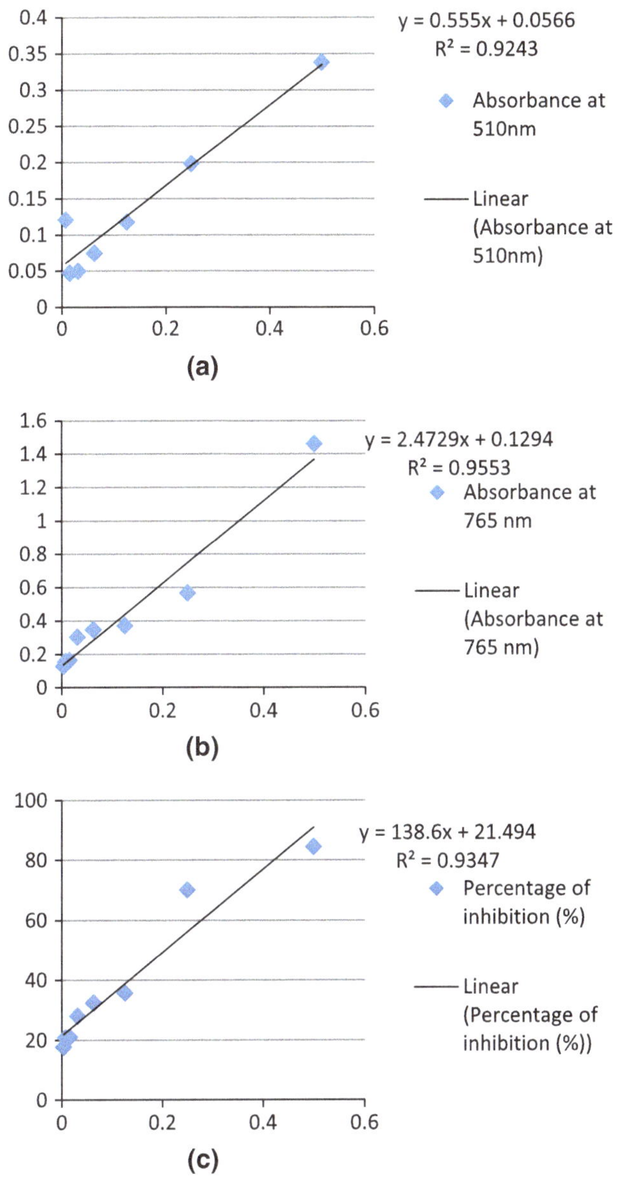

Fig. 3 **a** Standard curve of catechin hydrate as a reference for total flavonoid content. **b** Standard curve of gallic acid as a reference for total phenolic content. **c** Standard curve for DPPH free radical scavenging activity

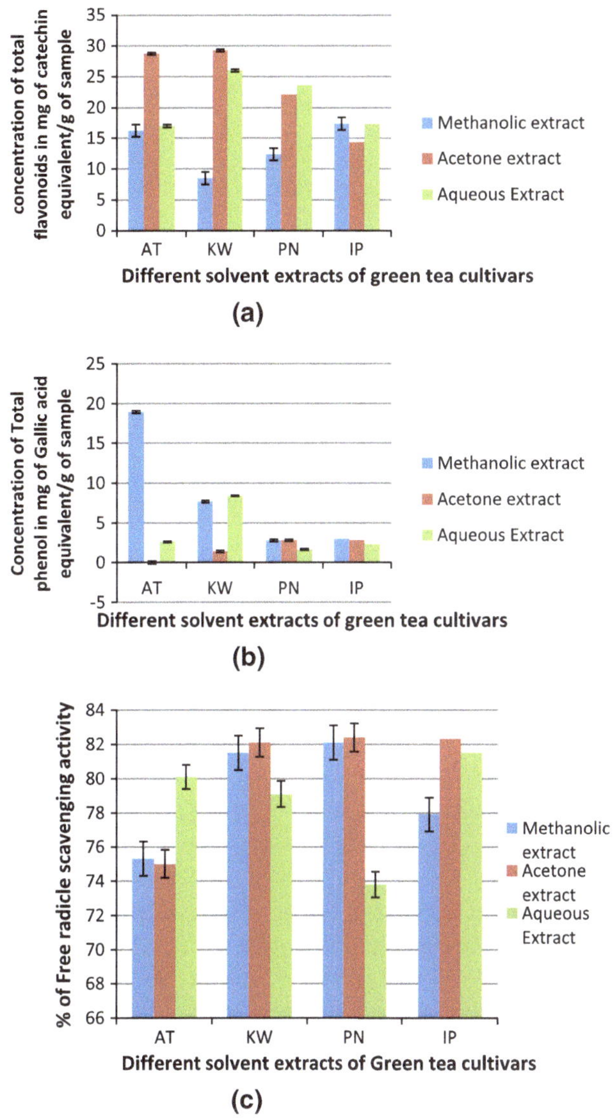

Fig. 4 **a** Comparative analysis total flavonoids in mg of catechin equivalent/g of sample among different extracts of green tea cultivars. **b** Comparative analysis total phenolic content in mg of Gallic equivalent/g of sample among different extracts of green tea cultivars. **c** Comparative analysis of DPPH free radical scavenging activity among different extracts of green tea cultivars

Significant difference in total phenolic content was observed among the investigated GT cultivars ($P \geq 0.05$) (Fig. 3b). Methanolic extract of AT sample with a concentration of 18.32 ± 0.357 mg of gallic acid equivalent/g of sample depicted the highest concentration with reference to standard curve (Table 4; Figs. 3b, 4b).

Table 3 Quantitative estimation of total flavonoids in aqueous and solvent extracts of different green tea cultivars

Extracts	Total flavonoids in green tea cultivars (mg of catechin equivalent/g of sample)			
	AT	KW	IP	PN
Methanolic	16.25 ± 0.030	8.5 ± 0.003	17.4 ± 0.003	12.35 ± 0.013
Acetone	28.75 ± 0.010	29.25 ± 0.015	14.35 ± 0.019	22.15 ± 0.005
Aqueous	17.05 ± 0.007	26.0 ± 0.077	17.3 ± 0.0007	23.6 ± 0.017

AT Assam green tea, *KW* Kashmir green tea, *IP* and *PN* Uttarakhand green tea

Table 4 Quantitative estimation of total phenols in aqueous and solvent extracts of different green tea cultivars

Extracts	Total phenols in green tea cultivars (mg of gallic acid equivalent/g of sample)			
	AT	KW	IP	PN
Methanolic	18.32 ± 0.357	7.68 ± 0.165	3.02 ± 0.007	2.81 ± 0.142
Acetone	0.79 ± 0.020	1.4 ± 0.061	2.86 ± 0.074	2.84 ± 0.028
Aqueous	2.62 ± 0.101	8.41 ± 0.208	1.68 ± 0.0499	2.34 ± 0.038

AT Assam green tea, *KW* Kashmir green tea, *IP* and *PN* Uttarakhand green tea

Table 5 Quantitative estimation of free radical scavenging activity by DPPH assay in aqueous and solvent extracts of different green tea cultivars

Extracts	Free radical scavenging activity in Green tea cultivars (% inhibition of DDPH radical)			
	AT	KW	IP	PN
Methanolic	75.30 ± 0.011	81.50 ± 0.002	77.90 ± 0.002	82.10 ± 0.016
Acetone	75.00 ± 0.053	82.10 ± 0.001	82.30 ± 0.0005	82.40 ± 0.004
Aqueous	80.10 ± 0.003	79.10 ± 0.003	81.50 ± 0.005	73.80 ± 0.152

AT Assam green tea, *KW* Kashmir green tea, *IP* and *PN* Uttarakhand green tea

There was significant difference in the antioxidant potentials of different extracts of GT cultivars ($P \geq 0.05$) which represents the variation in percentage of oxidant scavenging capacity as performed by DPPH free radical scavenging assay (Table 5; Figs. 3c, 4c). It was found to be significantly greater in acetone extract of PN, KW and IP GT sample i.e., between 82.4 ± 0.004 and 82.3 ± 0.005 %. The least free radical scavenging activated was estimated to be 73.80 ± 0.152 % in aqueous extract of PN sample.

The HPLC calibration curve revealed the presence of EGCg based on the retention time of identified peaks compared with standard (4.457 min) (Fig. 5a, b). The concentration of EGCg in purified residues of GT cultivars was estimated in the range of 0.65 ± 1.76 mg/g of EGCg to 13.78 ± 3.535 mg/g of EGCg with approximately 71 % percentage of EGCg in the purified residue of AT. The regression line expressed as correlation coefficient was linear ($r^2 = 0.9897$) built with standard EGCg (Figs. 6, 7).

Bases on Student's *t* test, ($P \geq 0.05$), a significant difference in the concentration of epigallocatechin gallate was depicted among Assam, Kashmir and Uttarakhand GT. However, there was insignificant difference in EGCg concentration between the GT cultivarscollected from Dehradun district.

Discussion

Previously, it has been found that GT possesses tannins, steroids, flavonoids, and alkaloids that have medicinal value as well as food value (Tariq and Reyaz 2012). The multipotency of GT is due to the presence of an active component namely catechin, a flavonoid that serves in antimicrobial defence mechanism, high antioxidant property, prevention against molecular damage in cancer, protection against cardiovascular damage, diabetes, etc. (Anand et al. 2012). There is a positive and highly significant relationship between the phenolics and flavonoids with antioxidant activity of GT (Hajimahmoodi et al. 2008). A plethora of evidence suggests strong antioxidant potentials of tea catechins in suppressing the production of excess free radicals which at the molecular level help prevent cellular damage from certain oxidation reactions in the body (Sabhapondit et al. 2012). The total phenolic, flavonoids, and antioxidants activity are the parameters depicting the quality of tea according to their biological properties (Azzahra et al. 2012). Total catechin content could be used to indicate the quality potential of tea, with high content being related to high quality (Obanda et al. 1997). To ascertain diversity, careful study of secondary metabolites, especially those which are major contributors to quality, is essential.

Solvents, such as methanol, ethanol, acetone, ethyl acetate, and their combinations have been used for the extraction of phenolics from plant materials, often with different proportions of water. Selecting the right solvent affects the amount and rate of polyphenols extracted (Xu and Chang 2007). In particular, methanol has been generally found to be more efficient in extraction of lower molecular weight polyphenols while the higher molecular weight flavanols are better extracted with aqueous acetone (Metivier et al. 1980; Labarbe et al. 1999; Prior et al. 2001; Guyot et al. 2001).

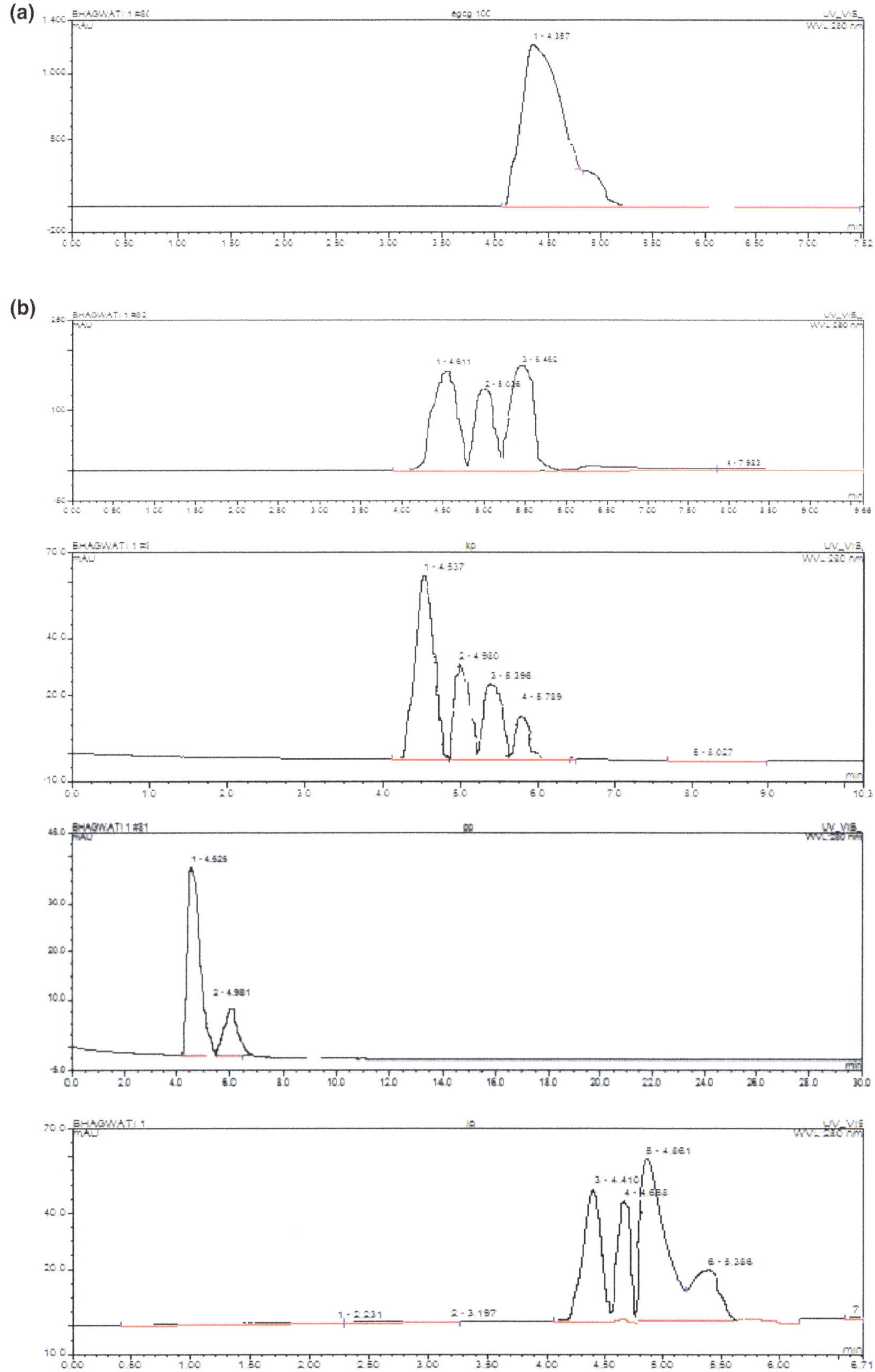

Fig. 5 **a** HPLC chromatogram of standard EGCg. **b** HPLC chromatogram of purified residues of AT, KW, PN and IP extract

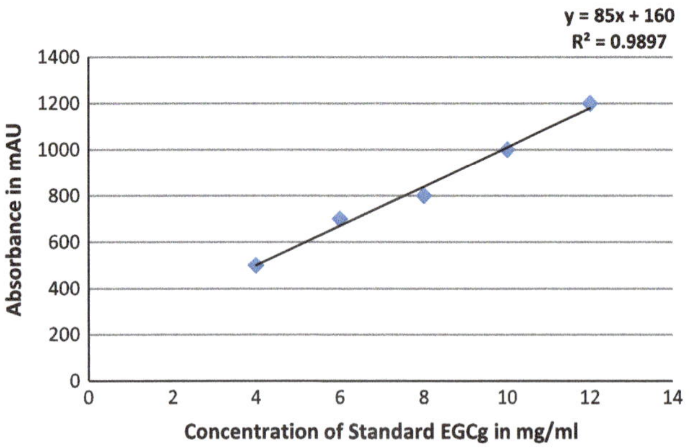

Fig. 6 Standard calibration curve of Epigallocatechin gallate (EGCg)

As per the preliminary observation, the GT cultivars showed the presence of phytoconstituents with varying percentage yield. The percentage yields of methanolic, acetone and aqueous extracts are different from one plant to another. These differences might be explained due to the differences in the nature of the secondary plant metabolism and their solubility in different solvents (Al-Younis and Abdulla 2009).

Kashmir (KW) and Assam GT (AT) cultivars have shown maximum flavonoids and phenolic contents in their respective acetonic and methanolic extracts where as higher free radical scavenging activity was determined among all the extracted cultivars under investigation. The present study suggests that the free radical scavenging activity of the cultivars may be due to the presence of antioxidants, i.e., phenolic and flavonoids compounds as detected by the phytochemical estimation. In our initial study, separation and characterization of the catechins among the purified GT extracts by HPLC suggests the remarkable difference in the concentration of varying catechins that can be attributed to the differences in the geographical distribution of the cultivars.

Conclusion

The phytochemical analysis of GT cultivars from Assam, Kashmir, and Uttarakhand region revealed the presence of terpenoids, tannins, alkaloids, phytosterols, glycosides, flavonoids, amino acid, and carbohydrates. The present comprehensive study demonstrates a marked influence of geographical location and the prevalent environmental specification (both agronomic and pedoclimatic) on the phytochemical characteristics of the GT cultivars. Owing to the significant interaction of the environment and catechin biosynthesis, the regional quality of GT cultivars can be enhanced by improvising the present expressions of the target genes characterized for catechin synthesis and hence will represent the regional cultivars as a marked herbal product with high nutraceutical application. Thus, further work is required to investigate these parameters to evaluate the diversity among the tea cultivars. Also, screening and characterization of the purified phytoconstituents by Prep-HPLC is essential to study the therapeutic potentials of GT and to confer it as a potent candidate for herbal drug development.

Fig. 7 Concentration and absorbance (280 nm) of EGCG among GT cultivars

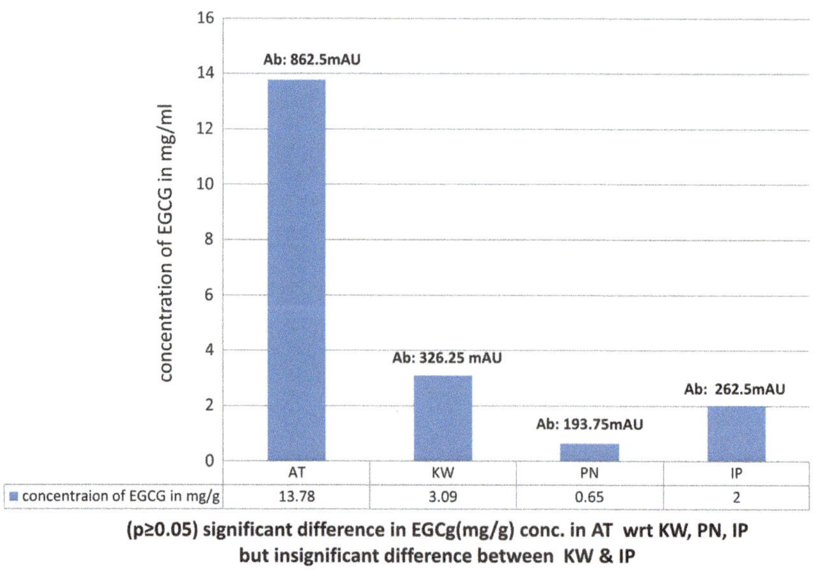

Acknowledgments We acknowledge Department of Biotechnology, Graphic Era University, Dehradun, for providing all laboratory facilities.

Conflict of interest The authors declare that they have no competing interest.

References

Anand J, Rai N, Kumar N, Gautam P (2012) Green tea: a magical herb with miraculous outcomes. Int Res J Pharm 3:139–147

Archana S, Abraham J (2011) Comparative analysis of antimicrobial activity of leaf extracts from fresh green tea, commercial green tea and black tea on pathogens. J App Pharm Sci 1:149–152

Azzahra LF, Fouzia H, Mohammed L, Noureddine B (2012) Antioxidant response of *Camellia sinensis* and *Rosmarinus officinalis* aqueous extracts toward H_2O_2 stressed mice. J App Pharm Sci 2:70–76

Baruah PK (1965) Classification of tea plants: species hybrids. Two Bud 13:14–16

Fujimura Y, Tachibana H, Maeda-Yamamoto M, Miyase T, Sano M, Yamada K (2002) Antiallergic tea catechin, (−)-epigallocatechin-3-O-(3-O-methyl)-gallate, suppresses Fc-epsilonRI expression in human basophilic KU812 cells. J Agric Food Chem 50:5729–5734

Gulati A, Rajkumar S, Karthigeyan S, Sud RK, Vijayan D, Thomas J, Rajkumar R, Das SC, Tamuly P, Hazarika M, Ahuja PS (2009) Catechin and catechin fractions as biochemical markers to study the diversity of Indian tea (*Camellia sinensis* (L.) O. Kuntze) germplasm. Chem Biodivers 6:1042–1052

Guyot S, Marnet N, Drilleau J (2001) Thiolysis-HPLC characterization of apple procyanidins covering a large range of polymerization states. J Agric Food Chem 49:14–20

Hajimahmoodi M, Hanifeh M, Sadeghi MR, Sadeghi N, Jannat B (2008) Determination of total antioxidant capacity of green teas by the ferric reducing antioxidant power assay. Iran J Environ Health Sci Eng 5:167–172

Hamilton-Miller JMT (2001) Anticarcinogenic properties of tea (*Camellia sinensis*). J Med Microbiol 50:299–302

Horzic D, Komes D, Belscak A, Ganic KK, Ivekovic D, Karlovic D (2009) The composition of polyphenols and methylxantine in teas and herbal infusions. Food Chem 115:441–448

Khan FA, Hussain I, Farroq S, Ahmad M, Arif M, Rehman I (2011) Phytochemical screening of some Pakistanian medicinal plants. Middle East J Sci Res 8:575–578

Koh GY, Chou GX, Liu Z (2009) Purification of a water extract of Chinese sweet tea plant (*Rubus suavissimus* S. Lee) by alcohol precipitation. J Agri Food Chem 57:5000–5006

Kyung HR, Yinzhe J (2006) Recovery of catechin compounds from Korean tea by solvent extraction. Short communication. Bioresour Technol 97:790–793

Labarbe B, Cheynier V, Brossaud F, Souquet JM, Moutounet M (1999) Quantitative fractionation of grape proanthocyanidins according to their degree of polymerization. J Agric Food Chem 47:2719–2723

Lister E, Wilson P (2001) Measurement of total phenolic and ABTS assay for antioxidant activity (personal communication). Crop Research Institute, Lincoln

Manach C, Scalbert, Morand AC, Rémésy C, Jiménez L (2004) Polyphenols: food sources and bioavailability. Am J Clin Nutr 79:727–747

Mcclanahan C (2012) Phytochemicals. Sigma Aldrich 7:4–12

Metivier RP, Francis FJ, Clydesdale FM (1980) Solvent extraction of anthocyanins from wine pomace. J Food Sci 45:1099–1100

Al-Younis NK, Abdullah AF (2009) Isolation and antibacterial evaluation of plant extracts from some medicinal plants in Kurdistan region. J Duhok Univ 12:250–255

Obanda M, Owuor PO, Taylor SJ (1997) Flavanol composition and caffeine content of green leaf as quality potential indicators of Kenyan black teas. J Sci Food Agri 74:209–215

Pandey N, Meena RM, Rai SK, Rai SP (2011) Medicinal plants derived nutraceuticals: a re-emerging health aid. Int J Pharm Biosci 2:419–441

Prior RL, Lazarus SA, Cao G, Muccitelli H, Hammerstone JF (2001) Identification of procyanidins and anthocyanins in blueberries and cranberries (*Vaccinium* spp.) using high-performance liquid chromatography/mass spectrometry. J Agric Food Chem 49:1270–1276

Sabhapondit S, Karak T, Bhuyan LP, Goswami BC, Hazarika M (2012) Diversity of catechin in northeast Indian tea cultivars. Sci World J 2012:485193

Sanderson GW, Kanapathipillai (1964) Further study on the effect of climate and clone on the chemical composition of fresh tea flush. Tea Q 35:222–229

Subhashini R, Mahadeva R, Sumathi P, Gunalan G (2010) A comparative phytochemical analysis of cocoa and green tea. Ind J Sci Tech 3:188–192

Tariq AL, Reyaz AL (2012) Phytochemical analysis of *Camellia sinensis* leaves. Int J Drug Dev Res 4:311–316

Tariq M, Naveed A, Barkat Ali K (2010) The morphology, characteristics, and medicinal properties of *Camellia sinensis*' tea. J Med Plants Res 4(19):2028–2033

Wight W (1962) Tea classification revised. Curr Sci 31:298–299

Xiong L, Li J, Li Y, Yuan L, Liu S, Huang J, Liu Z (2013) Dynamic changes in catechin levels and catechin biosynthesis-related gene expression in albino tea plants (*Camellia sinensis*). Plant Physiol Biochem 71:132–143

Xu BJ, Chang SK (2007) A comparative study on phenolic profiles and antioxidant activities of legumes as affected by extraction solvents. J Food Sci 72:159–166

Yamamoto T, Juneja LR, Chu DC, Kim M (1997) Chemistry and applications of green tea. CRC Press, Boca Raton, pp 109–121

Yang D, Liu Y, Sun M, Zhao L, Wang Y, Chen X, Wei C, Gao L, Xia T (2012) Differential gene expression in tea (*Camellia sinensis* L.) calli with different morphologies and catechin contents. J Plant Physiol 169:163–175

Yuegang Z, Hao C, Yiwei D (2000) Simultaneous determination of catechins, caffeine and gallic acids in green, oolong, black and pu-erh teas using HPLC with a photodiode array detector. Talanata 57:307–316

16

Statistical optimization of process parameters for the production of tannase by *Aspergillus flavus* under submerged fermentation

S. K. Mohan · T. Viruthagiri · C. Arunkumar

Abstract Production of tannase by *Aspergillus flavus* (MTCC 3783) using tamarind seed powder as substrate was studied in submerged fermentation. Plackett–Burman design was applied for the screening of 12 medium nutrients. From the results, the significant nutrients were identified as tannic acid, magnesium sulfate, ferrous sulfate and ammonium sulfate. Further the optimization of process parameters was carried out using response surface methodology (RSM). RSM has been applied for designing of experiments to evaluate the interactive effects through a full 31 factorial design. The optimum conditions were tannic acid concentration, 3.22 %; fermentation period, 96 h; temperature, 35.1 °C; and pH 5.4. Higher value of the regression coefficient ($R^2 = 0.9638$) indicates excellent evaluation of experimental data by second-order polynomial regression model. The RSM revealed that a maximum tannase production of 139.3 U/ml was obtained at the optimum conditions.

Keywords Tannase · Response surface methodology · Plackett–Burman design · Submerged fermentation

Introduction

Tannin acyl hydrolase (EC 3.1.1.20), commonly known as tannase, catalyses the hydrolysis of ester and depside bonds in hydrolysable tannins such as tannic acid, resulting glucose and gallic acid (Lekha and Lonsane 1997). Tannase is extensively used in the preparation of instant tea, beer, wine, coffee flavored soft drinks, etc. Tannase is also used in the production of gallic acid, a substrate for chemical synthesis of propyl gallate and trimethoprim which have application in food and pharmaceutical industries (Das Mohapatra et al. 2006; Lekha and Lonsane 1994). Many fungi, such as *Aspergillii, Penicillii, Fusaria,* and *Trichoderma* (Mohan et al. 2012; Batra and Saxena 2005; Murugan et al. 2007; Paranthaman et al. 2009), yeast like *Candida* sp., and *Saccharomyces cerevisiae* (Mondal et al. 2001a) have been reported as tannase producers. Few bacteria like *Bacilli*, *Corynebacterium* sp., *Lactobacillus* sp., and *Serratia* sp. are known to produce tannase (Rodríguez et al. 2008; Milva et al. 2010).

In industrial level tannase is mainly produced by *Aspergillus* species under submerged fermentation (SmF). The SmF is widely used for enzyme production because it offers many advantages like uniform process conditions, viz. concentration, temperature, pH, aeration and agitation in the bioreactors (Lekha and Lonsane 1997). In addition, the utilization of agro-industrial wastes, on one hand, provides alternative substrates and, on the other hand, helps to solve pollution problems by eliminating the need for disposal of the wastes. The nature of the substrate employed is the most important factor affecting fermentation processes, and its selection depends upon several factors mainly related to cost and availability and, thus, may necessitate the screening of several agro-industrial residues (Couto and Sanroman 2006).

The use of a sequential experimental design strategy is a useful tool for process optimization. Response surface methodology (RSM) provides important information regarding the optimum level of each variable along with its interactions with other variables and their effects on product yield. It reduces the number of experiments without neglecting the interactions among the parameters.

S. K. Mohan (✉) · T. Viruthagiri · C. Arunkumar
Department of Chemical Engineering, Annamalai University, Annamalainagar 608002, Tamilnadu, India
e-mail: skmohan75@gmail.com

This multivariate approach also improves statistical interpretation possibilities and evaluates the relative significance of several contributing factors even in the presence of complex interactions (Dilipkumar et al. 2011). RSM is widely used for multivariable optimization studies in several biotechnological processes such as the optimization of media, process conditions, hydrogen production, fermentation, sorption of dyes, etc. (Mannan et al. 2007; Pan et al. 2008). In the present work, optimization of tannase production by *Aspergillus flavus* sp. using tamarind seed powder as substrate was carried out using a sequential strategy of the experimental design. So far no work is available on the production of tannase using tamarind seed as substrate. Tamarind seed powder is a cheap substrate which is abundantly available in southern part of India. Hence, it was selected as substrate for the production of tannase.

Materials and methods

Microorganism and culture conditions

The tannase producing fungal culture, *A. flavus* (MTCC 3783) was obtained from IMTECH, Chandigarh and used for tannase production. The fungal culture was maintained on Czapek Dox minimal media agar slants supplemented with 1 % tannic acid as the sole carbon source. The fungal strain was sub-cultured periodically, grown at 30 °C for 7 days. The well grown culture was stored at 4 °C in a refrigerator and used for further sub-culturing.

The composition of the Czapek Dox minimal medium used for tannase enzyme production was: tannic acid, 10 g/l; sodium nitrate, 6 g/l; potassium dihydrogen orthophosphate, 1.52 g/l; magnesium sulfate, 0.52 g/l; potassium chloride, 0.52 g/l; ferrous sulfate, 0.01 g/l and zinc sulfate, 0.01 g/l. Cells were harvested from slants and used to inoculate liquid medium.

Production of tannase in submerged fermentation (SmF)

The spore suspension was inoculated in 250 ml Erlenmeyer flask containing 100 ml of Czapek Dox minimal medium. 3 gm of substrate (tamarind seed) was added separately to the above mentioned medium for studying their effect on the enzyme production. The cultures were grown at 30 °C, 140 rpm for 6 days in an incubator shaker. The samples were withdrawn at regular intervals of 24 h. The biomass was separated by filtering through Whatman No.1 filter paper. The cell free culture broth was assayed for the tannase activity.

Assay of tannase

0.1 ml of enzyme solution was incubated with 0.3 ml of 1.0 % (w/v) tannic acid and 0.2 M acetate buffer (pH 5.0) at 40 °C for 10 min and then the enzyme production was stopped by cooling to 0 °C by the addition of 2 ml bovine serum albumin (BSA) (1 mg/ml), which precipitates the remaining tannic acid. Simultaneously, a control without the enzyme was incubated and the samples were analyzed. The tubes were then centrifuged (5,000×*g*, 10 min) and the precipitate was dissolved in 2 ml of sodium dodecyl sulfate (SDS)—triethanolamine (1 % w/v SDS in 5 % v/v triethanolamine) solution and the absorbency was measured at 550 nm after addition of 1 ml of $FeCl_3$ (0.01 M $FeCl_3$ in 0.01 N HCl) (Mondal et al. 2001b). One unit of tannase enzyme was defined as the amount of enzyme required to hydrolyze one micro mole of ester linkage of tannic acid in 1 min at specific condition.

Plackett–Burman design

Plackett–Burman design, an efficient technique for medium component selection (Plackett–Burman 1946) was used to determine the factors that significantly influence the tannase production. Twelve variables (Table 1) were screened in 20 experimental runs (Table 2) and the insignificant variables were eliminated to obtain a smaller, more manageable set of factors. The low level (−1) and high level (+1) of each factor are listed in Table 1. The statistical software package 'Design Expert 7.1.5' was used for analyzing the experimental data.

Table 1 Nutrient screening using a Plackett–Burman design

Levels (g/ml)			
Nutrient code	Nutrient	Low (−1)	High (+1)
A	Tannic acid	1	5
B	Yeast extract	0.1	1
C	Magnesium sulfate	0.1	1
D	Ferrous sulfate	0.1	1
E	Ammonium nitrate	0.1	1
F	Ammonium chloride	0.1	1
G	Urea	0.1	1
H	Potassium chloride	0.1	1
I	Sodium nitrate	0.1	1
J	Potassium dihydrogen phosphate	0.1	1
K	Ammonium sulfate	0.1	1
L	Peptone	0.1	1

Box-Behnken design

The process variables with significant effects on enzyme production were optimized using Box-Behnken design. To determine the response pattern and synergy of variables the full 2^k composite design was performed giving $2^k + 2k + n_0$ combinations where k is the number of independent variables and n_0 is the number of replications of the experiments at center point. This provided 31 experimental runs performed with four factors at five coded levels (−2, −1, 0, +1 and +2) in duplicate, with central points in triplicate to determine the experimental error (Box and Draper 1987). The coded and actual values of the variables are presented in Table 3. The responses of the input variables were evaluated as a function of tannase activity, measured as the amount of tannase liberated and coded by Y (U/ml) (Table 4). A multiple regression analysis of the data was carried out to obtain an empirical model that relates the measured response to the independent variables. A second-order polynomial equation was obtained.

$$Y = \beta_0 + \sum \beta_i X_i + \sum \beta_{ii} X_i^2 + \sum \sum \beta_{ij} X_i X_j \quad (1)$$

where Y is the predicted response, β_0 is the intercept, β_i is the linear coefficient, β_{ii} is the quadratic coefficient and β_{ij} is the interaction coefficient. The relation between the coded forms of the input variable and the actual values of chosen variables is described as follows:

$$x_i = (X_i - X_0)/\Delta X_i \quad i = 1, 2, 3, \ldots, K \quad (2)$$

where x_i is the coded value, X_i is the actual value of an independent variable, X_0 is the value of X_i at center point and ΔX_i is the step change of the variable and $i = 1$–4. The analysis of results was performed with statistical and graphical analysis software (Design expert 7.1.5). The software was used for regression analysis of data obtained and to estimate the coefficient of regression equation. ANOVA (analysis of variance) for the final predictive equation and optimization values were also obtained using the same software package which indicates whether the variable was more significant or less significant.

Results and discussion

Screening of nutrients by PBD

Experiments were carried out based on Plackett–Burman design and the results obtained were given in Table 2. From the table, it was observed that the variation in tannase activity was 30–90 U/ml. From the Pareto chart (Fig. 1), the nutrients, tannic acid, magnesium sulfate, ferrous

Table 2 Plackett–Burman experimental design matrix for screening of important variables for tannase production using *Aspergillus flavus*

Run no.	*A*	*B*	*C*	*D*	*E*	*F*	*G*	*H*	*I*	*J*	*K*	*L*	Tannase (U/ml)
1	1	1	1	1	−1	−1	1	1	−1	1	1	−1	38.20
2	−1	1	−1	1	−1	1	1	1	1	−1	−1	1	65.40
3	−1	−1	−1	−1	1	−1	1	1	1	1	−1	−1	95.52
4	−1	1	1	−1	−1	−1	−1	1	−1	1	−1	1	80.82
5	1	−1	1	−1	1	1	1	1	−1	−1	1	1	39.67
6	1	−1	−1	−1	−1	1	−1	1	−1	1	1	1	58.80
7	−1	−1	−1	1	−1	1	−1	1	1	1	1	−1	42.80
8	1	−1	1	1	−1	−1	−1	−1	1	−1	1	−1	30.87
9	−1	1	−1	−1	1	−1	−1	−1	−1	−1	1	−1	50.89
10	−1	1	−1	−1	1	1	−1	1	1	−1	−1	−1	88.76
11	−1	−1	−1	−1	−1	−1	−1	−1	−1	−1	−1	−1	90.76
12	−1	1	1	1	1	−1	−1	1	1	−1	1	1	43.90
13	−1	−1	1	1	−1	1	1	−1	−1	−1	−1	1	57.90
14	−1	−1	−1	−1	1	−1	1	−1	1	1	1	1	86.54
15	1	−1	1	1	1	1	−1	−1	1	1	−1	1	38.00
16	−1	1	−1	1	1	1	1	−1	−1	1	1	−1	54.89
17	1	1	1	−1	−1	1	1	−1	1	1	−1	−1	47.02
18	1	1	−1	−1	−1	−1	1	−1	1	−1	1	1	60.76
19	1	−1	−1	1	1	−1	1	1	−1	−1	−1	−1	56.76
20	1	1	−1	1	1	−1	−1	−1	−1	1	−1	1	69.00

Table 3 Ranges of the independent variables used in RSM

Variables	Levels (U/ml)					
	Code	−2	−1	0	+1	+2
Tannic acid conc. (%)	X_1	1	2	3	4	5
Fermentation period (h)	X_2	48	72	96	120	144
Temperature (°C)	X_3	25	30	35	40	45
pH	X_4	4.5	5	5.5	6	6.5

Table 4 Central composite design (CCD) of factors in coded levels with enzyme activity as response

Run no.	X_1	X_2	X_3	X_4	Tannase activity (U/ml)	
					Experimental	Predicted
1	0	0	2	0	100.45	95.98
2	0	0	−2	0	94.6	90.37
3	0	0	0	0	137.54	134.71
4	0	−2	0	0	98.5	94.96
5	−1	1	−1	1	78.5	81.94
6	−1	−1	−1	1	85.5	84.56
7	1	1	−1	1	90.67	94.55
8	1	1	1	−1	95.67	98.75
9	0	0	0	2	90.54	85.72
10	−1	1	−1	−1	99.54	99.19
11	0	0	0	0	134.32	134.71
12	−1	−1	1	−1	100.43	98.91
13	1	−1	1	−1	95.7	98.61
14	0	0	0	0	134.23	134.71
15	0	0	0	0	134.13	134.71
16	1	−1	−1	1	95.34	97.73
17	0	0	0	0	134.22	92.49
18	0	0	0	0	134.31	134.71
19	1	1	1	1	97.43	97.73
20	0	2	0	0	97.65	92.49
21	1	1	−1	−1	103.21	105.65
22	0	0	0	0	134.23	134.71
23	−1	1	1	1	90.36	92.72
24	−1	−1	1	1	93.15	97.06
25	−1	1	1	−1	96.22	99.90
26	−1	−1	−1	−1	90.43	96.48
27	−2	0	0	0	106.4	102.44
28	1	−1	1	1	100.2	102.913
29	0	0	0	−2	102.56	98.67
30	2	0	0	0	119.5	114.75
31	1	1	1	−1	98.4	98.76

sulfate and potassium dihydrogen phosphate were found to be significant for the production of tannase by *A. flavus* sp. using tamarind seed powder as substrate.

The equation obtained was

$$Y = -14.35A - 2.268B - 14.49C - 14.81D - 0.310E - 10.348F + 3.276G + 2.0H - 0.212I - 5.062J - 17.86K + 5.80L$$

where A, tannic acid; B, yeast extract; C, magnesium sulfate; D, ferrous sulfate; E, ammonium nitrate; F, ammonium chloride; G, urea; H, potassium chloride; I, sodium nitrate; J, potassium dihydrogen phosphate; K, ammonium sulfate; L, peptone.

From the above equation, the negative coefficient for K, D, A and C indicates that increase in the nutrient concentration decreases the tannase activity. Hence their levels were maintained at their low (−1) levels in the optimization of process variables.

Optimization of process parameters by RSM

The levels of the factors (tannic acid concentration, fermentation period, temperature and pH) and the effect of their interactions on tannase production were determined by central composite design of RSM. Thirty-one experiments were performed with different combinations of the factors shown in Table 4. The predicted and observed responses along with the design matrix were presented in Table 5. The results were analyzed by ANOVA. The second-order regression equation provided levels of tannase activity as a function of tannic acid concentration, fermentation period, temperature and pH which can be presented in terms of coded factors as in the following equation:

$$\begin{aligned} Y = {} & 134.711 + 3.07708X_1 - 0.6183X_2 + 1.4033X_3 \\ & - 3.2354X_4 - 6.5288X_1^2 - 10.2475X_2^2 - 10.3850X_3^2 \\ & - 10.6288X_4^2 - 0.21X_1X_2 - 1.90X_1X_3 + 1.5381X_1X_4 \\ & - 0.4306X_2X_3 - 1.3325X_2X_4 + 2.5175X_3X_4 \end{aligned}$$

where Y is the tannase activity (U/ml) produced as a function of the coded levels of tannic acid concentration (X_1), fermentation period (X_2), temperature (X_3) and pH (X_4).

The ANOVA for tannase production was given in Table 5. A Model F value of 28.49 implies the model was significant. Values of "Prob>F" <0.05 indicate the model terms were significant. Values >0.1 indicate the model terms were not significant. In the present work, the linear effect of tannic acid concentration and pH, square effects of X_1, X_2, X_3 and X_4 and interactive effects of X_3 and X_4 were significant for tannase production. The coefficient of determination (R^2) for the tannase activity was calculated as 0.9638, which is close to 1 and can account for up to 96.38 % of the variability of the response. The predicted R^2 value of 0.7785 is in reasonable agreement with the adjusted R^2 value of 0.9299. An adequate precision value >4 is desirable. The adequate precision value of 16.64 indicates an adequate signal and suggests that the model can be used to navigate the design space. The coefficient of variation (CV) indicates the degree of precision with which the experiments are compared. Generally, the higher the value of the CV is the lower the reliability of the experiment. Here a lower value of CV (4.30) indicates greater reliability of the experiments performed. The above model can be used to predict the tannase production within the limits of the experimental factors.

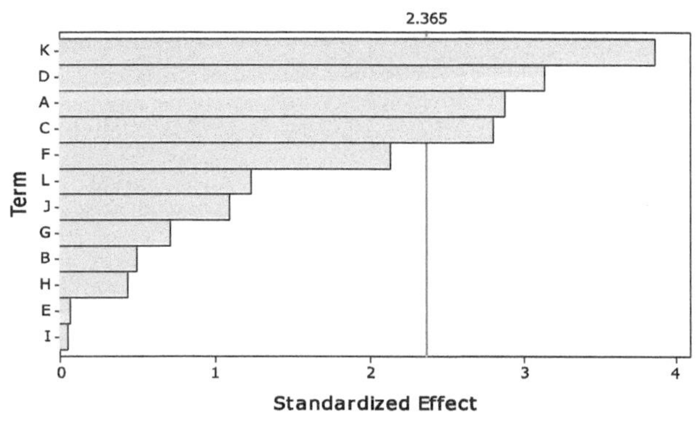

Fig. 1 Pareto chart showing the effect of media components on tannase activity using *Aspergillus flavus*

Response surface plots

The interactive effects of variables on tannase production were studied by plotting 3D surface curves against two independent variables, while keeping the other variable at its central (0) level. The 3D curves of the calculated tannase production and contour plots from the interactions between the variables were shown in Figs. 2, 3, 4 and 5. Figure 2 shows the dependency of tannase on tannic acid concentration and fermentation period. The tannase activity increased with an increase in tannic acid concentration to about 3.22 %. Tannic acid is an inducer for the production of tannase and hence tannic acid activity increases. But at higher concentration (above 3.22 %) tannase activity decreases. This may be due to inducer solubility, toxicity level and saturation which determine the optimum concentration (Pinto 2003).

From Fig. 2, it was also inferred that an increase in fermentation period resulted in an increase in tannase activity up to 96 h after that the tannase activity decreases. A decline in enzyme activity after 96 h of fermentation may be due to decrease in nutrient availability in the medium. These are in agreements with the results obtained by Beniwal and Chhokar (2010). Figures 4 and 5 show the dependency of tannase activity on temperature. Generally, temperature influences the metabolic activity of cells. At lower and higher temperatures, the tannase activity was found to be low. The maximum tannase activity was observed at a temperature of 35.1 °C. These results are well supported by Beniwal and Chhokar (2010). Similarly for pH, at low and higher levels the activity was low

Table 5 Analysis of variance (ANOVA) for response surface quadratic model for the production of tannase

Source	Coefficient factor	Sum of squares	*DF*	*F*	*P* > *F*
Model	134.79	8,045.48	14	28.49	<0.0001
X_1	3.08	209.76	1	10.40	0.0057
X_2	−0.62	8.64	1	0.43	0.5228
X_3	1.40	44.48	1	2.21	0.1582
X_4	−3.24	231.90	1	11.50	0.0040
$X_1 \times X_2$	−0.21	0.65	1	0.032	0.8604
$X_1 \times X_3$	−1.90	52.81	1	2.62	0.1265
$X_1 \times X_4$	1.54	33.68	1	1.67	0.2160
$X_2 \times X_3$	−0.43	2.64	1	0.13	0.7227
$X_2 \times X_4$	−1.33	25.97	1	1.29	0.2742
$X_3 \times X_4$	2.52	92.71	1	4.60	0.0483
$X_1 \times X_1$	−6.55	1,169.38	1	57.98	<0.0001
$X_2 \times X_2$	−10.27	2,874.51	1	142.53	<0.0001
$X_3 \times X_3$	−10.41	2,952.02	1	146.37	<0.0001
$X_4 \times X_4$	−10.65	3,091.94	1	153.31	<0.0001
Residual	–	302.52	15	–	–
Lack of fit	–	289.31	9	14.60	0.0020
Pure error	–	13.21	6	–	–
Cor total	–	8,348.00	29	–	–

Std. Dev. 4.49; R^2 0.9638; Mean 104.33; Adj R^2 0.9299; CV % 4.30; Pred R^2 0.7785; Adeq Precision 16.644

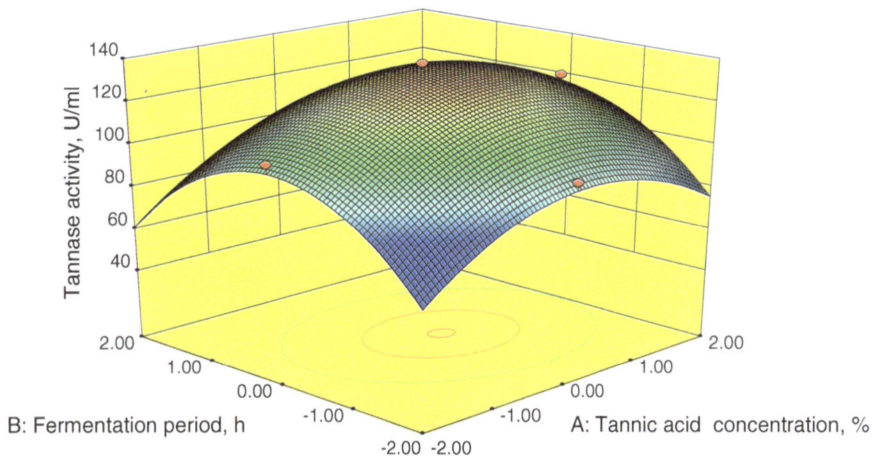

Fig. 2 3D plot showing the effect of tannic acid concentration and fermentation period on tannase activity

(Figs. 3, 5). Tannase activity was maximum in the acidic pH range, and its activity decreases in the alkaline range (Rodríguez et al. 2008; Raghuwanshi et al. 2011). The effect of pH on the activity of tannase may be attributed to its amino acids and active site which undergoes protonation or deprotonation; in addition, it may attributed to the conformational changes induced by amino acids ionization (Sabu et al. 2005).

The optimum conditions for the maximum production of tannase were determined by the response surface analyses and also estimated from the regression equation. The optimum conditions were tannic acid concentration, 3.22 %; fermentation period, 96 h; temperature, 35.1 °C; and pH 5.4. The predicted results were shown in Table 4. The predicted values from the regression equation closely agreed with the experimental values. Validation of the experimental model was tested by carrying out a batch experiment under optimal operating conditions. Three repeated experiments were performed and the results were compared. The tannase activity obtained from the experiments was very close to the response predicted by the regression model, which proves the validity of the model. At these optimized conditions the maximum tannase activity was found to be 139.3 U/ml.

Conclusions

The response surface methodology was effectively applied for the production of tannase from *A. flavus,* MTCC 3783

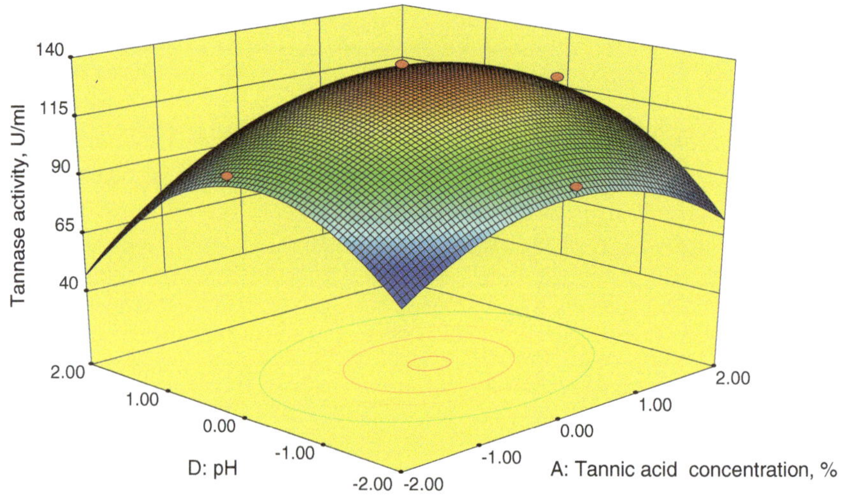

Fig. 3 3D plot showing the effect of tannic acid concentration and pH on tannase activity

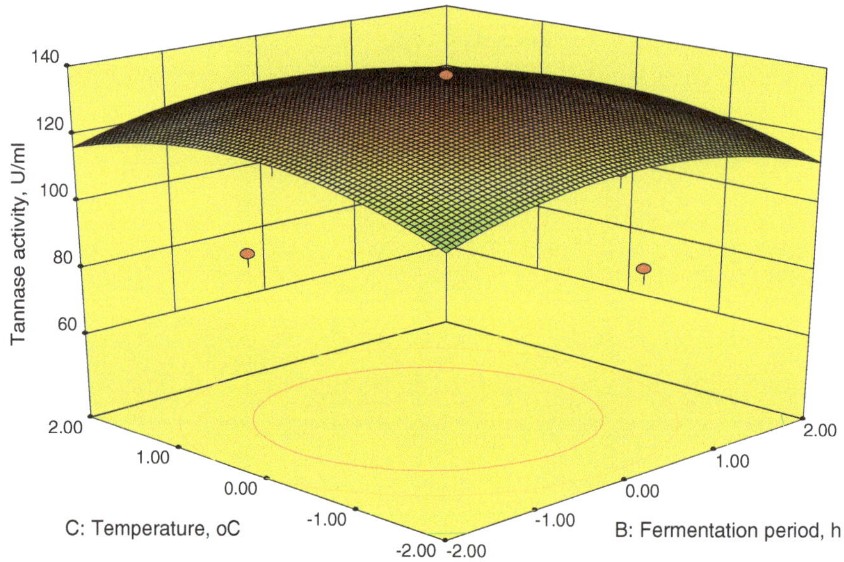

Fig. 4 3D plot showing the effect of fermentation period and temperature on tannase activity

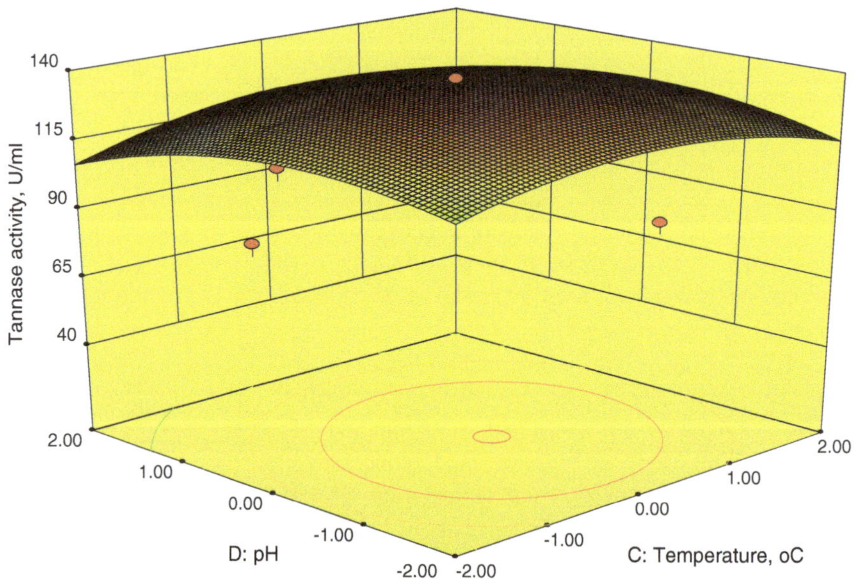

Fig. 5 3D plot showing the effect of temperature and pH on tannase activity

using tamarind seed powder as a substrate in submerged fermentation. Central composite design was employed to evaluate the effects of tannic acid concentration, fermentation period, temperature and pH on production of tannase by *A. flavus.* Using the above optimized nutrient solution, maximum tannase activity of 139.3 U/ml was obtained at the inducer tannic acid concentration of 3.22 %, temperature of 35.1 °C, pH of 5.4, and fermentation period of 96 h. The statistical design of experiment offers efficient methodology to identify the significant variables and to optimize the factors with minimum number of experiments for tannase production by *A. flavus.*

Acknowledgments The authors gratefully acknowledge the Bioprocess Research Laboratory, Department of Chemical Engineering, Annamalai University for providing the necessary facilities for the successful completion of this research work.

Conflict of interest The authors declare that they have no conflict of interest in the publication.

References

Batra A, Saxena RK (2005) Potential tannase producers from the genera *Aspergillus* and *Penicillium.* Process Biochem 40:1553–1557

Beniwal V, Chhokar V (2010) Statistical optimization of culture conditions for tannase production by *Aspergillus awamori* MTCC 9299 under submerged fermentation. Asian J. Biotechnol 2(1):46–52

Couto SR, Sanroman MA (2006) Application of solid-state fermentation to food industry—a review. J Food Eng 76(3):291–302

Das Mohapatra PK, Mondal KC, Pati BR (2006) Production of tannase through submerged fermentation of tannin containing plant extracts by *Bacillus licheniformis* KBR6. Pol J Microbiol 55(4):297–401

Dilipkumar M, Rajasimman M, Rajamohan N (2011) Application of statistical design for the production of inulinase by *Streptomyces* sp. using pressmud. Front Chem Sci Eng 5(4):463–470

Lekha PK, Lonsane BK (1994) Comparative titres, location and properties of tannin acyl hydrolase produced by *Aspergillus niger* PKL104 in solid-state, liquid surface and submerged fermentations. Process Biochem 29:497–503

Lekha PK, Lonsane BK (1997) Production and application of tannin acyl hydrolase: state of the art. Adv Appl Microbiol 44:215–260

Mannan S, Fakhruĺ-Razi A, Alam MZ (2007) Optimization of process parameters for the bioconversion of activated sludge by *Penicillium corylophilum* using response surface methodology. J Environ Sci 19(1):23–28

Milva P, Lucia RL, Roberto A, Alessandro E, Guido P, Monia R, Arianna L, Antonio F, Simone G, Silvano EF (2010) Tannic acid degradation by bacterial strains *Serratia* sp. and *Pantoea* sp. isolated from olive mill waste mixtures. Int J Biodeter Biodegr 64:73–80

Mohan SK, Viruthagiri T, Arunkumar C (2012) Optimization and production of tannase by *Aspergillus foetidus* through submerged fermentation. Asian J Bio Pharma Res 4(2):22–29

Mondal KC, Banerjee D, Jana M, Pati BR (2001a) Colorimetric assay method for determination of tannin acyl hydrolase (EC.3.1.1.20) activity. Anal Biochem 295:168–171

Mondal KC, Banerjee D, Banerjee R, Pati B (2001b) Production and characterization of tannase from *Bacillus cereus* KBR9. J Gen Appl Microbiol 47:263–267

Murugan K, Saravanababu S, Arunachalam M (2007) Screening of tannin acyl hydrolase (E.C.3.1.1.20) producing tannery effluent fungal isolates using simple agar plate and SmF process. Bioresource Technol 98:946–949

Pan CM, Fan YT, Xing Y, Hou HW, Zhang ML (2008) Statistical optimization of process parameters on biohydrogen production from glucose by *Clostridium* sp. *Fanp2.* Bioresource Technol 99(8):3146–3154

Paranthaman R, Vidyalakshmi R, Murugesh S, Singaravadivel K (2009) Optimization of various culture media for tannase production in submerged fermentation by *Aspergillus flavus.* Adv Biol Res 3:34–39

Pinto GAS (2003) Production of tannase by *Aspergillus niger.* Ph.D. Thesis, Rio de Janeiro, p 213

Raghuwanshi S, Dutt K, Gupta P, Misra S, Saxena RK (2011) *Bacillus sphaericus*: the highest bacterial tannase producer with potential for gallic acid synthesis. J Biosci Bioeng 111:635–640

Rodríguez H, De las Rivas B, Gómez-Cordovés C, Munoz R (2008) Degradation of tannic acid by cell-free extracts of *Lactobacillus plantarum.* Food Chem 107:664–670

Sabu A, Kiran GS, Pandey A (2005) Purification and characterization of tannin acyl hydrolase from *Aspergillus niger* ATCC16620. Food Technol Biotechnol 43:133–138

17

Plant growth regulation of Bt-cotton through *Bacillus* species

Pavan Kumar Pindi · Tasleem Sultana · Praveen Kumar Vootla

Abstract Deccan plateau in India periodically experiences droughts due to irregular rain fall and the soil in many parts of the region is considered to be poor for farming. Plant growth promoting rhizobacteria are originally defined as root-colonizing bacteria, i.e., *Bacillus* that cause either plant growth promotion or biological control of plant diseases. The study aims at the isolation of novel *Bacillus* species and to assess the biotechnological potential of the novel species as a biofertilizer, with respect to their plant growth promoting properties as efficient phosphate-solubilizing bacteria. Seven different strains of *Bacillus* were isolated from cotton rhizosphere soil near boys' hostel of Palamuru University which belongs to Deccan plateau. Among seven isolated strains, *Bacillus* strain-7 has shown maximum support for good growth of eight cotton cultivars. This bacterial species is named *Bacillus* sp. PU-7 based on the phenotypic and phylogenetic analysis. Among eight cotton cultivars, Mahyco has shown high levels of IAA, proteins, chlorophyll, sugars and low level of proline. Efficacy of novel *Bacillus* sp. PU-7 with Mahyco cultivar has been checked experimentally at field level in four different cotton grown agricultural soils. The strains supported plant growth in almost all the cases, especially in the deep black soil, with a clear evidence of maximum plant growth by increased levels of phytohormone production and biochemical analysis, followed by shallow black soil. Hence, it is inferred that the novel isolate can be used as bioinoculant in the cotton fields.

P. K. Pindi (✉) · T. Sultana · P. K. Vootla
Department of Microbiology, Palamuru University, Mahabubnagar 509001, Andhra Pradesh, India
e-mail: pavankumarpindi@gmail.com

Keywords *Bacillus* sp. PU-7 · Cotton seed varieties · IAA · Proline · Proteins · Sugars

Introduction

A group of rhizosphere bacteria (rhizobacteria) that exert a beneficial effect on plant growth is referred to as plant growth promoting rhizobacteria (PGPR), belongs to several genera: e.g., *Agrobacterium, Alcaligenes, Arthrobacter, Actinoplanes, Azotobacter, Bacillus, Pseudomonas* sp., *Rhizobium*. Multiple species of *Bacillus* and *Paenibacillus* are known to promote plant growth. The principal mechanisms of growth promotion include production of growth stimulating phytohormones, solubilization and mobilization of phosphate, siderophore production, antibiosis, i.e., production of antibiotics, inhibition of plant ethylene synthesis, and induction of plant systemic resistance to pathogens (Richardson et al. 2009; Idris et al. 2007; Gutierrez-Manero et al. 2001; Whipps 2001).

Ability of *Bacillus* sp. to produce organic acid such as gluconic, citric and fumaric acids under P-limiting conditions may increase the solubility of poorly soluble phosphorus. Several soil bacteria and fungi notably species of *Pseudomonas*, *Bacillus* and *Aspergillus*, etc., secrete organic acids and lower the pH in their vicinity to bring about solubilization of bound phosphates in soil (Sundra Rao and Sinha 1963). The N_2-fixing bacterium associated with nonlegumes includes species of *Achromobacter*, *Alcaligenes*, *Arthrobacter*, *Acetobacter*, *Azomonas*, *Beijerinckia* and *Bacillus*. Among PGPR species, IAA producing bacteria are belonging to *Aeromonas* (Halda-Alija 2003), *Azotobacter* (Ahmad et al. 2008) and *Bacillus* (Swain et al. 2007).

Recently, Choudhary and Johri (2008) explicated the mechanisms and role of *Bacillus* species as inducers of systemic resistance in relation to plant–microbe interactions and demarcated the pathways involved in their regulation. Species of *Bacillus* are common inhabitants among the resident microflora of inner tissues of various species of plants, including cotton, grape, peas, spruce, and sweet corn, where they play an important role in plant protection and growth promotion (Berg et al. 2005; Shishido et al. 1999; Bell et al. 1995).

Phytostimulation

Root-colonizing species of *Bacillus* and *Paenibacillus* are well known for the enhancement of plant growth (Idris et al. 2007). Growth promoting effects of various PGPRs are due to bacterial production of plant growth regulators such as indole-3-acetic acid (IAA), gibberellins, and cytokinins (Bottini et al. 2004; Bloemberg and Lugtenberg 2001). 80 % of rhizosphere colonizing bacteria has been reported positive for IAA production. Idris et al. (2004) showed production of substances with auxin (IAA)-like bioactivity from strains of *Bacillus subtilis/B. amyloliquefaciens* including strain FZB42. Gutierrez-Manero et al. (2001) has confirmed the production of gibberellins from *B. pumilus* and *B. licheniformis*. IAA plays an important key component in shaping plant root architecture such as root vascular tissue differentiation, regulation of lateral root initiation, polar root hair positioning, and root gravitropism (Aloni et al. 2006). Production of IAA from Gram-positive bacterium *B. amyloliquefaciens* FZB42 was first demonstrated by Idris et al. (2007) and its production was enhanced when the bacterium was fed with tryptophan. Application of phosphate solubilizers alone or in combination with nitrogen fixers has been found beneficial for cotton and wheat fields (Zaidi and Khan 2005; Kundu and Gaur 1980).

Production of phytohormones by PGPR

The production of phytohormones by PGPR is the most important mechanisms by which many rhizobacteria promote plant growth (Spaepen et al. 2007). Phytohormones are signal molecules acting as chemical messengers and play a fundamental role as growth and development regulators in the plants. Phytohormones are organic compounds that in extremely low concentrations influence biochemical, physiological and morphological processes in plants, and their synthesis is finely regulated (Fuentes-Ramírez and Caballero-Mellado 2006). Numerous fungal and bacterial species can produce phytohormones (Tsavkelova et al. 2006). The phytohormone producing ability is widely distributed among bacteria associated with soil and plants. Studies have demonstrated that the PGPR can stimulate plant growth through the production of auxins (IAA) (Spaepen et al. 2008), gibberellins (Bottini et al. 2004) and cytokinins (Timmusk et al. 1999), or by regulating the high levels of endogenous ethylene in the plant (Glick et al. 1998).

Indole acetic acid (IAA) producing rhizobacteria

Many important plant–microbial interactions center on the production of auxins, IAA being the main plant auxin. It is responsible for the division, expansion and differentiation of plant cells and tissues and stimulates root elongation. IAA synthesis has been detected in many rhizobacteria as well as in pathogenic, symbiotic and free living bacterial species (Costacurta and Vanderleyden 1995; Tsavkelova et al. 2006).

Siderophore-producing rhizobacteria

Siderophores are produced by various types of bacteria in response to iron deficiency which occur in neutral to alkaline pH soils due to low iron solubility at elevated pH (Sharma and Johri 2003). Iron is essential for cellular growth and metabolism such that Fe acquisition through siderophore production plays an essential role in determining the competitive fitness of bacteria to colonize plant roots and to compete for iron with other microorganisms in the rhizosphere (Crowley and Gries 1994; Crowley 2006). Siderophore-producing organisms can prevent the proliferation of pathogenic microorganisms by sequestering Fe^{3+} in the area around the root (Siddiqui 2006).

Materials and methods

The experiments were carried out in the Department of Microbiology, Palamuru University, Mahabubnagar, Andhra Pradesh, India.

Sample collection

Cotton rhizosphere soil sample was collected near boys' hostel of Palamuru University and screened for *Bacillus* species by serial dilution of the sample and dilution of 10^{-4} was spread on Soya bean trypticase agar.

Preparation of standard inoculums of seven species of *Bacillus*

Inoculums of seven isolates of *Bacillus* sp. were prepared in selective medium. 150 ml broth medium was inoculated in 500-ml conical flask and incubated at 28 °C under shaking at 100–150 rpm for 3 days to give an optical density of 0.5. Broth culture of *Bacillus* sp. was inoculated in to peat (100 ml kg^{-1} of peat) which was sterilized at 121 °C and 15 psi pressure for 1 h. Peat-based *Bacillus* inoculums were incubated at 28 °C by adding 10 % sugar solution for 3–4 days to increase the population up to 10^8 CFU ml^{-1}. *Bacillus* inoculations having at 10^8 MPN bacterial cells per gram of peat were applied to cotton seed as seed coating for a period of 90 days in sterilized soil.

Preparation of sterilized soil

1:1 ratio of sand and red soil were mixed properly and sterilized in an autoclave at 121 °C, 15 lbs pressure.

Genomic DNA extraction

The genomic DNA from the bacterial cells was obtained using a modification of the method described by Sambrook et al. (1989). The bacterial cells from pure culture were harvested by centrifugation (12,000 rpm) for 2 min, and the cell pellets were mixed with 600 μl of lysis buffer [10 mM Tris–HCl, 1 mM EDTA (pH 7.5), 0.5 % SDS, 100 g ml^{-1} proteinase C] and incubated at 37 °C for 1 h after the addition of 100 μl 5 M NaCl, and 80 μl CTAB/NaCl buffer. Samples were incubated at 65 °C for 10 min and samples were cooled to room temperature, followed by extraction of the aqueous phase with an equal volume of chloroform:isoamyl alcohol (24:11, v/v) and then with an equal volume of phenol:chloroform:isoamylalcohol (25:24:1, v/v) which was centrifuged at 12,000 rpm and 4 °C for 10 min. Isopropanol (0.6×) was mixed with the aqueous phase, and centrifuged at 12,000 rpm and 4 °C for 10 min. The DNA pellets were vacuum dried, and then dissolved in Tris buffer [10 mM Tris–HCl, and 1 mM EDTA (pH 7.5)].

PCR analysis

The small subunit rRNA gene of each sample's culture DNA was amplified using 16S rRNA Universal primers. The PCR amplification reaction mixture of 50 μl contained 4 μl bacterial DNA (nearly 200 ng), 1 μl Taq-DNA polymerase, 5 μl of Taq buffer, 5 μl of 2 mM dNTP mix, 5 μl of forward primer (10 pM μl^{-1}) and 5 μl of reverse primer (10 pM μl^{-1}). Amplification was carried out in a Bio-Rad thermo cycler run for 30 cycles. In each cycle denaturation was done at 94 °C for 20 s, annealing at 48 °C for 20 s and extension was done at 72 °C for 40 s and a final extension was carried out for 5 min at 72 °C at the end of all 30 cycles. The amplified DNA fragment of approximately 1,542 bp was separated on a 1 % agarose gel and purified by Qiagen spin columns (Mullis 1990; Barlett and Stirling 2003).

16S rRNA gene sequencing and phylogenetic tree construction

For 16S rRNA gene sequencing, DNA was prepared using the Mo Bio microbial DNA isolation kit (Mo Bio Laboratories Inc., Solano Beach, CA, USA) and sequenced as described previously (Lane 1991). The 1,502 nucleotides of 16S rRNA gene sequence of the isolate were subjected to BLAST sequence similarity search (Altschul et al. 1990) and EzTaxon (Chun et al. 2007) to identify the nearest taxa. All the 16S rRNA gene sequences belonging to the family "*Bacillaceae*" were downloaded from the database (http://www.ncbi.nlm.nih.gov), aligned using the CLUSTAL_X program (Thompson et al. 1997) and the alignment corrected manually. Phylogenetic trees were constructed using two tree-making algorithms, the maximum likelihood (ML) using the PhyML program (Guindon and Gascuel 2003) and Neighbor joining method (Saitou and Nei 1987) using the PHYLIP package, version 3.5 (Felsenstein 1993) and the resultant tree topologies were evaluated by bootstrap analysis based on 1,000 resamplings using the SEQBOOT and CONSENSE programs in the PHYLIP package. Pairwise evolutionary distances were computed using DNADIST program with the Kimura 2-parameter model as developed by Kimura (1980).

Phenotypic characterization of novel *Bacillus* sp. PU-7

Cell morphology and motility were studied using a light microscope. Motility was assessed on TSA medium containing l^{-1} pancreatic digest of casein (17 g), papaic digest of soyabean meat (3 g), sodium chloride (5 g), dipotassium hydrogen phosphate (2.5 g), dextrose (2.5 g) and agar (0.4 g). Growth at different temperatures, salt tolerance, biochemical characteristics, carbon assimilation, H_2S production and the sensitivity of the culture to different antibiotics were determined by previously described methods (Lanyi 1987; Smibert and Krieg 1994). Biochemical characteristics were also double checked with Hi25™ Enterobacteriaceae identification kit (Cat #KB003) and HiCarbohydrate™ kit parts A, B and C (Cat #KB009 of HiMedia, Mumbai, India) according to the manufacturer's protocol. Growth of $PU1^T$ at different pHs was checked on NA medium buffered either with citric acid–NaOH (for pH 5 and 6), phosphate (for pH 7 and 8), glycine–NaOH (for pH 9 and 10) or Tris buffer (for pH 11 and 12).

Production of IAA from *Bacillus* sp. PU-7

IAA production was detected as described by Brick et al. (1991). *Bacillus* culture was grown for 48 h on the respective media at 36 °C. Well-grown cultures were centrifuged at 3,000 rpm for 30 min. The supernatant

(2 ml) was mixed with two drops of orthophosphoric acid and 4 ml of the Salkowski reagent (50 ml, 35 % of perchloric acid, 1 ml 0.5 M $FeCl_3$ solution). Development of pink color indicates IAA production.

Siderophore production from *Bacillus* sp. PU-7

Siderophore production was detected by Schwyn and Neilands (1987) using blue agar plates containing the dye chrome azurol S. An orange halo around the colony is indicative of siderophore production.

Phosphate solubilization of *Bacillus* sp. PU-7

For phosphate solubilization assay, a medium containing 2 g yeast extract, 20 g glucose, 2 g tri calcium phosphate, 60 mg actidione, and 15 g agar mixed with 1,000 ml water, adjusted to pH 7, was used. A loopful inoculum of strain *Bacillus* sp. PU-7 was placed in the center of petri dishes containing the media and incubated at 28 °C for 5 days. Bacterial colony forming clear zone was considered as phosphate solubilizer (Rosas et al. 2006).

Biochemical analysis of cotton plants inoculated with *Bacillus* sp. PU-7

Extraction and analysis of total protein

The total protein was extracted by homogenizing 0.5 g plant tissue in 10 ml of 0.2 M perchloric acid. The homogenate was centrifuged at 5,000*g* for 10 min at 24 °C. Ethanol–ether–chloroform (2:2:1; v/v/v) solvent mixture was used twice for the extraction of the pellet. To the residue, 0.2 M NaOH was added and left overnight. The supernatant was used for total protein estimation (Lowry et al. 1951).

Estimation of sugars

Total soluble sugar was analyzed by heating 1 g of the plant tissue with 0.2 % anthrone reagent and reading the intensity at 625 nm using UV–VIS spectrophotometer (Spectronic D20) (Mahadevan and Shridhar 1986).

Estimation of proline

Proline estimation was carried out as described by Bates et al. (1973). Fresh mass of 0.5 g leaf tissue was taken and homogenized in 5 ml of 3 % (w/v) sulphosalicylic acid. The residue was removed by centrifugation at 5,000 rpm for 10 min and the supernatant was filtered through Whatman # 2 filter paper. The filtrate was mixed with an equal volume of ninhydrin and glacial acetic acid and incubated at 95 °C for 1 h. The reaction was terminated by placing in an ice bath for about 30 min and then extracted with 4 ml toluene by mixing vigorously for 15 s. The toluene phase containing the chromophore was aspirated, warmed to room temperature for 10 min and the proline content was determined colorimetrically and expressed in mg g^{-1}.

Estimation and extraction of chlorophyll

Chlorophyll pigment was extracted from 1 g of cotton leaves inoculated with *Bacillus* sp. PU-7 in 80 % acetone and estimated as described by Harborne (1973). The extracts were filtered in dark. OD values of filtrate were measured at 650 and 663 nm in UV–VIS spectrophotometer. The amount of total chlorophyll was calculated by Arnon's formula.

Phytohormone production from cotton plants inoculated with Bacillus sp. PU-7

Estimation of IAA One gram of leaf sample was crushed with 1 ml of phosphate buffer. Samples were centrifuged and two drops of perchloric acid were added to the supernatant to make up the volume to 2 ml with Salkowski reagent (i.e., 2 % 0.5 M $FeCl_3$ in 35 % perchloric acid). OD values were taken after 25 min at 530 nm by UV–VIS spectrophotometer. Standard graph was prepared by plotting concentration of IAA in μg ml^{-1} vs optical density at 530 nm.

Collection of soil samples from different agricultural fields of cotton Four different soil samples were collected from cotton fields of Mahabubnagar District, viz., shallow black soil from Malleboinpally, red soil from Makthal, deep black soil from Kalwakurthy and sandy soil from Narayanpet.

Physico-chemical characteristics of soil samples of agricultural fields of cotton Soil available nitrogen was estimated by alkaline potassium permanganate method (Subbiah and Asija 1956), available phosphorous was determined by Bray and Kurtz (1945) and potassium determined by flame photometrically (Jackson 1973).

Statistical analysis

A statistical analysis was performed using a statistical software SPSS for Windows version 17.0. ANOVA, *t* test was applied on subjects to know the significance of multiple mean differences and mean difference, respectively. *P* values are significant at 5 % level (or 95 confidence level).

Results

Isolation of novel PGPR *Bacillus* sp. PU-7

The main aim of present investigation is to isolate novel and efficient *Bacillus* species from cotton rhizosphere soil near boy's hostel in Palamuru University. Seven species of *Bacillus* were isolated, among them *Bacillus* strain-7 has shown maximum growth of plant with 8 cultivars of cotton from 17 cultivars tested (Table 1). We subsequently named the isolated strain as *Bacillus* sp. PU-7 (Fig. 1) based on 16S rRNA gene sequence.

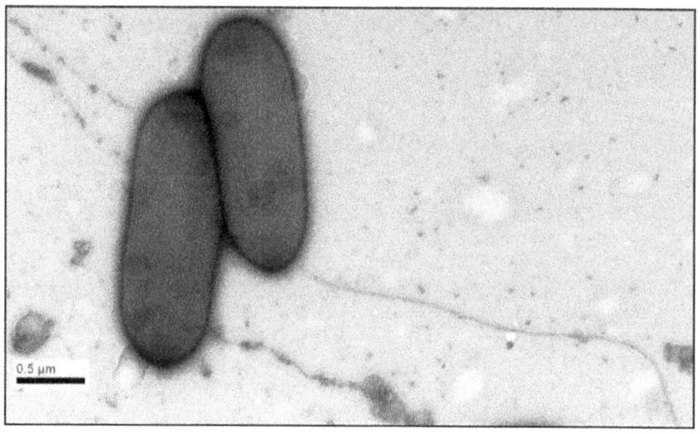

Fig. 1 Electron micrograph of negatively stained cells of *Bacillus* sp. PU-7. *Bar* 0.5 μm

Description of novel *Bacillus* sp. PU-7

Cells are Gram-positive, motile with single mono-polar flagellum, rod-shaped (0.6–0.7 μm in width and 1.6–2 μm in length) and occur singly and multiply by binary fission. Colonies on nutrient agar are circular, 1–2 mm in diameter, smooth, cream in color, opaque, crateri form and entire. Cells grow from 18 to 40 °C with an optimum temperature of 37 °C and tolerate up to 9.0 % NaCl (w/v). Growth occurs in a pH range of 7–10, designated strain $PU1^T$ was isolated from cotton rhizosphere soil near the boy's hostel of Palamuru University, Mahabubnagar district, Andhra Pradesh, India. Cells of the strain $PU1^T$ are positive for catalase, oxidase, phosphatase, lipase and urease and negative for gelatinase, amylase, protease, cellulase, lysine decarboxylase and ornithine decarboxylase (Table 2). The 16S rRNA gene sequence analysis indicated *Bacillus psychrodurans* and *Bacillus psychrotolerans* members of family "Bacillaceae" (phylum "Firmicutes") are the closest related species with a sequence similarity of 96.0–96.2 %. Other members of the family "Bacillaceae" had sequence similarities of <96.0 %. Based on the above-mentioned phenotypic and phylogenetic characteristics, strain PU-7 is proposed as the representative of a new species, novel *Bacillus* sp. PU-7 (Fig. 2).

Bacillus sp. PU-7 was tested for plant growth promoting characteristics (Table 3). Eight cultivars which have shown maximum growth with *Bacillus* sp. PU-7 (Fig. 3) were analyzed for biochemical characters and phytohormone production. Among eight cultivars of cotton, Mahyco has

Table 1 Plant growth parameters of 17 cotton cultivars (60 days old) with 7 different isolates of *Bacillus*

Strain/seed variety	Control	Strain-1	Strain-2	Strain-3	Strain-4	Strain-5	Strain-6	Strain-7
Mahyco	+	++	+	++	+	+	+	+++
Ajeet	+	+	+	−	++	+	+	++
Rashi	+	−	++	+	−	++	−	+++
Tulasi	+	++	+	+	+	−	+	++
Marvel	+	+	−	−	++	−	+	+++
Bunni	+	+	++	−	−	+	++	++
PCH-125	+	++	+	+	++	+	−	++
Nusun	+	++	−	+	+	++	++	+++
Kaveri	+	−	++	+	+	+	+	+++
Raj seeds	+	+	+	−	+	+	+	+++
Super seeds	±	+	+	+	+	+	++	+
Veda	+	+	−	++	+	+	+	++
Brahmaputra	±	+	+	+	+	+	+	+
S99Bt	±	+	+	+	+	+	−	++
Obama Bt	+	+	+	−	+	−	+	+++
Bunni seeds	+	−	++	++	−	++	+	++
Sunny (NCS-108)	+	+	+	++	+	++	+	+++

+++ plant growth above 40 cm, ++ plant growth below 30 cm, + plant growth below 20 cm, ± plant growth below 15 cm, − plant growth below 10 cm shown maximum levels of IAA, proteins sugars, chlorophyll, and minimum content of proline and Sunny (NCS-108) has shown minimum levels of IAA, proteins, sugars chlorophyll and high level of proline (Table 4).

Efficacy of novel *Bacillus* sp. PU-7 as a PGPR

Based on the above discussed results, it was apparent that *Bacillus* sp. PU-7 could be as used as a PGPR in bioformulation of poor soils that support least plant growth. We have collected four soil samples that were known to be poor for plant growth as described in "Materials and methods" and NPK and trace elements were determined (Table 5). These soil samples were inoculated with *Bacillus* sp. PU-7 and uninoculated sample was taken as control. Mahyco cultivar was grown in these 4 different soils for a period of 90 days in order to see the efficiency of novel isolate in natural conditions. Mahyco growth (physical growth parameters, phytohormone production and biochemical characteristics) was good in all soil types, but maximum growth was recorded in deep black soil followed by shallow black soil and minimum growth was observed in sandy soil followed by red soil (Tables 6, 7).

Discussion

Rhizosphere is defined as the soil influenced by roots, bacterial species that carry out functions which promote growth of plants. These bacteria are designated as PGPR (Martínez-Viveros et al. 2010). *Pseudomonas* and *Bacillus* genera are the most commonly investigated PGPR, and often the dominating bacterial groups in the rhizosphere (Morgan et al. 2005). *Bacillus* species have been reported to promote the growth of a wide range of plants (De Freitas et al. 1997; Kokalis-Burelle et al. 2002). Trials with rhizosphere-associated plant growth promoting N_2-fixing and phosphate solubilising *Bacillus* species indicated yield increase in sorghum (Broadbent et al. 1977), maize (Pal 1998), rice (Sudha et al. 1999), sugar beet (Cakmakci et al. 1999), barley (Sahin et al. 2004) and apples (Aslantas et al. 2007).

McSpadden Gardener (2004) and Ona et al. (2003) demonstrated the enhancement of plant growth by *Bacillus* and *Paenibacillus*. They promote plant growth by the solubilization of phosphorus and production of phytohormones, such as IAA (Lal and Tabacchioni 2009). Among PGPR species, *Azospirillum* is one of the best studied IAA

Table 2 Characteristics that differentiate the strain PU-7 with the closely related species of the genus *Bacillus*

Characteristic	PU-7	*Bacillus psychrodurans* DSM 11713^T	*Bacillus psychrotolerans* DSM 11706^T	*Bacillus insolitus* DSM 5^T
Cell morphology	Rods	Rods	Regular rods	Rods
Cell size (μm)	0.6–0.7 × 1.6–2	0.5–0.6 × 2–5	0.4–1 × 2–7	1.0–1.5 × 1.6–2.7
Nitrate reduction	–	+	–	–
Salinity tolerance (%)	9	5	3	2
Temperature range (°C)	18–40	−2 to 30	−2 to 30	0–25
Anaerobic growth	–	+	–	–
β-Galactosidase	–	w	–	+
Lysine decarboxylase	–	w	w	+
Ornithine decarboxylase	–	+	+	+
Nitrate reduction	–	+	–	+
Hydrolysis of				
DNA	–	+	+	+
Gelatin	–	+	–	+
Starch	–	+	+	–
Tween 60	+	w	+	–
Urea	+	–	–	w
Acid production from				
D-Glucose,	+	+	w	–
L-Arabinose,	–	w	w	–
D-Xylose	+	w	w	–
D-Mannitol	+	+	w	–
Utilization of				
Citrate	+	–	–	+

+ positive, – negative, *w* weak

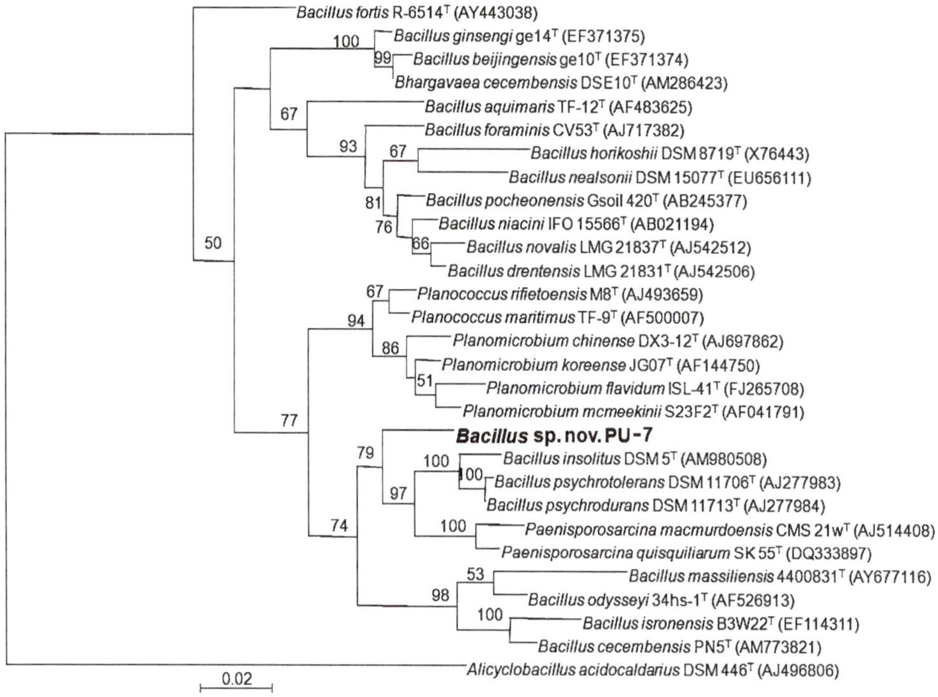

Fig. 2 Phylogenetic tree based on 16S rRNA gene sequences showing the relationship of *Bacillus* sp. PU-7 with the species of the genus *Bacillus*. Phylogenetic tree was constructed using the maximum likelihood method. *Numbers at nodes* are bootstrap values. The *bar* represents 0.02 substitutions per alignment position

Table 3 Plant growth promoting characteristics of novel *Bacillus* sp. PU-7

Characteristics	*Bacillus* sp. PU-7
Phosphate solubilization test	+
IAA production	+
Siderophore production	+

producers (Dobbelaere et al. 1999). Other IAA producing bacteria belonging to *Aeromonas* (Halda-Alija 2003), *Azotobacter* (Ahmad et al. 2008), *Bacillus* (Swain et al. 2007), *Burkholderia* (Halda-Alija 2003), *Enterobacter* (Shoebitz et al. 2009), *Pseudomonas* (Hariprasad and Niranjana 2009) and *Rhizobium* (Ghosh et al. 2008) genera have been isolated from different rhizosphere soils. Inoculation with IAA producing PGPR has been used to stimulate seed germination, to accelerate root growth and modify the architecture of the root system, and to increase the root biomass. In recent studies, Tsavkelova et al. (2007) have extended beyond individual strains as inoculants and reported an increase in the germination of orchid seeds (*Dendrobium moschatum*) inoculated with *Sphingomonas* sp. and IAA producing *Mycobacterium* sp. In addition to stimulating root growth, IAA producing bacteria can also be used to stimulate tuber growth. Swain et al. (2007) reported a positive effect of *B. subtilis* IAA producing strains on the edible tubercle *Dioscorea rotundata* L. in one of their studies. They applied a suspension of *B. subtilis* on the surface of the plants, which resulted in an increase in stem and root length, increased fresh weight of the stem and root, an increase in the root:stem ratio and increased numbers of sprouts as compared with non-inoculated plants. Phytohormones such as IAA may indirectly improve P acquisition by plants by increasing root growth

Fig. 3 Eight different cultivars of cotton inoculated with novel *Bacillus* sp. PU-7

Table 4 Phytohormone and biochemical production of 8 cotton cultivars (60 days old) inoculated with *Bacillus* sp. PU-7 (mean ± SE)

Cultivar	IAA (μg g^{-1})	Proteins (mg g^{-1})	Sugars (mg g^{-1})	Chlorophyll (mg g^{-1})	Proline (μg g^{-1})
Mahyco	749.67 ± 22.81	270 ± 11.55	129 ± 5.86	156.67 ± 6.67	340 ± 30.55
Rashi	657 ± 23.16	218.34 ± 15.90	140.66 ± 15.45	134 ± 8.72	350.67 ± 25.57
Marvel	658.34 ± 13.64	223 ± 4.62	118 ± 8.39	136.67 ± 14.53	330.00 ± 25.17
Nusun	634 ± 20.23	208 ± 15.62	122.67 ± 11.85	113.33 ± 6.67	325.33 ± 16.38
Kaveri	561 ± 10.69	175 ± 8.66	101.67 ± 10.14	113.33 ± 8.82	343.34 ± 29.06
Raj seeds	536.34 ± 14.84	165 ± 10.41	129 ± 12.42	89.67 ± 5.78	258.33 ± 13.64
Obama	502 ± 13.00	140 ± 2.31	126.67 ± 17.64	106.67 ± 14.53	240 ± 15.28
Sunny	431 ± 20.60	124 ± 3.06	103.33 ± 6.67	156.67 ± 8.82	277 ± 11.36
P value	0.0001**	0.0001**	0.24	0.05*	0.05*

* Significant; ** Highly significant

Table 5 Physico-chemical characteristics and available sulphur and micronutrients content of four agricultural soils of Mahabubnagar dist

GPS location	Soil type	pH	N	P	K	S	Fe	Mn	Zn	Cu
Malleboinpally lati 78.13388, long 16.77465	Shallow black soil	8.0	220.10	95.62	118.42	8.8	2.27	24.15	0.26	0.25
Makthal lati 77.527836, long 16.724264	Red soil	7.0	199.47	86.19	103.69	11.2	5.32	26.31	0.98	0.24
Kalwakurthy lati 78.492231, long 16.66739	Deep black	8.0	248.95	103.26	132.14	8.2	2.24	32.123	0.35	0.51
Narayanpet lati 77.49425, long 16.74231	Sandy soil	8.0	192.96	79.87	98.78	3.2	0.13	9.48	0.05	0.03

NPK in kg ha^{-1}, sulphur and micronutrients in ppm

Table 6 Plant growth parameters, phytohormone production and biochemical characteristics of Mahyco cotton cultivar in four different agricultural soils of Mahabubnagar district

Soil type	Location	Combination	Height of the plant (cm)		Plant fresh weight (g)		Plant dry weight (g)	
			S	*R*	*S*	*R*	*S*	*R*
Shallow black soil	Malleboinpally	Control	22.4 ± 0.00	11.3 ± 0.00	3.9 ± 0.00	1.45 ± 0.00	0.52 ± 0.00	0.4 ± 0.00
		M + BP	33.2 ± 0.52*	22.3 ± 0.34*	4.52 ± 0.014*	1.74 ± 0.04*	0.67 ± 0.028*	0.56 ± 0.014*
Red soil	Makthal	Control	20.2 ± 0.00	11.0 ± 0.00	3.85 ± 0.00	1.4 ± 0.00	0.5 ± 0.00	0.39 ± 0.00
		M + BP	30.7 ± 0.17*	20.3 ± 0.14*	4.24 ± 0.02*	1.66 ± 0.014*	0.61 ± 0.01*	0.61 ± 0.00*
Deep black soil	Kalwakurthy	Control	25.4 ± 0.00	12.7 ± 0.00	3.99 ± 0.00	1.89 ± 0.00	0.58 ± 0.00	0.48 ± 0.00
		M + BP	37.9 ± 0.02*	23.4 ± 0.14*	4.62 ± 0.014*	1.95 ± 0.025*	0.67 ± 0.02*	37.9 ± 0.02*
Sandy soil	Narayanpet	Control	20.1 ± 0.00	10.2 ± 0.00	3.45 ± 0.00	1.34 ± 0.00	0.45 ± 0.00	0.32 ± 0.00
		M + BP	30.16 ± 0.013*	20.16 ± 0.14*	4.20 ± 0.00*	1.62 ± 0.007*	0.61 ± 0.007*	0.51 ± 0.007*

M Mahyco cotton cultivar, *BP Bacillus* sp. PU-7, *S* shoot, *R* root

* Significant at 0.05

(Marschner et al. 2011). *Bacillus* and *Paenibacillus* are also able to produce endospores, which enhances their persistence and viability in soils (Lal and Tabacchioni 2009; Nicholson 2002). Yao et al. (2006) described the cotton plant crop response to *B. subtilis* FZB 24 experimental conditions which has increased the 31 % average yield (t ha^{-1}), 19 % bolls/plant mean number and 11 % mean plant height in cm.

In the present investigation *Bacillus* sp. PU-7 which was identified based on 16S rRNA gene sequence has shown positive results for plant growth promoting characteristics such as phosphate solubilization, IAA and siderophore production.

The ability of PGPRs to convert insoluble phosphorus (P) to an accessible form is an important trait for increasing the plant yields. The fact that certain microbes are capable of dissolving relatively insoluble phosphatic compounds has opened the possibility of inducing microbial solubilization of the phosphates in the soil (Zaidi et al. 2009). Rhizobacteria solubilize the mineral P in the rhizosphere and hence, provided soluble P to the plants. The cause of the mineral P solubilization could probably be due to

Table 7 Phytohormone production and biochemical characteristics of Mahyco cotton cultivar in four different agricultural soils (with mean ± SE) of Mahabubnagar district

Soil type	Location	Combination	IAA (μg g^{-1})	Proteins (mg g^{-1})	Sugars (mg g^{-1})	Chlorophyll (mg g^{-1})	Proline (μg g^{-1})
Shallow black soil	Malleboinpally	Control	382 ± 0.00	62 ± 0.00	77 ± 0.00	62.9 ± 0.00	520 ± 0.00
		M + BP	782.67 ± 4.24*	175.67 ± 1.41*	132.33 ± 1.41*	147.67 ± 2.12*	408.33 ± 10.21*
Red soil	Makthal	Control	650 ± 0.00	102 ± 0.00	92.00 ± 0.00	109 ± 0.00	680 ± 0.00
		M + BP	715.33 ± 14.14*	137.67 ± 0.71*	109.00 ± 1.41*	113.00 ± 1.41*	556.67 ± 15.28*
Deep black soil	Kalwakurthy	Control	700 ± 0.00	69.00 ± 0.00	87.00 ± 0.00	70.1 ± 0.00	410 ± 0.00
		M + BP	811.67 ± 10.61*	219.00 ± 1.41*	151.00 ± 1.42*	182.33 ± 3.54*	316.67 ± 5.77*
Sandy soil	Narayanpet	Control	350 ± 0.00	58.00 ± 0.00	68.00 ± 0.00	52.3 ± 0.00	740 ± 0.00
		M + BP	635.67 ± 5.66*	99.00 ± 3.54*	99.00 ± 4.95*	107.33 ± 3.54*	646.67 ± 30.33*

* Significant at 0.05

secretion of organic acids such as gluconic, 2-ketogluconic, oxalic, citric, acetic, malic, and succinic acid, etc. It is clear from the present results that *Bacillus* sp. PU-7 is a potential phosphate solubilizer.

Bacilli strains that can solubilizes P and produce siderophore and IAA have widely been reported (Raddadi et al. 2008; Trivedi and Pandey 2008) and have been shown to promote the growth of maize and wheat (Beneduzi et al. 2008; Trivedi and Pandey 2008).

Efficient and potential isolate *Bacillus* sp. PU-7 has shown maximum growth and high levels of IAA, proteins, sugars, chlorophyll and proline with Mahyco cultivar. This novel isolate has shown tremendous growth in different agricultural soil types, but maximum growth was observed in black soil in terms of plant growth, phytohormonal production and biochemical characteristics.

Conclusion

Novel *Bacillus* sp. PU-7 is an efficient isolate which is enhancing good growth of cotton plants in terms of physical growth parameters, phytohormonal, biochemical properties and supports plant growth in deep black soils followed by shallow black soils. Hence, this novel species with the biotechnological potential as a biofertilizer, with respect to their plant growth promoting properties can be used as bioinoculant in the cotton fields. It is concluded from present findings that efficient novel isolate can be used as a bioinoculant/PSB in different agricultural soils of cotton.

Acknowledgments The authors are thankful to Prof. G. Bagyanarayana, Vice Chancellor and Prof. K. Venkatachalam, Registrar, Palamuru University for the constant support and encouragement.

Conflict of interest The authors declare that they do not have any conflict of interest.

References

Ahmad F, Ahmad I, Khan MS (2008) Screening of free-living rhizospheric bacteria for their multiple plant growth promoting activities. Microbiol Res 163:173–181

Aloni R et al (2006) Role of cytokinin and auxin in shaping root architecture: regulating vascular differentiation, lateral root initiation, root apical dominance and root gravitropism. Ann Bot 97:883–893

Altschul S, Gish W, Myers EW, Lipman DJ (1990) Basic local alignment search tool. Mol Biol 215:403–410

Aslantas R, Cakmakci R, Sahin F (2007) Effect of plant growth promoting rhizobacteria on young apple trees growth and fruit yield under orchard conditions. Sci Hortic 111:371–377

Barlett JM, Stirling D (2003) A short history of the polymerase chain reaction. Methods Mol Biol 226:3–6

Bates LE, Waldren RP, Teare FD (1973) Rapid determination of free proline for water stress studies. Plant Soil 39:205–207

Bell CR et al (1995) Endophytic bacteria in grapevine. Can J Microbiol 41:46–53

Beneduzi A, Peres D, Da Costa PB, Bodanese Zanettini MH, Passaglia LM (2008) Genetic and phenotypic diversity of plant-growth-promoting bacilli isolated from wheat fields in southern Brazil. Res Microbiol 159:244–250

Berg G et al (2005) Impact of plant species and site on rhizosphere-associated fungi antagonistic to *Verticillium dahliae* Kleb. Appl Environ Microbiol 71:4203–4213

Bloemberg GV, Lugtenberg BFJ (2001) Molecular basis of plant growth promotion and biocontrol by rhizobacteria. Curr Opin Plant Biol 4:343–350

Bottini R, Cassan F, Picolli P (2004) Gibberellin production by bacteria and its involvement in plant growth promotion. Appl Microbiol Biotechnol 65:497–503

Bray R, Kurtz LT (1945) Determination of total organic and available forms of phosphorus in soils. Soil Sci 59:39–46

Brick JM, Bostock RM, Silverstone SE (1991) Rapid in situ assay for indole acetic acid production by bacteria immobilized on nitrocellulose membrane. Appl Environ Microbiol 57:535–538

Broadbent P, Bake KF, Franks N, Holland J (1977) Effect of *Bacillus* sp. on increased growth of seedlings in steamed and in non treated soil. Phytopathology 67:1027–1034

Cakmakci R, Kantar F, Algur OF (1999) Sugar beet and barley yields in relation to *Bacillus polymyxa* and *Bacillus megaterium* var. *phosphaticum* inoculation. J Plant Nutr Soil Sci 162:437–442

Choudhary DK, Johri BN (2008) Interactions of *Bacillus* sp. and plants-with special reference to induced systemic resistance (ISR). Microbial Res 164:493–513

Chun J et al (2007) Ez Taxon: a web based tool for the identification of prokaryotes based on 16S ribosomal RNA gene sequences. Int J Syst Evol Microbiol 57:2259–2261

Costacurta A, Vanderleyden J (1995) Synthesis of phytohormones by plant-associated bacteria. Crit Rev Microbiol 21:1–18

Crowley DE (2006) Microbial siderophores in the plant rhizosphere. In: Barton LL, Abadía J (eds) Iron nutrition in plants and rhizospheric microorganisms. Springer, Netherlands, pp 169–198

Crowley DE, Gries D (1994) Modeling of iron availability in the plant rhizosphere. In: Manthey JA, Crowley DE, Luster DG (eds) Biochemistry of metal micronutrients in the rhizosphere. Lewis Publishers, Boca Raton, pp 199–224

De Freitas JR, Banerjee MR, Germida JJ (1997) Phosphate-solubilizing rhizobacteria enhance the growth and yield but not phosphorus uptake of canola (*Brassica napus* L.). Biol Fertil Soil 24:358–364

Dobbelaere S, Croonenborghs A, Thys A, Vande Broek A, Vanderleyden J (1999) Phytostimulatory effect of *Azospirillum brasilense* wild type and mutant strains altered in IAA production on wheat. Plant Soil 212:155–164

Felsenstein J (1993) PHYLIP (phylogeny inference package), version 3.5.1. Department of Genome Sciences, University of Washington, Seattle

Fuentes-Ramírez LE, Caballero-Mellado J (2006) Bacterial biofertilizers. In: Siddiqui ZA (ed) PGPR: biocontrol and biofertilization. Springer, Netherlands, pp 143–172

Ghosh S, Sengupta C, Maiti TK, Basu PS (2008) Production of 3-indolylacetic acid in root nodules and culture by a *Rhizobium* species isolated from root nodules of the leguminous pulse *Phaseolus mungo*. Folia Microbiol 53:351–355

Glick BR, Penrose DM, Li J (1998) A model for the lowering of plant ethylene concentrations by plant growth-promoting bacteria. J Theor Biol 190:63–68

Guindon S, Gascuel O (2003) A simple, fast, and accurate algorithm to estimate large phylogenies by maximum likelihood. Syst Biol 52:696–704

Gutierrez-Manero FJ, Ramos B, Probanza A, Mehouachi J, Talon M (2001) The plant growth promoting rhizobacteria *Bacillus pumilus* and *Bacillus licheniformis* produce high amounts of physiologically active gibberellins. Physiol Plant 111:206–211

Halda-Alija L (2003) Identification of indole-3-acetic acid producing freshwater wetland rhizosphere bacteria associated with *Juncus effusus* L. Can J Microbiol 49:781–787

Harborne JB (1973) Phytochemical methods. Chapman and Hall Ltd, London

Hariprasad P, Niranjana SR (2009) Isolation and characterization of phosphate solubilizing rhizobacteria to improve plant health of tomato. Plant Soil 316:13–24

Idris EES et al (2004) Use of *Bacillus subtilis* as biocontrol agent. 6. Phytohormone action of culture filtrate prepared from plant growth promoting *Bacillus amyloliquefaciens* FZB24, FZB42, FZB45 and *Bacillus subtilis* FZB37. J Plant Dis Prot 111:583–597

Idris EES, Iglesias DJ, Talon M, Borriss R (2007) Tryptophan-dependent production of indole-3-acetic acid (IAA) affects level of plant growth promotion by *Bacillus amyloliquefaciens* FZB42. Mol Plant Microbe Interact 20:619–626

Jackson ML (1973) Soil chemical analysis. Prentice Hall of India Pvt. Ltd., New Delhi

Kimura M (1980) A simple method for estimating evolutionary rates of base substitutions through comparative studies of nucleotide sequences. J Mol Evol 16:111–120

Kokalis-Burelle N, Vavrina CS, Rosskopf EN, Shelby RA (2002) Field evaluation of plant growth promoting rhizobacteria amended transplant mixes and soil solarization for tomato and pepper production in Florida. Plant Soil 238:257266

Kundu BS, Gaur AC (1980) Effect of nitrogen fixing and phosphate solubilizing microorganisms as single and composite inoculants on cotton. Ind J Mirobiol 20:225–229

Lal S, Tabacchioni S (2009) Ecology and biotechnological potential of *Paenibacillus polymyxa*: a minireview. Ind J Mirobiol 49:2–10

Lane DJ (1991) 16S/23S rRNA sequencing. In: Stackebrandt E, Goodfellow M (eds) Nucleic acid techniques in bacterial systematics. Wiley, Chichester, pp 115–175

Lanyi B (1987) Classical and rapid identification methods for medically important bacteria. Methods Microbiol 19:1–67

Lowry OH, Rosebrough NJ, Farr AL, Randall RJ (1951) Protein measurement with the Folin-phenol reagent. J Biol Chem 193:265–275

Mahadevan A, Shridhar R (1986) Methods in physiological plant pathology. Sivakami publications, Madras, p 213

Marschner P, Crowley D, Rengel Z (2011) Rhizosphere interaction between microorganisms and plants govern iron and phosphorus acquisition along the root axis—model and research methods. Soil Biol Biochem 43:883–894

Martínez-Viveros O, Jorquera MA, Crowley DE, Gajardo G, Mora ML (2010) Mechanisms and practical considerations involved in plant growth promotion by rhizobacteria. J Soil Sci Plant Nutr 10:293–319

McSpadden Gardener BB (2004) Ecology of *Bacillus* and *Paenibacillus* spp. in agricultural systems. Phytopathology 94: 1252–1258

Morgan JA, Bending GD, White PJ (2005) Biological costs and benefits to plant–microbe interactions in the rhizosphere. J Exp Bot 56:1729–1739

Mullis KB (1990) The unusual origin of the polymerase chain reaction. Sci Am 262(4):56–65

Nicholson W (2002) Roles of *Bacillus* endospores in the environment. Cell Mol Life Sci 59:410–416

Ona O, Smets I, Gysegom P, Bernaerts K, Van Impre J, Prinsen E, Vanderleyden J (2003) The effect of pH on indole-3-acetic acid (IAA) biosynthesis of *Azospirillum brasilense* Sp7. Symbiosis 35:199–208

Pal SS (1998) Interaction of an acid tolerant strain of phosphate solubilizing bacteria with a few acid tolerant crops. Plant Soil 198:169–177

Raddadi N, Cherif A, Boudabous A, Daffonchio D (2008) Screening of plant growth promoting traits of *Bacillus thuringiensis*. Ann Microbiol 58:47–52

Richardson AE, Barea JM, McNeill AM, Prigent-Combaret C (2009) Acquisition of phosphorus and nitrogen in the rhizosphere and plant growth promotion by microorganisms. Plant Soil 321:305–339

Rosas SB, Andres JA, Rovera M, Correa NS (2006) Phosphate solubilizing *Pseudomonas putida* can influence rhizobia-legume symbiosis. Soil Biol Biochem 38:3502–3505

Sahin F, Cakmakci R, Kantar F (2004) Sugar beet and barley yields in relation to inoculation with N2-fixing and phosphate solubilizing bacteria. Plant Soil 265:123–129

Saitou N, Nei M (1987) The neighbor-joining method: a new method for reconstructing phylogenetic trees. Mol Biol Evol 4:406–425

Sambrook J, Fritschim EF, Maniatis T (1989) Molecular cloning: a laboratory manual. Cold spring Harbor Laboratory Press, New York

Schwyn B, Neilands JB (1987) Universal chemical assay for detection and determination of siderophore. Anal Biochem 160:47–56

Sharma A, Johri BN (2003) Growth promoting influence of siderophore-producing *Pseudomonas* strains GRP3A and PRS9 in maize (*Zea mays* L.) under iron limiting conditions. Microbiol Res 158:243–248

Shishido M, Breuil C, Chanway CP (1999) Endophytic colonization of spruce by plant growth promoting rhizobacteria. FEMS Microbiol Ecol 29:191–196

Shoebitz M, Ribaudo CM, Pardo MA, Cantore ML, Ciampi L, Curá JA (2009) Plant growth promoting properties of a strain of *Enterobacter ludwigii* isolated from *Lolium perenne* rhizosphere. Soil Biol Biochem 41:1768–1774

Siddiqui ZA (2006) PGPR: prospective biocontrol agents of plant pathogens. In: Siddiqui ZA (ed) PGPR: biocontrol and biocontrol. Springer, Netherlands, pp 112–142

Smibert RM, Krieg NR (1994) Phenotypic characterization. In: Gerhardt P, Murray RGE, Wood WA, Krieg NR (eds) Methods for general and molecular bacteriology. ASM, Washington, DC, pp 607–654

Spaepen S, Vanderleyden J, Remans R (2007) Indole-3-acetic acid in microbial and microorganism-plant signaling. FEMS Microbiol Rev 31:425–448

Spaepen S, Dobbelaere S, Croonenborghs A, Vanderleyden J (2008) Effects of *Azospirillum brasilense* indole-3-acetic acid production on inoculated wheat plants. Plant Soil 312:15–23

Subbiah BV, Asija GL (1956) A rapid procedure for the estimation of available nitrogen in soils. Curr Sci 25:259–260

Sudha SN, Jayakumar R, Sekar V (1999) Introduction and expression of the cry1Ac gene of *Bacillus thuringiensis* in a cereal associated bacterium, *Bacillus polymyxa*. Curr Microbiol 38:163–167

Sundra Rao WVB, Sinha MK (1963) Phosphate dissolving organisms in soil and rhizosphere. Indian J Agric Sci 33:272–278

Swain MR, Naskar SK, Ray RC (2007) Indole-3-acetic acid production and effect on sprouting of Yam (*Dioscorea rotundata* L.) minisetts by *Bacillus subtilis* isolated from culturable cowdung microflora. Pol J Microbiol 56:103–110

Thompson JD et al (1997) The CLUSTAL_X windows interface: flexible strategies for multiple sequence alignment aided by quality analysis tools. Nucl Acids Res 25:4876–4882

Timmusk S, Nicander B, Granhall U, Tillberg E (1999) Cytokinin production by *Paenibacillus polymyxa*. Soil Biol Biochem 31:1847–1852

Trivedi P, Pandey A (2008) Plant growth promotion abilities and formulation of *Bacillus megaterium* strain B 388 (MTCC6521) isolated from a temperate Himalayan location. Indian J Microbiol 48:342–347

Tsavkelova EA, Klimova SY, Cherdyntseva TA, Netrusov AI (2006) Microbial producers of plant growth stimulators and their practical use: a review. Appl Biochem Micro 42:117–126

Tsavkelova EA, Cherdyntseva TA, Klimova SY, Shestakov AI, Botina SG, Netrusov AI (2007) Orchid-associated bacteria produce indole-3-acetic acid, promote seed germination, and increase their microbial yield in response to exogenous auxin. Arch Microbiol 188:655–664

Whipps JM (2001) Microbial interactions and biocontrol in the rhizosphere. J Exp Bot 52:487–512

Yao AV, Bochow H, Karimov S, boturov U, Sanginboy S, Sharipov K (2006) Effect of FZB24 *Bacillus subtilis* as a biofertilizer on cotton yields in field tests. Arch Phytopathol Plant Prot 39:1–6

Zaidi A, Khan S (2005) Interactive effect of rhizospheric microorganisms on growth, yield and nutrient uptake of wheat. J Plant Nutr 28:2079–2092

Zaidi A, Khan MS, Ahemad M, Oves M (2009) Plant growth promotion by phosphate solubilizing bacteria. Acta Microbiol Immunol Hung 56:263–284

18

Simultaneous purification of nitrile hydratase and amidase of *Alcaligenes* sp. MTCC 10674

S. K. Bhatia · P. K. Mehta · R. K. Bhatia · T. C. Bhalla

Abstract *Alcaligenes* sp. MTCC 10674 has a bienzymatic system for the hydrolysis of nitriles. The nitrile hydratase and amidase have been purified simultaneously to homogeneity using a combination of $(NH)_4SO_4$ precipitation, ion exchange chromatography and gel permeation chromatography. Nitrile hydratase and amidase have molecular weight of 47 and 114 kDa, respectively and exist as heterodimer. Optimum temperatures for maximum activity of nitrile hydratase and amidase were 15 °C (2.4 U/mg protein) and 45 °C (2.3 U/mg protein), respectively. Nitrile hydratase showed maximum 7.8 U/mg protein at 50 mM acrylonitrile and amidase has 9.2 U/mg protein at 25 mM propionamide. Nitrile hydratase has V_{max} 10 μmol/min/mg and K_m 40 mM, while amidase has V_{max} 12.5 μmol/min/mg and K_m 45.5 mM, respectively. Heavy metal ions Hg^{2+}, Ag^{+}, Pb^{2+} and Cu^{2+} were strong inhibitors of nitrile hydratase and amidase activity.

Keywords *Alcaligenes* sp. MTCC 10674 · Bienzymatic · Nitrile hydratase · Amidase

Introduction

Nitrilase (EC 3.5.5.1), nitrile hydratase (EC 4.2.1.84) and amidase (EC 3.5.1.4) constitute an important class of nitrilase superfamily (Branner 2002). Nitrilase hydrolyzes the nitriles to acids and ammonia in one-step reaction (Sorokin et al. 2007). However, nitrile hydratase first hydrates the nitrile to corresponding amide which is subsequently hydrolyzed by amidases into organic acid and ammonia (Bhalla and Kumar 2005; Bhatia et al. 2013). There is considerable industrial interest in the enzymatic conversion of nitriles because of the increasing demand for conducting such conversions under mild conditions that are often compatible with the sensitive structures of many industrially important compounds. This environmental friendly bioconversion allows clean and mild synthesis with high selectivity and yield. Several bioprocesses have already been reported for the conversion of nitriles and amide compounds into their corresponding acids (Raj et al. 2006). Number of amides has been synthesized using nitrile hydratase, e.g., nicotinamide and butyramide (Raj et al. 2006; Prasad et al. 2007). Amidases are used as catalyst in effluent treatment and their acyltransferase activity is harnessed for the synthesis of pharmaceutically important compounds such as acetohydroxamic acids and benzohydroxamic acid (Prasad et al. 2007; Sharma et al. 2012; Bhatia et al. 2012). Despite immense potential of these hydrolyzing enzymes, these are not used for the commercial production of acids because of non availability of desired enzymes vis-à-vis their cost, selectivity and stability. *Alcaligenes* sp. MTCC 10674 has a nitrile hydratase and amidase bienzymatic system for the hydrolysis of nitriles. Nitrile hydratase and amidase system was used for the production of α-hydroxyisobutyric acid from α-hydroxyisobutyronitrile (Bhatia et al. 2013), which finds its use in the synthesis of polymethyl methacrylate and acrylic glass (Singh et al. 2006) and acyltransferase activity of amidase used for benzohydroxamic acid production (Bhatia et al. 2012). Whole cell has nitrile hydratase activity accompanied with amidase activity which led to carboxylic acid as side product during the conversion of nitriles into amide (Brady et al. 2004), and purified enzyme can be used to overcome this. Therefore, the objective of

S. K. Bhatia · P. K. Mehta · R. K. Bhatia · T. C. Bhalla (✉)
Department of Biotechnology, Himachal Pradesh University, Summer Hill, Shimla 171005, India
e-mail: bhallatc@rediffmail.com

present study was to develop simple steps for the purification of nitrile hydratase and amidase to reduce the cost of enzyme production and to characterize the nitrile hydratase and amidase using acrylonitrile and propionamide as substrate, respectively.

Materials and methods

Chemicals

All chemicals were of analytical grade and they were purchased from Alfa Aesar, Johnson Matthey Company and Sigma (India).

Medium and cultural conditions

Alcaligenes sp. MTCC 10674 previously isolated from the soil sample of orchid garden of Kinnaur District of Himachal Pradesh (India), was cultured in minimal salt media (MSM) having pH 7.0 and containing g/L, $Na_2HPO_4 \cdot 12H_2O$ 2.5 g, K_2HPO_4 2.0 g, $MgSO_4 \cdot 7H_2O$ 1.0 g, $FeSO_4 \cdot 7H_2O$ 0.1 g, $CaCl_2 \cdot 2H_2O$ 0.6 g, peptone 5.0 g at temperature 25 °C for 24 h. MSM was supplemented with 40 mM isobutyronitrile after 6 h of growth as nitrogen source.

Enzyme assay of nitrile hydratase and amidase

Nitrile hydratase assay

The assay mixture contained 0.125 M NaH_2PO_4/Na_2HPO_4 buffer (pH 8.0), 50 mM acrylonitrile and purified nitrile hydratase of *Alcaligenes* sp. MTCC 10674 at 15 °C for 20 min. Acrylamide production was measured spectrophotometrically at 230 nm. One unit of nitrile hydratase activity was defined as amount of enzyme that hydrates the acrylonitrile to release 1 μmol of acrylamide per minute under assay condition.

Amidase assay

Assay of purified amidase was carried out in 0.075 M NaH_2PO_4/Na_2HPO_4 buffer (8.0), 50 mM amide and purified amidase of *Alcaligenes* sp. MTCC 10674 at 45 °C for 20 min. Ammonia assay was performed for the amidase activity. One unit of amidase activity was defined as the amount of enzyme that hydrolyzes the propionamide to release 1 μmol of ammonia per minute under assay condition. Protein estimation was done according to Bradford method (Bradford 1976).

Purification of nitrile hydratase and amidase

Preparation of cell free extract

Alcaligenes sp. MTCC 10674 cells were cultured in 1 L minimal salt medium at 25 °C. After 24 h of incubation, cells were harvested from the culture broth by centrifugation at 10,000 *g* and washed twice with 0.1 M K_2HPO_4/KH_2PO_4 (pH 7.0) and suspended in the same buffer. Bacterial culture of *Alcaligenes* sp. MTCC 10674 (15 mg/ml) was disrupted using BeadBeater^{TM}. The resultant suspension was centrifuged at 10,000 g at 4 °C for 20 min to remove cell debris. The supernatant fluid was designated as the cell free extract (cell lysate) and stored at 4 °C.

Ammonium sulfate fractionation

The cell free extract was subjected to ammonium sulfate saturation (20–60 %) and the precipitates collected after centrifugation at 15,000 g (25 min at 4 °C) were suspended and dialyzed against the same buffer. The ammonium sulfate fractionate (ASF) having nitrile hydratase and amidase activity was further used for purification.

Ion exchange chromatography

The ASF having nitrile hydratase and amidase activity were subjected to DEAE-ion exchange chromatography. After loading the ASF onto the column, it was washed with potassium phosphate buffer pH 7.0 (0.05 M), until there was no further elution of protein. The column was subsequently eluted with a linear gradient of NaCl (from 0 to 0.5 M) in the same buffer.

Gel permeation chromatography

The protein fraction of ion exchange chromatography having amidase activity were pooled together and applied to gel permeation chromatography column. The gel permeation chromatography was performed using column (2.6 × 60 cm) packed with Sephacryl S-100 high resolution (GE Healthcare) matrix. The gel permeation column was pre-equilibrated with buffer and it was eluted with potassium phosphate buffer pH 7.0 (0.05 M) at a flow rate of 1.0 ml/min. The molecular weights of purified nitrile hydratase/amidase of *Alcaligenes* sp. MTCC 10674 were determined by SDS/Native-PAGE (Laemmli 1970).

Characterization of purified nitrile hydratase/amidase

Buffer pH

The activities of purified nitrile hydrolyzing enzymes of *Alcaligenes* sp. MTCC 10674 were assayed in different buffers, e.g., citrate buffer (4.0–6.0), sodium phosphate buffer (6.0–8.0), potassium phosphate buffer (6.0–8.0), borate buffer (7.0–9.0) and carbonate buffer (9.0–10.0). The 1.0 ml reaction mixture contained 0.1 M of specified buffer, and different substrates, acrylonitrile for nitrile hydratase and propionamide for amidase. Reaction was carried out at 30 °C for 1 h.

Buffer molarity and temperature

The activity of purified nitrile hydrolyzing enzyme was estimated in K_2HPO_4/KH_2PO_4. The effect of buffer molarity on the activities of the enzymes was studied by varying the buffer concentration from 0.025 to 0.125 M in the reaction. The temperature optimum was determined by varying reaction temperature from 25 to 55 °C.

Incubation time and stability

Incubation time for optimum activity of nitrile hydrolyzing enzyme was studied by varying the incubation time of reaction from 10 to 90 min. Thermal stability of the purified enzymes (nitrile hydratase/amidase) was investigated at 25 to 55 °C.

Effect of metals ion and other chemicals

The nitrile hydratase and amidase activity was assayed in the presence of various metal ions and chemicals ($CaCl_2$, $CdCl_2$, $CsCl_2$, $CoCl_2$, $CuSO_4$, $FeSO_4$, $HgCl_2$, $MgSO_4$, $MnCl_2$, $ZnSO_4$, KCl, NaCl, EDTA, DTT and urea) with a final concentration of 1 mM.

Results and discussion

Purification

Nitrile hydratase and amidase of *Alcaligenes* sp. MTCC 10674 were purified using different chromatography techniques. Acrylonitrile was used as substrate for nitrile hydratase characterization and propionamide was used as substrate for characterization of amidase. DEAE Sepharose ion exchange chromatography was performed and amidase got eluted with 0.125 M NaCl in fraction number 4–9 (Fig. 1) and further increase in NaCl concentration up to 0.15 M resulted in nitrile hydratase elution (Fig. 2). Fraction number 6, 7, 8, 9, 10 and 11 of 0.15 M NaCl elution contained purified nitrile hydratase and 6.3-fold purification was achieved in a single step (Table 1). Nitrile hydratase has been already purified from various organisms *Corynebacterium pseudodipthereticu* and *Rhodococcus rhodochrous* PA-34 up to 8.8 and 52-fold, respectively (Li et al. 1992; Prasad et al. 2009). Fraction 4, 5, 6, 7 and 8 obtained with 0.125 M NaCl showed amidase activity, were pooled together and applied in gel permeation column for further purification. Purified amidase was obtained in fraction number 14, 15 and 16 (Fig. 3), and 9.2-fold purification was achieved (Table 2), previously amidase has been purified from *Pseudonocardia thermophila* and *Delftia tsuruhatensis* CCTCCM 205114 up to 48 and 105-fold, respectively (Egorova et al. 2004; Wang et al. 2011). The purified amidase consists of two subunits of 52 and 49 kDa (Fig. 4a), while in native PAGE a single band of 114 kDa was observed (Fig. 4b). Nitrile hydratase also exists as heterodimer of 24 and 21 kDa (Fig. 5a), and in native PAGE a single band of 47 kDa was observed (Fig. 5b). Purified amidase exists as dimer (52 and 49 kDa) and has two subunits α and β as the amidase of *Pseudonocardia thermophila* (2*52 kDa),

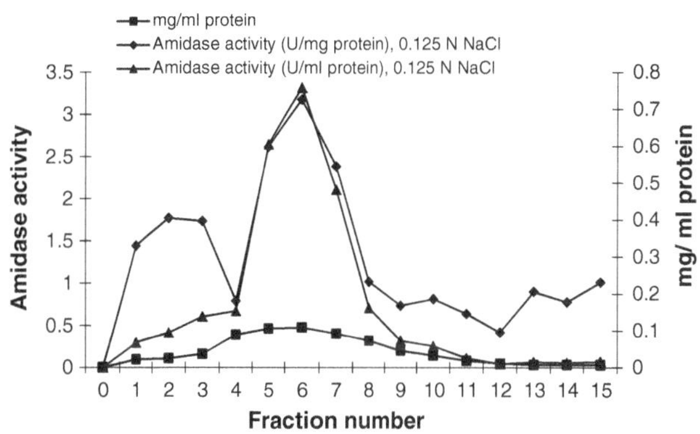

Fig. 1 Elution profile of protein and amidase activity with 0.125 N NaCl during DEAE-Sepharose chromatography

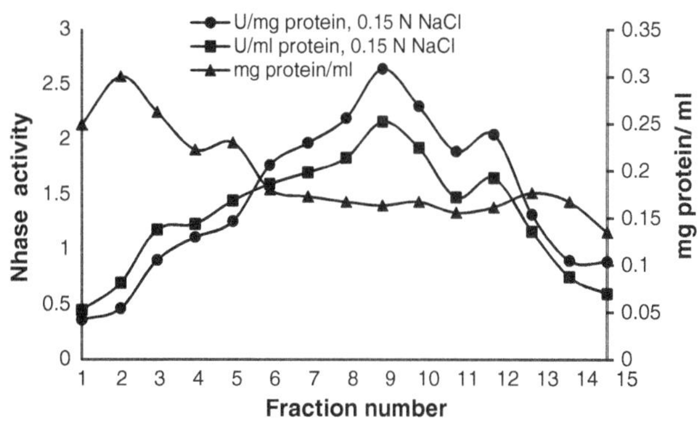

Fig. 2 Protein elution profile and Nhase activity profile in ion exchange chromatography with 0.15 N NaCl elution

Table 1 Purification summary of nitrile hydratase of *Alcaligenes* sp. MTCC 10674

Purification step	Volume (ml)	Protein (mg/ml)	Total protein (mg)	Total activity (U)	Specific activity (U/mg)	% Yield	Fold purification
Homogenate	40	3.3	132	47.2	0.35	100	1
20–60 %	12	6	72	35.0	0.48	74	1.4
DEAE-Sepharose	8	0.163	1.304	2.904	2.22	6	6.3

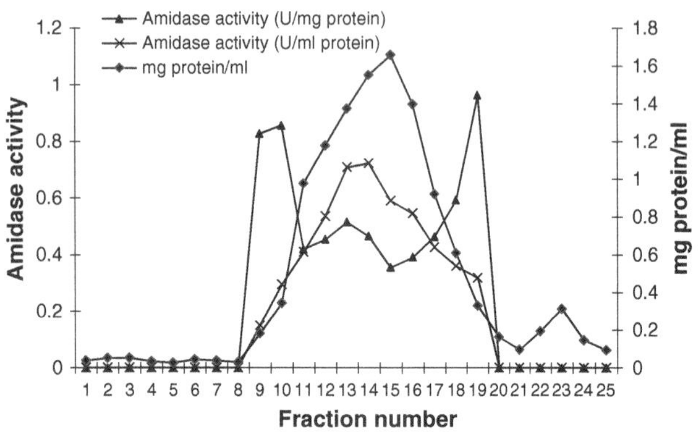

Fig. 3 Elution profile of protein and amidase activity during gel permeation chromatography

Brevibacterium sp. R312 (2*54.7 kDa) (Egorova et al. 2004; Baek et al. 2003). Nitrile hydratase was found as heterodimer of 24 and 21 kDa, as already reported in *Rhodococcus* sp. AJ270 (α-22.9, β-23.4) and *Rhodococcus equi* TG328-2 (α-23, β-24) (Song et al. 2007; Rzeznicka et al. 2010).

Buffer pH and molarity dependence of Nhase and amidase

Nhase and amidase have maximum hydrolysis activity, i.e., 1.75 U/mg protein and 1.2 U/mg protein, respectively in sodium phosphate buffer (pH 8.0). Nitrile hydratase requires 0.125 M buffer for optimum activity (0.14 U/mg protein) and maximum amidase activity was achieved in 0.075 M (1.95 U/mg protein). Both the enzymes are active at neutral pH, as already reported for the nitrile hydratase of *Corynebacterium* sp. C5 (Tani et al. 1989).

Effect of temperature

Amidase showed maximum activity at 45 °C (2.3 U/mg protein), while nitrile hydratase at 15 °C (2.4 U/mg protein). Deviation from the optimal temperature decreased the activity of both enzymes. Amidase has optimum activity at 45 °C same to the amidases of *Pseudonocardia thermophila*, *Bravibacillus* sp. R312, and *Sulfolobus solfataricus,* while hydratases require low temperature for its optimum activity as already reported in *Brevibacterium* sp. R312, *Rhodococcus* sp. AJ270 and *Rhodococcus equi* TG328-2 (Egorova et al. 2004; Baek et al. 2003; Song et al. 2007; Rzeznicka et al. 2010; Kotlova et al. 1999; d'Abusco et al. 2001).

K_m and V_{max} value

Amidase has V_{max} of 12.5 μmol/min/mg and K_m of 45.5 mM. K_m values using propionamide as substrate has been reported earlier from *Pseudonocardia thermophila* (7.4 mM), *Rhodococcus* sp. (1 mM), *Rhodococcus erythropolis* (1.3 mM), *Rhodococcus* sp. (2.6 mM), *Rhodococcus* sp. N-771 (34.2 mM) and *Brevibacterium* sp. (88 mM), respectively, while V_{max} for *Pseudomonas putida* was 0.019 (Egorova et al. 2004; Fournand et al. 1998; Park and Uhm 2008; Nawaz et al. 1994; Thiery et al. 1986; Wyndham and Slater 1986). Purified nitrile hydratase has K_m and V_{max} value for acrylonitrile as 40 mM and 10 μmol/min/mg, respectively. Nitrile hydratases of *Brevibacterium* sp. and *Nocardia* sp. (Kaakeh et al. 1991; Yu et al. 2006) have K_m 16.6, 16.7 and 82 mM, respectively, while V_{max} value for *Rhodococcus* sp. RHA1 and *Rhodococcus* sp. SHZ1 were respectively 2.8 and 12.3 μmol/min/mg (Wang et al. 2007).

Table 2 Purification summary of amidase of *Alcaligenes* sp. MTCC 10674

Purification step	Volume	Protein (mg/ml)	Total protein (mg)	Total activity (U)	Specific activity (U/mg protein)	% Yield	Fold purification
Homogenate	40	3.3	132	18	0.14	100	1.0
20–60 %	12	6	72	12	0.16	66	1.1
DEAE-Sepharose	10	1.4	14	5.5	0.40	30	2.9
Sephacryl-100	6	0.44	2.64	3.36	1.30	19	9.2

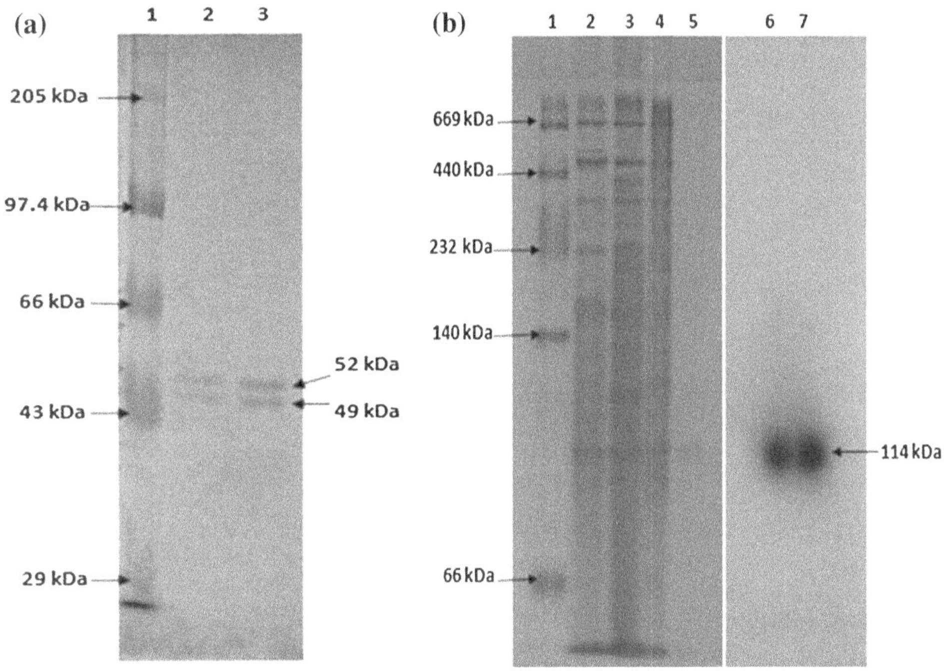

Fig. 4 **a** SDS-PAGE of purified fractions of amidase, lane (1) marker (2) and (3) purified fraction. **b** Native-PAGE analysis of purified amidase fraction, lane (1) Marker (2) crude (3) precipitate (4) pooled fraction of ion exchange chromatography (5) purified fraction of gel permeation (6) and (7) zymogram of purified fraction 6

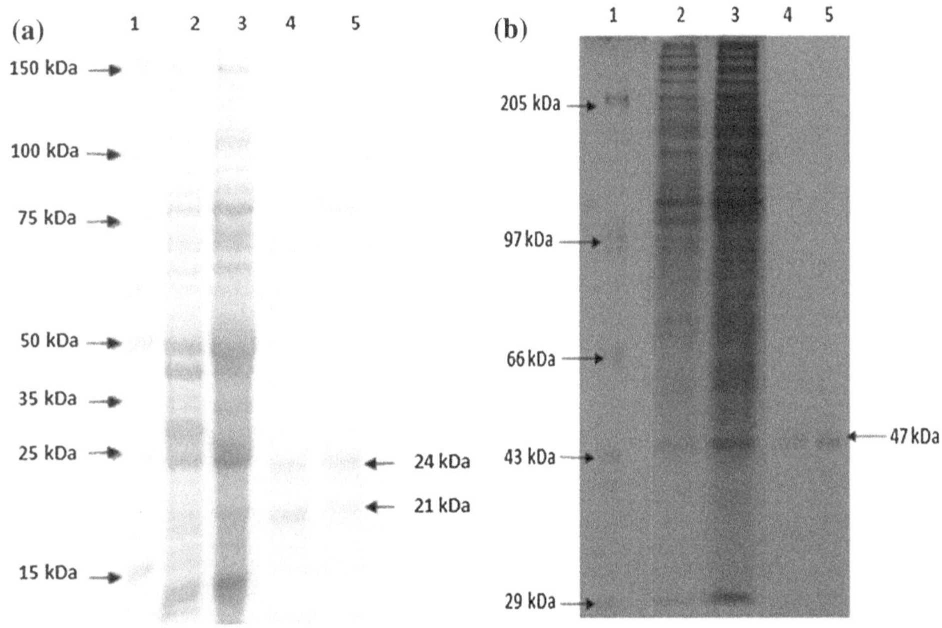

Fig. 5 SDS-PAGE of nitrile hydratase, **a** lane (1) marker (2) crude (3) precipitate (4) fraction number 14, 15. **b** Native-PAGE analysis of purified nitrile hydratase (1) marker (2) crude (3) precipitate (4) purified fraction 14 and 15

Incubation time and stability

Activities of nitrile/amide hydrolyzing enzymes were determined at different time interval (10–90 min). Nitrile hydratase and amidase gave maximum activity, respectively, 4.6 and 6.4 U/mg protein at 20 min of incubation. Amidase has half lives of 31 h, 23 h, 10 h, 4 h, 3 h and 2 h 30 min, respectively at 10, 15, 25, 35, 45 and 55 °C, while nitrile hydratase has half lives of 43, 31, 5, 3, 2.6 and 1 h, respectively at 10, 15, 25, 35, 45 and 55 °C. Both enzymes were stable at lower temperature, and increase in temperature resulted in rapid inactivation of enzyme. Amidase has half lives of 2 h 30 min at 55 °C, while amidase of *Rhodococcus erythropolis* No. 7 has a half life of 30 min at 55 °C (Park and Uhm 2008). Nitrile hydratase has half lives of 31 h at 15 °C and is more stable in comparison to nitrile hydratase of *Microbacterium imperiale* CBS 498-74 having half life of 6.5 h (Cantarella et al. 2004). Both enzymes are stable at lower temperature, increase in temperature resulted in rapid inactivation of enzyme.

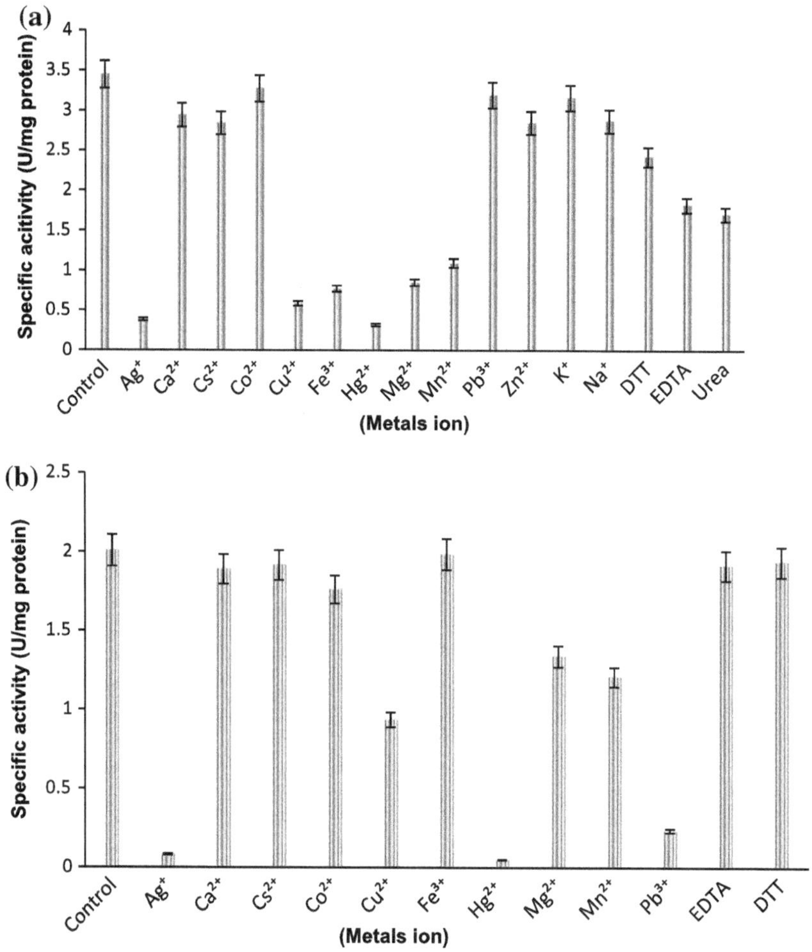

Fig. 6 **a** Effect of metal ions on amidase activity. **b** Effect of metals ion on nitrile hydratase activity

Effect of metal ions

Alcaligenes sp. MTCC 10674 nitrile/amide hydrolyzing enzymes were strongly inhibited by heavy metals ion. Addition of Hg^{2+}, Ag^{+}, Pb^{2+}, Cu^{2+}, Fe^{2+} and Mg^{2+} resulted into 92, 89, 88, 83, 77 and 75 % inhibition of amidase activity (Fig. 6a). In case of Nhase, addition of Hg^{2+}, Ag^{+}, Pb^{2+} and Cu^{2+} in reaction resulted into 97, 95, 88 and 55 % inhibition of activity, respectively (Fig. 6b). Other metal ions and chemical compounds, i.e., Ca^{2+}, Cs^{2+}, Co^{3+}, Zn^{2+}, Na^{+}, K^{+}, urea, EDTA and DTT showed no significant effect on nitrile hydratase and amidase activity. *Alcaligenes* sp. MTCC 10674 nitrile/amide hydrolyzing enzymes were strongly inhibited by heavy metal ions. Metal ions Hg^{2+}, Ag^{+}, Pb^{2+} and Cu^{2+} are strong inhibitors of nitrile hydrolyzing enzyme activity. Metal ions inhibition study revealed that enzymes contains sulfhydral (–SH), alcohol, or acid groups as part of their active sites, any reagent/metal ion which can react with these side groups of amino acid and act as an irreversible inhibitor. Heavy metals such as Ag^{+}, Hg^{2+} and Pb^{2+} have strong affinities for –SH groups and inhibit activity of nitrile metabolizing enzyme *Paracoccus* sp. M-1 (Shen et al. 2012). Several amidases and nitrile hydratase enzyme have been already purified, but simultaneous purification of these two enzymes, however, appear unique which makes this purification process simple and rapid.

Acknowledgments The authors acknowledge the Department of Biotechnology and University Grant Commission, India for financial support in the form of Senior Research Fellowship to Mr. Shashi Kant Bhatia, Praveen Kumar Mehta and Ravi Kant Bhatia. The computational facility availed at Bioinformatics Centre, H P University is also duly acknowledged.

References

Baek DH, Song JJ, Lee SG, Kwon SJ, Asano Y, Sung MH (2003) New thermostable D-methionine amidase from *Brevibacillus borstelensis* BCS-1 and its application for D-phenylalanine production. Enzyme Microb Tech 32:131–139

Bhalla TC, Kumar H (2005) *Nocardia globerula* NHB-2: a versatile nitrile-degrading organism. Can J Microbiol 51:705–708

Bhatia RK, Bhatia SK, Mehta PK, Bhalla TC (2012) Bench scale production of benzohydroxlic acid using acyl transferase activity from *Alcaligenes* sp. MTCC 10674. J Indus Microbiol Biotechnol.

Bhatia SK, Mehta PK, Bhatia RK, Bhalla TC (2013) An isobutyronitrile induced bienzymatic system of *Alcaligenes* sp. MTCC 10674 for the production of alpha-hydroxyisobutyric acid. Bioprocess Biosyst Eng 36(5):613–625

Bradford MM (1976) A rapid and sensitive method for the quantization of microgram quantities of protein utilizing the principle of protein-dye binding. Anal Biochem 72:248–254

Brady D, Beeton A, Zeevaart J, Kgaje C, Rantwijk FV (2004) Characterisation of nitrilase and nitrile hydratase biocatalytic systems. Appl Microbiol Biotechnol 64:76–85

Branner C (2002) Catalysis in the nitrilase superfamily. Curr Opin Stuct Biol 12:775–782

Cantarella M, Cantarella L, Gallifuoco A, Frezzini R, Spera A, Alfani F (2004) A study in UF-membrane reactor on activity and stability of nitrile hydratase from *Microbacterium imperiale* 498-74 resting cells for propionamide production. J Mol Catal B 29:105–113

d'Abusco SA, Ammendola S, Scandurra R, Politi L (2001) Molecular and biochemical characterization of the recombinant amidase from hyper thermophilic archaeon *Sulfolobus solfataricus*. Extremophiles 5:183–192

Egorova K, Trauthwein H, Verseck S, Antranikian G (2004) Purification and properties of an enantioselective and thermo active amidase from the thermophilic actinomycetes *Pseudonocardia thermophila*. Applied Microbiol Biotechnol 65:38–45

Fournand D, Arnaud A, Galzy P (1998) Acyl transfer activity of an amidase from *Rhodococcus* sp. strain R312: formation of a wide range of hydroxamic acids. J Mol Catal B 4:77–90

Kaakeh MR, Legras JL, Duran R, Chion ch KNCK, Arnaud A, Galzy P, Zentralbl (1991) Purification and properties of the nitrile hydratase of a new strain of *Rhodococcus* sp. Microbiol 146:89–98

Kotlova EK, Chestukhina CG, Astaurova OB, Leonova TE, Yanenko AS, Debaboy VG (1999) Isolation and primary characterization of an amidase from *Rhodococcus rhodochrous*. Biochemistry (Mosc) 64:384–390

Laemmli UK (1970) Cleavage of structural proteins during the assembly of the head of bacteriophage T4. Nature 227:680–685

Li WZ, Zhang YQ, Yang HF (1992) Formation and purification of nitrile hydratase from *Corynebacterium pseudodiphteriticum* ZBB-41. Appl Biochem Biotechnol 36:171–181

Nawaz MS, Khan AA, Seng JE, Leakey JE, Siitonen PH, Cerniglia CE (1994) Purification and characterization of an amidase from an acrylamide-degrading *Rhodococcus* sp. Appl Environ Microbiol 60:3343–3348

Park HJ, Uhm KN (2008) HKR-stereoselective amidase from *Rhodococcus erythropolis* No. 7 acting on 4-chloro-3-hydroxybutyramide. J Microbiol Biotechnol 18:552–559

Prasad S, Misra A, Jangir VP, Awasthi A, Raj J, Bhalla TC (2007) A propionitrile-induced nitrilase of *Rhodococcus* sp. NDB 1165 and its application in nicotinic acid synthesis. World J Microbiol Biotechnol 23:345–353

Prasad S, Raj J, Bhalla TC (2009) Purification of a hyperactive nitrile hydratase from resting cells of *Rhodococcus rhodochrous* PA-34. Indian J Microbiol 49(3):237–242

Raj J, Prasad S, Bhalla TC (2006) *Rhodococcus rhodochrous* PA-34: a potential biocatalyst for acrylamide synthesis. Process Biochem 41:1359–1363

Rzeznicka K, Schatzle S, Bottcher D, Klein J, Bornscheuer UT (2010) Applied Cloning and functional expression of a nitrile hydratase (NHase) from *Rhodococcus equi* TG328-2 in *Escherichia coli*, its purification and biochemical characterization. Appl Microbiol Biotechnol 85:1417–1425

Sharma M, Sharma NN, Bhalla TC (2012) Biotransformation of acetamide to acetohydroxamic acid at bench scale using acyl transferase activity of amidase of *Geobacillus pallidus* BTP-5x MTCC 9225. Indian J Microbiol 52(1):76–82

Shen W, Chen H, Kaizhi J, Jun N, Yan X, Li S (2012) Cloning and characterization of a novel amidase from *Paracoccus* sp. M-1, showing aryl acylamidase and acyl transferase activities. Appl Microbiol Biotechnol 4:1007–1013

Singh R, Sharma R, Tiwari N, Geetanjali Rawart DS (2006) Nitrilase and its application as a 'green' catalyst. Chem Biodivers 3: 1279–1287

Song L, Wang M, Yang X, Qian S (2007) Purification and characterization of the enantioselective nitrile hydratase from *Rhodococcus* sp. AJ270. Biotechnol J 2:717–724

Sorokin DY, Pelt S, Tourova TP, Takaichi S, Muyzer G (2007) Acetonitrile degradation under haloalkaline conditions by *Natronocella acetinitrilica* gen. nov. sp. nov. Microbiol 153: 1157–1164

Tani Y, Kurihara M, Nishise H (1989) Characterization of nitrile hydratase and amidase, which are responsible for the conversion of dinitriles to mononitriles, from *Corynebacterium* sp. Agric Biol Chem 53:3151–3158

Thiery A, Maestracci M, Arnaud A, Galzy P (1986) Acyltransferase activity of the wide spectrum amidase of *Brevibacterium* sp. R312. J Gen Microbiol 132:2205–2208

Wang C, Zhang G, Xu X, Li C (2007) Inducing expression and reaction characteristic of nitrile hydratase from *Rhodococcus* sp. SHZ-1. Chin J Chem Engin 15:573–578

Wang YS, Cheng F, Zheng RC, Wang YJ, Zheng YG (2011) Characterization of an enantioselective amidase with potential application to asymmetric hydrolysis of (R, S)-2, 2-dimethylcyclopropane carboxamide. World J Microbiol Biotechnol 27:2885–2892

Wyndham RC, Slater JH (1986) A comparative study of acquired amidase activity in *Pseudomaonas* sp. J General Microbiol 132:2185–2204

Yu H, Shi Y, Luo H, Tian Z, Zhu Y, Shen Z (2006) An over expression and high efficient mutation system of a cobalt-containing nitrile hydratase. J Mol Catal B 43:80–85

19

Chemical composition, antioxidant activity and in vitro antibacterial activity of *Achillea wilhelmsii* C. Koch essential oil on methicillin-susceptible and methicillin-resistant *Staphylococcus aureus* spp.

Seyedeh Mahsan Hoseini Alfatemi · Javad Sharifi Rad · Majid Sharifi Rad · Sasan Mohsenzadeh · Jaime A. Teixeira da Silva

Abstract The present study investigated the chemical composition of the essential oil (EO) from aerial parts (flowering stage) of *Achillea wilhelmsii* C. Koch by GC–MS. In addition, the antioxidant activity of the EO as well as its antimicrobial activity against methicillin-susceptible and methicillin-resistant *Staphylococcus aureus* (MRSA) strains was tested. Antioxidant activity was measured by the ability of the EO to scavenge 1,1-diphenyl-2-picrylhydrazyl (DPPH) radicals while the antimicrobial activity was assessed by the disc-diffusion method. In total, 52 compounds were recognized, accounting for 97.33 % of the EO. The main compounds in the EO were carvacrol (22.49 %), dihydrocarvone (13.23 %), linalool (12 %), 1,8-cineol (11.42 %), camphene (8.31 %), thymol (5.28 %), camphor (3.71 %), pulegone (2.82 %) α-terpineol (2.11 %), bornyl acetate (1.14 %), and farganol (1.01 %). The EC_{50} value of the EO was 0.01 and 0.08 mg/mL for the antioxidant and DPPH-scavenging ability, respectively. *A. wilhelmsii* EO affected methicillin-sensitive *Staphylococcus aureus* (MSSA) and MRSA, but the impact was more effective on MSSA.

Keywords *Achillea wilhelmsii* · Antioxidant activity · Essential oil (EO) · GC–MS · MRSA · MSSA · *Staphylococcus aureus*

Abbreviations

A. wilhelmsii	*Achillea wilhelmsii*
BHA	Butylated hydroxyanisole
EO	Essential oil
DPPH 1	1-Diphenyl-2-picrylhydrazyl
GC–MS	Gas chromatography–mass spectrometry
MRSA	Methicillin-resistant *Staphylococcus aureus*
MSSA	Methicillin-sensitive *Staphylococcus aureus*
S. aureus	*Staphylococcus aureus*

S. M. H. Alfatemi
Department of Bacteriology and Virology, Shiraz Medical School, Shiraz University of Medical Sciences, P.O. Box 71455-119, Shiraz, Iran

J. S. Rad
Zabol Medicinal Plants Research Center, Zabol University of Medical Sciences, P.O. Box 61615-585, Zabol, Iran

J. S. Rad
Department of Pharmacognosy, Faculty of Pharmacy, Zabol University of Medical Sciences, P.O. Box 61615-585, Zabol, Iran

J. S. Rad (✉)
Cereal Health Research Center of Zabol, Zabol University of Medical Sciences, P.O. Box 61615-585, Zabol, Iran
e-mail: javad.sharifirad@gmail.com

M. S. Rad
Department of Range and Watershed Management, Faculty of Natural Resources, University of Zabol, Zabol, Iran

M. S. Rad
Department of Rangeland Science, Gorgan University of Agricultural Sciences and Natural Resources, 49138-15739 Gorgan, Iran

S. Mohsenzadeh
Department of Biology, College of Sciences, Shiraz University, 71454 Shiraz, Iran

J. A. T. da Silva
P.O. Box 7, Miki Cho Post Office, 3011-2, Ikenobe, Kagawa-Ken 761-0799, Japan

Introduction

Herbal medicines are considered to be an important natural medicine for the treatment of health conditions and diseases. The excessive and repeated use of the same drugs used in modern medicine has led to the evolution of antibiotic-resistant microbes, including *Staphylococcus aureus* whose emergence of antibiotic-resistant strains reduces the number of antibiotics available to treat clinical infections caused by this bacterium (Parker and Jevons 1964). *S. aureus* is a highly versatile pathogen with considerable importance in human medicine. *S. aureus* is responsible for a wide range of hospital and community-acquired infections worldwide, from skin infections and food poisoning to life-threatening situations such as toxic-shock syndrome, endocarditis, pneumonia, bacteraemia and osteomyelitis (Kim et al. 2006; Akineden et al. 2008). Traditional medicine involving herbs (or the compounds within them) can solve health and medical problems caused by *S. aureus*. The essential oils (EOs) (primarily from leaves) of *Thymus vulgaris* and *Eucalyptus globulus* were tested against clinical isolates of methicillin-resistant *Staphylococcus aureus* (MRSA), both EOs to possessing antibacterial activity against MRSA, the former being more potent than the latter (Tohidpour et al. 2010). *Daucus crinitus* (a medicinal plant) EOs (derived from stems and leaves of wild plants) inhibited *S. aureus* (Bendiabdellah et al. 2013).

Achillea wilhelmsii C. Koch a perennial medicinal herb belonging to the Asteraceae family has a relatively wide distribution in different parts of Iran (Rechinger 1963; Mozaffarian 1966). It is native to Western Asia and Europe, although populations have also been discovered in North America, Australia and New Zealand (Dokhani et al. 2005). *A. wilhelmsii* has a wide range of reported biological activities, including antispasmodic (Yaeesh et al. 2006), antacid (Niazmand et al. 2010), antioxidant (Candan et al. 2003; Baris et al. 2006; Nemeth and Bernath 2008; Fathi et al. 2011), antihyperlipidemia (Asgary et al. 2000), antihypertensive (Niazmand et al. 2011) and antitumoral (Csupor-Loffler et al. 2009).

The hygiene industry utilizes *A. wilhelmsii* EO to make skin tender and soft and to treat skin inflammations using cream formulations (Pieroni et al. 2004). *A. wilhelmsii* is rich in sesquiterpenes, lactones flavonoids and monoterpenoids which have antioxidant activities (Jaimand and Rezaee 2001; Saeidinia et al. 2005).

The main purpose of the present study was to perform a biological examination on *A. wilhelmsii* C. Koch from Golmakan Khorasan Razavi, Iran by assessing the antioxidant and antimicrobial activity of the EO.

Materials and methods

Plant preparation and procedure

Aerial parts (stems, leaves and flowers) of the flowering stage of *A. wilhelmsii* C. Koch (Fig. 1a, b) were collected in June 2012 from Golmakan (36°28′44″N, 59°9′17″E), Khorasan Razavi, Iran. The plant was taxonomically identified by a botanist at the herbarium of Pharmacognosy, Department of the Faculty of Pharmacy affiliated to Shahid Beheshti University of Medical Sciences of Iran.

Extraction and isolation of the EO

Plant parts were air-dried in the shade at ambient temperature (18–25 °C) for 12 days. Dried aerial parts (100 g) were cut into small pieces and hydro-distilled for 4 h using Clevenger-type apparatus. The resulting EO was dried over

Fig. 1 *A. wilhelmsii* C. Koch in the flowering stage used for EO analyses in this study. **a** Wild population; **b** close-up of flowering capitulum

anhydrous sodium sulfate and stored at 4 °C until GC–MS analysis and bioassays.

GC–MS analysis

In this study, an HP 6890 N GC system coupled with an HP MSD5973 N quadruple mass spectrometer was utilized. The extracted compounds were separated on an HP-5MS capillary column (30 m length, 0.25 mm internal diameter, 0.25 mm film thickness). Split injection:sample ratio for distillation was 50:1. The column oven temperature was programmed to rise from an initial 40 to 150 °C at 4 °C/min, and then to 240 °C at 10 °C/min. Injection temperature and ion source temperature were 240 °C. Helium was used as the carrier gas with a flow rate of 1.2 mL/min. The ionizing energy was 70 eV. All data were obtained by collecting the full-scan mass spectra within the scan range 50–550 amu. Compounds were identified using the Wiley 7n.L Mass Spectral Library (Wiley, New York, NY, USA). Trace compounds were defined as those detected at <0.04 % of the EO.

Antioxidant activity

Antioxidant activity was measured by the paired diene method (Lingnert et al. 1979). The antioxidant activity measured is the ability of the EO to inhibit the peroxidation of linoleic acid in which the double bond is altered to a paired diene. Each EO sample (0.01–30 mg/mL) in methanol (100 μL) was mixed with 3 mL of 10 mM linoleic acid (Sigma Chemical Co., St. Louis, MO, USA) to form an emulsion in 0.2 M sodium phosphate buffer (pH 6.6) in test tubes and placed in the dark at 37 °C to quicken oxidation. After incubation for 17 h, 7 mL of 70 % methanol in deionized water was added, and the absorbance of the mixture was measured at 234 nm against a blank in a Hitachi U-2001 spectrophotometer (Tokyo, Japan). Antioxidant activity was quantified as follows: Antioxidant activity (%) = [(ΔA_{234} of control − ΔA_{234} of sample)/ΔA_{234} of control] × 100. Analyses were repeated three times. α-Tocopherol, butylated hydroxyanisole (BHA) and ascorbic acid (Sigma) were used as standard controls.

Scavenging ability on 1, 1-diphenyl-2-picrylhydrazyl radicals

The scavenging ability of 1,1-diphenyl-2-picrylhydrazyl (DPPH, Sigma) radicals, which was measured by the method of Shimada et al. (1992), is the ability of the EO to react quickly with DPPH radicals and to decrease most DPPH radical molecules. The assay was repeated three times. α-Tocopherol, BHA and ascorbic acid were used as standard controls. Each EO sample (0.5–30 mg/mL) in methanol (5 mL) was mixed with 1 mL of methanolic solution containing DPPH radicals, resulting in a final concentration of 0.2 mM DPPH. The mixture was shaken vigorously, left to stand for 45 min in the dark, and the absorbance was then measured at 517 nm against a blank. The scavenging ability was calculated as follows:

Scavenging ability (%) = [(ΔA_{517} of control − ΔA_{517} of sample)/ΔA_{517} of control] × 100.

EC_{50} value (mg/mL) is the efficient concentration at which the antioxidant activity was inhibited by 50 % and DPPH radicals were scavenged by 50 %, and was gained by interpolation from linear regression analysis.

Antibacterial study

Preparation of microorganisms

The *S. aureus* strains utilized in this study were clinical isolates from patients with *S. aureus*, obtained from the microbiological laboratory of the central hospital in Shiraz, Iran. This study was approved by the ethics committees of Zabol and Shiraz Universities of Medical Sciences. MRSA that were isolated were identified by screening tests on Mueller–Hinton agar (MHA, Torlak, Berlin, Germany) supplemented with 5 % NaCl and 1 mg/mL oxacillin-impregnated disc to isolate MRSA (Roberts et al. 2002). Finally, 10 MRSA strains and 5 methicillin-susceptible *S. aureus* (MSSA) strains were isolated from patients. In this study, two standards strains, ATTC 25923 (MRSA) and PTCC 1341 (MSSA), were used.

Disc-diffusion assay

Antimicrobial tests were performed by the disc-diffusion method using 100 μL of suspension (containing 2.0×10^8 CFU/mL of bacteria) dispersed evenly on MHA in sterilized Petri dishes (80 mm in diameter). To the discs (6 mm in diameter, HiMedia Laboratories Pvt. Ltd., Mumbai, India), 20, 50, 100 and 200 μL of EO and placed on the inoculated agar. The inoculated plates were maintained at 4 °C for 2 h and incubated at 37 °C for 24 h. Antimicrobial activity was evaluated by measuring the zone of inhibition (mm) against the test bacterial (MRSA and MSSA) strains.

Statistical analysis

The EO was prepared in triplicate for chemical characterization and for antioxidant and antibacterial assays. Data was subjected to analysis of variance following a completely random design to determine the least significant difference (LSD) at $P < 0.05$ using SPSS v. 11.5.

Results and discussion

The composition of *A. wilhelmsii* essential oil

The mass spectra and retention indices (RI) were used in this study to determine the EO composition of *A. wilhelmsii*. In total, 52 compounds were identified accounting for 97.33 % of the EO components (Table 1). The main compounds of the EO were (in decreasing order) carvacrol (22.49 %), dihydrocarvone (13.23 %), linalool (12 %), 1,8-cineol (11.42 %), camphene (8.31 %), thymol (5.28 %), camphor (3.71 %), pulegone (2.82 %), α-pinene (2.2 %), terpineol (2.11 %), bornyl acetate (1.14 %) and farganol (1.01 %). Some compounds were detected in trace amounts (not listed in Table 1): heptanal, isopentyl isovalerate, neo-3-thujanol, *cis*-jasmone, elemol. The main compound in the EO of *A. wilhelmsii* from the Golmakan Khorasan Razavi (Iran) area was carvacrol (22.42 %). Javidnia et al. (2004) also found 25.1 % carvacrol in *A. wilhelmsii* oil as the main compound. The major constituent of the EO of the flowers and leaves of *A. wilhelmsii* from Mazandaran (Iran) province was camphor, 21.2 % and 24.1 %, respectively (Azadbakht et al. 2003). Carvacrol and camphor have no harmful effects on humans and environment (Rajendran and Sriranjini 2008; Khani and Asgari 2012). The amount of camphor in *A. wilhelmsii* EO collected from Kerman was 9.0 % (Afsharypuor et al. 1996) and in the EO of aerial parts from Kazeroon (Iran, Fars) province was 2.2 % (Javidnia et al. 2004). Afsharypuor et al. (1996) reported that main compound in the EO of *A. wilhelmsii* was caryophyllene oxide (12.5 %), much higher than that reported in our study (0.08 %). 1,8-cineol, which was found at 3.32 % of the stem EO in this study was the major constituent of the oil of *A. wilhelmsii* from Egypt and Turkey (Javidnia et al. 2004; Baris et al. 2006). The variations in the qualitative and quantitative composition of the EOs from different locations within the same country or from different countries are likely caused by genetic variation, growth conditions, geographic variation and analytical protocols used to assess the EOs. Previous studies showed that monoterpenes, the main part of the *A. wilhelmsii* EO from Golmakan, have influential insecticidal effects against stored product insects (Papachristos et al. 2004; Rajendran and Sriranjini 2008). Consequently, the *A. wilhelmsii* EO from Golmakan could be an important optional phytochemical control strategy without undesirable effects such as direct toxicity to humans and environmental pollution (Rajendran and Sriranjini 2008; Khani and Asgari 2012). In addition, *A. wilhelmsii* EO contains sesquiterpenes, lactones and flavonoids, which have in the ability to lower blood lipid levels and hypertension (Asgary et al. 2000).

Table 1 Composition of the EO of *A. wilhelmsii*, identified by RI-MS*, relative to the literature

No.	Name of compound	RI	Relative % in EO
1	2-Methyl-butyl-acetate	885	0.43
2	Heptanal	910	0.01
3	α-Thujene	926	0.81
4	α-Pinene	935	2.22
5	Camphene	952	8.31
6	Sabinene	978	0.64
7	β-Pinene	984	0.78
8	3-Octanone	989	0.32
9	α-Phellandrene	1010	0.27
10	α-Terpinene	1015	0.73
11	*p*-Cymene	1028	0.81
12	1,8-Cineol	1034	11.42
13	Phenylacetaldehyde	1039	0.18
14	γ-Terpinene	1056	0.64
15	*cis*-Sabinene-hydrate	1064	0.16
16	*trans*-Linalool oxide	1076	0.29
17	Terpinolene	1086	0.31
18	α-Pinene oxide	1094	0.19
19	Linalool	1105	12.0
20	*cis*-Thujone	1108	0.21
21	Hotrienol	1111	0.25
22	Isopentyl isovalerate	1113	0.03
23	α-Thujone	1119	0.12
24	α-Campholenal	1123	0.32
25	Allo-Ocimene	1129	0.12
26	Isopinocarveol	1130	0.11
27	*trans*-Sabinol	1137	0.24
28	Camphor	1140	3.71
29	*p*-Menth-3-en-8-ol	1144	0.4
30	*neo*-3-Thujanol	1149	0.01
31	*trans*-β-Terpineol	1158	0.38
32	Borneol	1162	0.42
33	*n*-Nonanol	1166	0.01
34	4-Terpineneol	1170	0.37
35	*para*-Methyl-acetophenone	1175	0.12
36	*para*-Cymen-8-ol	1180	0.14
37	α-Terpineol	1185	2.11
38	Myrtenol	1190	0.21
39	Verbenone	1204	0.26
40	Farganol	1209	1.01
41	*trans*-Carveol	1218	0.12
42	Dihydrocarvone	1228	13.23
43	Pulegone	1248	2.82
44	Bornyl acetate	1252	1.14
45	Thymol	1288	5.28
46	Carvacrol	1328	22.49
47	Verbanol acetate	1369	0.25
48	*cis*-Jasmone	1398	0.04

Table 1 continued

No.	Name of compound	RI	Relative % in EO
49	Neryl isobutyrate	1467	0.38
50	Menthyl isovalerate	1532	0.42
51	Caryophyllene oxide	1549	0.08
52	Elemol	1568	0.01

RI retention indices relative to C6–C24 *n*-alkanes on the DB-5 column; *MS* mass spectrum (as indicated by the Wiley 7n.L Mass Spectral Library)

Table 2 EC_{50} values (mg/mL) of the *A. wilhelmsii* EO in two assays

	Antioxidant activity (mg/mL)	Scavenging ability (mg/mL)
A. wilhelmsii EO	0.01 ± 0.02 d	0.08 ± 0.08 d
α-Tocopherol	0.08 ± 0.01 b	0.11 ± 0.01 b
BHA	0.05 ± 0.01 c	0.07 ± 0.01 c
Ascorbic acid	4.03 ± 0.07 a	9.09 ± 0.06 a

Values are mean of ±SD of three replicates. Mean values with different letters within a column are significantly different ($P < 0.05$; LSD)

Table 3 Antibacterial activity of the EO of plants against MSSA standard, MSSA ($n = 5$), MRSA standard, and MRSA ($n = 10$) strains Diameter of the zone of inhibition (mm)

EO volume (μl)	MSSA standard	MSSA ($n = 5$)	MRSA standard	MRSA ($n = 10$)
25	14.01 ± 1.01	11.56 ± 1.04	6.25 ± 0.38	7.55 ± 0.28
50	16.01 ± 0.01	15.84 ± 0.89	12.1 ± 0.28	9.00 ± 0.33
100	14 ± 1.38	18.49 ± 1.23	12 ± 1.01	12.46 ± 1.07
200	27 ± 0.079	22.56 ± 0.45	19 ± 0.017	14.22 ± 0.11

Values are mean of ±SD of three replicates

Antioxidant activity and scavenging ability

The results for antioxidant activity and scavenging ability on DPPH radicals of the EOs assayed are summarized in Table 2. The efficiency of antioxidant activity and scavenging ability is inversely related with their EC_{50} values. The antioxidant activity EC_{50} values were 0.08, 0.05, 4.07 and 0.01 mg/mL for α-tocopherol, BHA, ascorbic acid and *A. wilhelmsii* EO, respectively. The scavenging ability EC_{50} values was 0.11, 0.07, 9.09 and 0.58 mg/mL for α-tocopherol, BHA, ascorbic acid and *A. wilhelmsii* EO, respectively. The antioxidant activity of *A. wilhelmsii* EO was stronger than three standards tested (Table 2), which could be used for the treatment of human diseases to remove free radicals (Dharmendra et al. 2009).

Antibacterial activity

The results of antibacterial activity of EOs (Table 3) showed that the maximum level of EO (200 μL) was inhibitory (largest zone of inhibition) against MSSA (22.56 mm) and MRSA (14.22 mm). The inhibitory activity against MSSA was greater than against MRSA. The EOs of *A. wilhelmsii* had a more negative impact on MSSA than MRSA (Table 3). Monoterpenes, which are rich in the EO of *A. wilhelmsii*, have powerful antibacterial effects (Unlu et al. 2002; Sokmen et al. 2004; Prabuseenivasan et al. 2006). In addition, phenolic and flavonoid compounds, also present in the EO of *A. wilhelmsii*, have antimicrobial activity (Stojanovic et al. 2005; Eleyinmi 2007; Yaghoubi et al. 2007; Mothana et al. 2009).

Conclusion

In this study, chemical composition, antioxidant activity and in vitro antibacterial activity of *A. wilhelmsii* L. Essential oil on methicillin-susceptible and methicillin-resistant *S. aureus* spp. was investigated. *A. wilhelmsii* C. Koch has emerged as an important medicinal plant. Its EO could be commercialized for its antioxidant, insecticidal and antibacterial applications, or used in the pharmaceutical, cosmetic or perfume industries.

Acknowledgments The authors are grateful to Professor Hildebert Wagner Center of Pharma-Research, Institute of Pharmacy LM-University of Munich, Germany, for his valuable comments on the manuscript.

Conflict of interest The authors declare that they have no conflict of interest.

References

Afsharypuor S, Asgaryand S, Lockwood GB (1996) Constituents of the essential oil of *Achillea wilhelmsii* from Iran. Planta Med 62(1):77–78.

Akineden O, Hassan AA, Schneide E, Usleber E (2008) Enterotoxigenic properties of *Staphylococcus aureus* isolated from goats' milk cheese. Int J Food Microbiol 124(2):211–216.

Asgary S, Naderi GH, Sarrafzadegan N, Mohammadifard N, Mostafavi S, Vakili R (2000) Antihypertensive and antihyperlipidemic effects of *Achillea wilhelmsii*. Drugs Exp Clin Res 26(3):89–93

Azadbakht M, Morteza-Semnani K, Khansari N (2003) The essential oil composition of *Achillea wilhelmsii* C. Koch. leaves and flowers. J Med Plants 2:55–59

Baris O, Culluce M, Sahin F, Ozer H, Kilic H, Ozkan H, Skmen M, Zbek T (2006) Biological activities of the essential oil and methanol extract of *Achillea biebersteinii* Afan. (Asteraceae). Turk J Biol 30:65–73

Bendiabdellah A, Amine Dib MEL, Meliani N, Muselli A, Nassim D, Tabti B, Costa J (2013) Antibacterial activity of *Daucus crinitus* essential oils along the vegetative life of the plant. J Chem 1:7–12.

Candan F, Unlu M, Tepe B, Daferera D, Polissiou M, Sokmen A, Akpulat HA (2003) Antioxidant and antimicrobial activity of the essential oil and methanol extracts of *Achillea millefolium* subsp. *millefolium* Afan. (Asteraceae). J Ethnopharmacol 87(2–3):215–220.

Csupor-Loffler B, Hajdu Z, Zupko I, Rethy B, Falkay G, Forgo P, Hohmann J (2009) Antiproliferative effect of flavonoids and sesquiterpenoids from *Achillea millefolium* S.L. on cultured human tumour cell lines. Phytother Res 23(5):672–676.

Dharmendra D, Prashant K, Jain SK (2009) *In-vitro* anti-oxidant activity of the ethyl acetate extract of gum guggul (*Commiphora Mukul*). Biological Forum 1(1):32–35

Dokhani S, Cottrell T, Khajeddin J, Mazza G (2005) Analysis of aroma and phenolic components of selected *Achillea* species. Plant Food Hum Nutr 6:55–62.

Eleyinmi AF (2007) Chemical composition and antibacterial of *Gongronema latifolium*. J Zhejiang Univ Sci B 8:352–358.

Fathi H, Lashtoo Aghaee B, Ebrahimzadeh MA (2011) Antioxidant activity and phenolic contents of *Achillea wilhelmsii*. Pharmacol Online 2:942–949

Jaimand K, Rezaee MB (2001) Comparative study of the essential oils of three *Achillea* species from Iran. J Essent Oil Res 13:354–356.

Javidnia K, Miri R, Sadeghpour H (2004) Composition of the volatile oil of *Achillea wilhelmsii* c. Koch from Iran. Daru 12:2–9

Khani A, Asgari J (2012) Insecticide activity of essential oils of *Mentha longifolia*, *Pulicaria gnaphalodes* and *Achillea wilhelmsii* against two stored product pests, the flour beetle, *Tribolium castaneum* and the *cowpea weevil*, *Callosobruchus maculatus*. J Insect Sci 12:102–107.

Kim JS, Song W, Kim HS, Cho HC, Lee KM, Choi MS, Kim EC (2006) Association between the methicillin resistance of clinical isolates of *Staphylococcus aureus*, their staphylococcal cassette chromosome mec *(SCCmec)* subtype classification and their toxin gene profiles. J Microbiol Infect Dis 56:289–295.

Lingnert H, Vallentin K, Eriksson CE (1979) Measurement of antioxidative effect in model system. J Food Process Preserv 3:87–103.

Mothana R, Lindequist U, Geraenert R, Bednarski P (2009) Studies of the in vitro anticancer, antimicrobial and antioxidant potentials of selected Yemeni medicinal plants from the island Soqotra. BMC Complement Altern Med 9:7–11.

Mozaffarian V (1966) A dictionary of Iranian plant names: Latin–English–Persian. Farhang Mo'aser

Nemeth E, Bernath J (2008) Biological activities of yarrow species (*Achillea* spp.). Curr Pharm Des 14(29):3151–3167.

Niazmand S, Khooshnood E, Derakhshan M (2010) Effects of *Achillea wilhelmsii* on rat's gastric acid output at basal, vagotomized, and vagal-stimulated conditions. Pharmacogn Mag 6(24):282–285.

Niazmand S, Esparham M, Rezaee SA, Harandizadeh F (2011) Hypotensive effect of *Achillea wilhelmsii* aqueous-ethanolic extract in rabbit. Aveicennal J Phytomed 1:51–56

Papachristos DP, Karamanoli KI, Stamopoulos DC, Menkissoglu-Spiroudi U (2004) The relationship between the chemical composition of three essential oils and their insecticidal activity against *Acanthoscelides obtectus* (Say). Pest Manag Sci 60(5):514–520.

Parker MT, Jevons MP (1964) A survey of methicillin resistance in *Staphylococcus aureus*. Postgrad Med J 40:170–178.

Pieroni A, Quave CL, Villanelli ML, Mangino P, Sabbatini G, Santini L, Boccetti T, Profili M, Ciccioli T, Rampa LG, Antonini G, Girolamini C, Cecchi M, Tomasi M (2004) Ethnopharmacognostic survey on the natural ingredients used in folk cosmetics, cosmeceuticals and remedies for healing skin diseases in the inland Marches. Central-Eastern Italy. J Ethnopharmacol 91(2–3):331–344.

Prabuseenivasan S, Jayakumar M, Ignacimuthu S (2006) *In vitro* antibacterial activity of some plant essential oils. BMC Complement Altern Med 6:39.

Rajendran S, Sriranjini V (2008) Plant products as fumigant for stored product insect control. J Stored Prod Res 44(2):126–135.

Rechinger KH (1963) Flora Iranica, Wien, Austria. Akademische Druke- U. Verlagsanstatt,

Roberts S, Young H, Faulkner S, Renshaw S, Morris AJ (2002) Value of broth cultures in detecting methicillin-resistant *Staphylococcus aureus*. NZ Med J 115:1–2

Saeidinia S, Gohari AR, Yassa N, Shafiee A (2005) Composition of the volatile oil of *Achillea conferta* DC. from Iran. Daru 13:34–36

Shimada K, Fujikawa K, Yahara K, Nakamura T (1992) Antioxidative properties of xanthan on the autoxidation of soybean oil in cyclodextrin emulsion. J Agric Food Chem 40:945–948.

Sokmen A, Sokmen M, Daferera D, Polissiou M, Candan F, Unlu M, Akpulat HA (2004) The in vitro antioxidant and antimicrobial activities of the essential oil and methanol extracts of *Achillea biebersteini* Afan. (Asteraceae). Phytother Res 18:451–456.

Stojanovic G, Radulovic N, Hashimoo T, Palic R (2005) In vitro antimicrobial activity of extracts of four *Achillea* species: the composition of *Achillea clavennae* L. (Asteraceae) extract. J Ethnopharmacol 101:185–190.

Tohidpour A, Sattari M, Omidbaigi R, Yadegar A, Nazemi J (2010) Antibacterial effect of essential oils from two medicinal plants against methicillin-resistant *Staphylococcus aureus* (MRSA). Phytomedicine 17(2):142–145.

Unlu M, Daferera D, Donmez E, Polissiou M, Tepe B, Sokmen A (2002) Compositions and the in vitro antimicrobial activities of the essential oils of *Achillea setacea* and *Achillea teretifolia* (Compositae). J Ethnopharmacol 83:117–121.

Yaeesh S, Jamal Q, Khan AU, Gilani AH (2006) Studies on hepatoprotective, antispasmodic and calcium antagonist activities of the aqueous-methanol extract of *Achillea millefolium*. Phytother Res 20(7):546–551.

Yaghoubi SMJ, Ghorbani GR, Soleimanianzad S, Satari R (2007) Antimicrobial activity of Iranian propolis and its chemical composition. Daru 15:24–48

20

Modulation of xylanase production from alkaliphilic *Bacillus pumilus* VLK-1 through process optimization and temperature shift operation

Lalit Kumar · Davender Kumar · Sushil Nagar · Rishi Gupta · Neelam Garg · Ramesh Chander Kuhad · Vijay Kumar Gupta

Abstract This study was aimed at enhancing the production of xylanase from an alkaliphilic *Bacillus pumilus* VLK-1 in submerged fermentation using wheat bran, a cheap and abundantly available agro-residue, through process optimization and to monitor the effect of temperature shift operation on it. The potential of xylanase in saccharification of wheat straw was also investigated. The results showed that optimization of the fermentation process by one variable approach increased the enzyme yield from 402 to 4,986 IU/ml. Subsequently, optimization of nitrogen and carbon sources through response surface methodology led to high level xylanase production (7,295 IU/ml) which was 1.46-fold greater than one variable approach after 56 h of cultivation at 30 °C. Temperature shift operation during fermentation resulted in maximum xylanase production in lesser duration (48 h instead of 56 h). Enzymatic hydrolysis of the alkali pretreated wheat straw with 500 IU xylanase alone released 173 ± 8 mg sugars/g whereas in combination with cellulase and β-glucosidase released 553 ± 12 mg sugars/g dry substrate in 6 h, indicating its potential in saccharification of the lignocellulosic substrate. Temperature shift operation is likely to be attractive for large scale industrial fermentation due to significant reduction in the operating cost. To our knowledge, this is the first report which showed the effect of temperature shift operation on xylanase production from bacteria. The xylanase production from *Bacillus* sp. in the present study is close to the highest titre reported in the literature. An enhanced xylanase production using wheat bran, a cheap and abundantly available agro-residue, will apparently reduce the enzyme cost, which would be beneficial for industry.

Keywords Wheat straw · Temperature shift · Saccharification · Response surface methodology · Xylanase

L. Kumar · D. Kumar · S. Nagar · V. K. Gupta (✉)
Department of Biochemistry, Kurukshetra University, Kurukshetra 136119, Haryana, India
e-mail: vkgupta59@rediffmail.com

N. Garg
Department of Microbiology, Kurukshetra University, Kurukshetra 136119, Haryana, India

R. Gupta · R. C. Kuhad
Lignocellulose Biotechnology Laboratory, Department of Microbiology, University of Delhi South Campus, Benito Juarez Road, New Delhi 110021, India

Introduction

Xylan, the major renewable hemicellulosic polysaccharide of plant cell walls, forms an interphase between lignin and other polysaccharides. It is a heteropolymer consisting of β-1,4-linked xylopyranose backbone with side-linked groups namely arabinofuranosyl, acetyl and glucuronosyl residues. Its complete hydrolysis requires the cumulative action of endo-β-1,4-xylanase (EC 3.2.1.8), β-xylosidase (EC 3.2.1.37), and a series of enzymes that degrade side chain groups. Among these, the most important enzyme is endo-β-1, 4-xylanase, which cleaves the glycosidic bonds in the xylan backbone to produce xylo-oligosaccharides of various lengths and xylose (Polizeli et al. 2005). Xylanase has gained immense interest due to its biotechnological potential in xylitol and ethanol production, paper industry, production of xylo-oligosaccharides, food industry, textile industry, animal feed industry, etc. (Dhiman et al. 2008; Kuhad and Singh 1993; Sharma and Kumar 2013). Xylanases used in industry are produced mainly from bacteria and fungi (Ahmed et al. 2009).

The wide-scale industrial applications of xylanase require cost-effective production of the enzyme to make the process economical. This can partly be achieved by using cheaply available agro-industrial residues such as wheat bran (Azeri et al. 2010; Sa-Pereira et al. 2002). The optimization of culture medium is a pre-requisite for successful use of microorganisms in industrial biotechnology because its composition can significantly affect the product yield. The optimization by one factor at a time approach requires a considerable effort and time, besides difficulty in analyzing the interactive effects of various factors. However, statistical design techniques can aid in optimization of culture medium, as they reveal the interactions among the process parameters at varying levels. In addition, the optimal level of each parameter for a given target can be calculated. Response surface methodology (RSM) is a widely used statistical approach to improve the enzyme yield though the magnitude of increase may vary (Bocchini et al. 2002; Riswan Ali et al. 2013).

The steady-state fermentation may not result in maximum productivity. The unsteady-state fermentation by changing the operational parameters such as pressure, temperature, flow rate, etc. via externally forced methods may enhance enzyme yield from microorganisms (Yuan et al. 2005). Therefore, it would be pertinent to investigate the effect of temperature shift during submerged fermentation (SmF) on xylanase yield.

Lignocellulosic biomass can be utilized to produce ethanol, a promising alternative energy source for the limited crude oil (Kuhad and Singh 1993). Wheat straw, an abundant low value by-product of wheat production worldwide, is an attractive lignocellulosic material for ethanol production. Its main constituents are 35–45 % cellulose, 20–30 % hemicelluloses and 8–15 % lignin (Saha et al. 2005) and its composition may vary depending on the wheat species, soil, climate conditions, etc. The first step in the conversion of wheat straw into ethanol is its saccharification to produce reducing sugars. Enzymatic saccharification of wheat straw is more promising compared to chemical hydrolysis as it requires low energy and is environment friendly. However, there is limited accessibility of these enzymes to their substrates in the lignocellulose complex due to the presence of lignin. It has been shown that pre-treatment of lignocellulosic materials with chemicals (dilute acid, alkali, sodium borohydrate), steam explosion, biological, or a combination of these, can significantly enhance the accessibility of the enzymes to cellulose and hemicelluloses present in these materials (Çöpür et al. 2012; Govumoni et al. 2013; Gupta et al. 2011; Mosier et al. 2005). Pre-treatment of lignocellulosic biomass is, therefore, crucial before enzymatic saccharification. As xylan is the major hemicellulose in secondary plant cell walls, addition of xylanase during hydrolysis of wheat straw is expected to increase the yield of monomeric sugars.

The present study was envisaged to modulate the xylanase production from an alkaliphilic *B. pumilus* VLK-1 (isolated from soil in our lab) using cheap and abundantly available agro-residue wheat bran through process optimization and temperature shift operation so as to obtain a high yield of the enzyme at low cost for industrial applications. The potential of this xylanase in saccharification of alkali pre-treated wheat straw was also evaluated.

Materials and methods

Raw materials and chemicals

Lignocellulosic substrates (wheat bran and wheat straw) were obtained from the local market, dried and powdered. The chemicals used in this study were purchased from Sigma Chemicals (St. Louis, MO), Hi-Media Laboratories, Mumbai (India) and Merck Laboratories, India.

Microorganism, maintenance and inoculum production

Bacillus. pumilus VLK-1 (GeneBank accession No. JQ350583), a potent xylanase producing alkaliphilic bacterial stain isolated in our laboratory from a soil sample collected from the vicinity of a plywood industry at Yamuna Nagar, Haryana, India. The bacterial strain was grown on xylan-agar medium containing (g/l): peptone 5.0, beef extract 3.0, xylan 1.0, and agar 20.0 (pH 7.0) followed by incubation at 37 °C for 24 h. The culture was stored at 4 °C and sub-cultured fortnightly. Inoculum was developed by inoculating the autoclaved nutrient broth in a conical flask with a loop full of the overnight grown culture of *B. pumilus* VLK-1 followed by incubation at 37 °C at 200 rpm.

Assay of xylanase

The activity of xylanase was assayed according to the method of Bailey et al. (1992) by measuring the amount of reducing sugars (xylose equivalent) liberated from xylan using 3,5-dinitrosalicylic acid (Miller 1959). The reaction mixture containing 490 μl of 1 % birch wood xylan (Sigma) as substrate (prepared in 0.05 M Tris–HCl buffer, pH 8.0) and 10 μl of appropriately diluted enzyme extract was incubated at 60 °C for 10 min. The reaction was then terminated by adding 1.5 ml of 3,5-dinitrosalicylic acid reagent. A control was run simultaneously that contained all of the reagents but the reaction was terminated prior to the addition of enzyme. The reaction mixture was heated in a boiling water bath for 10 min followed by cooling in ice-cold water. The absorbance of the resulting color was measured against the control at 540 nm in a spectrophotometer. The amount of xylose produced was measured from its standard curve. One international unit (IU) of xylanase activity was defined as the amount of enzyme required to release 1 μmol of xylose per min under the specified assay conditions.

Assay of cellulase

Cellulase activity [carboxymethyl cellulase (CMCase) and filter paper hydrolyzing activity (FPase)] was determined according to Ghosh (1987). One IU of cellulase activity was defined as the amount of enzyme required to release 1 μmol glucose per min under the specified assay conditions.

Optimization of xylanase production by one variable at a time approach

Xylanase production in SmF was optimized in flasks (250 ml) containing 50 ml of basal medium comprising of 0.2 g/l $MgSO_4{\cdot}7H_2O$, 1.0 g/l K_2HPO_4 and 2.0 % (w/v) wheat bran by varying one factor at a time. The enzyme production was carried out at 37 °C, pH 7.0, and 200 rpm with an inoculum size of 1.0 % (v/v). After 48 h of incubation, fermentation broth was centrifuged at 10,000×*g* for 15 min and the culture filtrate was collected for assay of xylanase activity as described above. The xylanase yield from *B. pumilus* VLK-1 under SmF conditions was optimized with respect to various parameters such as inoculum age (2–12-h-old culture), inoculum size (1.0–5.0 % v/v), cultivation time (0–72 h), cultivation temperature (30–45 °C), pH of the medium (4.0–11.0) and agitation rate (125–250 rpm) using one variable at a time approach. In addition to these factors, carbon and nitrogen source and additives were also optimized. Various carbon sources (1 % w/v) such as wheat bran, wheat straw, rice straw, birchwood xylan, oat spelt xylan, sugars (lactose, sucrose, xylose and glucose), carboxymethyl cellulose (CMC) and corn cob were tested as the sole substrate for xylanase production in the basal medium. The nitrogen source was optimized by supplementing the basal medium with different organic (yeast extract, tryptone, peptone, beef extract, and casein) and inorganic nitrogen (KNO_3, $NaNO_3$, $Ca(NO_3)_2$ and NH_4Cl) sources at 0.5 % (w/v) each. The optimum concentrations of the best carbon and nitrogen sources were also determined. The effect of different additives (Tween 20, Tween 40, Tween 80, olive oil, glycerol, EDTA and Triton X-100) on xylanase production was investigated by individually supplementing these in the basal production medium at a final concentration of 0.1 % (w/v or v/v) with flask lacking any additive as the control. All the experiments were performed in triplicate and the average values have been reported.

Optimization of xylanase production by statistical approach

The interaction among the four efficient factors, i.e. peptone (A), yeast extract (B), KNO_3 (C) and wheat bran (D), selected from one variable approach, was investigated on xylanase production from *B. pumilus* VLK-1 through RSM using statistical software package Design Expert 8.0.1, Stat-Ease, Inc., USA. A 2^4 full factorial central composite design (CCD) with 16 trials for factorial design, 8 trials for axial point, and 6 trials for replication of central point, leading to a set of 30 experiments was designed to optimize the concentrations of nitrogen source and carbon source for the production of xylanase from *B. pumilus* VLK-1. These 30 experiments were performed in triplicate and the average response value from each experiment was calculated.

Effect of temperature shift on xylanase production

Temperature shift during fermentation is likely to affect xylanase production because the optimum temperatures for growth and enzyme production by the bacterial strain may be different. So, the effect of temperature shift during fermentation was investigated on xylanase production by *B. pumilus* VLK-1. The organism was cultivated in the optimized SmF medium in shake flasks which were then incubated in an orbital shaker. One set of flasks was incubated at 30 °C and another set at 37 °C for a period of 56 h. Third set of flasks was given temperature shift treatment by initially incubating it at 37 °C for a period of 4–12 h and then shifting the cultivation temperature to 30 °C up to a total period of 56 h. Samples were harvested from all the three sets of flasks at different time intervals starting from 24 h for monitoring xylanase production. The experiment was performed in triplicates and the average values have been given.

Enzymatic saccharification of wheat straw

Wheat straw was washed, dried, milled and sieved to obtain a fine powder, which was used for enzymatic saccharification after alkali pre-treatment. The powdered wheat straw was treated with 0.1 N NaOH followed by incubation for 2 h at room temperature along with shaking. Thereafter, the wheat straw was thoroughly washed with tap water to neutral pH and filtered the contents through double-layered muslin cloth. The residue was dried overnight at 60 °C to a constant weight and used for saccharification. The alkali pre-treated wheat straw (1.0 g) was suspended at 2 % w/v consistency in 50 ml of 0.05 M sodium phosphate buffer pH 6.0, 1.0 ml of Tween 80, 500 IU xylanase (produced by *B. pumilus* VLK-1), 80 FPU of FPase and 160 U of β-glucosidase (Sigma) followed by incubation at 40 °C for 6 h in a rotary shaker at 100 rpm. An alkali untreated sample of wheat straw was also processed under identical conditions. A control was also run simultaneously in which no enzyme was added. After incubation, samples were centrifuged at 10,000×*g* for 15 min and the supernatants were analyzed for sugar content using 3,5-dinitrosalicylic acid. The efficiency of saccharification was evaluated by measuring the sugar yield (mg/g dry substrate). The saccharification was optimized with respect to different parameters.

Results and discussion

In this study, extracellular xylanase was produced in basal medium under SmF from *B. pumilus* VLK-1, an alkaliphilic strain. The xylanase titre under unoptimized conditions was 402 IU/ml, which increased to 7,295 IU/ml (18.1-fold increase) through optimization of fermentation parameters by one variable approach and RSM. The potential of this enzyme in saccharification of wheat straw was studied.

Optimization of xylanase production by one variable approach

Xylanase production was monitored using the basal medium (pH 7.0) inoculated with 1.0 % (v/v) of overnight grown secondary inoculum of *B. pumilus* VLK-1 and incubated at 37 °C under shaking at 200 rpm for 48 h by individually varying the factors viz. inoculum age and size, incubation time, cultivation temperature, initial pH of the production medium, carbon source, nitrogen source and additives.

Age and size of inoculum

The effect of inoculum age was studied by measuring the xylanase titre after inoculating 50 ml of the production medium with 1 % (v/v) of 2–12-h-old secondary inoculum. The highest enzyme production was recorded during the log phase reaching maximum (3,056 IU/ml) at 8 h followed by a decline. The profile of xylanase activity as a function of inoculum size (1.0–5.0 % v/v) revealed maximum enzyme yield at 1.0 % (Fig. 1a). An optimal inoculum level is necessary for maintaining the balance between proliferating biomass and available nutrients to obtain maximum enzyme yield. A lower enzyme yield at higher inoculum level could result from faster nutrient consumption. Moreover, a high level of inoculum is not favorable to its application in industry. Most researchers have reported the use of 1.0–2.0 % (v/v) inoculum for hyper production of xylanase in SmF (Nagar et al. 2012; Sanghi et al. 2009).

Cultivation period

Time course of xylanase production showed maximum enzyme activity after 56 h of incubation and thereafter, it remained more or less constant till 72 h (Fig. 1b). The optimum time resulting in maximum enzyme titre is likely to depend on several factors including the microbial strain. A survey of the literature revealed the highest enzyme production from *B. pumilus* SV-85 after 36 h (Nagar et al. 2010), *B. subtilis* ASH after 48 h (Sanghi et al. 2009), and *Bacillus* SSP-34 after 96 h (Subramaniyan and Prema 2000) of incubation. In the above reports, the activity of xylanase exhibited a decline after reaching a maximum value, which might be due to proteolysis of the enzyme. However, in the present study, though the incubation period for xylanase production from *B. pumilus* VLK-1 was longer than some other *Bacillus* sp. yet it did not decline after attaining the highest level. The rationale for this observation is not understood.

Cultivation temperature

A record of xylanase activity at various cultivation temperatures showed the highest enzyme production at 30 °C (Fig. 1c). The optimum temperature for enzyme production by an organism may vary since it is likely to affect growth of the organism. The present findings are in accordance with those of Ratto et al. (1992) who observed highest xylanase production by *Bacillus circulans* at 30 °C using xylan. Other *Bacillus* sp. have been reported to exhibit maximum xylanase titre at 37 °C (Nagar et al. 2012), 50 °C (Subramaniyan and Prema 2000) and 55 °C (Annamalai et al. 2009). Anand et al. (2013) reported the production of an alkali-tolerant and thermostable xylanase from *Geobacillus thermodenitrificans* at 60 °C, however, the activity was only 2.75 U/ml.

Cultivation pH

Each microorganism requires an optimum pH for its growth and activity. The initial pH of the medium may influence many enzymatic systems and the transport of enzymes across the cell membrane. Among a wide range of initial pH of the production medium tested, the optimum pH for xylanase production from *B. pumilus* VLK-1 was found to be 9.0 (Fig. 1d). A similar pH optimum was reported for xylanase production from *B. circulans* (Bocchini et al. 2002; Ratto et al. 1992), *Bacillus* sp. VI-4 (Yang et al. 1995) and *B. subtilis* (Annamalai et al. 2009). The optimum pH

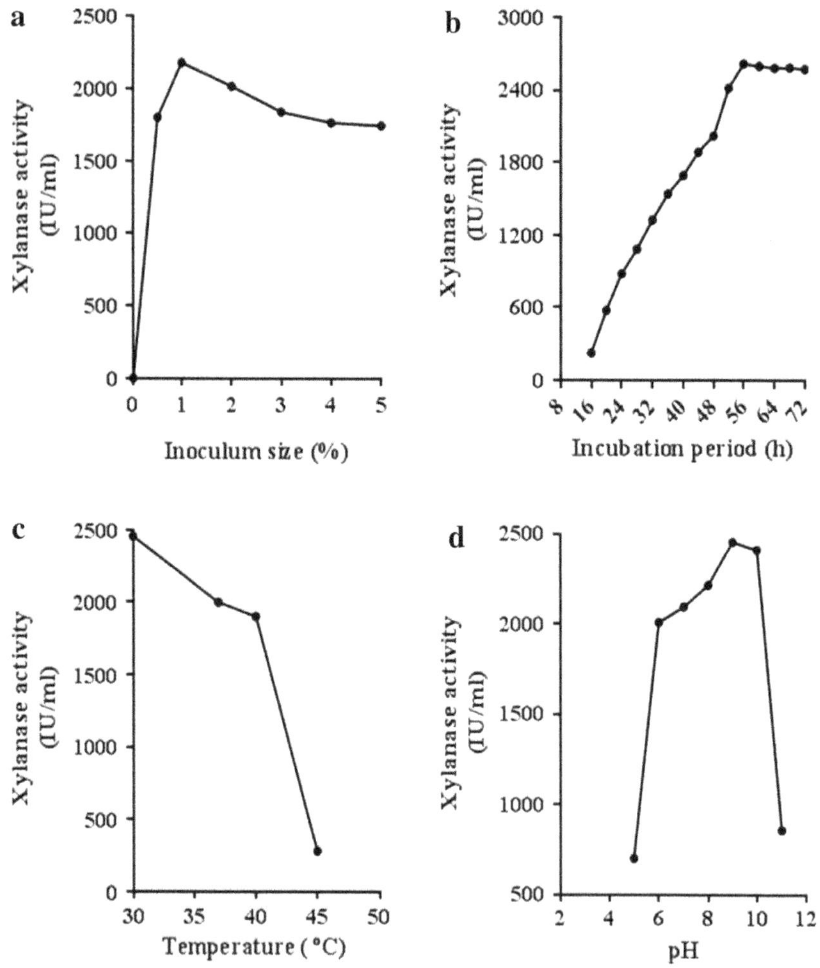

Fig. 1 Effect of **a** inoculum size, **b** incubation period, **c** cultivation temperature, and **d** pH of the production medium on xylanase production by *Bacillus pumilus* VLK-1 in SmF. The enzyme was produced in basal medium and the above mentioned factors were varied one at a time

documented by other researchers was 6.0 (Nagar et al. 2010), 7.0 (Archana and Satyanarayana 1998; Sanghi et al. 2009) and more than 9.0 (Nagar et al. 2012). Thus, the optimum pH for enzyme production varies from one bacterial strain to another. In the present study, the bacterial strain is alkaliphilic capable of growing in the pH range of 6.0–10.0.

Agitation rate

The highest xylanase production by *B. pumilus* VLK-1 was observed at an agitation rate of 200 rpm after 48 h of incubation. The enzyme yield was low if the agitation rate was increased or decreased beyond 200 rpm. This was consistent with earlier reports (Sanghi et al. 2009; Sepahy et al. 2011). The lower enzyme level under low agitation conditions may be attributed to the dissolved oxygen limitation for cell growth, improper mixing of media components and cell clumping.

Carbon source

Carbon source is an essential constituent of the fermentation medium, which affects cellular metabolism of the microorganism. The effect of various carbohydrates and agro-residues, each at a concentration of 1 % (w/v), was investigated on xylanase production by *B. pumilus* VLK-1. Among the different carbon sources tested, wheat bran supported the highest xylanase production followed by birchwood xylan, oat spelt xylan, wheat straw and rice straw (Fig. 2a). In control, the enzyme production was negligible. On varying the concentration of wheat bran from 1.0 to 8.0 % in the basal medium, the highest xylanase production was recorded at 4 %. The xylanase production was higher with wheat bran as carbon source as compared to pure xylan. It might be due to the fact that it contained 54 % carbohydrates (including xylan, pentoses and hexoses), 14 % protein, minerals, vitamins and amino acids (El-Sharnouby et al. 2012), which would support growth of the bacterium and hence, xylanase production. Wheat bran has been documented as the best carbon source for xylanase production from *Bacillus* species by several researchers (Archana and Satyanarayana 1998; Azeri et al. 2010; Irfan et al. 2012; Nagar et al. 2012; Sanghi et al. 2009). In contrast, higher production of a thermotolerant xylanase from *B. subtilis* in SmF was reported with cassava bagasse as compared to wheat bran as carbon source

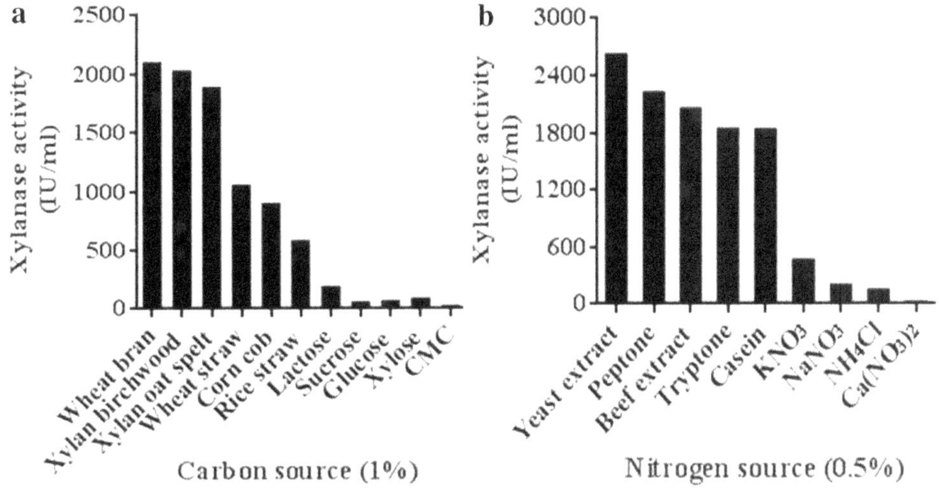

Fig. 2 Effect of different **a** carbon sources and **b** nitrogen sources on xylanase production by *Bacillus pumilus* VLK-1 in SmF

(Sugumaran et al. 2013). Anand et al. (2013) reported the production of an inducible xylanase with the same titre in the presence of xylan and wheat bran. Some researchers found birchwood or oat spelt xylan to support maximum enzyme titre (Annamalai et al. 2009; Giridhar and Chandra 2010; Sa-Pereira et al. 2002). However, it is uneconomical to use purified xylan as substrate for large scale production of xylanase. The use of wheat bran, a cheap and readily available agro-residue, is likely to reduce the cost of xylanase production, a desirable attribute for its application in industries.

Nitrogen source

The influence of several organic and inorganic nitrogen sources (0.5 % each) was examined on xylanase production by *B. pumilus* VLK-1. Addition of organic nitrogen sources such as peptone, yeast extract, and beef extract in the basal media resulted in higher enzyme titre as compared to inorganic compounds (Fig. 2b). The highest xylanase production in yeast extract might be due to the fact that it is a rich source of vitamins and minerals required for growth of the bacterium. The measurement of enzyme production using different concentrations of peptone, yeast extract and KNO_3 individually showed the highest enzyme titre at 0.2, 0.6 and 0.2 %, respectively. Further, the enzyme production was higher in the presence of a combination of 0.2 % peptone, 0.6 % yeast extract and 0.2 % KNO_3 as compared to individual nitrogen sources. There is a lot of variability with regard to optimal nitrogen source for xylanase production by different bacterial strains. The highest xylanase titre from *Bacillus* sp. SSP-34 was recorded with 0.25 % each of yeast extract and peptone (Subramaniyan et al. 2001) whereas tryptone was the best nitrogen source for *B. circulans* AB16 (Dhillon and Khanna 2000). In some cases, maximum xylanase production from *Bacillus* sp. was obtained with yeast extract (Annamalai et al. 2009; Sugumaran et al. 2013) or a combination of yeast extract and tryptone (Giridhar and Chandra 2010; Sepahy et al. 2011).

Effect of additives

The effect of various additives at a concentration of 0.1 % was examined on xylanase production. Addition of olive oil and Tween 80 individually to the basal medium stimulated the enzyme production by 37 and 38 %, respectively, whereas Tween 20 and Tween 40 did not show any significant effect as compared to the control (Table 1). Triton X-100 and EDTA were found to be inhibitory. These results are in line with earlier reports in which 0.1 % Tween 80 was found to enhance xylanase production (Giridhar and Chandra 2010). Nagar et al. (2010) observed stimulation of xylanase production from *B. pumilus* SV-85S by 0.2 % olive oil and 0.1 % Tween 80 but inhibition by a combination of these two additives.

Xylanase titre under optimized conditions through one variable approach

Xylanase production by *B. pumilus* VLK-1 increased from 402 to 4,986 IU/ml under the optimized conditions (1 % of 8-h-old secondary inoculum, incubation time 56 h, temperature 30 °C, pH 9.0, agitation rate 200 rpm, 4 % wheat bran as carbon source, a combination of 0.2 % peptone, 0.6 % yeast extract and 0.2 % KNO_3 as nitrogen source, 0.1 % olive oil and 0.1 % Tween 80) through one variable approach.

RSM for optimization of nitrogen and carbon source

The optimum combination of nitrogen source (peptone, yeast extract and KNO_3) and carbon source (wheat bran) for xylanase production from *B. pumilus* VLK-1 in SmF

Table 1 Effect of various additives on xylanase production by *B. pumilus* VLK-1

Additive	Xylanase activity (IU/ml)
Control	3,660 ± 21
Tween 20	3,661 ± 16
Tween 40	3,664 ± 11
Tween 80	3,980 ± 12
Olive oil	3,942 ± 10
Glycerol	2,939 ± 16
EDTA	2,927 ± 13
Triton X-100	567 ± 06

Table 2 Different coded level of variables of central composite design (CCD) for xylanase production by *B. pumilus* VLK-1 under submerged fermentation

Independent variables	Unit	Coded level of variables				
		−2	−1	0	+1	+2
Peptone	%	−0.17	0.05	0.28	0.5	0.73
Yeast extract	%	−0.25	0.1	0.45	0.08	1.15
KNO_3	%	−0.17	0.05	0.28	0.5	1.0
Wheat bran	%	−1.0	1.0	3.0	5.0	7.0

was determined through RSM. Each variable was used at five coded levels (Table 2) and all the variables were coded as '0' at the central point. The values of xylanase activity predicted by central composite design (CCD) and those observed experimentally are shown in Table 3. Responses of the CCD design were fitted with a polynomial quadratic equation. The overall polynomial equation for xylanase production was:

$$\begin{aligned} Y = {} & 6,667.3 - 194.06 \times A + 750.03 \times B - 40.79 \times C \\ & + 1,089.65 \times D - 492.89 \times A \times B - 552.15 \times A \times C \\ & + 398.16 \times A \times D + 217.75 \times B \times C + 623.76 \times B \times D \\ & - 464.97 \times C \times D - 783.55 \times A^2 - 681.5 \\ & \times B^2 1,270.83 \times C^2 - 1,257.15 \times D^2 \end{aligned}$$

where Y is xylanase activity, A is peptone, B is yeast extract, C is KNO_3 and D is wheat bran.

The analysis of variance (ANOVA) for this quadratic model clearly showed that the linear terms of yeast extract and wheat bran; square terms of peptone, yeast extract, KNO_3 and wheat bran, and all the interactive terms except of yeast extract and KNO_3 had a significant effect on enzyme production (on the basis of their p-values where $p \leq 0.0001$ is considered to be significant) (Table 4). The statistical significance of the model equation was evaluated by the F-test for analysis of variance (ANOVA) which showed that the regression was statistically significant with a 99 % confidence level. The model F-value of 117.58 (as shown by the Fisher's test) implied that the model was significant. A low probability value for the regression model [(Prob > F) < 0.0001] confirmed that the model was statistically significant for the optimization of nitrogen source and wheat bran for xylanase production. The determination coefficient (R^2) of the model was 0.99 indicating that the model could explain up to 99 % of the variability. Moreover, R^2 value was in reasonable agreement with adjusted R^2 (0.9825), which corrects the R^2 value for sample size and number of terms in the model. The predicted R^2 was 0.95. The Adequate precision measures the signal to noise ratio. A ratio >4 was desirable. The ratio of 33.72 indicated an adequate signal. So, the model was significant for the process. The response surface plots (3D and contour) were analyzed to determine interactions among the variables and the optimum value of each factor for maximum xylanase production. Each plot addressed the combined effect of two variables while the remaining variables were maintained at '0' (central) level.

Analysis of the 3D response surface plot between peptone and yeast extract (Fig. 3a) showed maximum xylanase activity at yeast extract concentration close to '+1' level and peptone concentration between '−1' and '0' level. An increase in the concentration of yeast extract from '−1' to '+1' level led to an enhancement of xylanase activity whereas increase in the concentration of peptone above the central level led to a decline in the activity of xylanase.

The interaction between wheat bran and peptone is shown by a contour plot (Fig. 3b). It revealed maximum xylanase activity when peptone concentration was at its '0' level and wheat bran concentration was between '0' and '+1' levels. The 3D plot between wheat bran and yeast extract (Fig. 3c) showed that both these factors at their '+1' levels supported the highest xylanase activity and any decrease from this level led to decline in enzyme activity. Analysis of the contour plot between KNO_3 and yeast extract (Fig. 3d) revealed maximum xylanase activity at the central level of KNO_3 and for yeast extract between '0' and '+1' levels. The 3D response plot between wheat bran and KNO_3 (Fig. 3e) was similar to KNO_3 and yeast extract. The simultaneous effect of all the four variables on xylanase activity can be seen in perturbation graph (Fig. 3f).

The optimum values of the four factors suggested by the response surface model were 0.29 g peptone, 0.27 g KNO_3, 0.77 g yeast extract and 4.53 g wheat bran for 100 ml of production medium. The concentrations of other medium components were kept same as in the original production medium. The model predicted 7,305.5 IU/ml of xylanase activity from the above optimized concentrations of the four factors. The predicted value was validated by determining xylanase production experimentally using the above optimized concentrations of the four factors and

Table 3 Full factorial design for xylanase production by *B. pumilus* VLK-1 under submerged fermentation

Run	Peptone	Yeast extract	KNO_3	Wheat bran	Xylanase activity (IU/ml)	
					Observed	Predicted
1	0.725	0.45	0.275	3	2,881.2	3,144.96
2	0.275	0.45	0.275	3	6,667.3	6,667.3
3	0.05	0.8	0.05	1	1,262	1,601.93
4	0.05	0.1	0.05	1	780	799
5	0.05	0.8	0.5	1	3,884	3,990
6	0.275	0.45	−0.175	3	1,576	1,640.39
7	0.5	0.1	0.5	1	956	1,013.28
8	0.275	0.45	0.275	3	6,667.3	6,667.3
9	0.5	0.1	0.5	5	1,988	1,811.44
10	0.275	−0.25	0.275	3	2,560	2,462.06
11	0.5	0.1	0.05	5	4,489.2	4,362.77
12	−0.175	0.45	0.275	3	4,328	3,921.23
13	0.275	0.45	0.275	3	6,667.3	6,667.3
14	0.05	0.8	0.05	5	5,240	5,162.37
15	0.5	0.8	0.5	5	4,048.2	4,008.74
16	0.275	0.45	0.275	3	6,667.3	6,667.3
17	0.05	0.1	0.05	5	1,828.68	1,864.5
18	0.275	0.45	0.275	3	6,667.3	6,667.3
19	0.275	0.45	0.275	3	6,667.3	6,667.3
20	0.275	0.45	0.725	3	1,734.96	1,502.39
21	0.5	0.8	0.5	1	588	715.55
22	0.05	0.1	0.5	5	1,549.44	1,521.79
23	0.5	0.1	0.05	1	1,604	1,704.73
24	0.275	0.45	0.275	7	3,404.4	3,818
25	0.275	0.45	0.275	−1	16	−567.94
26	0.05	0.1	0.5	1	1,818	2,316.27
27	0.5	0.8	0.05	5	6,024	5,689.09
28	0.5	0.8	0.05	1	528.72	536
29	0.275	1.15	0.275	3	5,465.6	5,441.36
30	0.05	0.8	0.5	5	5,628	1,521.79

other parameters as in one variable approach. The experimental value of xylanase activity (7,295 IU/ml) was close to the predicted value. The xylanase production using RSM was 1.46-fold higher than one variable approach. A comparison of xylanase production from *B. pumilus* VLK-1 under SmF conditions revealed that it was higher than the enzyme titre from most of the microbial sources but close to the highest value reported from *Bacillus* sp. in the literature (Table 5). Riswan Ali et al. (2013) recorded xylanase production of 4.12 IU/ml from *Cellulomonas fimi* in SmF using tapioca stem in RSM optimized medium components.

Thus, optimization experiments revealed maximum xylanase activity (7,295 IU/ml) after 56-h incubation at 30 °C with agitation of 200 rpm in the medium (pH 9.0) containing 4.53 % wheat bran, 0.29 % peptone, 0.77 % yeast extract, 0.27 % KNO_3, 0.1 % olive oil, and 0.1 % Tween 80 and inoculated with 1 % of 8-h-old secondary inoculum.

Table 4 Result of regression analysis for response surface quadratic model

Source	F-value	p value (Prob > F)	
Model	117.58	<0.0001	Significant
A: Peptone	9.89	0.0067	
B: Yeast extract	147.74	<0.0001	
C: KNO_3	0.44	0.5186	
D: Wheat bran	311.83	<0.0001	
AB	42.53	<0.0001	
AC	53.38	<0.0001	
AD	27.76	<0.0001	
BC	8.30	0.0114	
BD	68.82	<0.0001	
CD	37.85	<0.0001	
A^2	184.27	<0.0001	
B^2	139.40	<0.0001	
C^2	484.74	<0.0001	
D^2	474.36	<0.0001	

Modulation of xylanase production through temperature shift

The profile of xylanase production by *B. pumilus* VLK-1 under steady-state fermentation at 30 and 37 °C showed maximum enzyme activity after incubation for 56 h at both the temperatures; however, it was considerably higher at 30 °C than 37 °C (Table 6). This observation indicated that the optimum temperature for growth of the bacteria and enzyme production might be different. Hence, the effect of unsteady-state fermentation, which involved initial cultivation at 37 °C for different time periods (4,8 and 12 h) and subsequently shifting the temperature to 30 °C for the remaining time of fermentation up to a total duration of 56 h, was examined on xylanase production and the results are shown in Table 6. The data clearly showed that initial cultivation at 37 °C for 8 h and subsequently at 30 °C for the remaining period of fermentation led to nearly the same xylanase activity (7,106 IU/ml) after 48 h as that obtained (7,128 IU/ml) after 56 h under steady-state fermentation at 30 °C.

As a consequence of temperature shifting from 37 to 30 °C during cultivation, the time for maximum xylanase production was reduced from 56 to 48 h without any adverse effect on its activity. This approach of xylanase production might be beneficial for industry due to reduction in the operating cost. These results are similar to those of Yuan et al. (2005) who reported that unsteady-state fermentation by shifting the temperature of *Aspergillus niger* from 33 to 27 °C reduced the xylanase production

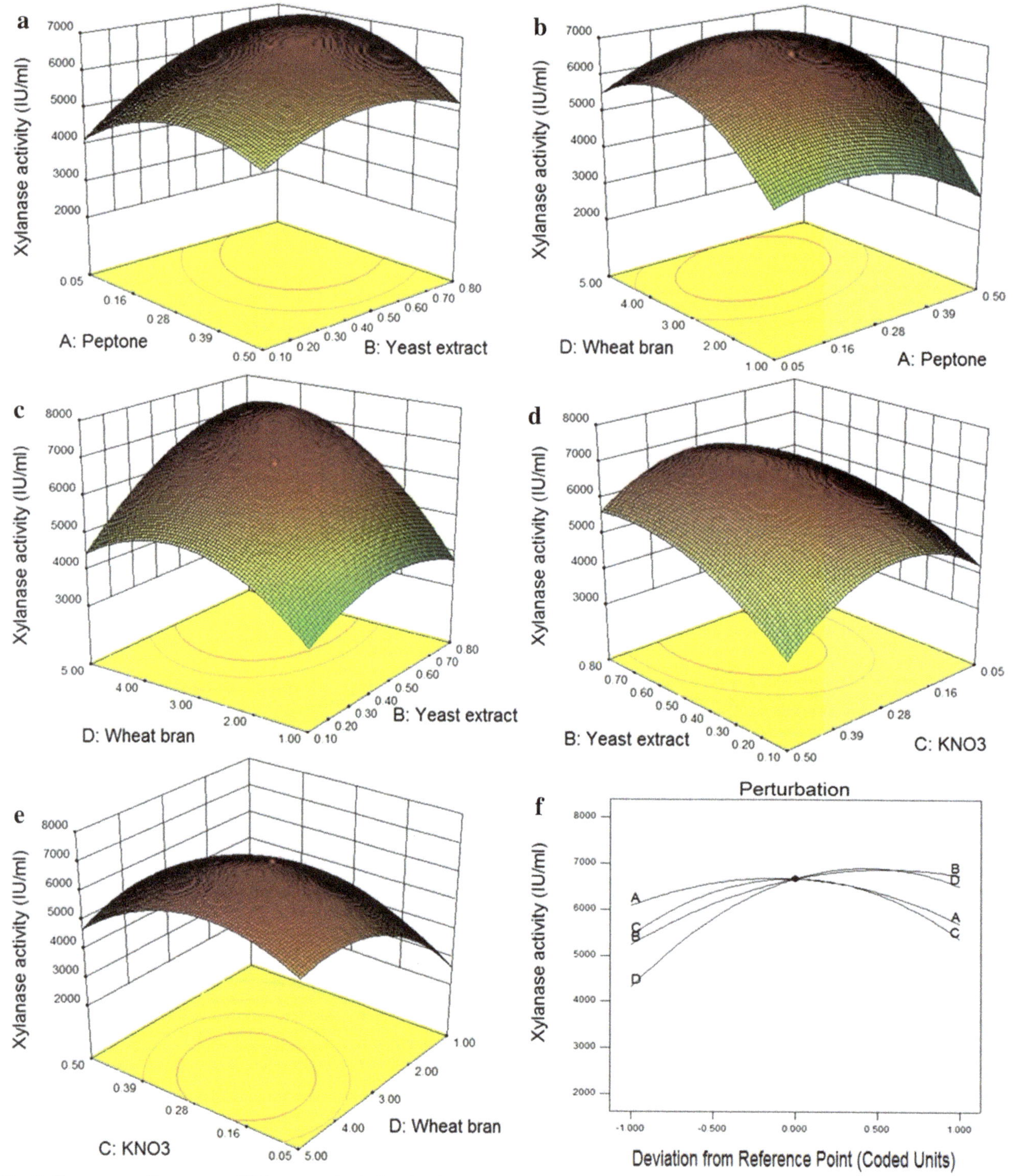

Fig. 3 Response surface curves (3D/contour plots) showing interactions between **a** peptone and yeast extract; **b** wheat bran and yeast extract; **c** KNO_3 and wheat bran; **d** peptone and wheat bran and **e** KNO_3 and yeast extract and **f** is the perturbation plot showing simultaneous interaction of all the four factors on xylanase activity

time by 16 h without any adverse effect on its activity. However, we could not find any report in the literature on enhancement of bacterial xylanase production through temperature shift. This approach would likely be advantageous for large scale enzyme production due to reduction in the operating cost of the fermentor.

Enzymatic saccharification of wheat straw

Enzymatic saccharification of alkali pre-treated wheat straw (1.0 g) was carried out using 0.05 M phosphate buffer (pH 6.0) at 2 % (w/v) consistency with xylanase alone or its combination with FPase and β-glucosidase. This process was optimized by varying the different parameters one at a time. The optimum values of the various parameters were found as 0.05 M phosphate buffer (pH 6.0), 500 IU of xylanase, 80 FPU of FPase, 160 IU β-glucosidase, incubation temperature 40 °C, time 6 h, and 1 % Tween 80 on the basis of release of maximum amount of reducing sugars per gram dry substrate under these conditions. Treatment of alkali pre-treated wheat straw with xylanase alone, mixture of FPase and β-glucosidase, and all the three enzymes (xylanase, FPase and β-glucosidase) for 6 h at 40 °C increased the amount of reducing sugars as a function of time attaining a peak at 6 h releasing 173 ± 8, 396 ± 10 and 553 ± 12 mg sugars/g dry substrate, respectively (Fig. 4). It implied that addition of xylanase during hydrolysis of wheat straw enhanced the amount of sugars in the hydrolyzate for conversion into ethanol. This fact is consistent with earlier reports (Govumoni et al. 2013; Kumar et al. 2012; Saha et al. 2005; Tabka et al. 2006). The optimum time of hydrolysis in the present study, however, was only 6 h which might be due to higher doses of enzymes used compared to other reports.

The sugar release from lignocellulosic substrates apparently depends on several factors including pre-treatment conditions, concentrations of enzymes, temperature and time. The amount of sugars released in the present study were in accordance with Saha et al. (2005) who had reported the release of 565 ± 10 mg/g of acid pre-treated wheat straw on its saccharification with a mixture of cellulase, β-glucosidase, xylanase, and esterase at 45 °C for 72 h. The maximum release of glucose from acidic steam-pre-treated wheat straw was reported with a mixture of cellulase, xylanase and feruloyl esterase (Tabka et al. 2006).

On hydrolysis of alkali untreated wheat straw at 40 °C for 6 h, the amount of reducing sugars released in the presence of only xylanase, mixture of cellulase and β-glucosidase, and all the three enzymes (xylanase, FPase and β-glucosidase) was 39 ± 2, 74 ± 3, and 119 ± 6 mg/g, respectively (data not shown). These values were much lower as compared to the sugars released from alkali pre-treated wheat straw signifying the importance of pre-treatment in hydrolysis of lignocellulosic substrate. The main goal of pre-treatment is to increase the enzyme accessibility and hence to improve digestibility of cellulose (Mosier et al. 2005). Alkali pre-treatment was reported to improve the enzymatic saccharification of wheat straw by removing its lignin from the structure (Çöpür et al. 2012; Govumoni et al. 2013; Gupta et al. 2011). Lignin degradation in alkali treatment could be explained by breakage of ester linkages between lignin

Table 5 Comparison of xylanase production by different *Bacillus* sp.

Organism	Carbon source	Xylanase activity (IU/ml)	References
Bacillus sp. AG20	Wheat bran	3.5	Azeri et al. (2010)
B. circulans	Xylan	400	Ratto et al. (1992)
B. circulans D1	Xylan	1,901	Bocchini et al. (2002)
B. licheniformis	Wheat bran	756	Archana and Satyanarayana (1998)
B. pumilus B20	Wheat bran	313	Geetha and Gunasekaran (2010)
B. pumilus SV34S	Wheat bran	3,454	Mittal et al. (2012)
B. pumilus SV-85S	Wheat bran	2,995 ± 200	Nagar et al. (2012)
B. pumilus SV-205	Wheat bran	7,382 ± 1,200	Nagar et al. (2010)
B. subtilis	Oat spelt xylan	128	Annamalai et al. (2009)
Bacillus sp. V1-4	Birchwood xylan	49	Yang et al. (1995)
B. subtilis	Oat spelt xylan	18	Sa-Pereira et al. (2002)
B. subtilis ASH	Wheat bran	410	Sanghi et al. (2009)
B. pumilus VLK-1	Wheat bran	7,295	Present study

Table 6 Effect of cultivation temperature shift on xylanase production by *B. pumilus* VLK-1

Cultivation temperature	Cultivation period			
	24 h	36 h	48 h	56 h
30 °C	1,858 ± 35	3,261 ± 121	5,126 ± 152	7,128 ± 142
37 °C	1,370 ± 42	2,721 ± 78	4,692 ± 134	5,696 ± 112
4 h at 37 °C, then shifted to 30 °C	1,697 ± 58	2,845 ± 81	4,994 ± 187	6,238 ± 124
8 h at 37 °C, then shifted to 30 °C	1,898 ± 46	3,429 ± 72	7,106 ± 172	7,122 ± 108
12 h at 37 °C, then shifted to 30 °C	1,544 ± 28	2,735 ± 62	4,748 ± 156	6,018 ± 161

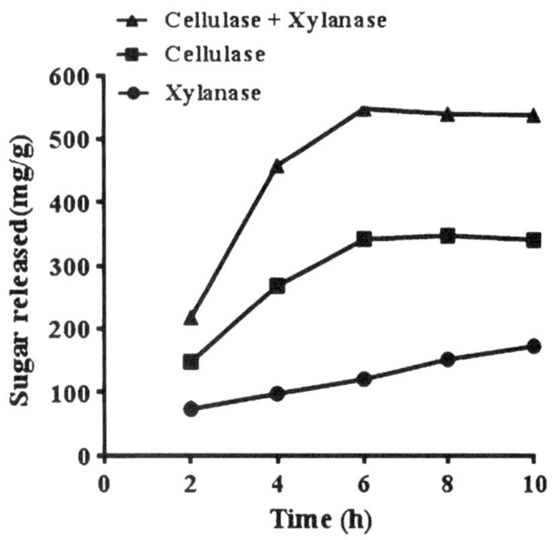

Fig. 4 Enzymatic saccharification of alkali pre-treated wheat straw with xylanase alone (500 IU), mixture of FPase (80 FPU) and β-glucosidase (160 IU), and all the three enzymes (xylanase, FPase and β-glucosidase) per gram dry substrate for 6 h at 40 °C

and xylan and deprotonation of lignin phenolic groups (Sun et al. 2005).

Conclusion

In this study, the enhanced levels of xylanase (7,295 IU/ml) were produced from the alkaliphilic *B. pumilus* VLK-1 by the process optimization employing both one variable and statistical approach using cost-effective agro-residue, which was 18.1-fold higher than under unoptimized conditions. Application of temperature shift during fermentation shortened the time duration from 56 to 48 h for maximum xylanase titre resulting in cost-effective enzyme production. Thus, xylanase production could be profitably modulated through process optimization and temperature shift for use in industry. The xylanase produced in the present study was found to be effective in saccharification of alkali pre-treated wheat straw.

Acknowledgments The authors feel obliged to Department of Biotechnology, Government of India for providing financial assistance in the form of a research project (BT/PR8665/PID/06/363/2007) for this work.

Conflict of interest The authors declare that they have no conflict of interest.

References

Ahmed S, Riaz S, Jamil A (2009) Molecular cloning of fungal xylanases: an overview. Appl Microbiol Biotechnol 84:19–35

Anand A, Kumar V, Satyanarayana T (2013) Characteristics of thermostable endoxylanase and β-xylosidase of the extremely thermophilic bacterium *Geobacillus thermodenitrificans* TSAA1 and its applicability in generating xylooligosaccharides and xylose from agro-residues. Extremophiles 17:357–366

Annamalai N, Thavasi R, Jayalakshmi S, Balasubramanian T (2009) Thermostable and alkaline tolerant xylanase production by *Bacillus subtilis* isolated from marine environment. Indian J Biotechnol 8:291–297

Archana A, Satyanarayana T (1998) Cellulase-free xylanase production by thermophilic *Bacillus licheniformis* A99. Indian J Microbiol 38:135–139

Azeri C, Tamer AU, Oskay M (2010) Thermoactive cellulase-free xylanase production from alkaliphilic *Bacillus* strains using various agro-residues and their potential in bioleaching of kraft pulp. Afr J Biotechnol 9:063–072

Bailey MJ, Biely P, Poutanen K (1992) Laboratory testing of method for assay of xylanase activity. J Biotechnol 23:257–270

Bocchini DA, Alves-Prado HF, Baida LC, Roberto IC, Gomes E, Da Silva R (2002) Optimization of xylanase production by *Bacillus circulans* D1 in submerged fermentation using RSM. Process Biochem 38:727–731

Çöpür Y, Tozluoglu A, Özyürek Ö (2012) Sodium borohydrate ($NaBH_4$) pretreatment for efficient enzymatic saccharification of wheat straw. Bioresour Technol 107:258–266

Dhillon A, Khanna S (2000) Production of a thermostable alkalitolerant xylanase from *Bacillus circulans* AB 16 grown on wheat straw. World J Microbiol Biotechnol 16:325–327

Dhiman SS, Sharma J, Battan B (2008) Industrial applications and future prospects of microbial xylanases: a review. Bio-Resources 3:1377–1402

El-Sharnouby GA, Aleid SM, Al-Otaibi MM (2012) Nutritional quality of biscuit supplemented with wheat bran and date palm fruits (*Phoenix dactylifora* L.). Food Nutr Sci 3:322–328

Geetha K, Gunasekaran P (2010) Optimization of nutrient medium containing agricultural waste for xylanase production by *Bacillus pumilus* B20. Biotechnol Bioprocess Eng 15:882–889

Ghosh TK (1987) Measurement of cellulase activity. Pure Appl Chem 59:257–268

Giridhar PV, Chandra TS (2010) Production of novel halo-alkali-thermo-stable xylanase by a newly isolated moderately halophilic and alkali-tolerant *Gracilibacillus* sp. TSCPVG. Process Biochem 45:1730–1737

Govumoni SP, Koti S, Kothagouni SY, Venkatesgwar S, Rao LV (2013) Evaluation of pretreatment methods for enzymatic saccharification of wheat straw fro bioethanol production. Carbohydr Polym 91:646–650

Gupta R, Khasa YP, Kuhad RC (2011) Evaluation of pretreatment methods in improving the enzymatic saccharification of cellulosic materials. Carbohydr Polym 84:1103–1109

Irfan M, Nadeem M, Syed Q, Baig S (2012) Effect of medium composition on xylanase production by *Bacillus subtilis* using various agricultural wastes. Am-Eurasian J Agric Environ Sci 12(5):561–565

Kuhad RC, Singh A (1993) Lignocellulose biotechnology: current and future prospects. Crit Rev Biotechnol 13:151–172

Kumar A, Gupta R, Shrivastava B, Khasa YP, Kuhad RC (2012) Xylanase production from an alkalophilic actinomycete isolate

Streptomyces sp. RCK-2010, its characterization and application in saccharification of second generation biomass. J Mol Catal B Enzyme 74:170–177

Miller GL (1959) Use of dinitrosalicylic acid reagent for determination of reducing sugar. Anal Chem 31:426–428

Mittal A, Nagar S, Gupta VK (2012) Production and purification of high levels of cellulase-free bacterial xylanase by *Bacillus* sp. SV-34S using agro-residue. Ann Microbiol.

Mosier N, Wyman C, Dale B, Elander R, Lee YY, Holtzapple M, Ladisch MR (2005) Features of promising technologies for pretreatment of lignocellulosic biomass. Bioresour Technol 96:673–686

Nagar S, Gupta VK, Kumar D, Kumar L, Kuhad RC (2010) Production and optimization of cellulase-free, alkali-stable xylanase by *Bacillus pumilus* SV-85S in submerged fermentation. J Ind Microbiol Biotechnol 37:71–83

Nagar S, Mittal A, Kumar D, Gupta VK (2012) Production of alkali tolerant cellulase free xylanase in high levels by *Bacillus pumilus* SV-205. Int J Biol Macromol 50:414–420

Polizeli ML, Rizzatti AC, Monti R, Terenzi HF, Jorge JA, Amorim DS (2005) Xylanases from fungi: properties and industrial applications. Appl Microbiol Biotechnol 67:577–591

Ratto M, Poutanen K, Viikari L (1992) Production of xylanolytic enzymes by an alkalitolernt *Bacillus circulans* stain. Appl Microbiol Biotechnol 37:470–473

Riswan Ali SB, Muthuvelayudham R, Viruthagiri T (2013) Statistical optimization of medium components for hemicellulase production using tapioca stem. J Microbiol Biotechnol Food Sci 2(6):2377–2382

Saha BC, Iten LB, Cotta MA, Wu YV (2005) Dilute acid pretreatment, enzymatic saccharification, and fermentation of wheat straw to ethanol. Process Biochem 40:3693–3700

Sanghi A, Garg N, Kuhar K, Kuhad RC, Gupta VK (2009) Enhanced production of cellulase-free xylanase by alkalophilic *Bacillus subtilis* ASH and its application in biobleaching of kraft pulp. Bio-Resources 4:1109–1129

Sa-Pereira P, Mesquita A, Duarte JC, Barros MRA, Costa-Ferreira M (2002) Rapid production of thermostable cellulase free xylanase by a strain of *Bacillus subtilis* and its properties. Enz Microb Technol 30:924–933

Sepahy AA, Ghazi S, Sepahy MA (2011) Cost-effective production and optimization of alkaline xylanase by indigenous *Bacillus mojavensis* AG137 fermented on agricultural waste. Enzyme Res.

Sharma M, Kumar A (2013) Xylanases: an overview. Br Biotechnol J 3910:1–28

Subramaniyan S, Prema P (2000) Cellulase-free xylanases from *Bacillus* and other microorganisms. FEMS Microbiol Lett 183:1–7

Subramaniyan S, Sandhia GS, Prema P (2001) Biotech control of xylanase production without protease activity in *Bacillus* sp. by selection of nitrogen source. Biotechnol Lett 23:369–371

Sugumaran KR, Kumar BK, Mahalakshmi M, Ponnusami V (2013) Cassava bagasse—a low cost substrate for thermotolerant xylanase production using *Bacillus subtilis*. Int J ChemTech Res 5(1):394–400

Sun JX, Xu F, Sun XF, Xiao B, Sun RC (2005) Physico-chemical and thermal characterization of cellulose from barley straw. Polym Degrad Stabil 88:521–531

Tabka MG, Herpoel-Gimbert I, Monod F, Sigoillot JC (2006) Enzymatic saccharification of wheat straw for bioethanol production by a combined cellulase, xylanase and feryloyl esterase treatment. Enz Microb Technol 39:897–902

Yang VW, Zhuang Z, Elegir G, Jeffries TW (1995) Alkaline active xylanase produced by an alkaliphilic *Bacillus* sp. isolated from kraft pulp. J Ind Microbiol 15:434–441

Yuan QP, Wang JD, Zhang H, Qian ZM (2005) Effect of temperature shift on production of xylanase by *Aspergillus niger*. Process Biochem 40:3255–3257

21

Preferential desulfurization of dibenzyl sulfide by an isolated *Gordonia* sp. IITR100

Abrar Ahmad · Ashok Kumar Chauhan · Hari Narayan Kushwaha · Saleem Javed · Ashwani Kumar

Abstract Several organosulfur compounds are present in the crude oil, and are required to be removed before its processing into transport fuel. For this reason, biodesulfurization of thiophenic compounds has been studied extensively. However, studies on the sulfide compounds are scarce. In this paper, we describe desulfurization of a model sulfidic compound, dibenzyl sulfide (DBS) by an isolated *Gordonia* sp. IITR100. The reaction was accompanied with the formation of metabolites dibenzyl sulfoxide, dibenzyl sulfone and benzoic acid. Studies with recombinant *E. coli* revealed that enzyme DszC of this isolate metabolizes DBS into dibenzyl sulfoxide and dibenzyl sulfone, but the reaction downstream to it is mediated by some enzyme other than its DszA. In reactions where DBS and dibenzothiophene (DBT) were present together, both IITR100 and recombinant *E. coli* exhibited preference for the desulfurization of DBS over DBT. The newly identified capability of IITR100 for desulfurization of both thiophenic and sulfidic compounds suggests its potential use in improved desulfurization of petroleum fractions.

Keywords Biodesulfurization · Dibenzyl sulfide · *Gordonia* sp. IITR100 · *dsz*ABC

A. Ahmad · A. K. Chauhan · A. Kumar (✉)
Environmental Biotechnology Division, CSIR-Indian Institute of Toxicology Research, M.G. Marg, Lucknow 226001, India
e-mail: ashwani.iitr26@gmail.com

A. Ahmad
e-mail: abrarbiochem@yahoo.co.in

A. K. Chauhan
e-mail: ashokchauhan44@gmail.com

H. N. Kushwaha
Pharmacokinetics and Metabolism Division, Central Drug Research Institute, Lucknow 226001, India
e-mail: harry2002_4u@yahoo.co.in

S. Javed
Department of Biochemistry, Faculty of Science, Jamia Hamdard, New Delhi 110062, India
e-mail: saleemjaved70@yahoo.co.in

Introduction

Several organosulfur compounds that include thiols, mercaptans, thiophenes and sulfides are present in crude petroleum (Peinder et al. 2010; Liu et al. 2010; Lam et al. 2012) which are required to be removed, prior to its use in automobiles and industry. For this reason, biodesulfurization of crude oil and its fractions has been studied extensively (Van Hamme et al. 2003a; Gupta et al. 2005; Kilbane 2006; Mohebali and Ball 2008; Xu et al. 2009). Most of these have been done on model thiophenic compound dibenzothiophene (DBT), and details of the desulfurization pathway, responsible enzymes and their genes have been reviewed (Denome et al. 1994; Piddington et al. 1995). Briefly, desulfurization of DBT is mediated by a '4S' pathway that includes serial activity of enzymes DszC, DszA and DszB, leading to the formation of hydroxybiphenyl as end product (Van Hamme et al. 2003a; Gupta et al. 2005; Mohebali and Ball 2008; Xu et al. 2009).

Studies on the desulfurization of sulfide compounds, however, are scarce (Van Hamme et al. 2003b, 2004; Kirkwood et al. 2005). The possible reason could be that the archetypal *Rhodococcus* sp. IGTS8, characterized extensively for the desulfurization of DBT (Mohebali and Ball 2008; Xu et al. 2009), was not found to be able to mediate the desulfurization of the model compound dibenzyl sulfide—DBS (Van Hamme et al. 2004). In contrast, *Rhodococcus* sp. strain JVH1 that was isolated for its ability for desulfurization of DBS does not utilize DBT as a sulfur source (Van Hamme et al. 2004). Consistent with this finding, genome sequence of JVH1 does not contain any genes that are similar to *dsz*ABC (Brooks and Van Hamme 2012). In this report, we describe the desulfurization of both DBT and DBS by an isolated strain *Gordonia* sp. IITR100.

Materials and methods

Chemicals

Dibenzothiophene (DBT), dibenzothiophene sulfone (DBTOO), dibenzyl sulfide (DBS), dibenzyl sulfone (DBSOO) and acetonitrile were purchased from Sigma-Aldrich (St. Louis, MO, USA). Dibenzyl sulfoxide was from MERCK-Schuchardt (Mumbai, India). All the other chemicals were of analytical grade.

Bacterium

A bacterium *Gordonia* sp. IITR100 (16S ribosomal RNA accession no; GU084407), obtained earlier by selective enrichment on dimethyl DBT from an oil-contaminated soil that was present around a refinery in Gujarat, India (Singh et al. 2011), was used. Nucleotide sequence of its *dsz*ABC genes (Accession no; KC693733.1) is >99 % identical to the corresponding genes of *Gordonia alkanivorans* strains 1B (Alves et al. 2007) and RIPI90.

Desulfurization by IITR100

For DBS desulfurization, 20 flasks, each containing 20 ml medium-1 (Na_2HPO_4, 2.0 g; KH_2PO_4, 1 g; $MgCl_2{\cdot}6H_2O$, 0.4 g; $(NH_4)_2C_2O_4$, 4.25 g; $Al(OH)_3$, 0.1 g; $SnCl_2{\cdot}2H_2O$, 0.5 g; KI, 0.05 g; LiCl, 0.01 g; $MnCl_2{\cdot}4H_2O$, 0.8 g; H_3BO_3, 0.05 g; $ZnCl_2$, 0.1 g; $CoCl_2{\cdot}6H_2O$, 0.1 g; $NiCl_2{\cdot}6H_2O$, 0.1 g; $BaCl_2$, 0.05 g; $(NH_4)_6Mo_7O_{24}{\cdot}4H_2O$, 0.05 g, 17.1 g Sucrose, per liter), along with 0.3 mM of DBS as sulfur source, were inoculated with IITR100. Three flasks for each time point, i.e., after 0, 2, 4, 6, and 8 days of incubation were removed. After estimation of growth (OD_{600}), the reaction was stopped by acidification to pH < 2.0, and levels of residual DBS and the formed metabolites were analyzed. Un-inoculated flasks were run in parallel and were processed likewise.

Similar experiment was set up for studying the desulfurization of DBT, except that 0.3 mM DBT was used as sulfur source. For experiments, where desulfurization of DBT and DBS was to be evaluated, when these were present together, similar set up was used except that 0.3 mM each of DBS and DBT was used as sulfur source.

Table 1 List of primers used in the present study

No.	Name	Sequence	Properties
1	*dsz*AF	GGAATTCCATATGGCTCAACGGCGACAACTGCATCTGGCCGGTTTC	Contains 1-36 bases of 5′end of *dsz*A and is preceded by site for *Nde*I
2	*dsz*AR	CCGCTCGAGGTGTGTCGAGGATGCCGGTATCAAGTTCTGAACCGG	Contains 1-33 bases of 3′end of *dsz*A and is preceded by site for *Xho*I
3	*dsz*CF	GGAATTCCATATGACTCTGTCCGTTGAAAAGCAGCACGTTCG	Contains 1-32 bases of 5′end of *dsz*C and is preceded by site for *Nde*I
4	*dsz*CR	CCGCCCAAGCTTCTAGGAGGTGAAGCCGGGAATCGGGTA	Contains 1-27 bases of 3′end of *dsz*C and is preceded by site for *Hind*III

Sequences were based on those of *dsz*ABC genes of IITR100 (Accession No.-KC693733.1)

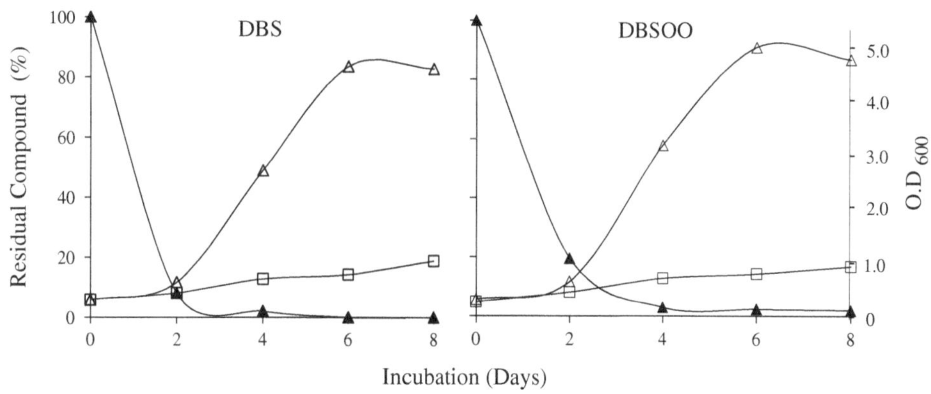

Fig. 1 Desulfurization of dibenzyl sulfide (DBS) and dibenzyl sulfone (DBSOO), in the presence of IITR100. Residual compound (*closed triangles*), and growth of IITR100 in presence of DBS/ DBSOO (*open triangles*) or their absence (*open squares*) are also shown

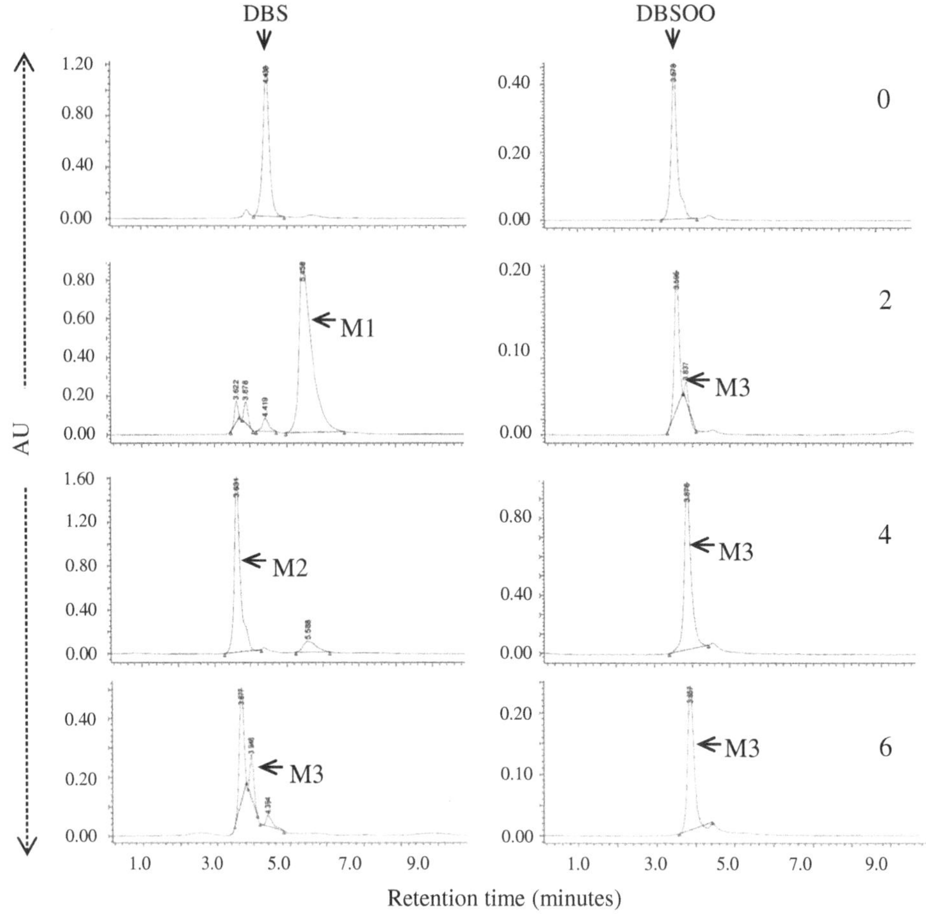

Fig. 2 Metabolism of dibenzyl sulfide (DBS) and dibenzyl sulfone (DBSOO) in the presence of IITR100, after 0, 2, 4 and 6 days of incubation. Retention times 5.62, 3.63 and 3.94 min of the formed metabolites M1, M2 and M3, respectively, were same as for dibenzyl sulfoxide (DBSO), dibenzyl sulfone (DBSOO) and benzoic acid

Desulfurization by recombinant *E. coli* cells

Recombinant *E. coli* DszA and *E. coli* DszC that were expressing DszA and DszC, respectively, were prepared by ligation of the amplified *dsz*A and *dsz*C genes (Table 1), respectively, with pET28a, and were followed by their cloning in *E. coli* BL21 as described earlier (Singh et al. 2007). After induction by IPTG, the cells from one liter culture were harvested by centrifugation at 3,500*g*, washed with 50 ml medium-1 and suspended in 100 ml of the same medium. For evaluation of the activity of DszC with the substrates, 0.3 mM of DBT, DBS or mixture of DBT and DBS were incubated with 10 ml suspension of the *E. coli* DszC. After incubation at 30 °C for different time periods, the reaction was stopped by acidification and residual substrate along with the formed metabolites was analyzed, as described below. Metabolism of DBTOO and DBSOO with *E. coli* DszA was also evaluated likewise.

Analytical methods

The acidified culture medium was extracted three times with equal volume of ethyl acetate. The residue obtained after complete evaporation of the solvent was dissolved in acetonitrile and a suitable aliquot was used for HPLC analysis, carried on a Waters instrument (Milford, MA, USA), equipped with PDA 996 detector and LiChrospher® 100 RP-18 columns (5 μm, 4.6 × 250 mm). It was run at 28 °C, using 100 % acetonitrile as mobile phase at the flow rate 0.50 ml/min.

Results

Desulfurization of dibenzyl sulfide by IITR100

The strain IITR100 was able to grow in liquid medium using either DBS or DBS sulfone as source of sulfur, whose levels declined progressively with increasing time periods (Fig. 1). The growth curves on both of these compounds were comparable, i.e., there was a lag period of 2 days and maximal growth (5.0 OD_{600}) was seen after 6 days of incubation. The growth with DBS was accompanied with the formation of three metabolites, i.e., M1, M2 and M3, which were subsequently identified as DBS sulfoxide, DBS sulfone, and benzoic acid, respectively, based on their co-migration on HPLC with authentic standards. Identity of M3 was further confirmed by its mass spectrum, which matched well with that of the authentic standard (data not shown). Similar to DBS, growth of IITR100 cells on DBS sulfone was accompanied with the formation of metabolite M3. Level of the formed M3 was highest after 4 days of incubation with IITR100, but decreased thereafter (Fig. 2), suggesting it to be an intermediary metabolite and not the end product. Based on the metabolites formed, desulfurization of DBS to benzoic acid appears to proceed via the formation of DBS sulfoxide and DBS sulfone (Fig. 3).

Transformation of DBT and DBS by recombinant *E. coli*

Incubation with the recombinant *E. coli* DszC that was harboring *dsz*C gene of IITR100 led to the oxidation of DBT and DBS to the corresponding sulfones (Fig. 4). A transient accumulation of DBS sulfoxides was observed after 24 h incubation, but was metabolized by 48 h. Incubation of DBT sulfone with recombinant *E. coli* DszA that was harboring *dsz*A gene of IITR100 led to its transformation into a metabolite DI (Fig. 5), possibly DBT sulfinite, but their incubation under the same conditions did not cause any metabolism of DBS sulfone.

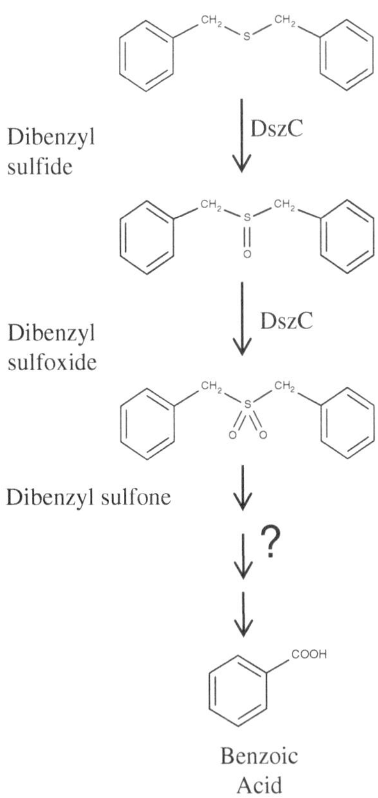

Fig. 3 Proposed pathway for desulfurization of DBS by IITR 100

Metabolism of DBT and DBS, when present alone or together

When the two chemicals were evaluated separately, desulfurization of DBS by IITR100 was marginally better than DBT, and both the chemicals were utilized completely by 3 and 4 days of incubation, respectively (Fig. 6a). But, when the two chemicals were present together, an enhanced preference for desulfurization of DBS was observed. Here, while nearly all of DBS was metabolized after 4 days of incubation, ~50 % of DBT was still present in the medium after the same period. Similarly, preferential metabolism of DBS was also observed when the metabolism of DBT and DBS by *E. coli* DszC cells was evaluated. Thus, when the two chemicals were present individually, ~nil and 10 % of DBS and DBT, respectively, remained in the reaction medium after 12 h of incubation (Fig. 6b). But when the two chemicals were present together, the difference in their metabolism became sharper and ~30 and 70 % of DBS and DBT remained after the same time period.

Discussion

The study was designed to study the desulfurization of a model sulfide compound dibenzyl sulfide (DBS) by an isolated strain *Gordonia* sp. IITR100, when it is present alone or in the presence of a thiophenic compound DBT. Results revealed that the desulfurization of DBS proceeds via the formation of metabolites DBS sulfoxide, DBS sulfone and benzoic acid. While the conversion of DBS to DBS sulfone is mediated by the activity of DszC, further metabolism of the formed DBS sulfone is mediated by some enzyme other than DszA. The results are in agreement with earlier studies, where formation of the same metabolites has been reported by the activity of *Rhodococcus* sp. strains JVH-1(Van Hamme et al. 2004) and K1bD (Kirkwood et al. 2005). The enzymes responsible for the activity, however, were not studied in these bacteria. Metabolism of DBS to DBS sulfone by DszC that was obtained from the strain IGTS8 has been shown earlier (Lei and Tu 1996), but the reaction was extremely slow and only 6 % of DBS was converted into DBS sulfone by the end of the reaction. Nearly 70 % of it remained as the intermediary metabolite DBS sulfoxide. In contrast, the metabolism of DBS to DBSOO by *E. coli* DszC in the present study was rapid and no significant accumulation of DBS sulfoxide was observed at the end of the reaction (Fig. 4).

Based on the analogy with DBT desulfurization, it has been proposed that the formed DBS sulfone might undergo further metabolism by enzyme DszA (Van Hamme et al.

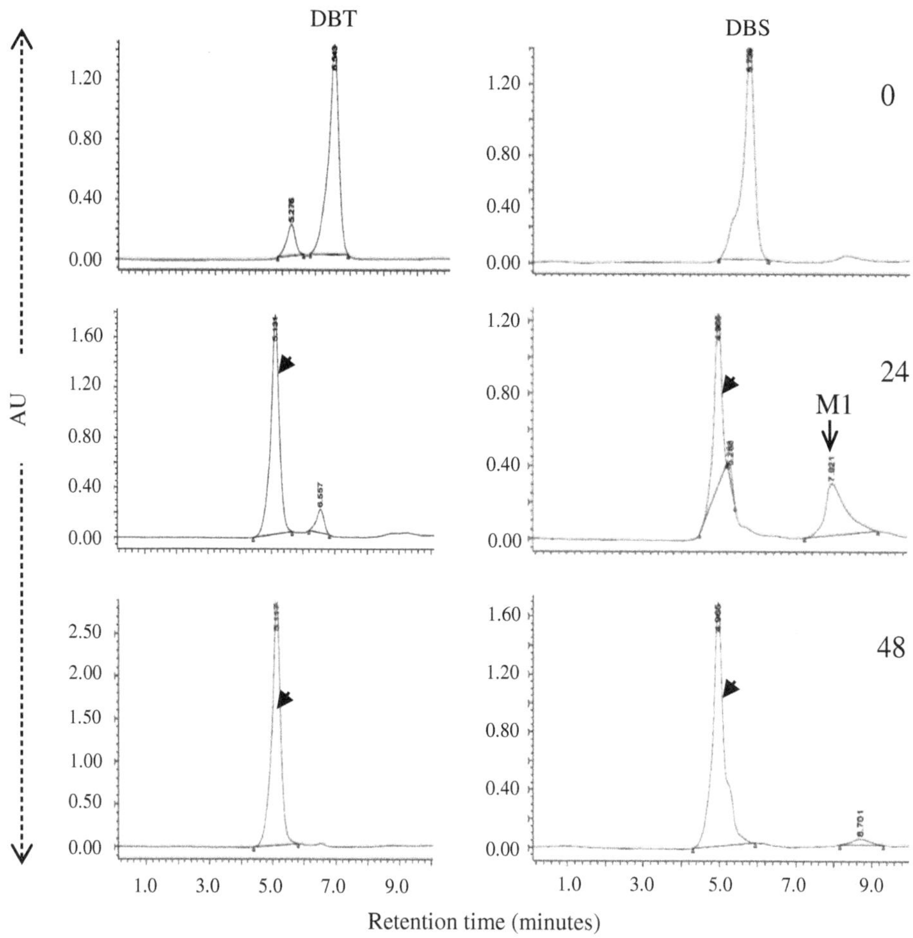

Fig. 4 Metabolism of DBT and DBS by recombinant *E. coli* DszC to their corresponding sulfones (shown by *arrowheads*), after 0, 24 and 48 h of incubation. Metabolite M1 is the intermediary dibenzyl sulfoxide

2004). Results of the present study, however, revealed that although IITR100 is able to cause desulfurization of DBS, its DszA has no activity towards the metabolism of DBS sulfone. It suggests that this reaction in IITR100 might proceed by some enzyme that is distinct from DszABC. The same might also be true for the previously characterized DBS-metabolizing strain *Rhodococcus* sp. JVH1, which does not carry *dsz*ABC genes (Brooks and Van Hamme 2012).

Interestingly, although IITR100 was isolated initially for its activity of DBT desulfurization, it exhibits preference for the desulfurization of DBS over DBT, possibly due to the similar preference by its DszC (Fig. 6). Similarly, a strain *Paenibacillus* sp. A11-2, isolated by enrichment on DBT, exhibits 2-threefold higher activity towards benzothiophene (Konishi et al. 2000). Thus, it is a possibility that several desulfurizing strains described in the literature, whose *dszABC* genes are different from the archetypal *Rhodococcus* IGTS 8, might actually have preference for other organosulfur compounds, which can be evaluated using their binary or bigger combination of the substrates.

Conclusion

- *Gordonia* sp. strain IITR100 mediates desulfurization of dibenzyl sulfide (DBS).
- Metabolism of DBS to DBS sulfone is mediated by enzyme DszC.
- Further metabolism of DBS sulfone takes place with some enzyme other than DszA.
- IITR100 as well as recombinant *E. coli* DszC exhibits preference for the desulfurization of DBS over DBT, when present together.

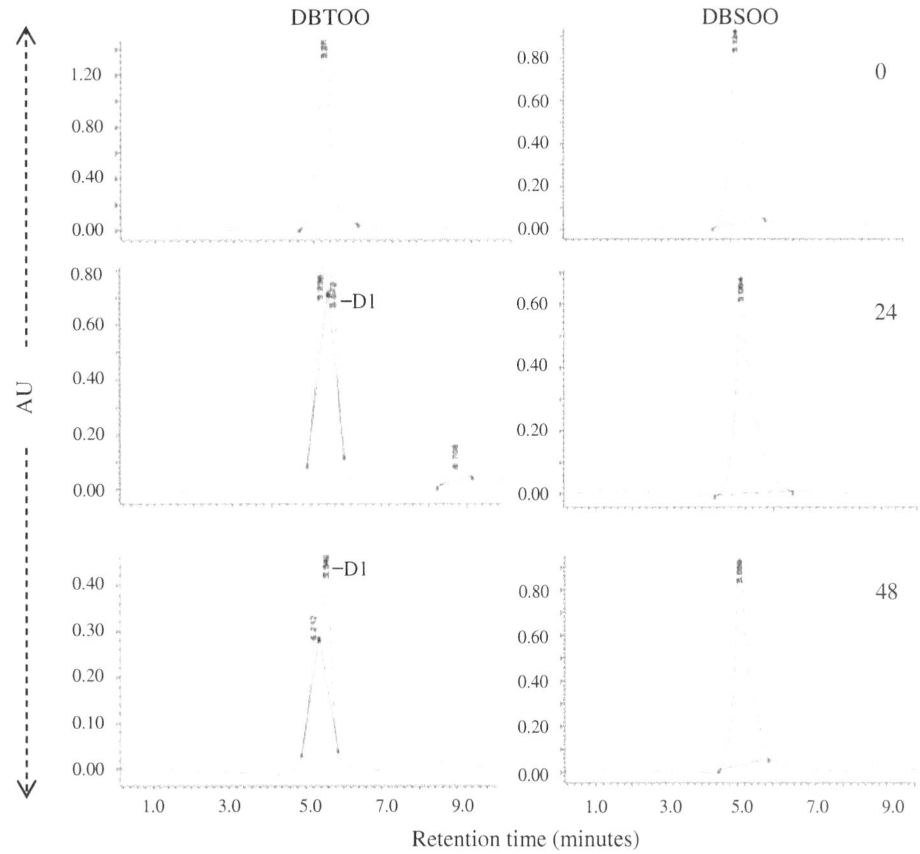

Fig. 5 Metabolism of DBT sulfone (DBTOO) and DBS sulfone (DBSOO) by *E. coli* DszA. Formation of a metabolite D1 after 24 and 48 h incubation with DBTOO, but no metabolism of DBSOO under the same conditions, is also shown

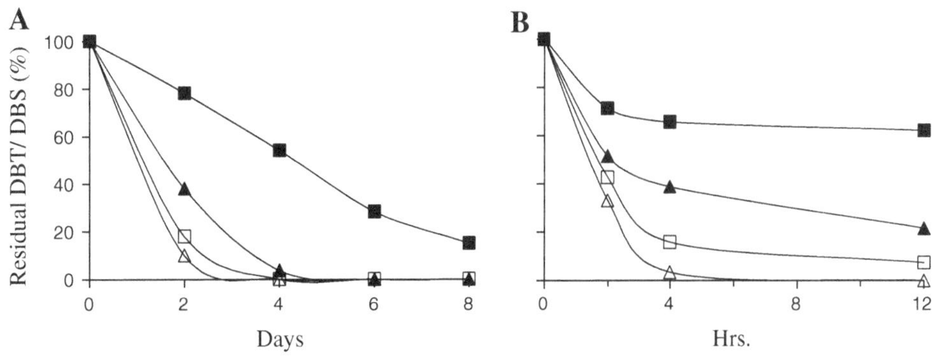

Fig. 6 Desulfurization/metabolism of DBS (*triangles*) and DBT (*squares*) with IITR100 (Panel **a**) and with recombinant *E. coli* DszC (Panel **b**), when present separately (*open*) or together (*closed*)

- IITR 100 has potential for desulfurizing both aliphatic and aromatic organosulfur compounds.
- This strain could be used for biodesulfurization of broad substrate range of organosulfur compounds in petroleum fractions.

Acknowledgments Financial assistance by a grant SIP-08 from Council of Scientific and Industrial Research (CSIR), India, is gratefully acknowledged. Authors AA and AKC sincerely thank Council of Scientific and Industrial Research (CSIR) and Indian Council of Medical Research (ICMR), India, respectively, for fellowship support. The funding sources had no role in study design; in the collection, analysis and interpretation of data. We are also thankful to our colleagues in analytical facility of Indian Institute of Toxicology Research, Lucknow, and Central Drug Research Institute, Lucknow, for their help in HPLC and GC–MS analysis.

Conflict of interest The authors declare that they have no conflict of interest in the publication.

References

Alves L, Melo M, Mendonca D, Simoes F, Matos J, Tenreiro R, Girio FM (2007) Sequencing, cloning and expression of the *dsz* genes required for dibenzothiophene sulfone desulfurization from *Gordonia alkanivorans* strain 1B. Enzyme Microb Technol 40:1598–1603

Brooks SL, Van Hamme JD (2012) Whole-genome shotgun sequence of rhodococcus species strain JVH1. J Bacteriol 194(19):5492–5493.

Denome SA, Oldfield C, Nash LJ, Young KD (1994) Characterization of the desulfurization genes from *Rhodococcus* sp. strain IGTS8. J Bacteriol 176:6707–6716

Gupta N, Roychoudhury PK, Deb JK (2005) Biotechnology of desulfurization of diesel: prospects and challenges. Appl Microbiol Biotechnol 66:356–366

Kilbane JJ 2nd (2006) Microbial biocatalyst developments to upgrade fossil fuels. Curr Opin Biotechnol 17:305–314

Kirkwood KM, Ebert S, Foght JM, Fedorak PM, Gray MR (2005) Bacterial biodegradation of aliphatic sulfides under aerobic carbon-or sulfur-limited growth conditions. J Appl Microbiol 99:1444–1454

Konishi J, Onaka T, Ishii Y, Suzuki M (2000) Demonstration of the carbon-sulfur bond targeted desulfurization of benzothiophene by thermophilic *Paenibacillus* sp. strain A11-2 capable of desulfurizing dibenzothiophene. FEMS Microbiol Lett 187(2):151–154

Lam V, Li G, Song C, Chen J, Fairbridge C, Hui R, Zhang J (2012) A review of electrochemical desulfurization technologies for fossil fuels. Fuel Process Technol 98:30–38

Lei B, Tu SC (1996) Gene overexpression, purification, and identification of a desulfurization enzyme from *Rhodococcus* sp. strain IGTS8 as a sulfide/sulfoxide monooxygenase. J Bacteriol 178(19):5699–5705

Liu P, Shi Q, Chung KH, Zhang Y, Pan N, Zhao S (2010) Molecular characterization of sulfur compounds in Venezuela crude oil and its SARA fractions by electrospray ionization Fourier transform ion cyclotron resonance. Energy Fuels 24(9):5089–5096

Mohebali G, Ball AS (2008) Biocatalytic desulfurization (BDS) of petrodiesel fuels. Microbiology 154:2169–2183

Peinder P, Visser T, Wagemans R, Blomberg J, Chaabani H, Soulimani F, Weckhuysen BM (2010) Sulfur speciation of crude oils by partial least squares regression modeling of their infrared spectra. Energy Fuels 24(1):557–562

Piddington CS, Kovacevich BR, Rambosek J (1995) Sequence and molecular characterization of a DNA region encoding the dibenzothiophene desulfurization operon of *Rhodococcus* sp. strain IGTS8. Appl Environ Microbiol 61:468–475

Singh AK, Chaudhary P, Macwan AS, Diwedi UN, Kumar A (2007) Selective loss of *lin* genes from hexachlorocyclohexane-degrading *Pseudomonas aeruginosa* ITRC-5 under different growth conditions. Appl Microbiol Biotechnol 76(4):895–901

Singh P, Ahmad A, Faraz S, Macwan A, Kumar A, Srivastava P (2011) Desulfurization of recalcitrant organosulfur compound 4, 6-DMDBT by the isolated strain IITR100 (Abstract). In: Proceedings of the 80th meeting of 'Society of Biological chemists (India)' held at Lucknow, India

Van Hamme JD, Singh A, Ward OP (2003a) Recent advances in petroleum microbiology. Microbiol Mol Biol Rev 67:503–549

Van Hamme JD, Wong ET, Dettman H, Gray MR, Pickard MA (2003b) Dibenzyl sulfide metabolism by white rot fungi. Appl Environ Microbiol 69(2):1320–1324

Van Hamme JD, Fedorak PM, Foght JM, Gray MR, Dettman HD (2004) Use of a novel fluorinated organosulfur compound to isolate bacteria capable of carbon-sulfur bond cleavage. Appl Environ Microbiol 70(3):1487–1493

Xu P, Feng J, Yu B, Li F, Ma C (2009) Recent developments in biodesulfurization of fossil fuels. Adv Biochem Eng Biotechnol 113:255–274

22

Nanoemulsion: an advanced mode of drug delivery system

Manjit Jaiswal · Rupesh Dudhe · P. K. Sharma

Abstract An advanced mode of drug delivery system has been developed to overcome the major drawbacks associated with conventional drug delivery systems. This review gives a detailed idea about a nanoemulsion system. Nanoemulsions are nano-sized emulsions, which are manufactured for improving the delivery of active pharmaceutical ingredients. These are the thermodynamically stable isotropic system in which two immiscible liquids are mixed to form a single phase by means of an emulsifying agent, i.e., surfactant and co-surfactant. The droplet size of nanoemulsion falls typically in the range 20–200 nm. The main difference between emulsion and nanoemulsion lies in the size and shape of particles dispersed in the continuous phase. In this review, the attention is focused to give a basic idea about its formulation, method of preparation, characterization techniques, evaluation parameters, and various applications of nanoemulsion.

Keywords Nanoemulsion · Drug delivery · Emulgents · High-pressure homogenization

Introduction

Nanoemulsions are a colloidal particulate system in the submicron size range acting as carriers of drug molecules. Their size varies from 10 to 1,000 nm. These carriers are solid spheres and their surface is amorphous and lipophilic with a negative charge. Magnetic nanoparticles can be used to enhance site specificity. As a drug delivery system they enhance the therapeutic efficacy of the drug and minimize adverse effect and toxic reactions. Major application includes treatment of infection of the reticuloendothelial system (RES), enzyme replacement therapy in the liver, treatment of cancer, and vaccination. An emulsion is a biphasic system in which one phase is intimately dispersed in the other phase in the form of minute droplets ranging in diameter from 0.1 to 100 μm. It is a thermodynamically unstable system, which can be stabilized by the presence of an emulsifying agent (emulgent or emulsifier). The dispersed phase is also known as internal phase or the discontinuous phase while the outer phase is called dispersion medium, external phase or continuous phase. The emulsifying agent is also known as intermediate or interphase. The term 'nanoemulsion' also refers to a miniemulsion which is fine oil/water or water/oil dispersion stabilized by an interfacial film of surfactant molecule having droplet size range 20–600 nm. Because of small size, nanoemulsions are transparent. There are three types of nanoemulsion which can be formed: (a) oil in water nanoemulsion in which oil is dispersed in the continuous aqueous phase, (b) water in oil nanoemulsion in which water droplets are dispersed in continuous oil phase, and (c) bi-continuous nanoemulsions.

Advantages of nanoemulsion

(a) It may be used as substitute for liposomes and vesicles (Bouchemal et al. 2004a, b).
(b) It improves the bioavailability of drug (Kim et al. 2001; Wagner et al. 1996).
(c) It is non-toxic and non-irritant in nature.
(d) It has improved physical stability.
(e) Nanoemulsions have small-sized droplets having greater surface area providing greater absorption.

M. Jaiswal (✉) · R. Dudhe · P. K. Sharma
Department of Pharmacy, School of Medical and Allied Sciences, Galgotias University, G. B. Nagar, Uttar Pradesh, India
e-mail: manjitj2@gmail.com

(f) It can be formulated in variety of formulations such as foams, creams, liquids, and sprays.
(g) It provides better uptake of oil-soluble supplements in cell culture technology.
(h) It helps to solubilize lipophilic drug.
(i) Helpful in taste masking.
(j) Less amount of energy is required.

Components of nanoemulsion

The main components of nanoemulsion are oil, emulsifying agents, and aqueous phases (Gasco et al. 1991; Kriwet and Müller-Goymann 1995; Trotta 1999). Oils can be of any type like castor oil, corn oil, coconut oil, evening primrose oil, linseed oil, mineral oil, olive oil, peanut oil, etc. A mixture of oil and water may yield a crude temporary emulsion, which upon standing, will separate in two distinct phases due to the coalescence of the dispersed globules. Emulgents or emulsifying agents can impart stability to such systems. Emulgents are broadly classified as surfactants like spans and tweens, hydrophilic colloids such as acacia and finely divided solids, e.g., bentonite and veegum. An emulgent, in addition to its emulsifying properties, should be nontoxic and its taste, odour and chemical stability should be compatible with the product. Some of the desirable properties of an emulgent are: (1) it should be able to reduce the surface tension to below 10 dynes/cm, (2) it should be adsorbed rapidly around dispersed phase globule to form a complete and coherent film to prevent coalescence, (3) it should help in building up an adequate zeta potential and viscosity in the system so as to impart optimum stability, and (4) it should be effective in a fairly low concentration. Emulgents form monomolecular, multimolecular or particulate films around the dispersed globules (Sharma and Jain 1985).

Monomolecular films

Surfactant type of emulgents stabilizes a nanoemulsion by forming a monolayer of adsorbed molecules or ions at the interface reducing interfacial tension. In modern day practice, combination of emulgents is preferred over single emulgent. The combination consists of a predominantly hydrophilic emulgent in the aqueous phase and a hydrophobic agent in the oily phase to form a complex film at the interface.

Multimolecular films

Hydrated lyophilic colloids form multimolecular films around globules of dispersed oil. Hydrated colloids do not cause any appreciable lowering of surface tension and their ability to form strong, coherent multimolecular films. Their tendency to increase the viscosity of the continuous phase enhances the stability of emulsion.

Solid particulate films

The emulgents forming particulate films are small solid particles that are wetted to some degree by both aqueous and non-aqueous liquid phases. They are concentrated at the interface where they produce a film around the dispersed globules thus preventing coalescence.

Formulation aspects and method of preparation of nanoemulsion

Formulation of nanoemulsion includes active drug, additive and emulsifier. The various methods for the preparation of nanoemulsion include two methods: (a) high-energy emulsification and (b) low-energy emulsification. The high-energy emulsification method includes high-energy stirring, ultrasonic emulsification, high-pressure homogenization, microfluidization, and membrane emulsification (Tiwari and Amiji 2006; Perdiguer et al. 1997; Banker et al. 2002). The low-energy emulsification method includes phase inversion temperature, emulsion inversion point, and spontaneous emulsification (Ahuja et al. 2008). Using a combined method, which includes the high-energy and low-energy emulsification, it is possible to prepare reverse nanoemulsion in a highly viscous system.

Ultrasonic emulsification

Ultrasonic emulsification is very efficient in reducing droplet size. In ultrasonic emulsification, the energy is provided through sonotrodes called as sonicator probe. It contains piezoelectric quartz crystal which can expand and contract in response to alternating electric voltage. As the tip of sonicator contacts the liquid, it produces mechanical vibration and cavitation occurs. Cavitation is the formation and collapse of vapour cavities in liquid. Thus, ultrasound can be directly used to produce emulsion; it is mainly used in laboratories where emulsion droplet size as low as 0.2 micrometer can be obtained.

High-pressure homogenization

The preparation of nanoemulsion requires high-pressure homogenization. This technique makes use of high-pressure homogenizer/piston homogenizer to produce nanoemulsion of extremely low particle size (up to 1 nm) (Asua 2002; Anton et al. 2008).

Microfluidization

Microfluidization is a patented mixing technology, which makes use of a device called microfluidizer. This device uses high pressure which forces the drug product through the interaction chamber resulting in a very fine particle of submicron range. The process is repeated several times to obtain a desired particle size to produce uniform nanoemulsion.

Phase inversion temperature

This method involves change in phase by applying a higher temperature to a microemulsion (El-Aasser et al. 1986; Pouton 1997).

Spontaneous emulsification

It involves three steps: (a) preparation of homogeneous organic solution consisting of oil and lipophilic surfactant in water miscible solvent and hydrophilic surfactant, (b) the organic phase is injected in aqueous phase under continuous magnetic stirring, o/w emulsion is formed, and (c) the aqueous phase is removed by evaporation under reduced pressure (Solans et al. 2005; Tadros et al. 2004).

Factors to be considered during preparation of nanoemulsion

1. Surfactant must be selected carefully such that an ultralow interfacial tension may be achieved which is a primary requirement to produce nanoemulsion.
2. Concentration of surfactant must be high enough to stabilize the microdroplets to produce nanoemulsion.
3. The surfactant must be flexible or fluid enough to promote the formation of nanoemulsion.

Characterization of nanoemulsion

A stable nanoemulsion is characterized by the absence of the internal phase, absence of creaming, absence of deterioration by microorganisms, and maintenance of elegance in respect of appearance, colour, odour and consistency (Sharma and Jain 1985). Hence the instability of emulsion can be classified as follows:

Flocculation and creaming

Flocculation consists of the joining together of globules to form large clumps or floccules, which rise or settle in the emulsion more rapidly than the individual globules. The rising up or settling down of dispersed globules to give a concentrated layer is known as creaming. Thus flocculation leads to creaming.

Cracking

Cracking of an emulsion refers to separation of the dispersed phase as a layer. Whereas a creamed emulsion may be reconstituted by shaking or agitation, a cracked emulsion cannot be corrected. Cracking represents permanent instability. Cracking of the emulsion may be due to: (1) addition of an emulgent of opposite nature, (2) decomposition or precipitation of emulgent, (3) addition of a common solvent in which both oily and aqueous phases are miscible, (4) extremes of temperature, (5) microorganisms, (6) creaming.

Miscellaneous instability

Emulsions may deteriorate if stored under extremely high or low temperature or in the presence of light. Hence emulsions are usually packed in air-tight, coloured containers and stored at moderate temperature.

Phase inversion

It is the change in the type of emulsion from o/w to w/o and vice versa. It is the physical process. Phase inversion may be brought about by varying the phase volume ratio, addition of electrolytes, and temperature changes.

Evaluation parameters of nanoemulsion

Droplet size analysis

Droplet size analysis of nanoemulsion is measured by a diffusion method using a light-scattering, particle size-analyzer counter, LS 230. It is also measured by correlation spectroscopy that analyzes the fluctuation in scattering of light due to Brownian motion. Droplet size analysis of nanoemulsion can also be performed by transmission electron microscopy (TEM) (Bouchemal et al. 2004a, b; Alka et al. 2007; Farhan et al. 2008).

Viscosity determination

The viscosity of nanoemulsion is measured by using Brookfield-type rotary viscometer at different shear rates at different temperatures.

Dilution test

Dilution of a nanoemulsion either with oil or with water can reveal this type. The test is based on the fact that more of the continuous phase can be added into a nanoemulsion without causing the problem of its stability. Thus, an o/w nanoemulsion can be diluted with water and a w/o nanoemulsion can be diluted with oil.

Drug content

Preweighed nanoemulsion is extracted by dissolving in a suitable solvent, extract is analyzed by spectrophotometer or HPLC against standard solution of drug (Singh and Vingkar 2008; Chen et al. 2008).

Polydispersity

It indicates the uniformity of droplet size in nanoemulsion. The higher the value of polydispersity, lower will be uniformity of droplet size of nanoemulsion. It can be defined as the ratio of standard deviation to mean droplet size. It is measured by a spectrophotometer.

Dye test

If a water-soluble dye is added in an o/w nanoemulsion the nanoemulsion takes up the colour uniformly. Conversely, if the emulsion is w/o type and the dye being soluble in water, the emulsion takes up the colour only in the dispersed phase and the emulsion is not uniformly coloured. This can be revealed immediately by microscopic examination of the emulsion.

Refractive index

Refractive index of nanoemulsion is measured by Abbes refractometer.

pH

The pH of nanoemulsion can be measured by pH meter.

Zeta potential

Zeta potential is measured by an instrument known as Zeta PALS. It is used to measure the charge on the surface of droplet in nanoemulsion (Erol and Hans-Hubert 2005).

Fluorescence test

Many oils exhibit fluorescence when exposed to UV light. When a w/o nanoemulsion is exposed to a fluorescence light under a microscope, the entire field fluoresces. If the fluorescence is spotty, the nanoemulsion of o/w type.

Percentage transmittance

Percentage transmittance of nanoemulsion is measured by a UV-visible spectrophotometer.

Conductance measurement

The conductance of nanoemulsion is measured by a conductometer. In this test a pair of electrodes connected to a lamp and an electric source is dipped into an emulsion. If the emulsion is o/w type, water conducts the current and lamp gets lit due to passage of current between the electrodes. The lamp does not glow when the emulsion is w/o: oil being in external phase does not conduct the current.

Filter paper test

This test is based on the fact that an o/w nanoemulsion will spread out rapidly when dropped onto filter paper. In contrast, a w/o nanoemulsion will migrate only slowly. This method should not be used for highly viscous creams (Sharma and Jain 1985).

Conclusion

Nanoemulsions are widely used in pharmaceutical systems. Nanoemulsion formulation offers several advantages such as delivery of drugs, biological or diagnostic agents. The most important application of nanoemulsion is for masking the disagreeable taste of oily liquids. Nanoemulsion may also protect the drugs, which are susceptible to hydrolysis and oxidation. Nowadays, nanoemulsions are used for targeted drug delivery of various anticancer drugs, photo sensitizers or therapeutic agents. Nanoemulsion can also provide prolonged action of the medicaments. Overall all nanoemulsion formulation may be considered as effective, safe and with increased bioavailability. It is expected that further research and development will be carried out in the future regarding nanoemulsion.

Conflict of interest The authors hereby want to declare that there is no conflict of interest whatsoever in this publication.

References

Ahuja A, Ali J, Baboota S, Faisal MS, Shakeell F, Shafiq S (2008) Stability evaluation of Celecoxib nanoemulsion containing Tween 80. Thai J Pharm Sci 32:4–9

Alka AJA, Baboota S, Shakeel F, Shafiq S (2007) Design development and evaluation of novel nanoemulsion formulations for transdermal potential of Celecoxib. Acta Pharm 57:315–332.

Anton N, Benoit JP, Saulnier P (2008) Design and production of nanoparticles formulated from nano-emulsion templates-a review. J Control Release 128:185–199

Asua JM (2002) Miniemulsion polymerization. Prog Polym Sci 27:1283–1346

Banker GS, Lieberman HA, Rieger MM (2002) Pharmaceutical dosage forms. Disperse Syst Marcel Dekker 2(3):339–340

Bouchemal K, Briancon S, Fessi H, Perrier E (2004a) Nano-emulsion formulation using spontaneous emulsification: solvent, oil and surfactant optimization. Int J Pharm 280:242

Bouchemal K, Briancon S, Fessi H, Perrier E (2004b) Nano-emulsion formulation using spontaneous emulsification: solvent, oil and surfactant optimization. Int J Pharm 280:243

Chen H, Du D, Mao CMD, Wan J, Xu H, Yang X (2008) Hydrogel-thickened nanoemulsion system for topical delivery of lipophilic drugs. Int J Pharm 353:272

El-Aasser MS, Lack CD, Vanderhoff JW, Fowkes FM (1986) Miniemulsification process-different form of spontaneous emulsification. Coll Surf 29:103–118

Erol Y, Hans-Hubert B (2005) Design of a phytosphingosine-containing, positively-charged nanoemulsion as a colloidal carrier system for dermal application of ceramides. Eur J Pharm Biopharm 60:93

Farhan AJ, Mushir A, Faiyaz S, Cushman T, Roop KK, Sheikh S (2008) Investigation of nanoemulsion system for transdermal delivery of domperidone: ex-vivo and in vivo studies. Curr Nanosci 4(4):381–390

Gasco MR, Gallarate M, Pattarino F (1991) In vitro permeation of azelaic acid from viscosized microemulsions. Int J Pharm 69:193–196

Kim CK, Cho YJ, Gao ZG (2001) Preparation and evaluation of biphenyl dimethyl dicarboxylate microemulsions for oral delivery. J Control Release 70:149–155

Kriwet K, Müller-Goymann C (1995) Diclofenac release from phospholipid drug systems and permeation through excised human stratum corneum. Int J Pharm 125:231–242

Perdiguer AC, Dachs FJG, Carreras N, Valdivia (1997) Nanoemulsion of the oil water type, useful as an ophthalmic vehicle and process for the preparation thereof Assignee: Laboratorios Cusi, S.A. (Barcelona, ES)

Pouton CW (1997) Formulation of selfemulsifying drug delivery systems. Adv Drug Deliv Rev 25:47–58

Sharma SN, Jain NK (1985) A text book of professional pharmacy. Vallabh Prakashan, 1st edn, p 201

Singh KK, Vingkar SK (2008) Formulation, antimalarial activity and biodistribution of oral lipid nanoemulsion of primaquine. Int J Pharm 347:138

Solans C, Izquierdo P, Nolla J, Azemar N, Garcia-Celma MJ (2005) Nanoemulsions. Curr Opin Coll Interface Sci 10:102–110

Tadros T, Izquierdo P, Esquena J, Solans C (2004) Formation and stability of nano-emulsions. Adv Coll Interface Sci 108:303–318

Tiwari SB, Amiji MM (2006) Nanoemulsion formulations for tumor-targeted delivery. Nanotech Cancer Therapy. Taylor and Francis Group Editors, pp 723–739

Trotta M (1999) Influence of phase transformation on indomethacin release from microemulsions. J Control Release 60:399–405

Wagner JG, Gerrard ES, Kaiser DG (1996) The effect of the dosage form on serum levels of indoxole. Clin Pharmacol Ther 7:610–619

23

Phosphate-solubilizing bacteria-assisted phytoremediation of metalliferous soils: a review

Munees Ahemad

Abstract Heavy metal pollution of soils is of great concern. The presence of the toxic metal species above critical concentration not only harmfully affects human health but also the environment. Among existing strategies to remediate metal contaminates in soils, phytoremediation approach using metal accumulating plants is much convincing in terms of metal removal efficiency, but it has many limitations because of slow plant growth and decreased biomass owing to metal-induced stress. In addition, constrain of metal bioavailability in soils is the prime factor to restrict its applicability. Phytoremediation of metals in association with phosphate-solubilizing bacteria (PSB) considerably overcomes the practical drawbacks imposed by metal stress on plants. This review is an effort to describe mechanism of PSB in supporting and intensifying phytoremediation of heavy metals in soils and to address the developmental status of the current trend in application of PSB in this context.

Keywords Bioremediation · Heavy metals · Hyperaccumulator plants · Phosphate-solubilizing bacteria · Phytoremediation · Rhizobacteria

Introduction

Soil is one of the most important natural resource on which lives of all plants, animals and microorganisms directly or indirectly dependent. In soils, different microorganisms thrive on abundantly present nutrients therein and through various interactions play a pivotal role in cycling of nutrients and pedogenesis (Ahemad and Khan 2013). Alteration or disturbance in soil ecosystem by added pollutants leads to substantial changes in functional activities of these important soil microorganisms (Swain and Abhijita 2013). Among pollutants, enormous amounts of toxic heavy metals such as chromium, cadmium, copper, zinc, mercury and lead contaminate soils through various geogenic, anthropogenic and technogenic activities (Ahemad 2012; Liu et al. 2013; Waterlot et al. 2013; Chodak et al. 2013). Due to non-biodegradable nature, metals in soils persist longer and pose a risk to human health through food chain because of their carcinogenicity, mutagenicity and teratogenicity (Ahemad and Malik 2011; Ali et al. 2013; Ahemad and Kibret 2013a). In addition, metals exceeding threshold limit affect microbial diversity and soil fertility (Huang et al. 2009). Thus, remediation of such metal-stressed soils is of paramount significance as they are rendered inappropriate for agricultural application.

Many physicochemical technologies are already in practice to clean up the metal-contaminated soils (Hashim et al. 2011). However, these conventional technologies are generally too costly to be applied to decontaminate the metal-polluted sites. Moreover, they generally adversely affect the texture and organic components, which are important to sustain the fertility of soils (Rajkumar et al. 2010). In view of sustainability issues and environmental ethics, bioremediation, the exploitation of biological processes for the cleanup of contaminated sites, is a promising, benign and ecologically sound alternative to chemical technologies (Hashim et al. 2011; Gillespie and Philp 2013). Among different bioremediation approaches, phytoremediation (utilizing metal accumulating plants to detoxify and extract contaminants in polluted soils) is

M. Ahemad (✉)
Department of Agricultural Microbiology, Faculty of Agricultural Sciences, Aligarh Muslim University, Aligarh 202002, UP, India
e-mail: muneesmicro@rediffmail.com

gaining wide acceptance due to being cheap and environmentally safe but the major drawback of this technique is that it is time-consuming and high levels of metals decease the remediating efficiency of plants (Ali et al. 2013). Interestingly, interactions between plant and metal resistant bacteria have shown better remediation of heavy metals, and this synergism not only expedites the remediation process by ameliorating phytostabilization (reduction in metal toxicity through metal immobilization) and phytoextraction (metal accumulation as a result of metal mobilization) of metal species but also accelerate the plant growth and development (Khan et al. 2009).

Since last decades, several phosphate-solubilizing bacteria (PSB) exhibiting both heavy metal detoxifying traits and plant growth promoting activities have been explored and have been implicated in phytoremediation of metalliferous soils (He et al. 2010; Misra et al. 2012; Oves et al. 2013; Ahemad and Kibret 2013b). This review is an effort to emphasize how the beneficial association between plants and PSB can be used to remediate the metal-stressed soils efficiently. In this review, mechanism of PSB mediation in supporting and intensifying phytoremediation process is discussed in detail.

Phytoremediation: an overview

Currently, several physicochemical and biological techniques are in practice to remediate the metal-contaminated soils. Of them, remediation processes based upon the physicochemical parameters are very costly and also affect the soil properties, biodiversity and fertility. Different remediation technologies have been compared in Table 1 in terms of cost. It is obvious from the enlisted remediation approaches that phytoextraction (phytoremediation type) is one of the most cost effective method to remediate the metal-polluted soils (Padmavathiamma and Li 2007). Phytoremediation occurs only at marginal cost, which is due to harvesting and field management, e.g., weed control. In addition, the resulting biomass of phytoremediating plants can be used for heat and energy production in specialized facilities (Peuke and Rennenberg 2005). Unlike physicochemical processes, phytormediation is an eco-friendly and comprehensive strategy having no side effects on soil texture and health (Suresh and Ravishankar 2004).

Table 1 Cost of different remediation technologies

Process	Cost (US$/ton)	Other factors
Vitrification	75–425	Long-term monitoring
Land filling	100–500	Transport/excavation/monitoring
Chemical treatment	100–500	Recycling of contaminants
Electrokinetics	20–200	Monitoring
Phytoextraction	5–40	Disposal of phytomass

Source: Glass (1999)

In phytoextraction, metals are accumulated in plant biomass from moderately contaminated soils. On the other hand, phytostabilization is a long-term, in situ approach applicable in polymetallic soils wherein concentration and the area of metal contaminant are so extensive that phytoextraction cannot work; therefore, metals are not allowed to enter plants but are captured in situ through biosorption, precipitation or reducing toxicity (de-Bashan et al. 2012). Plants selected for phytostabilization must be metal-tolerant (metallophytes) and should not accumulate metals into root tissues. Despite having mechanisms to evade metal translocation in shoot tissues, considerable amount of metals may be found in the shoot parts. Metal accumulation in plants can measured in terms of (1) bioconcentration factor (BF)/accumulation factor (AF) which is the ratio of metal concentration in the shoot tissue to the soils and (2) translocation factor (TF) which is the ratio of metals in shoot to those in root. For better phytostabilization, these values should preferably be $\ll 1$ while the values must be $\gg 1$ in addition to higher plant biomass for an ideal phytoextracting plants (Peuke and Rennenberg 2005; Mendez and Maier 2008).

Phosphate-solubilizing bacteria

After nitrogen, phosphorus (P) is the second essential macronutrient for plant growth and development. Generally, substantial amount of phosphorus (P) occurs in soil ranging from 400 to 1,200 mg/kg of soil, either in mineral forms, e.g., apatite, hydroxyapatite and oxyapatite, or organic forms such as, inositol phosphate (soil phytate), phosphomonoesters, phosphodiesters and phosphotriesters (Ahemad et al. 2009). However, the concentration of soluble forms of P in soil is usually ~ 1 mg/kg or less (Goldstein 1994). In addition, it has a very limited bioavailability to growing plants due to high reactivity of phosphate ions in soils. To circumvent this deficiency, phosphatic fertilizers are applied in soils. But most of the applied P in the forms of fertilizers is precipitated; consequently, a very small fraction is available for absorption by plants. As an eco-friendly and economical alternative to provide substantial amount of soluble P to plants for growth promotion is the exploitation of P solubilization and mineralization traits of PSB.

Additionally, PSB not only protect plants from phytopathogens through the production of antibiotics, HCN, phenazines and antifungal metabolites, etc. (Upadhayay and Srivastava 2012; Singh et al. 2013), but also promote plant growth through N_2 fixation (He et al. 2010),

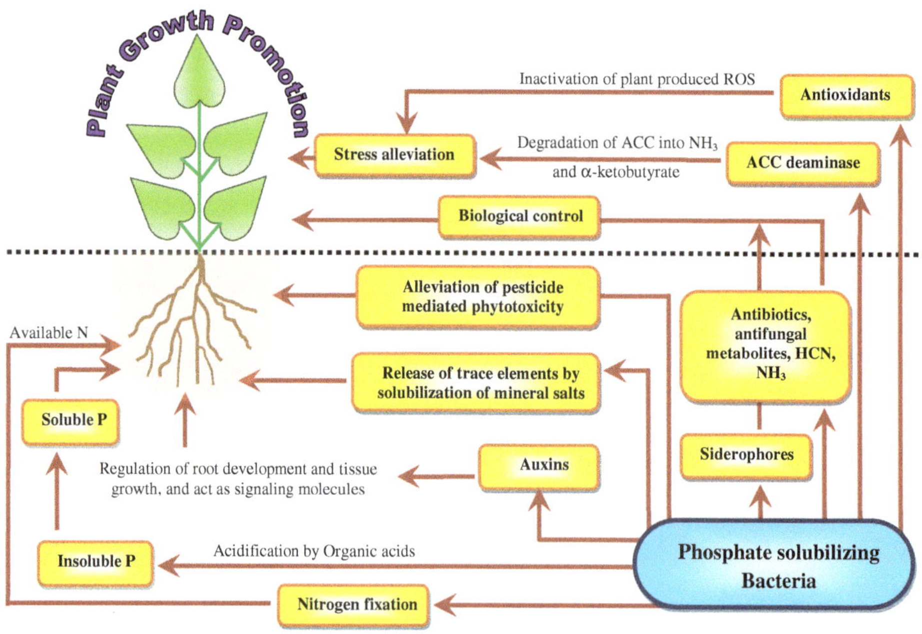

Fig. 1 Mechanisms of phosphate-solubilizing bacteria-mediated plant growth promotion. *ROS* reactive oxygen species, *ACC* 1-aminocyclopropane-1-carboxylate, *NH_3* ammonia, *HCN* hydrogen cyanate, *IAA* indole-3-acetic acid, *P* phosphate

siderophore production (Ahemad and Khan 2012a, b), phytohormone secretion (Misra et al. 2012; Oves et al. 2013) and lowering ethylene levels (Jiang et al. 2008; Kumar et al. 2009) (Fig. 1).

Mechanisms of PSB-assisted metal phytoremediation

Although decontamination of metal-polluted soils using plants (phytoextraction/phytostabilization) has shown encouraging results, this approach has limitations in case of the polluted sites wherein metal concentration is extremely elevated (Gamalero and Glick 2012). Under high metal stress, their physiological activities are hampered; growth and development are severely impeded; and resistance mechanisms are weakened, and in turn, they become prone to phyto-pathogen attacks (Ma et al. 2011c). Further, their metal phytoremediating efficiency is depressingly affected, and the process of metal decontamination is proportionally impeded depending upon several factors (Martin and Ruby 2004). Intended to overcome the noxious level of metals that significantly decline the plant growth, PSB with multiple plant growth promoting traits (Table 2; Fig. 1) and concurrent metal detoxifying potentials (Fig. 2) may increase the phytoremediation competence of plants by promoting their growth and health even under hazardous levels of different metals. As adjuncts with plants, PSB remediate metal-contaminated soils largely through facilitating either phytostabilization (decreasing metal toxicity by transforming metal species into immobile forms) or phytoextraction (metal mobilization and accumulation in plant tissues) (Fig. 3). Various plant growth promoting traits of PSB, such as organic acid production, secretion of siderophores, IAA production and ACC deaminase activity, contribute in enhancing the phytoremediation capability of plants.

Organic acids

In most of the metalliferous soils, metals are strongly adhered to soil particles; therefore, they are not easily available for uptake by phytoextracting plants (Gamalero and Glick 2012). In this context, PSB are very promising agents since they solubilize the insoluble and biologically unavailable metals such as Ni (Becerra-Castro et al. 2011), Cu (Li and Ramakrishna 2011) and Zn (He et al. 2013) by secreting low molecular weight organic acids; thus, they facilitate metal bioavailability for plant uptake (Becerra-Castro et al. 2011). A number of organic acids such as lactic, citric, 2-ketogluconic, malic, glycolic, oxalic, malonic, tartaric, valeric, piscidic, succinic and formic have been identified, which have chelating properties (Panhwar et al. 2013). Moreover, metal bioavailability in metal-stressed soils can be further increased by inoculating biosurfactant producing PSB as the bacterial biosurfactants aid in metal release from soil particles (Gamalero and Glick 2012; Singh and Cameotra 2013).

Table 2 Plant growth promoting substances released by phosphate-solubilizing bacteria

PGPR	Plant growth promoting traits	References
Pseudomonas aeruginosa strain OSG41	IAA, siderophores	Oves et al. (2013)
Pseudomonas sp.	IAA, HCN	Singh et al. (2013)
Acinetobacter haemolyticus RP19	IAA	Misra et al. (2012)
Pseudomonas putida	IAA, siderophores, HCN, ammonia	Ahemad and Khan (2011c, 2012b, c)
Pseudomonas fluorescens strain Psd	IAA, siderophores, HCN, antibiotics, biocontrol activity	Upadhayay and Srivastava (2012)
Bacillus thuringiensis	IAA	Sandip et al. (2011)
Pseudomonas aeruginosa	IAA, siderophores, HCN, ammonia	Ahemad and Khan (2010c, 2011a, e, 2012d)
Pseudomonas sp. TLC 6-6.5-4	IAA, siderophore	Li and Ramakrishna (2011)
Bacillus sp.	IAA, HCN	Karuppiah and Rajaram (2011)
Klebsiella sp.	IAA, siderophores, HCN, ammonia	Ahemad and Khan (2011b, d, 2012a)
Enterobacter asburiae	IAA, siderophores, HCN, ammonia	Ahemad and Khan (2010a, b)
Bacillus species PSB10	IAA, siderophores, HCN, ammonia	Wani and Khan (2010)
Arthrobacter sp. MT16, *Microbacterium* sp. JYC17, *Pseudomonas chlororaphis* SZY6, *Azotobacter vinelandii* GZC24, *Microbacterium lactium* YJ7	ACC deaminase, IAA, siderophore	He et al. (2010)
Pseudomonas sp.	IAA, siderophore, HCN, biocontrol potentials	Tank and Saraf (2009)
Enterobacter aerogenes NBRI K24, *Rahnella aquatilis* NBRI K3	ACC deaminase, IAA, siderophore	Kumar et al. (2009)
Enterobacter sp.	ACC deaminase, IAA, siderophore	Kumar et al. (2008)
Burkholderia	ACC deaminase, IAA, siderophore	Jiang et al. (2008)
Pseudomonas aeruginosa	ACC deaminase, IAA, siderophore	Ganesan (2008)

ACC 1-aminocyclopropane-1-carboxylate, *HCN* hydrogen cyanate, *IAA* indole-3-acetic acid

Siderophores

Generally, iron occurs mainly as Fe^{3+} and forms insoluble hydroxides and oxyhydroxides, thus is not easily available to both plants and microorganisms (Ahemad and Kibret 2013b). Under iron-limiting conditions to acquire iron, bacteria secret low molecular weight siderophores, which are iron chelators with exceptionally strong affinity for ferric iron (Fe^{3+}) (Schalk et al. 2011). Despite their preferential affinity for Fe^{3+}, they can also chelate several other metals such as, magnesium, manganese, chromium (III), gallium (III), cadmium, zinc, copper, nickel, arsenic and lead, and radionuclides, including plutonium (IV) with variable affinities (Nair et al. 2007; Rajkumar et al. 2010; Schalk et al. 2011). Supply of iron to growing plants under heavy metal pollution becomes more important as bacterial siderophores help to minimize the stress imposed by metal contaminants (Gamalero and Glick 2012). For instance, siderophore overproducing mutant NBRI K28 SD1 of phosphate-solubilizing bacterial strain *Enterobacter* sp. NBRI K28 not only increased plant biomass but also enhanced phytoextraction of Ni, Zn and Cr by *Brassica juncea* (Indian mustard) (Kumar et al. 2008).

Indole acetic acid

Phytohormone, indole-3-acetic acid (IAA) whose biosynthesis requires L-tryptophan as a precursor, is the most important auxin, which regulates several morphological and physiological functions in plants (Glick 2012). Although it has been implicated in stimulation of root growth, alleviation of salt stress, plant-pathogen interactions, legume-rhizobia interactions and eliciting induced systemic resistance against various diseases, it primarily is involved in stimulating the proliferation of lateral roots in plants, thereby root surface area is increased and they absorb more water and soil minerals (Egamberdieva 2009; Lugtenberg and Kamilova 2009, Ahemad and Kibret 2013b). Many phosphate-solubilizing bacterial genera (He et al. 2010; Ahemad and Khan 2011a, 2012b; Misra et al. 2012; Oves et al. 2013) in soils have been reported to secret IAA that is absorbed by plant roots to increase the endogenous pool of plant IAA (Glick et al. 2007). However, effects of variable IAA concentrations vary among different plant species. Moreover, optimum concentration of bacterial IAA has stimulatory effect, while high concentration (supra-optimal) of those is inhibitory to root growth (Glick 2012).

Generally, bacterial IAA facilitates adaptation of host plants in metal-contaminated sites through triggering

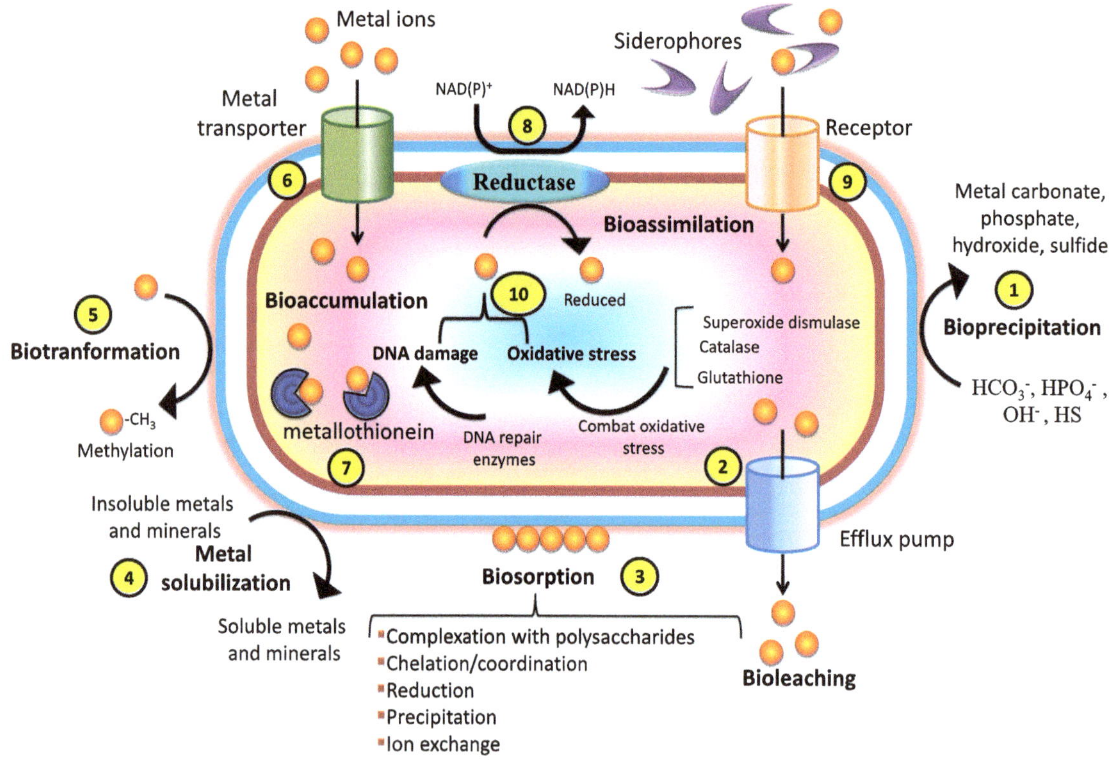

Fig. 2 Various bacterial interactions with heavy metals in metal-polluted soils: *1* precipitation/crystallization of metals occurs due to bacteria-mediated reactions or as a result of the production of specific metabolites. *2* Plasmid-DNA-encoded efflux transporters (e.g., ATP-ase pumps or chemiosmotic ion/proton pumps) expel the accumulated metals outside the cell. *3* Metals bind to the anionic functional groups (e.g., sulfhydryl, carboxyl, hydroxyl, sulfonate, amine and amide groups) of extracellular materials present on cell surfaces. *4* Organic acids secreted by bacteria solubilize the insoluble metal minerals. *5* Some bacteria utilize methylation as an alternative for metal resistance/detoxification mechanism, which involves the transfer of methyl groups to metals and metalloids. *6* Metals enter the bacterial cell by chromosomal DNA-encoded metal transporters either through ATP hydrolysis or as a result of chemiosmotic gradient across the cytoplasmic membrane. *7* Bacterial cell also accumulate substantial concentration of metals by the synthesis of low molecular mass cysteine-rich metal-binding proteins, metallothioneins having high affinities for several metals. *8* Membrane-embedded metal reductases, generally encoded by chromosomal DNA, reduce metals in the presence of electron donors. *9* Siderophore secretion decreases metal bioavailability by binding metal ions having chemistry similar to iron. *10* Superoxide dismutase, catalase and glutathione are activated to combat oxidative stress produced by the reactive oxygen species (ROS), and DNA repair system is activated to repair the DNA damaged due to various metal interactions within cell

physiological changes in plant cell metabolism under metal stress so that the growing plants can withstand high concentrations of heavy metals (Glick 2010). However, Hao et al. (2012) determined that bacterial IAA had a larger impact on the growth of host plants under metal stress rather than bacterial metal resistance through transposon mutagenesis in phosphate-solubilizing *Agrobacterium tumefaciens* CCNWGS0286.

1-Aminocyclopropane-1-carboxylate (ACC) deaminase

Another phytohormone, ethylene, modulates many important physiological activities of growing plants including root growth and development. Under both biotic (e.g., phyto-pathogen attacks) and abiotic (e.g., heavy metals, drought, flooding and salinity) stresses, plant produces ethylene up to the level that is inhibitory to root growth (Khalid et al. 2006; Arshad et al. 2007; Nadeem et al. 2007, 2009; Chen et al. 2013). Since phytoremediation approach to decontaminate the metal-spiked soils is largely reliant on the profuse growth of roots and the efficient uptake and mobilization of heavy metal ions via prolific root system to different plant parts, stress-induced ethylene at supra-optimal concentration leads to reduced root growth in turn, limiting the proficiency of metal remediating plants (Arshad et al. 2007; Gamalero and Glick 2012).

To counter this physiological crisis, an enzyme ACC deaminase (EC 4.1.99.4) produced by many soil microflora including PSB (Kumar et al. 2009; He et al. 2010), degrades ACC (an immediate precursor for ethylene in plants) into 2-oxobutanoate and ammonia hence decreases the ethylene biosynthesis in plant tissues (Saleem et al. 2007; Shaharoona et al. 2007; Zahir et al. 2009). Ammonia

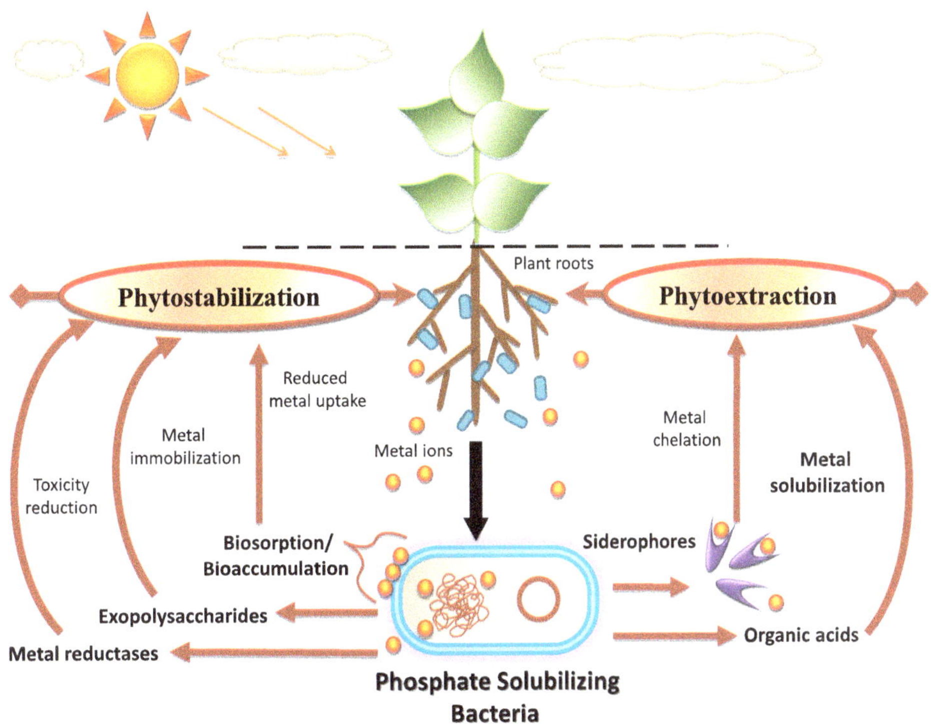

Fig. 3 Schematic portrayal of the role of metal resistant phosphate-solubilizing bacteria in alleviation of heavy metal toxicity, phytoextraction and phytostabilization

released in this way is utilized by ACC deaminase-expressing organisms as nitrogen source for growth (Glick 2005). In addition, while attached with the plant roots, ACC deaminase-containing bacteria act as a sink for ACC ensuring that ethylene level may not increased to the point where root growth and development is impaired (Glick 1995). Thus, bacterial ACC deaminase-induced extensive root proliferation in metal remediating (hyperaccumulator) plants results into efficient phytoremediation processes in metal-polluted soils (Arshad et al. 2007). Several species of ACC deaminase-containing PSB have been isolated and successfully improved the plant growth under metal stress (Ganesan 2008; Jiang et al. 2008; Sun et al. 2009).

Exploiting PSB in phytoremediation of metal-stressed soils

In various studies, growth promoting effects of PSB are well established both in unpolluted and polluted soils when used as inoculants (Ma et al. 2011a, b; Oves et al. 2013). However, degree of their impact on different plants varies depending upon plant species, bacterial species, soil types and environmental factors. In metalliferous soils, several authors have studied phytoremediation using PSB as bio-inoculants to remove different heavy metals from soils. Worldwide, the research in this direction is currently being carried out considering various aspects to overcome hurdles which impede the efficient removal of metal contaminants. In Table 3, various phytoremediation studies have been listed to show effects of different PSB using different plant species and metals. Many insights can be drawn following analyses of these studies:

1. Most of the laboratory or green house studies have employed plants of *Brassicaseae* family in conjunction with PSB because plant species of this family (hyper-accumulator plants) have been reported to accumulate substantial amount of metals in their tissues.
2. Diverse species of PSB have been used in these metal phytoremediation studies. However, species like *Pseudomonas aeruginosa*, being an opportunistic human pathogen, poses a challenge to be released in the soil environment (Walker et al. 2004). Hence, ethical and juridical considerations are needed if they are to be used as inoculants in fields.
3. Among environmentally toxic metals, only few metals such as, Ni, Cu, Zn and Cr have been studied extensively while other toxicologically important metals, e.g., As, Cd, Hg and Pb are least considered. Therefore, phytoremediation studies concerning other metals would reveal new challenges, insights and problems leading to pave ways for further research in this course.
4. Both approaches of metal remediation, phytostabilization and phytoextraction have been implicated in these studies. As the plants growing in metal-stressed soils are weakened due to metal-induced physiological

Table 3 Phosphate-solubilizing bacteria (PSB) mediated metal remediation and plant growth promotion

PSB	Plant	Heavy metals	Conditions	Role of PSB (Mode of metal remediation)	References
Pseudomonas aeruginosa strain OSG41	Chickpea (*Cicer arietinum*)	Cr	Pots	Increased the dry matter, symbiotic traits, grain yield and grain protein of chickpea plants in the presence of chromium and decreased the uptake of chromium by 36, 38 and 40 % in roots, shoots and grains, respectively (Phytostabilization)	Oves et al. (2013)
Acinetobacter haemolyticus RP19	Pearl millet (*Pennisetum glaucum*)	Zn	Pots	Increased significantly root length, shoot length, fresh weight and root biomass (Phytostabilization)	Misra et al. (2012)
Pseudomonas sp. A3R3	*Alyssum serpyllifolium*, *Brassica juncea*	Ni	Pots	Increased significantly the biomass (*B. juncea*) and Ni content (*A. serpyllifolium*) in plants grown in Ni-stressed soil (Phytoextraction)	Ma et al. (2011a)
Pseudomonas sp. TLC 6-6.5-4	*Zea mays*, *Helianthus annuus*	Cu	Pots	Significantly increased copper uptake by plants and also enhanced the biomass of maize (Phytoextraction)	Li and Ramakrishna (2011)
Psychrobacter sp. SRS8	*Ricinus communis*, *Helianthus annuus*	Ni	Pots	Stimulated plant growth and Ni accumulation in both plant species with increased plant biomass, chlorophyll and protein content (Phytoextraction)	Ma et al. (2011b)
Arthrobacter sp. MT16, *Microbacterium* sp. JYC17, *Pseudomonas chlororaphis* SZY6, *Azotobacter vinelandii* GZC24, *Microbacterium lactium* YJ7	*Brassica napus*	Cu	Gnotobiotic condition	Increased (16–41 %) root length (Phytoextraction)	He et al. (2010)
Bacillus species PSB10	Chickpea (*Cicer arietinum*)	Cr	Pots	Significantly improved growth, nodulation, chlorophyll, leghemoglobin, seed yield and grain protein and reduced the uptake of chromium in roots, shoots and grains (Phytostabilization)	Wani and Khan (2010)
Pseudomonas sp. SRI2, *Psychrobacter* sp. SRS8, *Bacillus* sp. SN9	*Brassica juncea*, *Brassica oxyrrhina*	Ni	Pots	Increased the biomass of the test plants and enhanced Ni accumulation in plant tissues (Phytoextraction)	Ma et al. (2009a)
Psychrobacter sp. SRA1, *Bacillus cereus* SRA10	*Brassica juncea*, *Brassica oxyrrhina*	Ni	Pots	Enhanced the metal accumulation in plant tissues by facilitating the release of Ni from the non-soluble phases in the soil (Phytoextraction)	Ma et al. (2009b)
Achromobacter xylosoxidans strain Ax10	Brassica juncea	Cu	Pots	Significantly improved Cu uptake by plants and increased the root length, shoot length, fresh weight and dry weight of plants (Phytoextraction)	Ma et al. (2009c)
Pseudomonas sp.	Chickpea	Ni	Pots	Enhanced fresh and dry weight of plants even at 2 mM nickel concentration (Phytostabilization)	Tank and Saraf (2009)

Table 3 continued

PSB	Plant	Heavy metals	Conditions	Role of PSB (Mode of metal remediation)	References
Enterobacter aerogenes NBRI K24, *Rahnella aquatilis* NBRI K3	*Brassica juncea*	Ni, Cr	Pots	Increased plant root length, dry weight, leaf protein and chlorophyll content with Ni and Cr uptake (Phytostabilization)	Kumar et al. (2009)
Bacillus weihenstephanensis strain SM3	*Helianthus annuus*	Ni, Cu, Zn	Pots	Increased plant biomass and the accumulation of Cu and Zn in the root and shoot systems, also augmented the concentrations of water soluble Ni, Cu and Zn in soil with their metal mobilizing potential (Phytoextraction)	Rajkumar et al. (2008)
Pseudomonas aeruginosa strain MKRh3	Black gram	Cd	Pots	Plants showed lessened cadmium accumulation, extensive rooting, and enhanced plant growth (Phytostabilization)	Ganesan (2008)
Enterobacter sp. NBRI K28, mutant NBRI K28 SD1	*Brassica juncea*	Ni, Cr, Zn	Pots	Improved plant growth parameters such as biomass, chlorophyll and protein and increased Ni, Cr and Zn uptake (Phytoextraction)	Kumar et al. (2008)
Burkholderia sp. J62	*Lycopersicon esculentum*	Pb, Cd	Pots	Increased root and shoot dry weight as well as Pb and Cd uptake (Phytoextraction)	Jiang et al. (2008)
Bacillus subtilis SJ-101	*Brassica juncea*	Ni	Growth chamber	Facilitated Ni accumulation (Phytoextraction)	Zaidi et al. (2006)
Pseudomonas sp., *Bacillus* sp.	Mustard	Cr	Pots	Stimulated plant growth and decreased Cr(VI) content (Phytostabilization)	Rajkumar et al. (2006)
Pseudomonas fluorescens	Soybean	Hg	Greenhouse	Increased plant growth (Phytostabilization)	Gupta et al. (2005)
Pseudomonas sp.	Soybean, mungbean, wheat	Ni, Cd, Cr	Pots	Promotes growth of plants (Phytostabilization)	Gupta et al. (2002)

damage, they become prone to diseases and pests attack. In practicing phytoextraction strategy, it is paramount important that selected plants must exhibit resistance to phytopathogens in order to smoothly function in metal-stressed soils. Moreover, further exploration and application of PSB strains, possessing additional traits which confer resistance to plants against various diseases, would be a better choice for metal phytoextraction.

Conclusions

Efficiency of phytoremediation of metal-polluted soils is chiefly determined by the metal bioavailability which in turn increases the metal uptake by plants. Hence, PSB compared with other plant growth promoting bacteria would be marvelous alternatives to boost this process as organic acids, and bio-surfactants secreted by these organisms solubilize sparingly soluble metal complexes, consequently increase bioavailability of metals and nutrient supply to soils. Thus, PSB with multifunctional activities (such as production of siderophore, IAA, ACC deaminase, organic acids and anti-pathogen metabolites) are better choice in assisting the phytoremediation process in metal-contaminated soils.

Conflict of interest The authors declare that there is no conflict of interests.

References

Ahemad M (2012) Implications of bacterial resistance against heavy metals in bioremediation: a review. IIOABJ 3:39–46

Ahemad M, Khan MS (2010a) Influence of selective herbicides on plant growth promoting traits of phosphate solubilizing *Enterobacter asburiae* strain PS2. Res J Microbiol 5:849–857

Ahemad M, Khan MS (2010b) Plant growth promoting activities of phosphate-solubilizing *Enterobacter asburiae* as influenced by fungicides. Eur Asian J Biosci 4:88–95

Ahemad M, Khan MS (2010c) Phosphate-solubilizing and plant-growth-promoting *Pseudomonas aeruginosa* PS1 improves greengram performance in quizalafop-p-ethyl and clodinafop amended soil. Arch Environ Contam Toxicol 58:361–372

Ahemad M, Khan MS (2011a) Toxicological assessment of selective pesticides towards plant growth promoting activities of phosphate solubilizing *Pseudomonas aeruginosa*. Acta Microbiol Immunol Hung 58:169–187

Ahemad M, Khan MS (2011b) Effects of insecticides on plant-growth-promoting activities of phosphate solubilizing rhizobacterium *Klebsiella* sp. strain PS19. Pestic Biochem Physiol 100:51–56

Ahemad M, Khan MS (2011c) Assessment of plant growth promoting activities of rhizobacterium *Pseudomonas putida* under insecticide-stress. Microbiol J 1:54–64

Ahemad M, Khan MS (2011d) Toxicological effects of selective herbicides on plant growth promoting activities of phosphate solubilizing *Klebsiella* sp. strain PS19. Curr Microbiol 62:532–538

Ahemad M, Khan MS (2011e) *Pseudomonas aeruginosa* strain PS1 enhances growth parameters of greengram [*Vigna radiata* (L.) Wilczek] in insecticide-stressed soils. J Pest Sci 84:123–131

Ahemad M, Khan MS (2012a) Biotoxic impact of fungicides on plant growth promoting activities of phosphate-solubilizing *Klebsiella* sp. isolated from mustard (*Brassica compestris*) rhizosphere. J Pest Sci 85:29–36

Ahemad M, Khan MS (2012b) Effect of fungicides on plant growth promoting activities of phosphate solubilizing *Pseudomonas putida* isolated from mustard (*Brassica compestris*) rhizosphere. Chemosphere 86:945–950

Ahemad M, Khan MS (2012c) Evaluation of plant growth promoting activities of rhizobacterium *Pseudomonas putida* under herbicide-stress. Ann Microbiol 62:1531–1540

Ahemad M, Khan MS (2012d) Alleviation of fungicide-induced phytotoxicity in greengram [*Vigna radiata* (L.) Wilczek] using fungicide-tolerant and plant growth promoting *Pseudomonas* strain. Saudi J Biol Sci 19:451–459

Ahemad M, Khan MS (2013) Pesticides as antagonists of rhizobia and the legume-*Rhizobium* symbiosis: a paradigmatic and mechanistic outlook. Biochem Mole Biol 1:63–75

Ahemad M, Kibret M (2013a) Recent trends in microbial biosorption of heavy metals: a review. Biochem Mol Biol 1:19–26

Ahemad M, Kibret M (2013b) Mechanisms and applications of plant growth promoting rhizobacteria: current perspective. J King Saud Univ Sci.

Ahemad M, Malik A (2011) Bioaccumulation of heavy metals by zinc resistant bacteria isolated from agricultural soils irrigated with wastewater. Bacteriol J 2:12–21

Ahemad M, Zaidi A, Khan MS, Oves M (2009) Biological importance of phosphorus and phosphate solubilizing microbes. In: Khan MS, Zaidi A (eds) Phosphate solubilizing microbes for crop improvement. Nova Science Publishers Inc., New York, pp 1–14

Ali H, Khan E, Anwar SM (2013) Phytoremediation of heavy metals-concepts and applications. Chemosphere 91:869–881

Arshad M, Saleem M, Hussain S (2007) Perspectives of bacterial ACC deaminase in phytoremediation. Trends Biotechnol 25:356–362

Becerra-Castro C, Prieto-Fernández A, Alvarez-Lopez V, Monterroso C, Cabello-Conejo MI, Acea MJ, Kidd PS (2011) Nickel solubilizing capacity and characterization of rhizobacteria isolated from hyperaccumulating and non-hyperaccumulating subspecies of *Alyssum serpyllifolium*. Int J Phytoremediat 1:229–244

Chen L, Dodd IC, Theobald JC, Belimov AA, Davies WJ (2013) The rhizobacterium *Variovorax paradoxus* 5C-2, containing ACC deaminase, promotes growth and development of *Arabidopsis thaliana* via an ethylene-dependent pathway. J Exp Bot.

Chodak M, Gołębiewski M, Morawska-Płoskonka J, Kuduk K, Niklińska M (2013) Diversity of microorganisms from forest soils differently polluted with heavy metals. Appl Soil Ecol 64:7–14

de-Bashan LE, Hernandez JP, Bashan Y (2012) The potential contribution of plant growth-promoting bacteria to reduce environmental degradation—a comprehensive evaluation. Appl Soil Ecol 61:171–189

Egamberdieva D (2009) Alleviation of salt stress by plant growth regulators and IAA producing bacteria in wheat. Acta Physiol Plant 31:861–864

Gamalero E, Glick BR (2012) Plant growth-promoting bacteria and metals phytoremediation. In: Anjum NA, Pereira ME, Ahmad I, Duarte AC, Umar S, Khan NA (eds) Phytotechnologies: remediation of environmental contaminants. CRC Press, Boca Raton, pp 361–376

Ganesan V (2008) Rhizoremediation of cadmium soil using a cadmium-resistant plant growth-promoting rhizopseudomonad. Curr Microbiol 56:403–407

Gillespie IMM, Philp JC (2013) Bioremediation, an environmental remediation technology for the bioeconomy. Trends Biotechnol.

Glass DJ (1999) U.S. and international markets for phytoremediation, 1999–2000. D. Glass Associates, Needham, p 266

Glick BR (1995) The enhancement of plant growth by free-living bacteria. Can J Microbiol 41:109–117

Glick BR (2005) Modulation of plant ethylene levels by the bacterial enzyme ACC deaminase. FEMS Microbiol Lett 251:1–7

Glick BR (2010) Using soil bacteria to facilitate phytoremediation. Biotechnol Adv 28:367–374

Glick BR (2012) Plant growth-promoting bacteria: mechanisms and applications. Hindawi Publishing Corporation, Scientifica

Glick BR, Todorovic B, Czarny J, Cheng Z, Duan J, McConkey B (2007) Promotion of plant growth by bacterial ACC deaminase. Crit Rev Plant Sci 26:227–242

Goldstein AH (1994) Involvement of the quinoprotein glucose dehydrogenase in the solubilization of exogenous phosphates by gram-negative bacteria. In: Torriani-Gorini A, Yagil E, Silver S (eds) Phosphate in microorganisms: cellular and molecular biology. ASM, Washington, DC, pp 197–203

Gupta A, Meyer JM, Goel R (2002) Development of heavy metal resistant mutants of phosphate solubilizing *Pseudomonas* sp. NBRI4014 and their characterization. Curr Microbiol 45:323–332

Gupta A, Rai V, Bagdwal N, Goel R (2005) In situ characterization of mercury resistant growth promoting fluorescent pseudomonads. Microbiol Res 160:385–388

Hao X, Xie P, Johnstone L, Miller SJ, Rensing C, Weia G (2012) Genome sequence and mutational analysis of plant-growth-promoting bacterium *Agrobacterium tumefaciens* CCNWGS0286 isolated from a zinc-lead mine tailing. Appl Environ Microbiol 78:5384–5394

Hashim MA, Mukhopadhyay S, Sahu JN, Sengupta B (2011) Remediation technologies for heavy metal contaminated groundwater. J Environ Manag 92:2355–2388

He LY, Zhang YF, Ma HY, Su LN, Chen ZJ, Wang QY, Meng Q, Fang SX (2010) Characterization of copper resistant bacteria and assessment of bacterial communities in rhizosphere soils of copper-tolerant plants. Appl Soil Ecol 44:49–55

He H, Ye Z, Yang D, Yan J, Xiao L, Zhong T, Yuan M, Cai X, Fang Z, Jing Y (2013) Characterization of endophytic *Rahnella* sp. JN6 from *Polygonum pubescens* and its potential in promoting growth and Cd, Pb, Zn uptake by Brassica napus. Chemosphere 90:1960–1965

Huang S, Peng B, Yang Z, Chai L, Zhou L (2009) Chromium accumulation, microorganism population and enzyme activities in soils around chromium-containing slag heap of steel alloy factory. Trans Nonferrous Met Soc China 19:241–248

Jiang CY, Sheng XF, Qian M, Wang QY (2008) Isolation and characterization of a heavy metal resistant *Burkholderia* sp. from heavy metal-contaminated paddy field soil and its potential in promoting plant growth and heavy metal accumulation in metal polluted soil. Chemosphere 72:157–164

Karuppiah P, Rajaram S (2011) Exploring the potential of chromium reducing *Bacillus* sp. and there plant growth promoting activities. J Microbiol Res 1:17–23

Khalid A, Akhtar MJ, Mahmood MH, Arshad M (2006) Effect of substrate-dependent microbial ethylene production on plant growth. Microbiology 75:231–236

Khan MS, Zaidi A, Wani PA, Oves M (2009) Role of plant growth promoting rhizobacteria in the remediation of metal contaminated soils. Environ Chem Lett 7:1–19

Kumar KV, Singh N, Behl HM, Srivastava S (2008) Influence of plant growth promoting bacteria and its mutant on heavy metal toxicity in *Brassica juncea* grown in fly ash amended soil. Chemosphere 72:678–683

Kumar KV, Srivastava S, Singh N, Behl HM (2009) Role of metal resistant plant growth promoting bacteria in ameliorating fly ash to the growth of *Brassica juncea*. J Hazard Mater 170:51–57

Li K, Ramakrishna W (2011) Effect of multiple metal resistant bacteria from contaminated lake sediments on metal accumulation and plant growth. J Hazard Mater 189:531–539

Liu M, Huang B, Bi X, Ren Z, Shenga G, Fu J (2013) Heavy metals and organic compounds contamination in soil from an e-waste region in South China. Environ Sci Process Impacts 15:919–929

Lugtenberg B, Kamilova F (2009) Plant growth-promoting rhizobacteria. Annu Rev Microbiol 63:541–556

Ma Y, Rajkumar M, Freitas H (2009a) Isolation and characterization of Ni mobilizing PGPB from serpentine soils and their potential in promoting plant growth and Ni accumulation by *Brassica* spp. Chemosphere 75:719–725

Ma Y, Rajkumar M, Freitas H (2009b) Improvement of plant growth and nickel uptake by nickel resistant-plant-growth promoting bacteria. J Hazard Mater 166:1154–1161

Ma Y, Rajkumar M, Freitas H (2009c) Inoculation of plant growth promoting bacterium *Achromobacter xylosoxidans* strain Ax10 for the improvement of copper phytoextraction by *Brassica juncea*. J Environ Manag 90:831–837

Ma Y, Prasad MNV, Rajkumar M, Freitas H (2011a) Plant growth promoting rhizobacteria and endophytes accelerate phytoremediation of metalliferous soils. Biotechnol Adv 29:248–258

Ma Y, Rajkumar M, Luo Y, Freitas H (2011b) Inoculation of endophytic bacteria on host and non-host plants-effects on plant growth and Ni uptake. J Hazard Mater 195:230–237

Ma Y, Rajkumar M, Vicente JA, Freitas H (2011c) Inoculation of Ni-resistant plant growth promoting bacterium *Psychrobacter* sp. strain SRS8 for the improvement of nickel phytoextraction by energy crops. Int J Phytoremediat 13:126–139

Martin TA, Ruby MV (2004) Review of in situ remediation technologies for lead, zinc, and cadmium in soil. Remediat Summer.

Mendez MO, Maier RM (2008) Phytostabilization of mine tailings in arid and semiarid environments-an emerging remediation technology. Environ Health Perspect 116:278–283

Misra N, Gupta G, Jha PN (2012) Assessment of mineral phosphate-solubilizing properties and molecular characterization of zinc-tolerant bacteria. J Basic Microbiol 52:549–558

Nadeem SM, Zahir ZA, Naveed M, Arshad M (2007) Preliminary investigations on inducing salt tolerance in maize through inoculation with rhizobacteria containing ACC deaminase activity. Can J Microbiol 53:1141–1149

Nadeem SM, Zahir ZA, Naveed M, Arshad M (2009) Rhizobacteria containing ACC-deaminase confer salt tolerance in maize grown on salt-affected fields. Can J Microbiol 55:1302–1309

Nair A, Juwarkar AA, Singh SK (2007) Production and characterization of siderophores and its application in arsenic removal from contaminated soil. Water Air Soil Pollut 180:199–212

Oves M, Khan MS, Zaidi A (2013) Chromium reducing and plant growth promoting novel strain *Pseudomonas aeruginosa* OSG41 enhance chickpea growth in chromium amended soils. Eur J Soil Biol 56:72–83

Padmavathiamma PK, Li LY (2007) Phytoremediation technology: hyper-accumulation metals in plants. Water Air Soil Pollut 184:105–126

Panhwar QA, Jusop S, Naher UA, Othman R, Razi MI (2013) Application of potential phosphate-solubilizing bacteria and organic acids on phosphate solubilization from phosphate rock in aerobic rice. Sci World J.

Peuke AD, Rennenberg H (2005) Phytoremediation. EMBO Rep 6:497–501

Rajkumar M, Nagendran R, Kui JL, Wang HL, Sung ZK (2006) Influence of plant growth promoting bacteria and Cr(VI) on the growth of Indian mustard. Chemosphere 62:741–748

Rajkumar M, Ma Y, Freitas H (2008) Characterization of metal-resistant plant-growth promoting *Bacillus weihenstephanensis* isolated from serpentine soil in Portugal. J Basic Microbiol 48:500–508

Rajkumar M, Ae N, Prasad MNV, Freitas H (2010) Potential of siderophore-producing bacteria for improving heavy metal phytoextraction. Trends Biotechnol 28:142–149

Saleem M, Arshad M, Hussain S, Bhatti AS (2007) Perspective of plant growth promoting rhizobacteria (PGPR) containing ACC deaminase in stress agriculture. J Ind Microbiol Biotechnol 34:635–648

Sandip B, Subrata P, Swati RG (2011) Isolation and characterization of plant growth promoting *Bacillus Thuringiensis* from agricultural soil of West Bengal. Res J Biotech 6:9–13

Schalk IJ, Hannauer M, Braud A (2011) New roles for bacterial siderophores in metal transport and tolerance. Environ Microbiol 13:2844–2854

Shaharoona B, Jamro GM, Zahir ZA, Arshad M, Memon KS (2007) Effectiveness of various *Pseudomonas* spp. and *Burkholderia* caryophylli containing ACC-deaminase for improving growth and yield of wheat (*Triticum aestivum* L.). J Microbiol Biotechnol 17:1300–1307

Singh AK, Cameotra SS (2013) Rhamnolipids production by multi-metal-resistant and plant-growth-promoting rhizobacteria. Appl Biochem Biotechnol 170:1038–1056

Singh Y, Ramteke PW, Shukla PK (2013) Isolation and characterization of heavy metal resistant *Pseudomonas* spp. and their plant growth promoting activities. Adv Appl Sci Res 4:269–272

Sun L, He L, Zhang Y, Zhang W, Wang Q, Sheng X (2009) Isolation and biodiversity of copper-resistant bacteria from rhizosphere soil of *Elsholtzia splendens*. Wei Sheng Wu Xue Bao 49:1360–1366

Suresh B, Ravishankar GA (2004) Phytoremediation-a novel and promising approach for environmental clean-up. Crit Rev Biotechnol 24:97–124

Swain H, Abhijita S (2013) Nitrogen fixation and its improvement through genetic engineering. J Global Biosci 2:98–112

Tank N, Saraf M (2009) Enhancement of plant growth and decontamination of nickel-spiked soil using PGPR. J Basic Microbiol 49:195–204

Upadhayay A, Srivastava S (2012) Evaluation of multiple plant growth promoting traits of an isolate of *Pseudomonas fluorescens* strain Psd. Indian J Exp Biol 48:601–609

Walker TS, Bais HP, Deziel E, Schweizer HP, Rahme LG, Fall R, Vivanco JM (2004) *Pseudomonas aeruginosa*-plant root interactions. Pathogenicity, biofilm formation and root exudation. Plant Physiol 134:320–331

Wani PA, Khan MS (2010) *Bacillus* species enhance growth parameters of chickpea (*Cicer arietinum* L.) in chromium stressed soils. Food Chem Toxicol 48:3262–3267

Waterlot C, Bidar G, Pelfrêne A, Roussel H, Fourrier H, Douay F (2013) Contamination, fractionation and availability of metals in urban soils in the vicinity of former lead and zinc smelters, France. Pedosphere 23:143–159

Zahir ZA, Ghani U, Naveed M, Nadeem SM, Asghar HN (2009) Comparative effectiveness of *Pseudomonas* and *Serratia* sp. containing ACC-deaminase for improving growth and yield of wheat (*Triticum aestivum* L.) under salt-stressed conditions. Arch Microbiol 191:415–424

Zaidi S, Usmani S, Singh BR, Musarrat J (2006) Significance of *Bacillus subtilis* strain SJ 101 as a bioinoculant for concurrent plant growth promotion and nickel accumulation in *Brassica juncea*. Chemosphere 64:991–997

Aquatic model for engine oil degradation by rhamnolipid producing *Nocardiopsis* VITSISB

Suki Roy · Shreta Chandni · Ishita Das · Loganathan Karthik · Gaurav Kumar · Kokati Venkata Bhaskara Rao

Abstract The present study was focused on isolation, screening, characterization and application of biosurfactant producing marine actinobacteria. Twenty actinobacteria were isolated from marine water sample and were primarily screened for biosurfactant production using hemolytic activity method. Among the 20 isolates, six showed positive result for hemolytic activity and those were taken for further secondary screening tests such as oil collapse method, oil spreading method and emulsification method. From the results of secondary screening analysis, two isolates (SIS-3 and SIS-20) were selected and further used to carry out biosurfactant characterization test such as pH, density, surface tension and viscosity determination. Comparing biosurfactant characterization results, SIS-3 was chosen for further analysis and application. FT-IR and GC–MS were carried out for analysis of biosurfactant from isolate SIS-3 and the compound detected was rhamnolipid. The isolate (SIS-3) was identified as *Nocardiopsis* using 16S rRNA gene sequencing and named as '*Nocardiopsis* VITSISB' (KC958579) which was further applied for immobilizing whole cells for engine oil degradation by constructing an aquatic model and using natural products such as soybean meal, sugarcane juice as nutrient source. The oil was efficiently degraded by rhamnolipid producing *Nocardiopsis* VITSISB (KC958579) within 25 days which indicated that the strain can act as a natural candidate for the bioremediation of oil spill in ocean.

Keywords Actinobacteria · Bioremediation · Biosurfactant · Rhamnolipid

Introduction

Disasters make headlines, but a hundreds and million gallon of oil quietly ends up in the sea every year which is not considered seriously. It has been analyzed that a large percentage of ocean pollution occurs due to oil leakage in marine water. When it comes to oil pollution, ocean is the worst sufferer in terms of loss of aquatic life (Murali et al. 2010). There are many incidents that have occurred worldwide. One of the recent oil spill incident reported by "NY Daily News", that in 27 July 2013, oil spilling occurred in the sea of Rayong near eastern Thailand due to the leakage of a pipe line of Oil and Natural Gas Company "Petroleum Authority Of Thailand Public Limited Company" (PTT Plc), buried deep into the ocean up to 20 km which spread nearly 50,000 tons of oil up to 984 feet near the eastern cost of Thailand. It caused lot of damage to the environment. Apart from this, many other incidences had occurred in all over the world due to many different causes such as oil leaked into the sea, accident of oil transporting ships from one place to another, which also affect valuable resources of the natural environment. Research is going on to find out the relevant solutions with respect to factors such as eco friendly and cost effective methods.

Biodegradation of oil by biosurfactant producing strain plays an important role in restoring oil-polluted

S. Roy · S. Chandni · I. Das · L. Karthik · G. Kumar · K. V. Bhaskara Rao (✉)
Molecular and Microbiology Research Laboratory, School of Bio Sciences and Technology, VIT University, Vellore 632014, Tamil Nadu, India
e-mail: kokatibhaskar@yahoo.co.in

L. Karthik
Department of Biotechnology, Shri JJT University, Jhunjhunu, Rajasthan, India

environments and habitats. Some microorganisms can produce amphoteric surface-active-agents which emulsify hydrocarbons in a growth medium and reduce the surface and interfacial tensions, which increase the solubility, and emulsification of the two immiscible phases also provides an insoluble substrate for microorganisms (Banat et al. 2000). These are known as biosurfactants. These are a structurally diverse group of surface-active molecules which are eco-friendly. It has been characterized that, biosurfactants having advantages of low toxicity, high biodegradability, better environmental compatibility, high foaming capability, higher selectivity, specific activity at extreme temperature, pH, salinity, can be analyzed for its applications for the specified issue.

Biosurfactants have been broadly used in environmental protection purpose; such methods are enhanced oil recovery (EOR), control of oil spills, detoxification of industrial effluents and soil contaminated with oil. Chemical dispersants that are mainly used to dissipate oil slicks lead to the increase in toxic hydrocarbon levels in fish by a factor of up to 100 and may kill fish eggs. Dispersed oil droplets infiltrate into deeper water and can lethally contaminate coral. Research indicates that some dispersants are poisonous to corals. A 2012 study found that Corexit dispersant had increased the toxicity of oil by up to 52 times. On the other hand, biosurfactants have an amplified adaptability in contrast to numerous chemical surfactants; therefore it may be appropriate for bioremediation (Borjana et al. 2001).

There is a dependence of biodegradation of a given hydrocarbon on its dispersion site (Mattei et al. 1986). However, minute information on either biosurfactants active in saline or biosurfactant produced by marine microorganisms has been reported so far. Marine biosurfactants produced by some marine microorganisms have been rewarded more interest, particularly for the bioremediation of the ocean polluted by crude oil (Maneerat 2005). It is not surprising that marine actinobacteria are found to be a valuable resource since they are different from terrestrial organisms due to their survival in adverse conditions of marine environment such as alkalinity, temperature etc. Marine biotechnology has opened up unexpectedly new horizons for discovering novel organisms and trapping their potential resources such as bioactive compounds which are widely used in different applications such as pharmaceutical, textile, cosmetic industries etc. (Fiedler et al. 2005; Blunt et al. 2009). The actinobacteria are distinguished by their gram-positive character and high guanine-plus-cytosine (G + C) content in their genomes (Okami 1986).

The present study was to isolate and characterize biosurfactant producing actinobacteria from marine water and to evaluate their application in engine oil degradation.

Materials and methods

Medium, chemicals and engine oil

All the media and chemical ingredients used in the current experiment were bought from Hi-Medium, Mumbai, India and SRL Pvt. Ltd., Mumbai, India. Engine oil used in this study was obtained from local automobile workshop (Vellore, India) and was sterilized by filtration.

Isolation of marine actinobacteria

The marine water sample was collected from Marina beach (13.05°N, 80.28°E), Chennai, Tamil Nadu, India. A tenfold serial dilution of the sample was prepared. 100 μl of serially diluted sample was inoculated into starch casein agar medium (pH 7.2) prepared with 50 % seawater to sustain the growth of actinobacteria. To avoid the growth of fungal and bacterial contaminant, potassium dichromate (50 μg ml^{-1}) and nalidixic acid (15 μg ml^{-1}) were supplemented. The plates were incubated at room temperature (28 °C) and monitored periodically over 3 months for actinomycetes growth. The pure isolates of the actinobacteria were transferred to ISP2 (Isolation streptomyces project medium No. 2) slants and preserved at 4 ± 2 °C (Karthik et al. 2010; Kathiresan et al. 2005).

Primary screening method

Biosurfactant producing marine actinobacteria were primarily screened using hemolytic assay. All the isolates were streaked into blood agar plate and incubated at 37 °C for 7 days. The appearance of clear zone indicated the production of biosurfactant. The results were interpreted as 'γ' for no hemolysis, 'α' for incomplete hemolysis and 'β' for complete hemolysis (Mulligan and Gibbs 1989).

Production media for biosurfactant

To select an appropriate growth media for biosurfactant producing actinobacteria, two different media were used. These are SS medium (soluble starch 25 g/l, glucose 10 g/l, yeast extract 2 g/l, calcium carbonate 3 g/l, trace salt solution 1 ml ($FeSO_4$, $CuSO_4$, $ZnSO_4$ and $MnCl_2$) and SS medium with oil (SS Medium with 0.5 % oil). One loop full culture of the isolates were inoculated into two media and incubated at 37 °C for 7 days. After 7 days, emulsification index was calculated for each medium.

Secondary screening method

Biosurfactant producing marine actinobacteria were screened by drop collapsing method, oil spreading method and emulsification index. Sodium lauryl sulphate was taken as positive control for all the methods (Chandran and Das 2010).

Drop collapse method

In drop collapse method, 2 μl of mineral oil was added to 96-well microtiter plate and 5 μl of the culture was added to the surface of the oil. After 1 min, the observation of shape of the drop on the surface of the oil was done. The interpretation of result was '+' to '++' corresponding to partial to complete spreading on the oil surface. Those cultures that gave rounded drops were scored as negative '−' indicative of the lack of biosurfactant production (Bodour and Miller-Maier 1998).

Oil spread method

In Petri plate, 40 ml of distilled water was added followed by 10 μl of engine oil on the surface of the water. 10 μl of culture was then added to the surface of oil. The diameter of the clearing zone on the oil surface and correlation of surfactant activity were measured (Walter et al. 2000).

Emulsification index (E_{24} %)

Emulsification activity was measured by calculating E_{24} described by Cooper and Goldenberg (1987). Kerosene was added to culture supernatant (1:2 v/v), mixed in a high speed vortex for 2 min and allowed to stand for 24 h. After 24 h, measurement of height of the stable emulsion layer was executed. The emulsion index E_{24} is calculated as the ratio of the height of the emulsion layer and the total height of liquid.

$$E_{24} = (\text{total height of the emulsified layer} / \text{total height of the liquid layer}) \times 100\,\% \quad (1)$$

Production and extraction of biosurfactant

Biosurfactant production was carried out in SS Medium with 0.5 % oil. One loop full culture of the isolates were inoculated into media and incubated in a shaker at 120 rpm for a period of 7 days at 37 °C. Centrifugation of the culture broth was carried out at 10,000 rpm for 10 min followed by extraction of solvents with chloroform and methanol (2:1 v/v). The solvent removal was done by rotary evaporation and biosurfactant was obtained as the resultant residue (Chandran and Das 2010).

Biosurfactant characterization

Biosurfactant characterization was conducted by determination of pH, density, surface tension and viscosity. In all the cases, sodium lauryl sulphate was taken as positive control and distilled water as negative control.

pH and density determination

The pH and density of the biosurfactant solution (1 mg/ml) was determined using pH meter and hydrometer, respectively.

Surface tension measurement

The capillary rise method was used to determine the surface tension of the culture free supernatant. A small bore capillary was inserted into the liquid whose surface tension was to be determined. The liquid level raised to a certain height in the tube was proportionally varying with the surface tension. The surface tension was calculated using the following formula (Harkins and Jordan 1991):

$$\gamma = dghr/2 \quad (2)$$

where, γ is surface tension, d is the density of the fluid, g is the acceleration due to gravity, h is the height of liquid rising and r is the radius of capillary's bore.

Viscosity determination

High viscosity of the oil limits its mobility. So, decrease in the viscosity of heavy oil is a way to improve oil recovery. The culture-free supernatant was used to determine the viscosity of the biosurfactant with the help of Ostwald viscometer. Distilled water was used as reference fluid. The formula used to determine viscosity was:

$$\eta_s = (\eta_w t_s \rho_s)/t_w \rho_w \quad (3)$$

η_s is the viscosity of sample, η_w is the viscosity of water, t_s and t_w are the time taken by sample and water, respectively, ρ_s and ρ_w are the density of sample and water, respectively (Cooper et al. 2013).

Analysis of biosurfactant

Orcinol assay

The nature of the biosurfactant was determined using orcinol assay. In this assay, 100 μl of cell-free supernatant was added to 900 μl of orcinol reagent and incubated at 80 °C for 30 min. Production of a blue-green color determined the presence of glycolipid in the supernatant. The mixture was allowed to come down to room temperature. The absorbance was measured at 421 nm (Harkins et al. 2010).

Stability analysis

To evaluate the temperature effect, 10 ml of the culture-free supernatant was exposed at 5, 50, 70 and 100 °C for 1 h and cooled to room temperature. To examine the pH effect, the pH of the culture supernatant was adjusted to 2.0, 4.0, 8.0 and 12.0. The effect of sodium chloride concentration was detected by adding sodium chloride into the medium to achieve the concentration of 2.0, 4.0, 7.0 and 10.0 % (w/v). Stability study was conducted by surface tension and emulsification index determination.

Fourier transforms infrared spectroscopy (FT-IR) analysis

One milligram of extracted biosurfactant was mixed with 20 mg of potassium bromide and pressed to obtain translucent pellets. The infrared spectra were performed on Mattson 1,000 FT-England FTIR system within the range of 500–4,000 cm^{-1} wave number (Chander et al. 2012).

Gas chromatography–mass spectrometry (GC–MS) analysis

The chloroform:methanol extract of SIS-3 was analyzed by gas chromatography–mass spectrometry (GC–MS) using a Thermo GC—Trace Ultra VER: 5.0, Thermo MS DSQ II with a Elite-5MS (30.0 m, 0.25 mm ID, 250 μm df). The temperature of the column was programmed from initial temperature 60 °C for 2 min, ramp 10 °C/min to 300 °C with a holding of 6 min, and the injector or detector temperature for the analysis was about 250 °C. Helium was used as the carrier gas at a flow rate of 1.0 ml/min. The identification of the chemical constituents was based on correlation with the existing mass spectra data with those obtained from the Wiley8.LIB and NIST08.LIB library spectrum provided by the software on a GC–MS system (Arino et al. 1996).

Strain identification

The potent strain showing significant biosurfactant activity was molecularly characterized with 16S rRNA gene sequencing. The actinobacterial DNA was isolated by Bacterial DNA Mini Spin Kit (2 μl) (Amnion, India) and PCR amplification was done using an initial denaturation step at 94 °C for 5 min followed by 35 cycle of 1 min at 94 °C, 30 s at 55 °C and 1:30 min at 72 °C and final extension at 72 °C for 5 min cooled at 4 °C, the Taq polymerase was used as 10× AMTaq Polymerase buffer: 5 μl AMTaq, Polymerase enzyme: 3 U and PCR grade water to make the volume up to 50 μl (Amnion, India) as per user manual (Saiki et al. 1985). The forward and reverse primers used are 5′-CWG RCC TAN CAC ATG SAA GTC-3′ (100 ng) and 5′-GRC GGW GTG TAC NAG GC-3′ (100 ng), respectively. The resemblance or homogeneity of the16S rRNA partial gene sequence was analyzed with the analogous sequences already present in the records bank-National Center for Biotechnology Information (NCBI) via BLAST search. The applied DNA sequences were allied and phylogenetic tree was created with the function of neighbor-joining technique, the used online software is known as ClustalW (Saitou and Nei 1987). The topologies of the resulted tree were calculated through bootstrap analyses of the neighbor-joining technique based on 1,000 re-samplings. The phylogenetic analysis was conceded using the TREEVIEW software (version 1.3.3).

Immobilization of whole cells of SIS-3 in calcium alginate beads

Immobilization was carried by suspending 100 ml of cultures in sodium alginate solution (2 % w/v). The mixture obtained was added dropwise through a syringe into calcium chloride solution (3.5 % w/v). Beads were formed when alginate got solidified upon contact with calcium chloride, entrapping the cells. The spherical beads (~2 mm diameter) got hardened in 30 min and then were washed with distilled water to eliminate excess calcium ions and cells (Taha et al. 2013).

Degradation of engine oil by designing aquatic model

Oil degradation ability of the biosurfactant was determined by setting up an aquatic model. It was constructed with an open glass chamber provided with 2 l of marine water, 100 ml of engine oil and limestone. To degrade the oil, whole-cell immobilization of *Nocardiopsis* VITSISB (KC958579) in calcium alginate was conducted. The immobilized cells were exposed to oil present in the aquatic model along with 100 ml of sugarcane juice and 50 mg of soya meal which acted as nutrient for the cells. The model was kept at room temperature for 25 days. Oil sample from the aquatic model was centrifuged at 5,000 rpm for 5 min. Water layer was discarded and the oil supernatant was taken. The oil was then extracted by means of chloroform (1:1) in separating funnel. Chloroform and supernatant was vortexed for 5 min and kept still for 10 min. The lower aqueous layer and top layer of chloroform with engine oil were taken out in another flask. A little amount of anhydrous sodium sulphate was used to eliminate the residual moisture. Solvent hexane was then evaporated at room temperature. The remaining oil left was then diluted with chloroform in the ratio of 1:9 (v/v) (Kumari and Abraham 2011). The oil degradation was analyzed using High Performance Liquid Chromatography (Empower Company, India). Acetonitrile:water (75:25 %) was used as mobile phase. The HPLC analysis was performed at a flow rate of 1 ml min^{-1}, with UV detection at a wavelength of at 244 nm.

Statistical analysis

All experiments were carried out in triplicates. Data are presented as mean ± standard deviation (SD). To prove significant differences, results were evaluated statistically, at 95 % confidence level ($p < 0.05$), using the Analysis of Variance (ANOVA) available in Graphpad Prism v.6.00 (La, Jolla, CA, USA).

Results

Total 20 marine actinobacteria were isolated from marine water sample. Out of 20 isolates, six isolates showed positive result in hemolytic activity on blood agar plates. The results are presented in Table 1. Two media (SS, SS + OIL) were used for the production of biosurfactant by marine actinobacteria. The emulsification index for the isolates in two media is mentioned in Fig. 1. The result of secondary screening of biosurfactant by six isolates is provided in Table 2. Biosurfactant characterization test were conducted and the results are summarized in Table 3. The surface tension was observed as 62.808 dyne/cm for SIS-3, 62.808 dyne/cm for SIS-20, and 68.6 dyne/cm for water. With reference to biosurfactant viscosity measurement result, SIS-3 showed 10.7×10^{-3} poise, SIS-20 showed 12.6×10^{-3} poise and sodium lauryl sulphate showed 10.1×10^{-3} poise, so lower the viscosity more would be the positive effect on emulsion stability of biosurfactant. Density measurement result showed SIS-3 as 1.16 g/cm^3, SIS-20 as 1.18 g/cm^3 and water as 68.6 g/cm^3. The orcinol assay showed positive result by formation of blue-green color. The constancy of biosurfactant produced by SIS-3 was examined over an extensive variation of temperatures.The surface tension reduction (dyne/cm) and percentage of emulsification activities were highest at the temperature of 50 °C as shown in Fig. 2. The biosurfactant showed relatively high efficiency at pH 12 where the emulsification activity showed almost 66 % emulsion and the surface tension reduced to 56.8 dyne/cm as shown in Fig. 3. The effect of sodium chloride addition on biosurfactant produced from SIS-3 was studied as shown in Fig. 4. Optimum constancy of biosurfactant was observed at 10 % NaCl concentration. At higher concentration of sodium chloride, the biosurfactant showed 76 % of the emulsification activity and more reduction in surface tension than in lower concentration of NaCl. Molecular compositions of biosurfactant produced by SIS-3 were evaluated by FT-IR as shown in Fig. 5. The broad band observed in biosurfactant was at 3,444 cm^{-1} which corresponded to the hydroxy group, H-bonded OH stretch in glycolipid. The peak at 1,689 cm^{-1} corresponded to C=O stretching of carbonyl groups which is characteristic for ester compounds. The band 1,396 cm^{-1} confirmed the presence of alkyl groups. The C–O–C stretching in the rhamnose was found at 1,083 cm^{-1} and the stretching of aromatic C–H in-plane bend occurred at 991 cm^{-1}. The GC–MS analysis of chloroform:methanol extracts of SIS-3 is shown in Fig. 6. The results showed that the major chemical constituent of chloroform:methanol extracts of SIS-3 was phenol, 3,5-bis(1,1-dimethylethyl)- by comparison of mass spectral data and retention times. The other major constituents present in the extract were benzene, 1,3-bis(1,1-dimethylethyl),dibutyl phthalate,2-tert-butyl-4,6-bis(3,5-di-tert-butyl-4 hydroxybenzyl) phenol and dodecane, 1-fluoro. The major chemical constituent phenol,3,5-bis(1,1-dimethylethyl) was responsible for surfactant activity. The molecular weight of phenol, 3,5-bis(1,1-dimethylethyl) ($C_{14}H_{22}O$) was 206. The taxonomic position of *Nocardiopsis* VITSISB (KC958579) was determined based on 16S rRNA partial gene sequencing. A phylogenetic tree was constructed based on the neighbor-joining method using Treeview software as shown in Fig. 7. The oil layer in the aquatic model was reduced after duration of 25 days as shown in Fig. 8. HPLC Chromatogram of degraded oil showed more number of peaks as compared to control engine oil as shown in Fig. 9.

Table 1 Haemolysis assay of 20 isolates of marine actinobacteria

Isolate name	Haemolysis activity	Size of zone of clearance (mm)
SIS-1	α	9 ± 0.5
SIS-2	β	18 ± 0.3
SIS-3	β	15 ± 0.5
SIS-4	α	8 ± 0.6
SIS-5	α	6 ± 0.6
SIS-6	α	5 ± 0.3
SIS-7	β	5 ± 0.6
SIS-8	α	6 ± 0.5
SIS-9	γ	
SIS-10	γ	
SIS-11	β	17 ± 0.6
SIS-12	α	3 ± 0.5
SIS-13	γ	-
SIS-14	α	16 ± 0.6
SIS-15	α	7 ± 0.5
SIS-16	α	4 ± 0.5
SIS-17	γ	-
SIS-18	β	16 ± 0.3
SIS-19	α	10 ± 0.2
SIS-20	α	13 ± 0.4

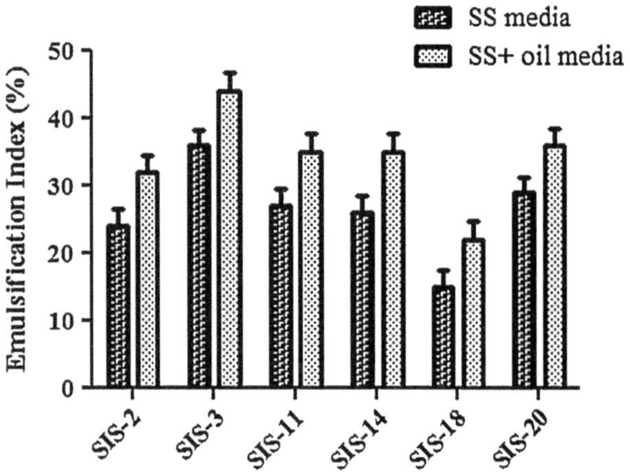

Fig. 1 Emulsification index (E_{24}) of the isolates in two media. Data are given as mean ± SD ($n = 3$). Results were considered significant at $p < 0.05$

Table 2 Secondary screening of biosurfactant produced by 6 isolates

Isolate name	Oil collapse method	Oil spread method Size of zone (mm)	Emulsification index (%)
SIS-2	+[a]	62 ± 0.2	32 ± 0.4
SIS-3	+	74 ± 0.5	44 ± 0.2
SIS-11	-[c]	64 ± 0.3	35 ± 0.3
SIS-14	+	75 ± 0.2	35 ± 0.1
SIS-18	+	52 ± 0.4	22 ± 0.2
SIS-20	++[b]	68 ± 0.3	36 ± 0.2
SLS	++	83 ± 0.1	82 ± 0.1

[a] Partial spreading on the oil surface
[b] Complete spreading on the oil surface
[c] Lack of biosurfactant production

Table 3 Characterization of biosurfactant produces by two strain SIS-20 and SIS-3

Properties	SIS-20	SIS-3	SLS [a]	H_2O
pH	6.95	7.24	7.13	7.0
Density (g/cm^3)	1.18	1.16	1.01	1
Surface tension (dyne/cm)	69.808	62.524	59.393	68.6
Viscosity (poise)	12.6×10^{-3}	10.7×10^{-3}	10.1×10^{-3}	9.1×10^{-3}

[a] Sodium Lauryl Sulphate

Discussion

As marine actinobacteria are promising organisms to get high yielding of bioactive compounds, they are efficiently used in different fields such as bioremediation, drug development and cosmetic products. Hence, with that concern, we had paid more concern in marine actinobacteria. In this present study, almost 20 isolates were collected from marine water which are further designated as SIS-1 to SIS-20.

Strain screening

The results of the present study reveal that, the isolates SIS-2, SIS-3, SIS-11, SIS-14, SIS-18 and SIS-20 showed very good hemolytic activity. Bernheimer and Avigad (1970) used this method to screen the biosurfactant producing microorganism. Carrillo et al. (1996) described the hemolytic activity as a primary technique to screen the biosurfactant activity. The zone of inhibition was directly proportional to the concentration of surfactant. In another study, Noha et al. (2004) reported that 16 % of false positive results in hemolytic action for biosurfactant selection are due to microorganism virulence issue, as well as that the biosurfactants that are poorly diffusible may not lyse the blood cells. Hence, hemolytic activity was recommended as a primary screening method. The hemolytic activity of the remaining 14 strains was ranged from (0 to 140 mm) and these 14 strains exhibited poor results for the drop collapsing method, oil spreading method (50–80 mm) and emulsification index (20–50 %) method.

In the drop collapse method, SIS-3 and SIS-20 showed positive results. The drop collapse technique may be considered as chief screening method for biosurfactant production. Bodour et al. (2003) reported that the drop collapse efficient technique may be employed for detection of biosurfactant producing microorganisms in natural environment. The oil spread method is an indirect dimension of surface action of a surfactant which is tested against oil; a higher surface action of biosurfactant is represented with a larger diameter of clearing zone (Rodrigues et al. 2006). Plaza et al. (2006) and Youssef et al. (2004) verified that the oil spreading method is an effective method to identify biosurfactant production by various microorganisms. The assay was also used for screening by Huy et al. (1999). Among the six isolates, SIS-3 and SIS-20 showed highest emulsification index. The emulsification index is a technique widely used to identify and quantify biosurfactants in cultures. Emulsification is an important process in alkane biodegradation (Thavasi et al. 2011). The diffusion of a liquid phase as microscopic droplets in an additional liquid continuous phase results in the development of emulsion (Desai and Banat 1997). This property is useful for making water/oil emulsions for food and cosmetics.

In all these studies, among the six isolates which are screened for primary test, only SIS-3 and SIS-20 showed maximum clear of zone, emulsification index and good result in drop collapse, hence these two isolates were studied further.

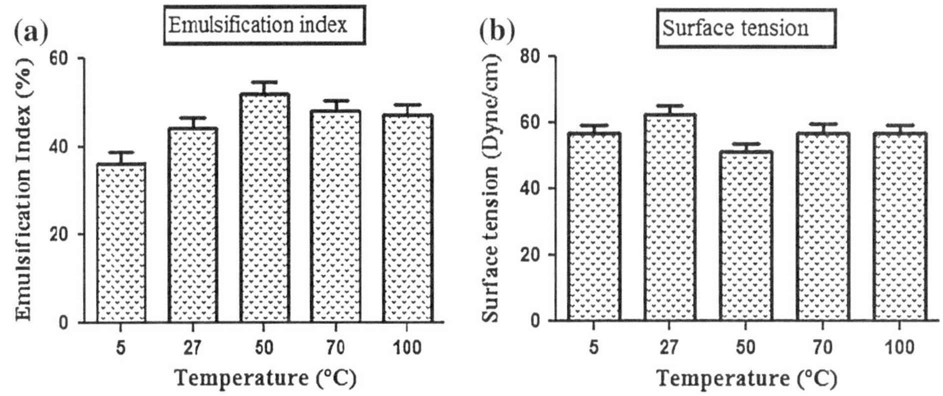

Fig. 2 Temperature stability of biosurfactant produced by SIS-3: **a** emulsification index, **b** surface tension. Data are given as mean ± SD ($n = 3$). Data are significant where $p < 0.05$

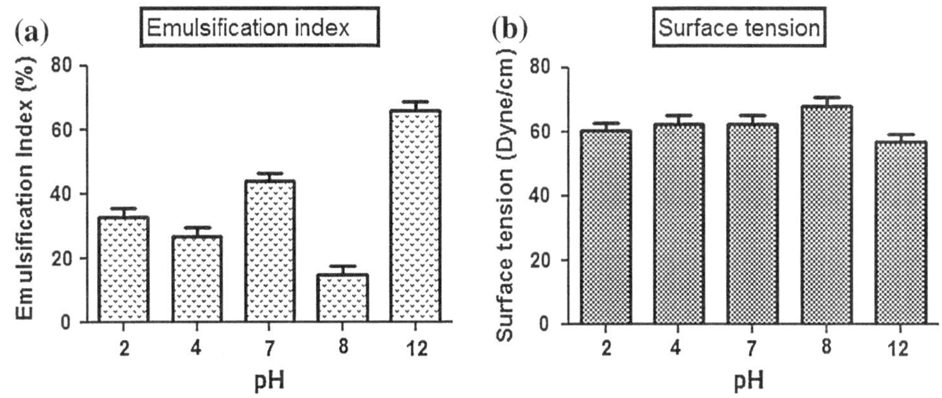

Fig. 3 pH stability of biosurfactant produced by SIS-3: **a** emulsification index, **b** surface tension. Data are given as mean ± SD ($n = 3$). Results were considered significant at $p < 0.05$

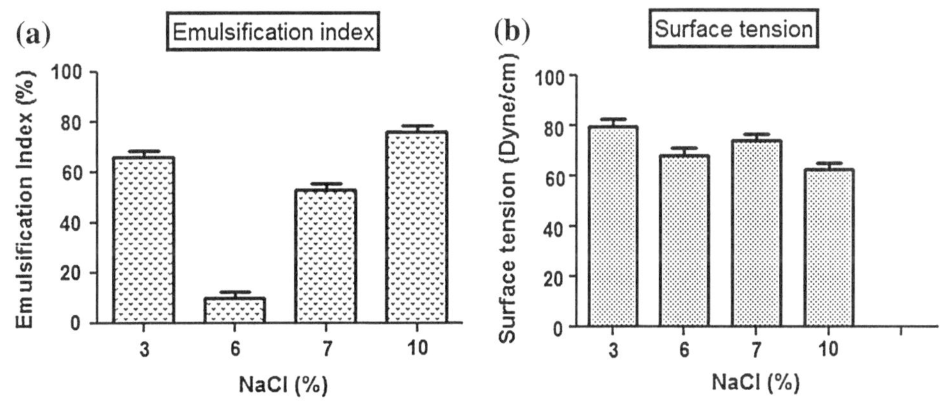

Fig. 4 Salinity stability of biosurfactant produced by SIS-3: **a** emulsification index, **b** surface tension. Data are given as mean ± SD ($n = 3$). Results were considered significant at $p < 0.05$

Production media for biosurfactant

For all the isolates, SS + oil media showed more emulsification index than SS media. The use of vegetable oil as carbon sources to produce biosurfactants was found to be an effective and low-cost alternative (Abouseoud et al. 2008). Hence, SS + oil media were chosen for further production of biosurfactant from marine actinobacteria.

Biosurfactant characterization

Attwood et al. (1970) already discussed that the higher the surface tension of biosurfactant, the higher will be the micelle forming activity which shows higher oil degradation activity. In the cosmetic industry, the cosmetic product quality including product viscosity and product stability get affected by different factors for example, foaming, emulsifying, water binding capability, dispersal and wetting features (Gharaei-Fathabad 2011). The variety of cleanup efforts varies extensively and depends on a number of issues, such as density of biosurfactant, the capacity of the spilling, weather as well as speed of reactions. The biosurfactant from SIS-3 was found to have more pH than SIS-20. Hence, the biosurfactant from SIS-3 can well accommodate in oceanic environment with respect to stability. Graham (2010) already mentioned that the "rise rate" of

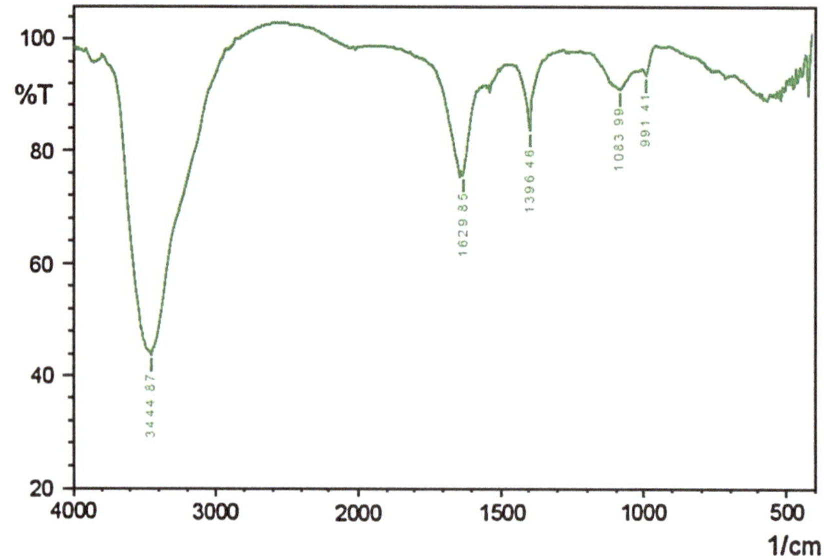

Fig. 5 FT-IR spectra of biosurfactant produced by SIS-3

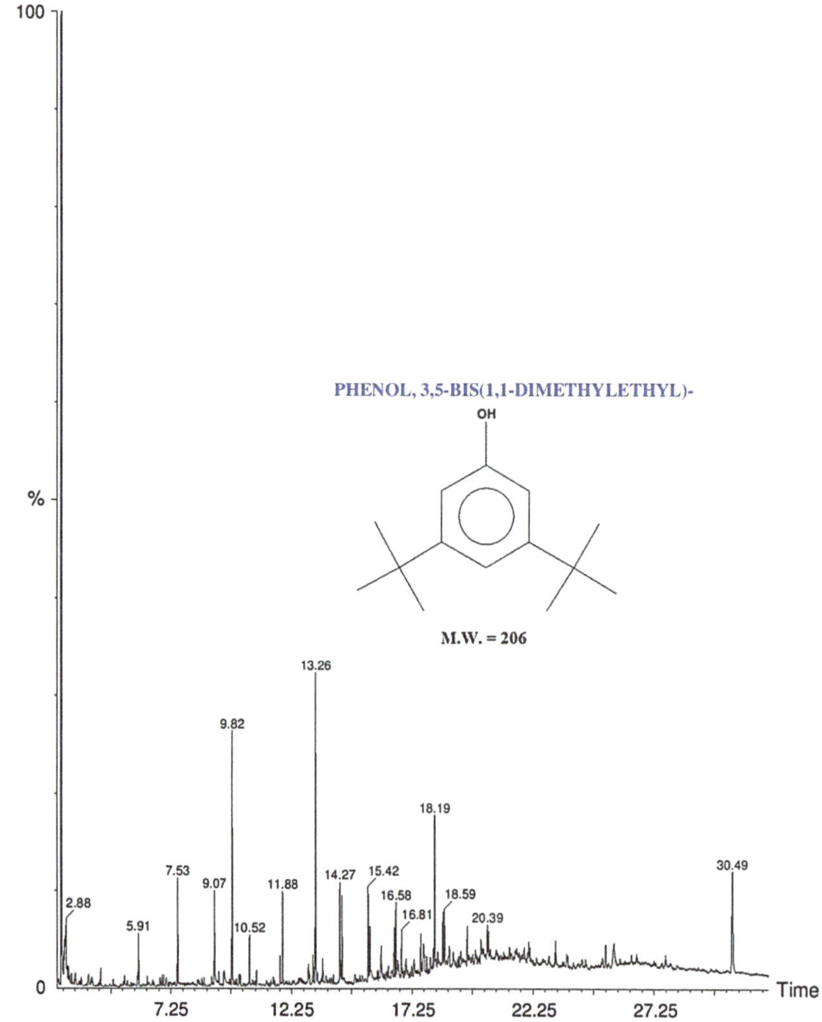

Fig. 6 GC–MS profile of chloroform:methanol extract of SIS-3

oil globules (their vertical velocity) is the rapidity at which oil particles move in the direction of the panel surface which is because of the dissimilarity in densities of oil and water. The closer the density of the biosurfactant in compared water, the lesser will be the time acquired for separation. Since SIS-3 showed best result with respect to pH,

Fig. 7 Phylogenetic tree of *Nocardiopsis* VITSISB (KC958579)

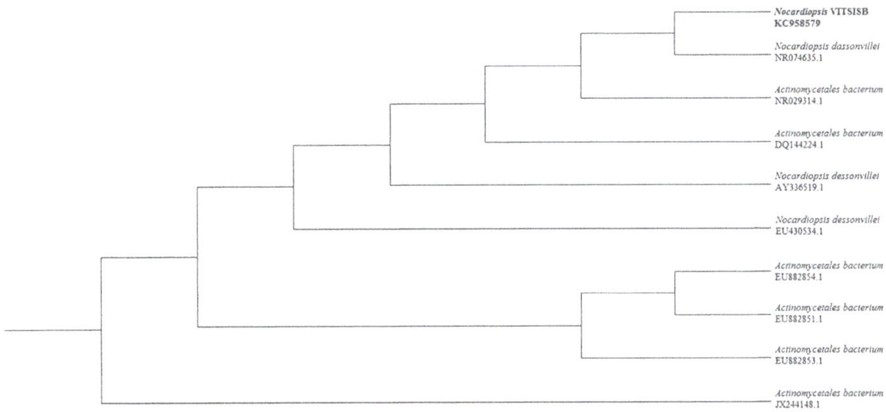

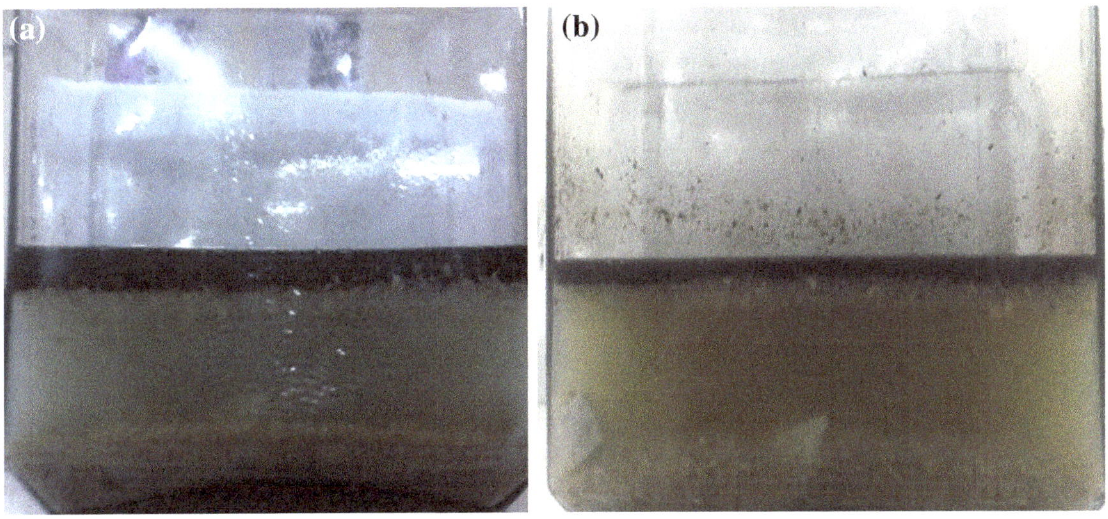

Fig. 8 Aquatic Model **a** day 1 and **b** day 25

density, viscosity and surface tension, it was selected for further analysis and application.

Orcinol assay

Production of a blue-green color after addition of orcinol indicated the presence of glycolipid in the biosurfactant produced by SIS-3 (Ahmad Mohammad et al. 2011). This was the first report on the production of rhamnolipid by marine *Nocardiopsis* sp.

Stability test

The isolate SIS-3 maintained its best biosurfactant properties at the temperature of 50 °C as shown. This activity indicates the significance of the biosurfactant in pharmaceutical, food and cosmetic industries, where leading to achieve sterility is of paramount importance (Abouseoud et al. 2008; Mulligan et al. 1984). The results indicated that the increase in pH had a positive effect on emulsification activity and surface tension. This might be because of a better constancy of fatty acid surfactant micelles in the occurrence of sodium hydroxide and the precipitation of secondary metabolite at high pH value (Abouseoud et al. 2008; Abu-Ruwaida et al. 1991). With result to the effect of salinity on biosurfactant, the result suggested that the biosurfactant might be useful in marine environments and other systems where salt concentration is above physiological level. There are reports on some biosurfactants produced by bacteria showing stability in the presence of high salt concentration (Chandran and Das 2010).

Based on the results obtained from the present study, the biosurfactant produced by SIS-3 had its best activity at high temperature, alkaline pH and high salinity which could be useful for bioremediation of spills in marine environment. In contrast to the report of Khopade et al. (2012), moderately thermostable

Fig. 9 HPLC profile of engine oil: **a** before degradation, **b** after degradation (25 days)

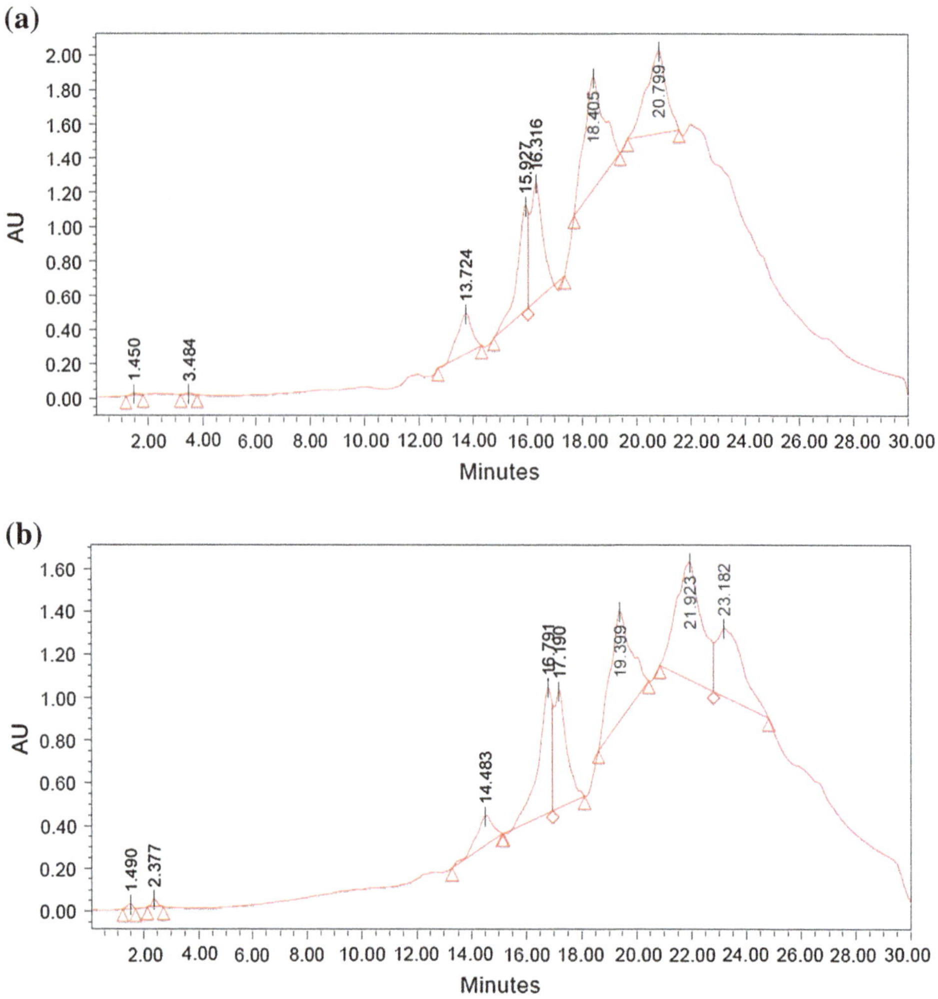

biosurfactant from marine *Nocardiopsis* sp. had been identified.

Fourier transforms infrared spectroscopy (FT-IR) analysis

The FT-IR analysis result was correlated with previous work of Chander et al. (2012) and Yan-Ping et al. (2009). From the FT-IR data, it is evident that rhamnolipid form of biosurfactant was dominant in the biosurfactant produced by SIS-3.

Gas chromatography–mass spectrophotometry (GC–MS) analysis

The major chemical constituent found in the present study is phenol, 3, 5-bis (1,1-dimethylethyl) which was responsible for surfactant activity. Hence, it can be interpreted that aromatic phenolic compounds might be responsible for the dissociation of hydrocarbon compounds present in the crude oil into many different simpler compounds (Kuo 1994).

Strain identification

Nocardiopsis VITSISB (KC958579) was found to be closely related to *Nocardiopsis dassonvillei* (NR074635) with 98 % similarity.

Degradation of engine oil by immobilized cell in aquatic model

In the aquatic model, sugarcane juice and soya meal were used as carbon and nitrogen source, respectively. Sugarcane juice is rich in monosaccharides, such as glucose, fructose; disaccharide such as sucrose (81–87 %), oligosaccharides (6 %) and polysaccharides (8 %). It is also enriched with various inorganic salts such as potassium, sodium, calcium, magnesium, iron, aluminium, copper, zinc, manganese, cobalt, silicon, chloride, phosphate and sulphate. The product that remains after extracting most of the oil from whole soya beans is the soybean meal. Soybean meal is characterized of having higher content of crude protein, about 40–49 %. Proteins of soybean meal are characterized and they contain higher quantity of lysine, tryptophane, isoleucine, valine and threonine (Poultry Feeding Standards) (Ensminger et al. 1990). Hence, combination of sugarcane juice and soya meal can act as complete nutrient formulation for the growth of actinobacteria. *Nocardiopsis* VITSISB (KC958579) was applied as immobilized cells to degrade engine oil. The immobilized microorganism is now swiftly used in the laboratory. Rather than using biosurfactant alone, immobilized living cells are more potential in the biodegradation of hydrocarbons. The use of immobilized biosurfactant producing organism is a rapid, nontoxic, inexpesnsive, versatile method (Siemann and Wagner 1993). The aquatic model provided similar oceanic environment of oil spill to the strain to degrade the oil so that the result of oil degradation can be correlated as of marine environment. In HPLC chromatogram, relative abundance was taken on *Y* axis and retention time was taken on *X* axis. The increase in the number of peaks after oil degradation indicated the breakdown of heavy hydrocarbons into several simpler compounds. The result of HPLC confirmed the degradation of engine oil by the isolated *Nocardiopsis* VITSISB (KC958579).

Conclusion

The marine actinobacteria, *Nocardiopsis* VITSISB (KC958579) isolated from marine water sample, was found to be the proficient producer of rhamnolipid as well as potent degrader of engine oil under oceanic condition with reference to high pH, temperature and salinity. The biosurfactant having elevated emulsification activity, extra surface tension reduction, low density and higher viscosity, can be credibly used towards biodegradation of oil spill in ocean as compared to chemical dispersants. The biosurfactant from *Nocardiopsis* VITSISB (KC958579) offered an extra advantage of its production on renewable sources. Hence, this specific property of the biosurfactant may replace chemical surfactant in various fields. It can be concluded that the immobilized culture of *Nocardiopsis* VITSISB (KC958579) can act as a new scheme for oil spill cleanup in ocean with respect to reusability and cost effectiveness.

Acknowledgments We are very thankful to the management of VIT University for providing well-furnished lab and facilities to carry out our project work. We are also thankful to VIT-SIF Lab, SAS and Chemistry Division for providing GC–MS facility.

Conflict of interest The authors declare that they have no conflict of interest in the publication.

References

Abouseoud M, Maachi R, Amrane A, Boudergua S et al (2008) Evaluation of different carbon and nitrogen sources in production of biosurfactant by *Pseudomonas fluorescens*. Desalination 223:143–151

Abu-Ruwaida AS, Banat IM, Haditirto S, Salem A et al (1991) Isolation of biosurfactant producing bacterial product characterization and evaluation. Acta Biotechnol 11:315–324

Ahmad Mohammad AM, Rudolf H, Francois L, Muller M et al (2011) Rhamnolipids: detection, analysis, biosynthesis, genetic regulation and bioengineering of production. Biosurfactants 20:13–55.

Arino S, Marchal R, Vandecasteele JP (1996) Identification and production of a rhamnolipidic biosurfactant by a *Pseudomonas* species. Appl Microbiol Biotechnol 45:162–168

Attwood D, Elworthy PH, Kayne SB (1970) Surfactant systems and their chemistry, pharmacy and biology. J Phys Chem 74:3529

Banat IM, Makkar RS, Cameotra SS (2000) Potential commercial applications of microbial surfactants. Appl Microbiol Biotechnol 53:495–508

Bernheimer AW, Avigad LS (1970) Nature and properties of a cytolytic agent produced by *Bacillus subtilis*. J Gen Microbiol 61:361–369

Blunt JW, Copp BR, Hu WP, Munro MHG et al (2009) Marine natural products. Nat Prod Rep 26:170–244

Bodour AA, Miller-Maier RM (1998) Application of a modified drop-collapse technique for surfactant quantitation and screening of biosurfactant-producing microorganisms. J Microbiol Methods 32:273–280

Bodour AA, Drees KP, Maier RM (2003) Distribution of biosurfactant-producing bacteria in undisturbed and contaminated arid southwestern soils. Appl Environ Microbiol 69:3280–3287

Borjana KT, George RI, Nelly EC (2001) Biosurfactant production by a new *Pseudomonas putida* strain. Am J Sci Ind Res 57:356–360

Carrillo PG, Mardaraz C, Pitta-Alvarez SJ et al (1996) Isolation and selection of biosurfactant-producing bacteria. World J Microbiol Biotechnol 12(82–84):19

Chander CR, Lohitnath T, Mukesh Kumar DJ, Kalaichelvan PT (2012) Production and characterization of biosurfactant from *Bacillus subtilis* MTCC441 and its evaluation to use as bioemulsifier for food bio-preservative. Adv Appl Sci Res 3:1827–1831

Chandran P, Das N (2010) Biosurfactant production and diesel oil degradation by yeast species *Trichosporon asahii* isolated from petroleum hydrocarbon contaminated soil. Int J Eng Sci Technol 2:6942–6953

Cooper DG, Goldenberg BG (1987) Surface-active agents from two *Bacillus* species. Appl Environ Microbiol 53:224–229

Cooper J, Solomon R, Elrod S, Barnes T, Crawford C et al (2013) Application of laser trap as a viscometer. Proc SPIE.

Desai JD, Banat IM (1997) Microbial production of surfactants and their commercial potential. Microbiol Mol Biol Rev 61:47–64

Ensminger ME, Oldfield JE, Heinemann WW (1990) Feeds and Nutrition. The Ensminger Publishing Company, USA, pp 593–666

Fiedler V, Dal Maso M, Boy M, Aufmhoff H et al (2005) The contribution of sulphuric acid to atmospheric particle formation and growth: a comparison between boundary layers in Northern and Central Europe. Atmos Chem Phys 5:1773–1785

Gharaei-Fathabad E (2011) Biosurfactants in pharmaceutical industry: a mini- review. Am J Drug Discov Dev 1:58–69

Graham P (2010) Deep sea oil spill cleanup techniques: applicability, trade-offs and advantages. ProQuest discovery guides. http://www.csa.com/discoveryguides/discoveryguides-main.php. Accessed Aug 2006

Harkins WD, Jordan FE (1991) The determination of surface tension (free surface energy) and the weight of falling drops. J Am Chem Soc 41:499

Harkins JP, Perfumo A, Marchant R, Banat IM (2010) Isolation and analysis of low molecular weight microbial glycolipids. In: Timmis KN (ed) Handbook of hydrocarbon and lipid microbiology, 5th edn. Springer, Berlin, pp 3705–3723

Huy N, Jin S, Amada K et al (1999) Characterization of petroleum-degrading bacteria from oil-contaminated sites in Vietnam. J Biosci Bioeng 88:100–102

Karthik L, Kumar G, Bhaskara Rao KV (2010) Diversity of marine actinomycetes from nicobar marine sediments and its antifungal activity. Int J Pharm Pharm Sci 2(1):199–203

Kathiresan K, Balagurunathan R, Manilamani Selvan M (2005) Fungicidal effect of marine actinomycetes against phytopathogenic fungi. IJBT 4:271–276

Khopade A, Biao R, Liu X, Mahadik K et al (2012) Production and stability studies of the biosurfactant isolated from marine *Nocardiopsis sp.* B4. Desalination 285:198–204

Kumari M, Abraham J (2011) Biodegradation of diesel oil using yeast *Rhodosporidium toruloides*. Res J Environ Toxicol 5:369–377

Kuo LC (1994) An experimental study of crude oil alteration in reservoir rocks by water washing. Org Geochem 21:465–479

Maneerat S (2005) Biosurfactant from marine microorganisms. J Sci Technol 27:1263–1272

Mattei G, Rambeloarisoa E, Giusti G, Rontani JF et al (1986) Fermentation procedure of a crude oil in continuous culture on seawater. Appl Microbiol Biotechol 23:302–304

Mulligan CN, Gibbs BF (1989) Correlation of nitrogen metabolism with biosurfactant production by *Pseudomonas aeruginosa*. Appl Environ Microbiol 55:3016–3019

Mulligan CN, Cooper DG, Neufeld RJ (1984) Selection of microbes producing biosurfactants in media without hydrocarbons. J Ferment Technol 62:311–314

Murali M, Kumar R, Vethamony (2010) Taluka level Environmental Sensitivity Index (ESI) and vulnerability mapping for oil spills: a pilot study from Goa state, India. Second International Conference on Coastal Zone Engineering and Management, Arabian Coast

Noha HY, Kathleen ED, David PN et al (2004) Comparison of methods to detect biosurfactant production by diverse microorganisms. J Microbiol Methods 56:339–347

Okami Y (1986) Marine microorganisms as a source of bioactive agents. Microb Ecol 12:65–78

Plaza G, Zjawiony I, Banat I (2006) Use of different methods for detection of thermophilic biosurfactant-producing bacteria from hydrocarbon-contaminated bioremediated soils. J Pet Sci Eng 50:71–77

Rodrigues LR, Teixeira JA, Van der Mei HC, Oliveira R (2006) Physiochemical and functional characterization of a biosurfactant produced by *Lactococcus lactis* 53. Colloids Surf B 49:79–86

Saiki RK, Scharf S, Faloona F, Mullis KB et al (1985) Enzymatic amplification of beta-globin genomic sequences and restriction site analysis for diagnosis of sickle cell anemia. Science 230:1350–1354

Saitou N, Nei M (1987) The neighbor-joining method: a new method for reconstructing phylogenetic trees. Mol Biol Evol 4:406–425

Siemann M, Wagner F (1993) Prospects and limits for the production of biosurfactant using immobilized biocatalysts. In: Kosaric N (ed) Surfactant science series – Biosurfactants, vol 48. Marcel Dekker, New York, pp 99–133

Taha TH, Alamri SA, Mahdy HM, Hafez E (2013) The effects of various immobilization matrices on biosurfactant production using hydrocarbon (HC) degrading marine bacteria via entrapment technique. J Biol Sci 13:48–57

Thavasi R, Jayalakshmi S, Banat IM (2011) Effect of biosurfactant and fertilizer on biodegradation of crude oil by marine isolates of *Bacillus megaterium*, *Corynebacterium kutscheri* and *Pseudomonas aeruginosa*. Bioresour Technol 102:772–778

Walter V, Christoph S, Rudolf H (2000) Screening concepts for the isolation of biosurfactant producing microorganisms. Madame Curie bioscience database, Austin

Yan-Ping G, Yong-You H, Roy R, Lin GH (2009) Characterization and micellization of rhamnolipidic fractions and crude extracts produced by *Pseudomonas aeruginosa* mutant MIG-N146. J Colloid Interface Sci 331:356–363

Youssef N, Duncan K, Nagle D et al (2004) Comparison of methods to detect biosurfactant production by diverse microorganisms. J Microbiol Methods 56:339–347

25

Growth of *Cyanobacterium aponinum* influenced by increasing salt concentrations and temperature

Dominik Winckelmann · Franziska Bleeke · Peter Bergmann · Gerd Klöck

Abstract The increasing requirement of food neutral biofuels demands the detection of alternative sources. The use of non-arable land and waste water streams is widely discussed in this regard. A Cyanobacterium was isolated on the area of a possible algae production side near a water treatment plant in the arid desert region al-Wusta. It was identified as *Cyanobacterium aponinum* PB1 and is a possible lipid source. To determine its suitability of a production process using this organism, a set of laboratory experiments were performed. Its growth behavior was examined in regard to high temperatures and increasing NaCl concentrations. A productivity of 0.1 g L^{-1} per day was measured at an alga density below 0.75 g L^{-1}. *C. aponinum* PB1 showed no sign of altered growth behavior in media containing 70 g L^{-1} NaCl or less. Detection of a negative effect of NaCl on the growth using Pulse-Amplitude-Modulation chlorophyll fluorescence analysis was not more sensitive than optical density measurement.

Keywords PAM · Bio fuel · *Cyanobacterium aponinum* · Evaporation · Open pond · Arid desert

Introduction

Microalgae and cyanobacteria were considered to be a good choice as lipid source for bio diesel production, for CO_2 sequestration, as nutritional supplement and for numerous other applications (Chisti 2008; Grobbelaar 2012). Recent studies consider microalgae and cyanobacteria to be not able to be used as a stand-alone commercial production process, but to be a means to utilize waste streams of other processes or to be a solution for a small-scale niche products (Van Beilen 2010; Chisti 2013). During the crude oil production in the Sultanate of Oman 800,000 m^3 of contaminated production water accumulates per day (Breuer and Al-Asmi 2010). Part of the water is treated in a reed bed water treatment plant. Different options for further use of the treated water are under investigation. One possible option is to establish an algae growth facility for which the possible location would be in the arid desert region al-Wusta, Oman.

The use of non-arable land in a desert region has the advantage of sustainable production without competing with possible crop production (Singh et al. 2011), while it raises some possible difficulties. The maximum light exposure, annual temperature regime, cloud coverage, availability of water and the general infrastructure are considered to be major factors influencing the production in large-scale algae cultures (Grobbelaar 2009). Furthermore a high evaporation rate and the high consumption of fresh water to keep the salinity of the growth medium under a certain threshold is an important economic factor (Guieysse et al. 2013). The use of indigenous species as production organisms, either in axenic cultures or as mixed cultures, is said to circumvent some of the difficulties associated with the location (Mutanda et al. 2011; Odlare et al. 2011).

In preparation of the establishment of an algae-production site in an arid desert, the effect of warm temperature and increasing salt content on the growth of a domestic specie and the possible productivity were determined.

D. Winckelmann · F. Bleeke
School of Engineering and Science, Jacobs-University Bremen, Campus Ring 1, 28759 Bremen, Germany

D. Winckelmann · F. Bleeke · P. Bergmann · G. Klöck (✉)
University of Applied Sciences Bremen, Am Neustadtswall 30, 28199 Bremen, Germany
e-mail: Gerd.Kloeck@hs-bremen.de

Therefore, the first step was to isolate different algae species from the water treatment plant. Thereby a strain of *Cyanobacterium aponinum* was found. *C. aponinum* is said to be able to grow in fresh and seawater media at temperatures of up to 45 °C (Moro et al. 2007) and might be used as lipid source (Karatay and Dönmez 2011).

The goal during the development of an algae growth process has to be to provide a process, which is able to deliver a stable biomass and product outcome, while being robust against outer circumstances. Grobbelaar (2000) gave an overview about the different fields and possibilities of improvement to optimize mass algal culture. He stated that process control should be improved to allow exploitation of physiological properties of the species to be grown. It was emphasized that carbon fixation and metabolism, and maximum photosynthetic rate are important to determine the overall efficiency of the photo capture and net carbon fixation, and possible damage to the photosystem reaction centers.

PAM chlorophyll fluorescence analysis were performed in laboratory studies to estimate CO_2 fixation, light acclimation states (Behrenfeld et al. 2004; Campbell et al. 1998), and the influence of salt stress on photosynthesis (Lu and Vonshak 1999) in cyanobacteria and green algae using the pulse modulation principle (Schreiber et al. 1986). But the large-scale cultivation of algae opens up some new, so far not existing challenges and raise the need for real time online measurement methods (Havlik et al. 2013).

Chlorophyll fluorescence analysis was performed during this study using a phyto-PAM, which was designed to distinguish between fluorescence excited by light of four different preset wavelengths (Kolbowski and Schreiber 1995) and thereby enable an estimate of the relative pigment-content associated with those wavelengths.

Methods

Culture media

With regard to future industrial sized applications a simple media containing the commercially available fertilizer WUXAL® Universaldünger liquid plant fertilizer (8 % N, 8 % P_2O_5, 6 % K2O, 0.01 % B, 0.004 % Cu, 0.02 % Fe, 0.012 % Mn, 0.004 % Zn; Wilhelm Haug GmbH & Co. KG, Germany) was used. The NaCl concentration added to the media was chosen corresponding to the salinity of the natural environment *C. aponinum* was isolated from. Wuxal liquid medium (WM) contained tap water with 0.002 % w/v $MgSO_4$, 1 % w/v NaCl and 0.05 % v/v Wuxal. Wuxal was sterile filtrated and added after autoclaving.

Isolation of algae

Environmental samples were taken in a reed bed system near Nimr, Sultanate of Oman (+18° 34′ 14.98″ N, +55° 49′ 5.99″ E). Initial growth was conducted in Erlenmeyer flasks containing WM soil media (WM with 3 % v/v soil extract (Watanabe 2005)) for 7 days at room temperature and natural light. After biomass increase was visible, the flasks were transferred into a light incubator (RUMED Licht Thermostat Typ 1301, Rubarth Apparate GmbH, Germany) and cultivated at 40° C and 110 ± 2.5 PAR m^{-2} s^{-1}. The flasks were shaken constantly at 150 rounds per minute (orbital shaker 3005, GFL—Gesellschaft für Labortechnik mbH, Germany). Single-celled algae were isolated using the 13-streak method on WM- soil plates containing 1.5 % (w/v) Agar. The agar plates were grown in the light incubator until growth was visible. Colonies containing only one algae species were transferred to liquid media.

Identification of algae

Genomic DNA was extracted from freshly grown *C. aponinum* PB1. Cells were broken-up via beat mill and DNA was extracted using a DNA extraction kit (DNeasy Plant Mini kit, Qiagen, Germany).The 16 s rRNA gene was amplified by PCR using the primers CYA106F, and a mixture of CYA781R (a) and CYA781R (b) (Boutte et al. 2006); (GenBank: JN584264.1)) and the internal transcribed spacer between the 16S and 23S rRNA genes (ITS) was amplified using the primer pair ULR and CSIF (without the G-C clamp (Janse et al. 2003); GenBank: KF982001). The PCR product was purified using the QIAquick PCR Purification Kit (Qiagen, Germany). Sequencing was performed using the Sanger sequencing technique by GATC Biotech AG (Germany).

Characterization of algae

The optical density was measured at 700 nm (OD700) using a Thermo Spectronic Genesys 20 Model 4001/4 (Thermo Electron Corporation, USA). For the determination of bio dry weight, part of a culture was washed and filtered. The filter was dried at 105 °C until the weight was constant.

Fluorescence analysis was performed using a phyto-PAM with phyto-EDF (Walz, Germany). The samples were transferred in a 1.5 ml cup and the perspex-rod sensor tip was 0.2 mm immersed into the algae suspension. Everything was covered and the fluorescence was measured (Settings: gain was adjusted to F 645 nm values above 200 and below 400, measuring light intensity 0, saturation light intensity 10, saturation pulse length 400 ms; For introduction, see (Schreiber et al. 1995)).

Each chlorophyll fluorescence measurement performed delivered four light-adapted steady state fluorescence values ($F'x$) and four maximum fluorescence values ($F'mx$), which were assigned to one of the four different used excitation wave length ($x = 470, 520, 645$ or 665 nm). The effective quantum efficiency of charge separation at photosystem II (yield) in actinic light was calculated according to Genty et al. (Kaplan et al. 1989) for each wavelength used (yield x).

$$\text{Yield} = \frac{F'm - \text{F}}{F'm}$$

To determine a shift in light utilization the ratio of steady-state fluorescence was calculated. The maximum signal of $F'x$ was set as 1, the lower values were normalized against it resulting in three values called rel.Fx.

Microalgae cultivation

Stock cultures of *C. aponinum* PB1 were grown in Erlenmeyer flasks at 24 ± 2 °C and illuminated with white fluorescent light (60 ± 2.5 PAR m^{-2} s^{-1}, measured using Meteon Irradiance meter with a PAR Lite sensor (Kipp and Zonen, Netherlands)). Temperature was measured using a Hobo® Pendant temp #UA-001-64 (Onset Computer Corporation, USA).

The growth experiment was performed at 38 ± 2 °C. All cultures were illuminated with white fluorescent light (110 ± 2.5 PAR m^{-2} s^{-1} for 14 h per day, in 49 cm high glass tubes with an inner diameter of 49 mm and a total liquid volume of 750 ml, and aerated with fresh air at a rate of 100 l/h. Initial inoculum was 1 × 10^6 cells per ml. The OD700, cell concentration (determined using a Zeiss Axiostar Plus light microscope (Carl Zeiss, Germany) and a Thoma Neu chamber with a depth of 0.1 mm (Paul Marienfeld GmbH & Co. KG, Germany), pH (using a wtw pH 315i (WTW Wissenschaftlich-Technische Werkstätten GmbH, Germany)) with an InLab Micro pH electrode (Mettler Toledo GmbH, Germany)), and chlorophyll fluorescence were determined twice a week before nine in the morning. Evaporated water was replaced with tap water or a NaCl solution (10 % w/v NaCl in tap water) and salinity was measured twice a week (using a conductivity meter LF 340 with a TetraCon® 325 (WTW Wissenschaftlich-Technische Werkstätten GmbH, Germany).

Results

Characterization of *C. aponinum* PB1

The sequence for the 16S rRNA gene covered 412 bases and the sequence gained for the ITS region covered 441 bases. After BLAST (Altschup et al. 1990) analysis a similarity of 99 % with an E value of 0.0 to *C. aponinum* PCC 10605 (GenBank: CP003947.1), respectively, 100 % and E value of 0.0 for the 16S rRNA gene alone (NR_102443.1), and a similarity of 98 % with an E value of 0.0 to *C. aponinum* ETS-03 (GenBank: AM238427.1) were found.

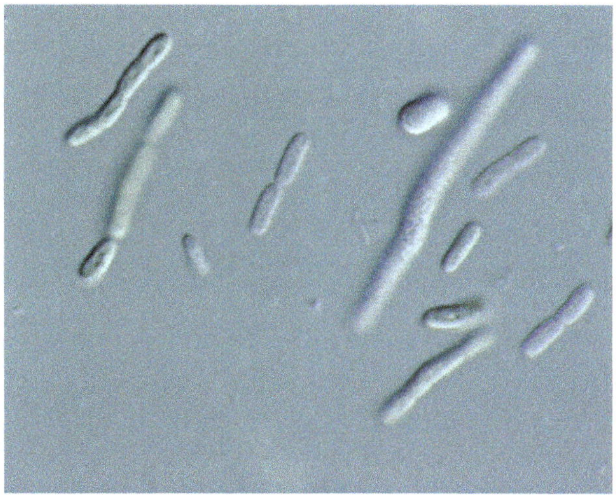

Fig. 1 Picture of *C. aponinum* PB1 (Culture was grown for 14 days in WM), Picture was made by Gerhard Kauer with light microscope and 630-fold magnification

Cyanobacterium aponinum PB1 grow in peanut-like or cylindrical shape with a length of 3–4 μm and a diameter of 2–3 μm. They appear solitary or during cell division as pairs. Grown under nutrient-limited conditions they show elongated cell shapes (Fig. 1). They appear blue–green in color.

Cyanobacterium aponinum PB1 is able to grow at 45 °C and 110 PAR m^{-2} s^{-1} (14 h light and 10 h dark) in WM with 10 g L^{-1} NaCl. Grown using the same experimental set-up at 28 °C and 180 PAR m^{-2} s^{-1} (14 h light and 10 h dark) exponential growth was achieved over a period of 14 days with a growth rate μ per day of 0.3 (SD 0.006) and a maximum light-adapted yield 645 nm of 0.45 (data not shown). The constant bio dry weight of washed cultures with an OD700 of 1 was 0.51 g L^{-1} (SD 0.0075). Cultures grown at RT showed no DNA fragmentation (Nedelcu 2006) after incubation at 42 °C in the dark (Fig. 2).

Growth in WM with increasing NaCl content

Cyanobacterium aponinum PB1 was grown in media with increasing NaCl concentrations (referred to as treated culture), namely 10, 13, 20, 26, 35, 58, 70, 92, and 140 g L^{-1}, and stable NaCL concentration (10 g L^{-1}, referred to as control culture). Growth was exponential for both cultures till day 6, when the NaCl content was increased to 20 g L^{-1}, and their growth rate μ per day was comparable

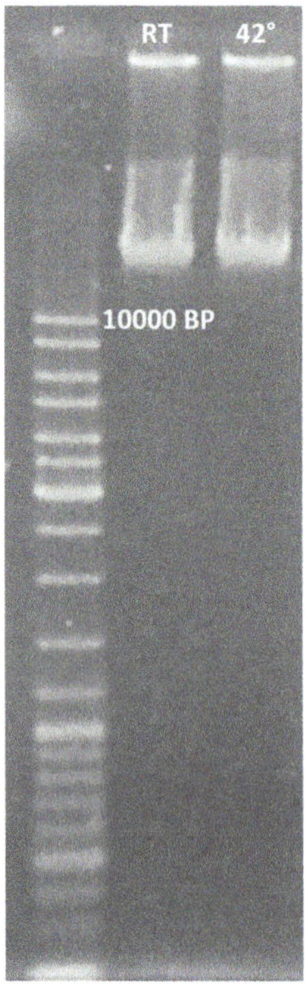

Fig. 2 No indication of *C. aponinum* genomic DNA fragmentation after 2 h incubation at 42° C (*right lane*) compared to 2 h incubation at room temperature (RT, *left lane*). DNA fragmentation was performed like described in Nedelcu 2006

with 0.35 for the treated culture and 0.33 for the control. Afterwards, the growth was linear with an increase of 1.3×10^6 cells per ml per day for the treated and 0.9×10^6 cells per ml per day for the control culture till day 16, after which the NaCl concentration was increased to 92 g L^{-1}. The cell density of treated culture decreased from 3×10^7 to 1.8×10^7 cells/ml during the growth in WM containing 92 g L^{-1} NaCl and even further to 1.5×10^7 cells/ml after the NaCl content was increased to 140 g L^{-1}. The cell content in the control culture stayed stable at 3×10^7 cells per ml over the whole time (Fig. 3). The optical density showed a similar course like the cell count (Fig. 4). The coloration became less green and paler after the NaCl content was 92 g L^{-1} or higher. The pH rose in the first 2 days from 7.8 to 9 on day 6, when the NACL content was increased to 20 g L^{-1}, but decreased till day 16, after which the NaCl concentration was increased to 92 g L^{-1}, to 5.8 for the treated culture and 5.5 for the control. Over the whole course the pH of the treated culture was slightly higher. This changed after the NaCl content was increased to 92 g L^{-1} when the pH of both cultures was at 5.6. The pH of the treated culture stayed stable during growth in 140 g L^{-1} Nacl while the pH of the control culture slightly increased (Fig. 4).

The yield of the treated culture increased from 0.12 in media containing 10 g L^{-1} NaCl to 0.4 in media containing 58 g L^{-1} NaCl and decreased stepwise to 0.12 when the culture was grown in media containing 140 g L^{-1} NaCl. The yield of the control culture increased from 0.22 to 0.31 and stabilized at 0.28 (Fig. 5).

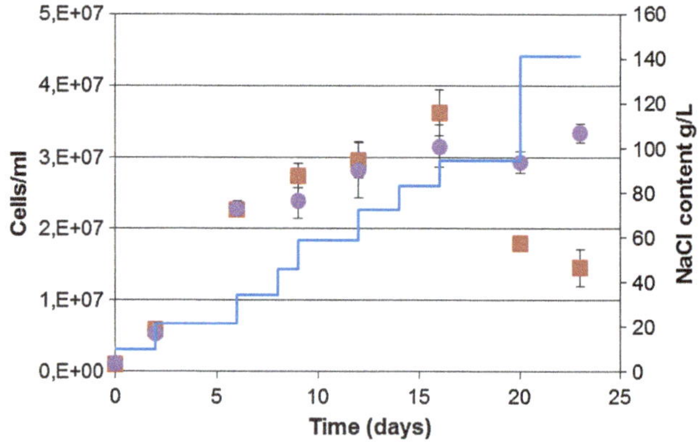

Fig. 3 Growth of *C. aponinum PB1* in WM with increasing NaCl content (filled squares; NaCl content as line) over time. As control a *C. aponinum* PB1 culture growing in WM with 10 g L^{-1} NaCl is shown (*filled circles*). *Error bars* represent three times the standard deviation

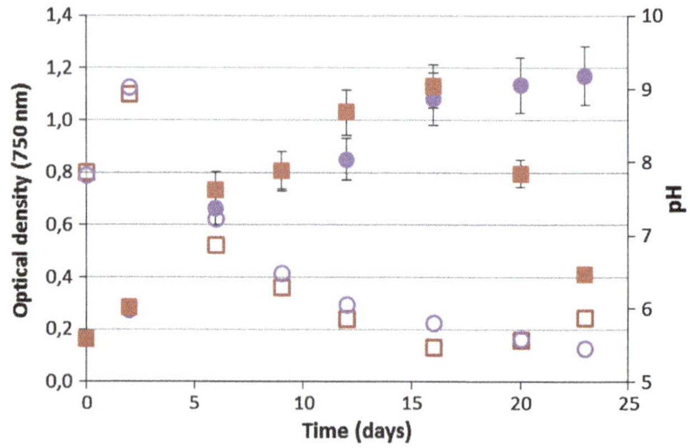

Fig. 4 pH change and growth as measured by optical density of *C. aponinum PB1* with increasing NaCl content (pH: *open squares*; OD: *filled squares*) over time. As control a *C. aponinum* PB1 culture growing in WM with 10 g L^{-1} NaCl is shown (pH: *open circles*; OD: *filled circles*). *Error bars* represent three times the standard deviation

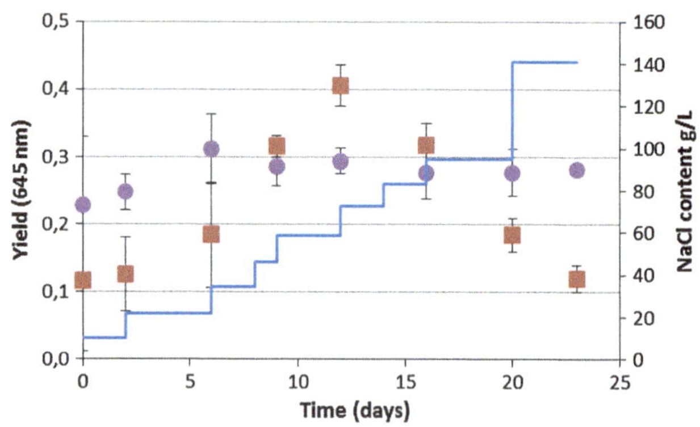

Fig. 5 Photosynthetic yield of *C. aponinum PB1* with increasing NaCl content (*filled squares*; NaCl content as *line*) over time. As control a *C. aponinum* PB1 culture growing in WM with 10 g L^{-1} NaCl is shown (*filled circles*). *Error bars* represent three times the standard deviation

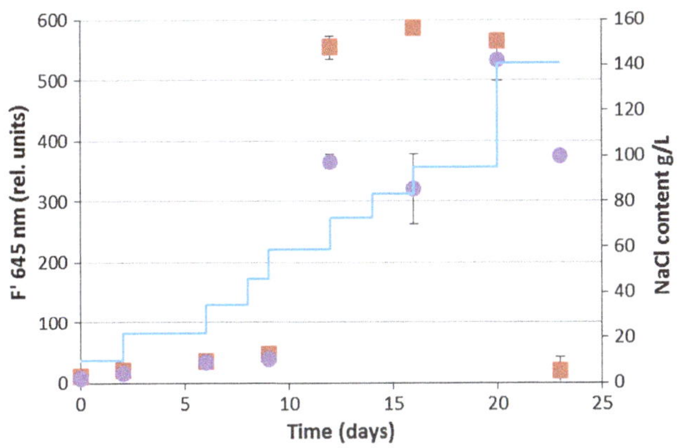

Fig. 6 Change of light-adapted steady state fluorescence (*F'*) of *C. aponinum PB1* with increasing NaCl content (*filled squares*; NaCl content as *line*) over time. As control a *C. aponinum* PB1 culture growing in WM with 10 g L^{-1} NaCl is shown (*filled circles*). *Error bars* represent three times the standard deviation

The light-adapted fluorescence ($F' = 645$ nm) of the treated and the control culture increases from 10 at a NaCl content of 10 g L^{-1} to 50 rel. units after growth in media containing 35 g L^{-1} NaCl. The fluorescence values increase steeply during the growth in WM with 58 g L^{-1} NaCl for both cultures, but while it stays stable for the treated culture till the NaCl content was increased to 140 g L^{-1} NaCl, after which the fluorescence values drop to undetectable values, it decreases for the control culture on day 16, increases till day 20 and declines till day 23 (Fig. 6).

The relative, normalized fluorescence values of the control culture stayed stable from day 6 to 16, after which the NaCl concentration was increased to 92 g L^{-1}. Thereafter, RelF 520 nm and RelF 665 nm increased slightly. The relF 520 nm of the stressed culture was slightly higher than the control value and decreased after the NaCl content was increased to 58 g L^{-1}. RelF 665 nm of the stressed culture showed similar values like the control culture declined after day 10.

Discussion

The isolate was assigned to be *C. aponinum* because of the molecular similarities, similar shape and color, and the ability to survive in BG11 media containing 34 g L^{-1} salt. All those features are used to distinguish *C. aponinum* ETS-03 from other Cyanobacteria and to establish the *C. aponinum* species (Moro et al. 2007). Our isolate was found in microbial mats in a warm water pond and able to be free living like *C. aponinum* ETS-03 (Moro et al. 2007). *C. aponinum* PB1 showed a change in color during temperature shift or nutrient stress (data not shown) and Moro reported unknown carotenoids in the *C. aponinum* ETS-03. *C. aponinum* was found to accumulate up to 45 % lipids per dry weight (Karatay and Dönmez 2011). Because of the ability to most likely survive the temperature conditions to be found at the test field area and the prospect of possible products made us chose *C. aponinum* PB1 for further tests in the laboratory.

Cyanobacterium aponinum PB1 showed no difference in growth until day 8 when the NaCl content was increased to 92 g L^{-1}. The control had reached the stationary phase at that time, while the other culture declined in OD and in cell concentration. *C. aponinum* PB1 could be part of the functional group of moderately halotolerant cyanobacteria, as defined by Reed and Stewart ((Reed and Stewart 1985). This group is able to synthesize Glucosylglycerol [O-a-D-glucopyranosyl-(1– >2)-glycerol] as their major compatible solute as response to increasing salt concentrations. Members of this group were found in sea and fresh water and were able to survive in media containing salt concentrations of double the amount of salt in sea water and even more (Reed and Stewart 1985). Hagemann defined the NaCl tolerance limit of moderately halotolerant strains as 1.7 M, which coincides with our own findings (1.7 M NaCl ~99 g L^{-1}).

It was shown that cyanobacteria alkalize the growth media when not provided with inorganic carbon (IC) (Kaplan et al. 1989). Cyanobacteria store CO_2 in form of IC in carboxysomes (Reinhold et al. 1991). The initial increase of pH during this experiment indicates an insufficient internal IC pool. This pool is filled over time by the CO_2-concentrating mechanism (Kupriyanova et al. 2013) using the excess CO_2 from the air leading to a decrease in pH. *C. aponinum* PB1 alkalized the medium to pH values of up to 10.5 during experiments using an insufficient aeration rate (data not shown). Cyanobacteria are able to survive salt stress by a number of different physiological changes. The decrease of pH until day 20 of the culture grown in WM with increasing NaCl compared to the control culture could be because of an increase in Na^+/H^+-antiporter synthesis during growth in elevated salt levels (Allakhverdiev et al. 1999). It was shown that the artificial induction of a Na^+/H^+-antiporter overexpression enabled a freshwater cyanobacterium to survive and grow in seawater (Waditee et al. 2002). Another response to increased salt levels is an increase in unsaturation of membrane lipids leads to a change in plasma membrane mobility (Allakhverdiev et al. 2000a). This might depress the functionality of K^+ channels, which are also point of entry for Na^+ molecules and of water channels (Allakhverdiev et al. 2000b). While in the same moment the synthesis of membrane bound Na^+/H^+-antiporter and H^+ATPase-enzymes might be increased (Allakhverdiev et al. 2001). All those processes are thought to decrease the Na^+ content in the cytosol and thereby protecting the photosystems against salt-induced inactivation (Singh et al. 2002). The counteracting mechanisms of increased number of enzymes and depression of functionality limit the survival under stress to a certain threshold, which is thought to be reached by the culture growing in increasing salt levels on day 20, like it is shown in change in pH, growth rate and photosynthetic activity. The mechanisms of lipid unsaturation and solute biosynthesis are said to be not only responsible for salt tolerance but also for tolerance to osmotic or high and low temperatures (Klähn and Hagemann 2011; Singh et al. 2002). Modern proteomic analyses try to explain the mechanisms of heat shock and salt-induced stress in more detail (Castielli et al. 2009; Ludwig 2012).

The photosynthetic yield 645 nm was below the maximum yield of the stock cultures grown at room temperature. The cultures photosynthetic capability was limited over the whole course of the experiment. The highest yield 645 nm was measured after growth in media with 58 g L^{-1} with 0.4. At the same day, the OD was significantly higher compared to the control culture. Afterwards, the yield decreased steadily and a decrease in growth could not be seen before a decrease in cell concentration or OD was measurable. Also the other fluorescence parameters were not reliable to indicate NaCl stress. Under controlled laboratory conditions, fluorescence values of dark-adapted and biomass-standardized samples were found to be reliable indicators for various stressors (Lu and Vonshak 1999). But the use of fluorescence values under less controlled conditions was found to be not reliable (Kruskopf and Flynn 2006).

The utilization of pH measurements as control parameter for culture stability could be used to indicate CO_2 depletion. A different method or changes in the measurement procedure have to be found to use PAM as control parameter for culture stability. A bypass with rest time for dark adaption might be an answer for semi-continuous online measurement.

The possible growth in temperatures above 30 °C in easy-made liquid media in combination with its free-living growth form predestine *C. aponinum* PB1 for field tests in the Omanian desert.

The mean conductivity of the feed water was measured with 14 ms cm^{-1} (personal communication). The maximum natural evaporation on the Arabic peninsula is difficult to measure and values of up to 14 mm per day are reported (Alsharhan et al. 2001). Evaporation would be even higher because of the by circulation enlarged surface of the algae culture. We assumed a maximum evaporation of 20 mm per day. The theoretical maximal conductivity, while harvesting every third day, due to evaporation would be 24 ms cm^{-1}, which correlates to the conductivity of a NaCl solution with 14 g L^{-1} (Lewis 1980). Therefore, as long as the main salt in the feed water is NaCl (Nübel et al. 2000) *C. aponinum* PB1 should be able to grow. The salt in the feed water might have a positive impact on the lipid content because higher salt levels were shown to increase lipid accumulation in cyanobacteria (Araujo et al. 2011; Bhadauriya et al. 2008).

Assuming a theoretical exponential growth with an average growth rate of 0.3 μ per day in a 10 cm deep open pond a biomass yield of 22,500 kg per hectare and year (expecting 300 working days) could be achieved. With a caloric value of 20 kJ per 10 g dry weight, like it was published for other cyanobacteria (Quintana et al. 2011), this would correspond to an energy content of 50,000 Mj. Similar and higher growth rates were reported for other cyanobacteria (Markou and Georgakakis 2011). A possible lipid yield of up to 10,125 kg of lipids per hectare and year if 45 % lipids per dry weight might be achieved (Karatay and Dönmez 2011). The usability as biodiesel highly depends on the lipid composition (Quintana et al. 2011), which is influenced by the growth stage, growth temperature and light intensity (Liu et al. 2005; Vargas et al. 1998; Walsh et al. 1997).

Presume exponential growth at dry weight concentrations below 0.37 g L^{-1} and harvesting two-third of the whole volume every three days 65,000 m^3 have to be moved per year. Assuming an average evaporation rate of 14 mm per day additional 140,000 m^3 of water have to be pumped. This accumulates to an energy consumption of 20,000 Mj for liquid movement alone.

To be able to give a final statement about the economic feasibility of an algae production process using *C. aponinum* the actual biomass composition achievable in large scale and possible synergies with established procedures have to be evaluated (Patel et al. 2012).

Acknowledgments We would like to thank Gerhard Kauer (Univeristy of Applied Sciences Emden) for his help. The project underlying this report was funded by the Federal Ministry of Education and Research of Germany under grant number FHprofUnt 17003X11 promoted. The responsibility for the content of this publication lies with the author.

References

Allakhverdiev SI, Nishiyama Y, Suzuki I, Tasaka Y, Murata N (1999) Genetic engineering of the unsaturation of fatty acids in membrane lipids alters the tolerance of Synechocystis to salt stress. Proc Natl Acad Sci U S A 96(10):5862–5867. Retrieved from http://www.pubmedcentral.nih.gov/articlerender.fcgi?artid=21951&tool=pmcentrez&rendertype=abstract

Allakhverdiev SI, Sakamoto A, Nishiyama Y, Inaba M, Murata N (2000a) Ionic and osmotic effects of NaCl-induced inactivation of photosystems I and II in Synechococcus sp. Plant Physiol 123(3):1047–1056. Retrieved from http://www.pubmedcentral.nih.gov/articlerender.fcgi?artid=59068&tool=pmcentrez&rendertype=abstract

Allakhverdiev SI, Sakamoto A, Nishiyama Y, Murata N (2000b) Inactivation of photosystems I and II in response to osmotic stress in Synechococcus. Contribution of water channels. Plant Physiol 122(4):1201–1208. Retrieved from http://www.pubmedcentral.nih.gov/articlerender.fcgi?artid=58955&tool=pmcentrez&rendertype=abstract

Allakhverdiev SI, Kinoshita M, Inaba M, Suzuki I, Murata N (2001) Unsaturated fatty acids in membrane lipids protect the photosynthetic machinery against salt-induced damage in Synechococcus. Plant Physiol 125(4):1842–53. Retrieved from http://www.pubmedcentral.nih.gov/articlerender.fcgi?artid=88840&tool=pmcentrez&rendertype=abstract

Alsharhan AS, Rizk ZA, Nairn AEM, Bakhit DW, Alhajari SA (2001) Physical geography of the Arabian Peninsula. In: AS Alsharhan, ZA Rizk, AEM Nairn, DW Bakhit, SA Alhajari (eds.) Hydrogeology of an arid region: the arabian gulf and adjoining areas, Elsevier Science BV, Amsterdam, pp. 7–53

Altschup SF, Gish W, Miller W, Myers EW, Lipman DJ (1990) Basic local alignment search tool. J Mol Biol 215:403–410

Araujo GS, Matos LJBL, Gonçalves LRB, Fernandes FAN, Farias WRL (2011) Bioprospecting for oil producing microalgal strains: evaluation of oil and biomass production for ten microalgal strains. BioResour Technol 102(8):5248–5250.

Behrenfeld MJ, Prasil O, Babin M, Bruyant F (2004) In search of a physiological basis for covariations in light-limited and light-saturated photosynthesis 1. J Phycol 40(1):4–25.

Bhadauriya P, Gupta R, Singh S, Bisen PS (2008) NaCl induced metabolic changes in the diazotrophic cyanobacterium Anabaena cylindrica. World J Microbiol Biotechnol 25(2):341–345.

Boutte C, Grubisic S, Balthasart P, Wilmotte A (2006) Testing of primers for the study of cyanobacterial molecular diversity by DGGE. J Microbiol Method 65(3):542–550.

Breuer R, Al-Asmi S (2010) NIMR water treatment project–up scaling a reed bed trail to industrial. In: SPE international conference on health, safety and environment in oil and gas exploration and production, vol 1. pp 286–296

Campbell D, Hurry V, Clarke AK, Gustafsson P, Öquist G (1998) Chlorophyll fluorescence analysis of cyanobacterial photosynthesis and acclimation chlorophyll fluorescence analysis of cyanobacterial photosynthesis and acclimation. Microbiol Mol Biol Rev 62(3):667–683

Castielli O, De la Cerda B, Navarro JA, Hervás M, De la Rosa MA (2009) Proteomic analyses of the response of cyanobacteria to different stress conditions. FEBS Lett 583(11):1753–1758.

Chisti Y (2008) Biodiesel from microalgae beats bioethanol. Trends Biotechnol 26(3):126–131.

Chisti Y (2013) Constraints to commercialization of algal fuels. J Biotechnol 167(3):201–214.

Grobbelaar JU (2000) Physiological and technological considerations for optimising mass algal cultures. J Appl Phycol 12:201–206

Grobbelaar JU (2009) Upper limits of photosynthetic productivity and problems of scaling. J Appl Phycol 21(5):519–522.

Grobbelaar JU (2012) Microalgae mass culture: the constraints of scaling-up. J Appl Phycol 24(3):315–318.

Guieysse B, Béchet Q, Shilton A (2013) Variability and uncertainty in water demand and water footprint assessments of fresh algae cultivation based on case studies from five climatic regions. Bioresour Technol 128:317–323.

Havlik I, Lindner P, Scheper T, Reardon KF (2013) On-line monitoring of large cultivations of microalgae and cyanobacteria. Trends Biotechnol 31(7):406–414.

Janse I, Meima M, Kardinaal WEA, Zwart G (2003) High-resolution differentiation of cyanobacteria by using rRNA-internal transcribed spacer denaturing gradient gel electrophoresis high-resolution differentiation of cyanobacteria by using rRNA-internal transcribed spacer denaturing gradient gel electroph. Appl Environ Microbiol 69(11):6634–6643.

Kaplan A, Scherer S, Lemer M (1989) Nature of the light-induced H + Efflux and Na + Uptake in, 1220–1225

Karatay SE, Dönmez G (2011) Microbial oil production from thermophile cyanobacteria for biodiesel production. Appl Energy 88(11):3632–3635.

Klähn S, Hagemann M (2011) Compatible solute biosynthesis in cyanobacteria. Environ Microbiol 13(3):551–562.

Kolbowski J, Schreiber U (1995) Computer-controlled phyto- plankton analyzer based on a 4-wavelength PAM Chl fluoro meter. In: Mathis P (ed) Photosynthesis: from light to biosphere. Kluwer Academic Publishers, Dordrecht, Boston, London

Kruskopf M, Flynn KJ (2006) Chlorophyll content and fluorescence responses cannot be used to gauge reliably phytoplankton biomass, nutrient status or growth rate. New Phytol 169:525–536.

Kupriyanova EV, Sinetova MA, Cho SM, Park Y-I, Los DA, Pronina NA (2013) CO2-concentrating mechanism in cyanobacterial photosynthesis: organization, physiological role, and evolutionary origin. Photosynth Res 117(1–3):133–146.

Lewis E (1980) The practical salinity scale 1978 and its antecedents. IEEE J Ocean Eng 5.

Liu X-J, Jiang Y, Chen F (2005) Fatty acid profile of the edible filamentous cyanobacterium *Nostoc flagelliforme* at different temperatures and developmental stages in liquid suspension culture. Process Biochem 40(1):371–377.

Lu C, Vonshak A (1999) Characterization of PSII photochemistry in salt-adapted cells of cyanobacterium Spirulina platensis. New Phytol 141:231–239

Ludwig M, Bryant DA (2012) Synechococcus sp. strain PCC 7002 transcriptome: acclimation to temperature, salinity, oxidative stress, and mixotrophic growth conditions. Front Microbiol 3:354.

Markou G, Georgakakis D (2011) Cultivation of filamentous cyanobacteria (blue-green algae) in agro-industrial wastes and wastewaters: a review. Appl Energy 88:3389–3401.

Moro I, Rascio N, Di Bella M, Andreoli C, La Rocca N (2007) *Cyanobacterium aponinum*, a new Cyanoprokaryote from the microbial mat of Euganean thermal springs (Padua, Italy). Arch Hydrobiol Suppl Algol Stud.

Mutanda T, Ramesh D, Karthikeyan S, Kumari S, Anandraj A, Bux F (2011) Bioprospecting for hyper-lipid producing microalgal strains for sustainable biofuel production. BioResour Technol 102(1):57–70.

Nedelcu AM (2006) Evidence for p53-like-mediated stress responses in green algae. FEBS Lett 580(13):3013–3017.

Nübel U, Garcia-Pichel F, Clavero E, Muyzer G (2000) Matching molecular diversity and ecophysiology of benthic cyanobacteria and diatoms in communities along a salinity gradient. Environ Microbiol 2(2):217–226. Retrieved from http://www.ncbi.nlm.nih.gov/pubmed/11220307

Odlare M, Nehrenheim E, Ribé V, Thorin E, Gavare M, Grube M (2011) Cultivation of algae with indigenous species—potentials for regional biofuel production. Appl Energy 88(10):3280–3285.

Patel B, Tamburic B, Zemichael FW, Dechatiwongse P, Hellgardt K (2012) Algal biofuels: a credible prospective? ISRN Renew Energy.

Quintana N, Van der Kooy F, Van de Rhee MD, Voshol GP, Verpoorte R (2011) Renewable energy from Cyanobacteria: energy production optimization by metabolic pathway engineering. Appl Microbiol Biotechnol 91(3):471–490.

Reed RH, Stewart WDP (1985) Osmotic adjustment and organic solute accumulation in unicellular cyanobacteria from freshwater and marine habitats. Mar Biol 88:1–9

Reinhold L, Kosloff R, Kaplan A (1991) A model for inorganic carbon fluxes and photosynthesis in cyanobacterial carboxysomes. Can J Bot.

Schreiber U, Schliwa U, Bilger W (1986) Continuos recording of photochemical and non-photochemical chlorophyll fluorescence quenching with a new type of modulation fluorometer. Photosynth Res 10:51–62

Schreiber U, Endo T, Mi H, Asada K (1995) Quenching analysis of chlorophyll fluorescence by the saturation pulse method: particular aspects relating to the study of eukaryotic algae and cyanobacteria. Plant Cell 36(5):873–882

Singh SC, Sinha RP, Häder D (2002) Role of lipids and fatty acids in stress tolerance in cyanobacteria. Acta Protozool 41:297–308

Singh A, Nigam PS, Murphy JD (2011) Renewable fuels from algae: an answer to debatable land based fuels. BioResour Technol 102(1):10–16.

Van Beilen JB (2010) Why microalgal biofuels won't save the internal combustion. Biofuels Bioprod Biorefining 4:41–52.

Vargas MA, Rodríguez H, Moreno J, Olivares H, Del Campo JA, Rivas, JGuerrero MG (1998) Biochemical composition and fatty acid content of filamentous nitrogen-fixing cyanobacteria. J Phycol 817:812–817

Waditee R, Hibino T, Nakamura T, Incharoensakdi A, Takabe T (2002) Overexpression of a Na+/H+ antiporter confers salt tolerance on a freshwater cyanobacterium, making it capable of growth in sea water. Proc Natl Acad Sci USA 99(6):4109–4114.

Walsh K, Jones GJ, Dunstan RH (1997) Effect of irradiance on fatty acid, carotenoid, total protein composition and growth of *Microcystis aeruginosa*. Phytochemistry 44(5):817–824

Watanabe MM (2005) Freshwater culture media. In: A Anderson, Robert A (ed.) Algal culturing techniques. Elsevier Academic Press, Amsterdam, Boston, Heidelberg, London, New York, Oxford, Paris, San Diego, San Francisco, Singapore, Sydney, Tokyo, pp. 13–20

26

Purification and biochemical characterization of a newly produced yellow laccase from *Lentinus squarrosulus* MR13

Mainak Mukhopadhyay · Rintu Banerjee

Abstract A novel yellow laccase was produced from *Lentinus squarrosulus* MR13 under solid state fermentation. The yellow laccase was purified by a factor of 12.67-fold by ammonium sulfate precipitation, anion exchange chromatography and gel filtration chromatography to a specific activity of 3,772.86 IU mg^{-1}. Its molecular mass was determined by SDS-PAGE and found to be 66 kDa. The activity of the enzyme was measured with 2,2′-azinobis-(3-ethylbenzothiazoline-6-sulfonic acid) as substrate and found to be stable in a broad range of pH (pH 4–9). The optimum temperature of the enzyme was 40 °C. The enzyme was stable at temperatures between 25 and 55 °C and decreased rapidly when the temperature was above 65 °C. Circular dichroism spectra also supported the temperature stability of the enzyme. The K_m and V_{max} values of the purified yellow laccase were 0.0714 mM and 0.0091 mM min^{-1}, respectively.

Keywords *Lentinus squarrosulus* MR13 · Yellow laccase · Circular dichroism spectroscopy · SDS-PAGE · Enzyme kinetics

Introduction

Laccase was first identified and reported by Yoshida (1883) from *Rhus vernicifera*, the Japanese lacquer tree. It was characterized as a copper-containing oxidase by Bertrand (1985). Fungal laccases possess various physiological roles. Laccases from the white-rot fungi, e.g. *Trametes versicolor* and *Pycnoporus cinnabarinus*, contribute to lignin biodegradation by oxidizing the phenolic subunits of lignin (Bourbonnais et al. 1995; Eggert et al. 1996; Thurston 1994). This multipurpose enzyme is used for various purposes which include dye decolorization (Abadulla et al. 2000), fruit juice stabilization (Minussi et al. 2002), pulp bioleaching (Arias et al. 2003) and phenol and aromatic compounds degradation (Crecchio et al. 1995). In fungal physiology, laccases are involved in plant pathogenesis, pigmentation, detoxification and lignin degradation.

Yellow laccase isolated from *Panus tigrinus* 8/18 could oxidize veratryl alcohol and nonphenolic lignin as model compounds without any synthetic mediator in the reaction mixture. Yellow laccase produced by different fungi did not show the characteristic absorption spectrum at 610 nm which is a typical characteristic feature of blue laccase due to the presence of type-I copper atom. The EPR and circular dichroism (CD) spectrum are also different for this group of laccases (Leontievsky et al. 1997; Rodakiewicz-Nowak et al. 1999). Yellow laccase produced during submerged fermentation is capable of oxidizing non-phenolic compounds in the absence of any mediators. This which is a unique characteristic that differentiates yellow laccases from blue laccase (Leontievsky et al. 1997). Information on the purification and characteristics of yellow laccases is extremely limited and their catalytic properties are seldom investigated or reported. Information on the catalytic properties of these enzymes could serve as a basis for the development of bio-preparations and creation of an effective technology for their application in bioremediation (Bezalel et al. 1996) and delignification of potential lignocellulosic substrate for biofuel generation (Baldrian 2004).

M. Mukhopadhyay · R. Banerjee (✉)
Microbial Biotechnology and Downstream Processing Laboratory, Agricultural and Food Engineering Department, Indian Institute of Technology, Kharagpur 721 302, India
e-mail: rb@iitkgp.ac.in; rb@agfe.iitkgp.ernet.in

M. Mukhopadhyay
e-mail: mainak.mukhopadhyay12@gmail.com

In this present investigation, a new yellow laccase from *Lentinus squarrosulus* MR13 was purified and characterized biochemically. The fundamental characteristics of the purified enzyme such as molecular mass, effect of pH, temperature on the enzyme activity and stability, isoelectric point, effect of different additives and inhibitors on enzyme activity were also studied.

Materials and methods

Chemicals

2,2′-Azinobis-(3-ethylbenzothiazoline-6-sulfonic acid) (ABTS) was purchased from Sigma-Aldrich Company (USA). Other chemicals were of reagent grade.

Microorganisms and enzyme production conditions

Lentinus squarrosulus (MR13) was collected from decayed wood in local field at Kharagpur and maintained in potato dextrose agar. Sub culturing was done in every 15 days to maintain its viability.

Substrate (rice straw) for the yellow laccase production (particle size 1 cm) was taken and Czapek-dox media was added to it in the ratio of 1:3. The substrate was mixed thoroughly, sterilized and inoculated with the inoculum prepared on wheat grain. Fermentation was carried out at 30 °C for 9 days.

After incubation the fermented biomass was soaked with water in 1:1 for 4 h at room temperature (32 °C) for leaching of the extracellular enzyme. Extraction of the enzyme from the biomass was performed using cotton cheese cloth. Maximum amount of extract was collected by extraction with pressure. The extract was centrifuged at 10,000 rpm at 4 °C for 10 min. The clear supernatant was used for subsequent studies.

Enzyme assay

Laccase activity was determined spectrophotometrically with ABTS as substrate and the oxidation was monitored at 436 nm ($\varepsilon = 29{,}300\ M^{-1}\ cm^{-1}$). One unit of enzyme activity (IU) is defined as the amount of enzyme that released 1 μmol of oxidized product per minute.

Protein estimation

The protein content of the culture filtrate was estimated by Lowry's method (Lowry et al. 1951) with bovine serum albumin as the internal standard.

Enzyme purification

Crude enzyme broth was concentrated by ammonium sulfate precipitation. The precipitated enzyme was dialysed and monitored at 280 nm followed by activity assay. The concentrated enzyme was loaded into a DEAE cellulose column (10 mm × 40 mm). The purified fraction was further subjected to Sephadex G 100. The final fraction of enzyme was further concentrated with Amicon ultrafiltration system having 10 kDa molecular mass cut off membrane. The purified enzyme was subsequently used for further characterization. SDS-PAGE was carried out to determine the homogeneity and apparent molecular mass of the purified laccase.

Determination of isoelectric point

Isoelectric pH of the purified laccase was determined using Rotofor (BioRad). Focusing was carried out at a constant power of 12 W for 2 h. Twenty samples were collected and analyzed for enzyme activity and pH. The ampholyte used was in the pH range of 3.0–10.0.

UV/visible absorption studies

Enzyme preparation was adjusted to 0.5 mg mL^{-1} in 0.1 M phosphate buffer (pH-6.8). The ratio A_{280}/A_{610} was used for the characterization of laccase.

Effect of pH and temperature on yellow laccase activity and stability

To estimate the optimum pH value, the activity of the purified enzyme was studied over a pH range of 3.0–10. For the determination of the pH stability, enzyme was kept at 4 °C for 1 h in different buffers (0.1 M) and the residual yellow laccase activity determined under standard assay conditions.

Effect of different temperatures on yellow laccase activity was measured at 25–65 °C by standard enzyme assay. To evaluate the temperature effect on the stability of yellow laccase, 100 μL protein solution was prepared in 0.1 M phosphate buffer of pH 6.8, kept at different temperatures ranging 25–65 °C for 60 h and the residual activity tested by standard enzyme assay measured.

Circular dichroism (CD) spectra of yellow laccase

CD spectra of the purified laccase was recorded at various temperature ranging 30–65 °C in 0.1 M sodium phosphate buffer (pH 6.8) using a CD Spectropolarimeter (JASCO J-810) (Yang et al. 1986).

Effect of metal ions on yellow laccase activity

To study the effect of various metal ions on enzyme activity, the enzyme was incubated with 1 mM of $CaCl_2$, $CuSO_4$, $CoCl_2$, $CuCl_2$, $(NH_4)_2SO_4$.$FeSO_4$, $FeCl_3$, $FeCl_2$, K_2SO_4, $NiSO_4$, KCl, $MnCl_2$, $MnSO_4$, NaCl, $MgSO_4$, Na_2SO_4, Na_2S, $MgCl_2$, Hg_2Cl_2 and $HgCl_2$ at 35 °C for 1 h. Then the required volume of enzyme was taken for assay by standard enzyme assay protocol.

Effect of inhibitors and different additives

Six potential laccase inhibitors were selected to evaluate their effect on the purified laccase. The enzyme was incubated with various inhibitors for 1 h at 35 °C and laccase activity was measured.

Effect of additives like sodium azide, dithiothreitol, L-cysteine, EDTA, sodium thioglycolate, β-mercaptoethanol, SDS, urea and hydrogen peroxide on activity of purified enzyme was thereafter evaluated. Enzyme solution was incubated with different concentrations of β-mercaptoethanol, SDS and urea for 1 h at 35 °C and enzyme activities were calculated.

Assay of substrate specificity and enzyme kinetics of yellow laccase

For the determination of substrate specificity of the enzyme, different aromatic substrates were added to the assay media by replacing ABTS. The substrates used were resorcinol, gallic acid, catechol, pyrogallol, syringaldazine, ferulic acid, vanillic acid and caffeic acid at 1 mM concentration during the assay. The substrate oxidation rate was followed by measuring the absorbance change with the molar extinction coefficient (ε) obtained from the literature (Wolfenden and Wilson 1982). Relative activity of ABTS was taken as 100 %.

Kinetic constants of the enzyme for the most common substrate ABTS were determined. The rate of substrate oxidation was determined by spectrophotometric method, using reported molar extinction coefficient ε (29,300 M^{-1} cm^{-1}) (Wolfenden et al. 1982). Concentration range of ABTS was 40–600 mM. Reactions were conducted at 35 °C. K_m and V_{max} values were calculated using the Lineweaver–Burk plot of the Michaelis–Menten equation.

Application of yellow laccase on lignocellulosic substrate for lignin degradation

Bambusa bamboos was selected for the application of yellow laccase. It is a potential substrate for the production of second generation bioethanol. To achieve higher accessibility in the cellulosic layer of the lignocellulosics, lignin must be degraded by pre-treatment process. In this experiment, crude yellow laccase was used to degrade lignin. After each cycle of the incubation of lignocellulosics with enzyme the residual lignin was estimated (Hussain et al. 2002).

Results

Purification of yellow laccase

The crude extract of the yellow laccase produced from *L. squarrosulus* MR13 was purified to homogeneity using a three-step purification procedure as summarized in Table 1. The first step of ammonium sulfate precipitation (60 %) resulted in an increase in specific activity to 827.27 IU mg^{-1} protein having 2.77-fold purification. The concentrated fraction was further subjected to ion exchange chromatography where a 7.43-fold increase in purification was achieved (Fig. 1a). The eluted fraction from ion exchange chromatography was pooled and used for gel filtration chromatography. The specific activity of laccase was found to be further increased to 3,772.86 IU mg^{-1} protein having 12.67-fold purification (Fig. 1b).

Determination of molecular mass

The purified *L. squarrosulus* MR13 yellow laccase showed a single band on SDS-PAGE with a mobility equivalent to the molecular mass of 66 kDa as visualized by Coomassie brilliant blue staining (Fig. 2).

UV/visible absorption studies

Absorption maxima of the purified enzyme were observed at 465 nm (Fig. 3). The absorbance spectra of purified enzyme showed no significance absorbance at 610 nm. The A_{280}/A_{610} ratio of purified enzyme was 312.5. This value indicated the absence of type III copper atom in the purified enzyme. From this observation it was concluded that the enzyme belongs to yellow laccase.

Determination of p*I*

Isoelectric focusing is an extremely powerful, non denaturing technique for localization of the proteins at their respective isoelectric points without any phase change. The absence of any phase change, separation of proteins in their native state and efficient temperature control in Rotofor® system leads to minimum denaturation of the enzyme. Isoelectric focusing of laccase revealed that there are only one laccase present which had p*I* at pH 8.2.

Table 1 Purification profile of yellow laccase of *Lentinus squarrosulus* MR13 produced by solid state fermentation

Purification step	Protein concentration (mg mL^{-1})	Enzyme activity (IU mL^{-1})	Specific activity (IU mg^{-1})	Purification (fold)	% Yield
Crude extract	1.985	593.8	299.14	01.00	100
$(NH_4)_2SO_4$ precipitation	2.515	2,080.6	827.27	02.77	81
DEAE cellulose	0.062	137.83	2,223.06	07.43	64
Sephadex G 100	0.021	79.23	3,772.86	12.61	45

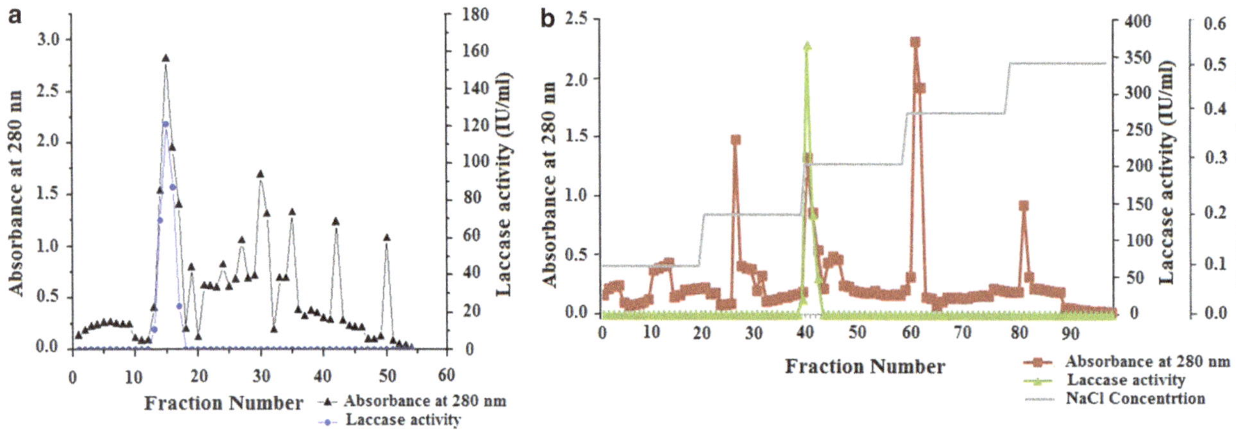

Fig. 1 **a** Ion exchange, **b** gel filtration chromatography profile of yellow laccase

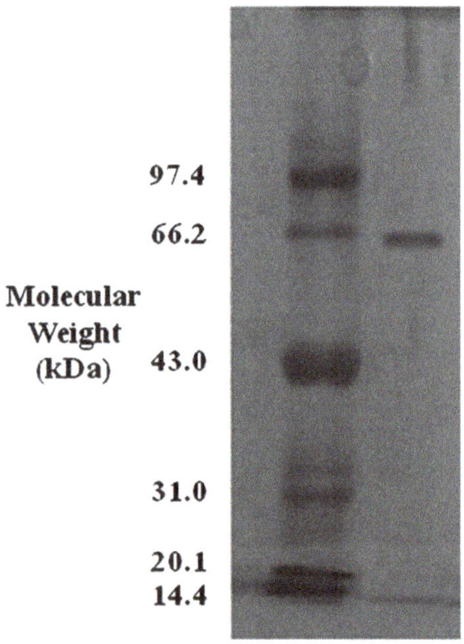

Fig. 2 SDS-PAGE for determination of purity and calculation of molecular weight of the yellow laccase (molecular weight 66 kDa)

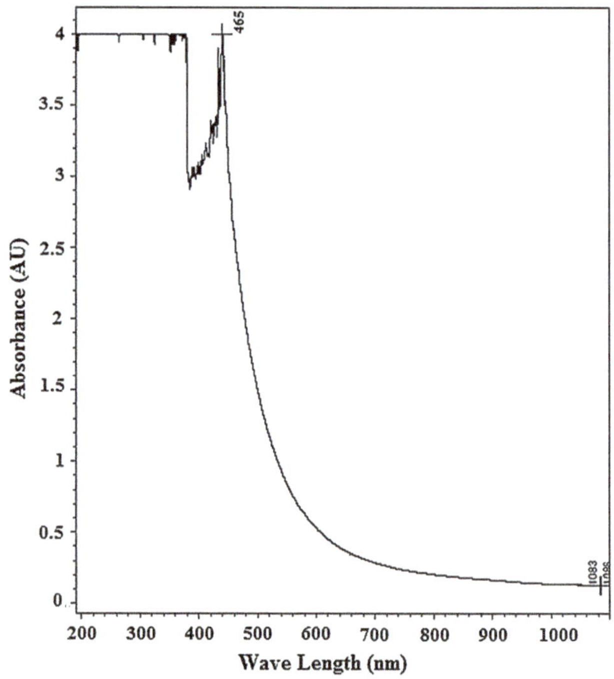

Fig. 3 UV/vis absorption spectra of purified yellow laccase

Effect of pH on yellow laccase activity and stability

The effect of pH on enzyme activity was analyzed by carrying out enzyme assays at different pH. The highest activity was observed at pH 4.5. Enzyme was able to hold back its activity in a long range of pH, from 4.0 to 9.0 (Fig. 4a). At pH 3.5 and 9.5 enzyme was able to retain its

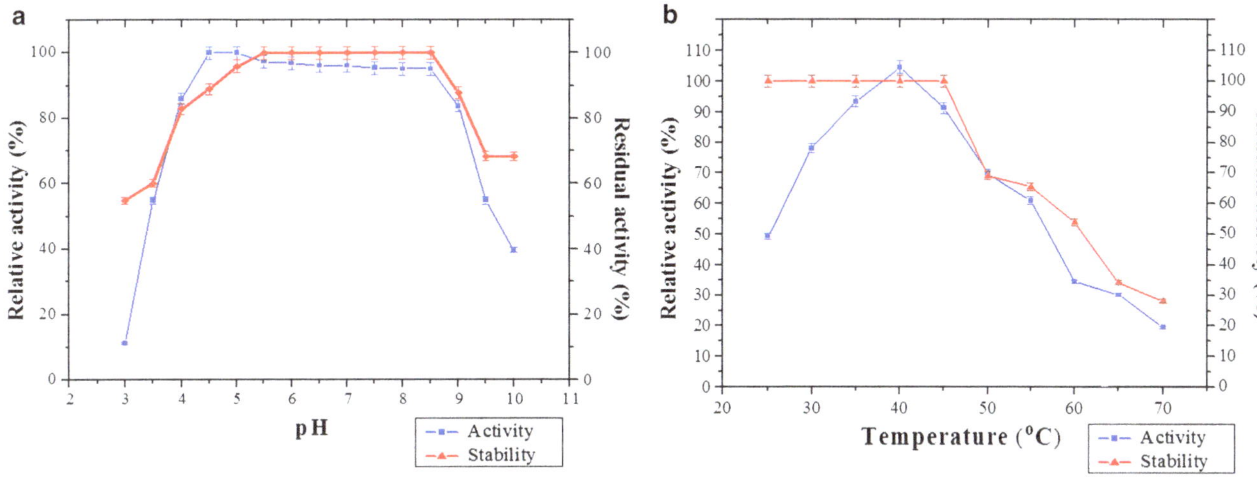

Fig. 4 **a** Effect of pH and **b** temperature on the activity and stability of purified yellow laccase

half activity. The purified enzyme exhibits maximum stability at the pH range 5.0–8.5. Comparable stability was also observed at pH 4.0, 4.5 and 9.0.

Effect of temperature on yellow laccase activity and stability

The optimum temperature of laccase activity was obtained at 40 °C, beyond which the activity of the enzyme started decreasing (Fig. 4b). From the obtained results the profile of enzyme activity can be shown as $35 > 40 > 45 > 30 > 50 > 55 > 60 > 65$ °C. Results for the effect of temperature on stability of yellow laccase indicated that the enzyme under evaluation has been considered to be stable in moderate to higher temperature. It could withstand the temperature up to 55 °C, but at 65 °C its activity was reduced to 34 %.

CD spectra of yellow laccase at different temperature

At lower temperature, yellow laccase showed more α-helix and minimum random coil. At higher temperature percentage of α-helix has decreased, which indicated that yellow laccase was more active at lower to moderate temperature (Fig. 5). In the temperature range 30–50 °C yellow laccase showed almost equivalent amount of α helix than β sheet (Table 2), which indicated the normal activity of yellow laccase. The spectral features of the enzyme at the temperature range of 30–50 °C remained almost constant with slight variations in magnitude.

Effect of metal ions

The stability of enzyme activity against different metal ions was studied and it was observed that the yellow laccase retains its normal activity in the presence of $(NH_4)_2SO_4$, whereas the activity was increased ~2 % in presence of $CuCl_2$ and 5 % in presence of $CuSO_4$. All the other metal ions acted as the inhibitor of the enzyme in 1 mM concentration (Table 3).

Effect of inhibitors

Effect of different inhibitors on enzyme activity was observed using sodium azide, dithiothreitol, L-cysteine, EDTA, sodium thioglycolate and H_2O_2 (Table 4). It was observed that 1 mM concentration of sodium azide, sodium thioglycolate and L-cysteine completely inhibited the enzyme activity. EDTA and H_2O_2 were able to inhibit the enzyme completely in higher concentrations, 2 and 5 mM, respectively. Dithiothreitol acted as a potent inhibitor which was able to inhibit the enzyme completely at 0.1 mM concentration.

Effect of different additives

The effect of additives like β-mercaptoethanol, SDS and urea, on activity of purified enzyme was evaluated. Results exhibited that β-mercaptoethanol, SDS and urea at their higher concentration were to inhibit the enzyme completely (Table 5). β-mercaptoethanol and urea at 0.1 mM concentration were able to inhibit only 14.29 and 12.5 % of enzyme activity, respectively. At 10 mM concentration of SDS the enzyme was completely inhibited.

Substrate specificity

The data on substrate specificity of the enzyme are summarized in Table 6. Syringaldazine showed highest relative activity (116.69 %), whereas other substrates like resorcinol, catechol, tannic acid, ferulic acid, vanillic acid and caffeic acid have nearly similar specificity towards the purified yellow laccase.

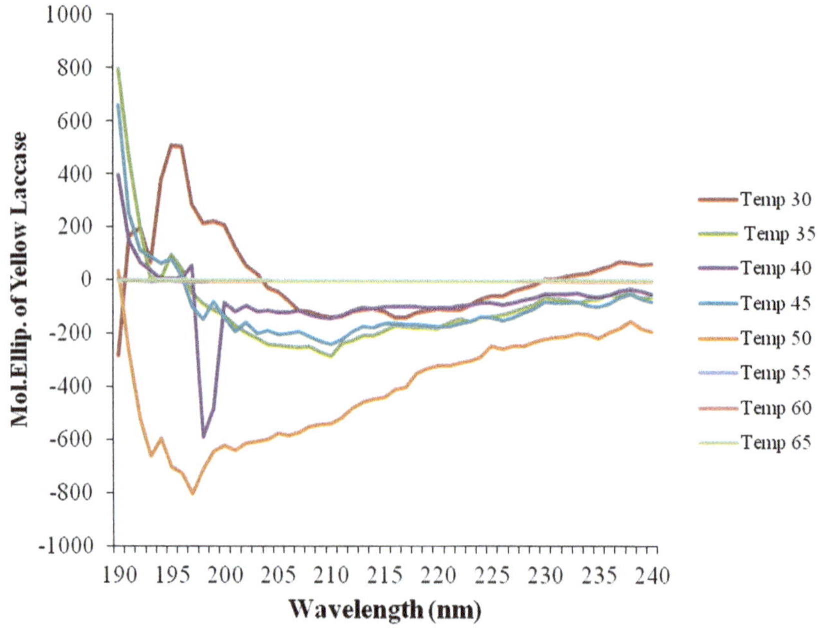

Fig. 5 CD spectra of yellow laccase at different temperature

Table 2 α Helix and β sheet comparison of yellow laccase at different temperature

Temperature (°C)	α Helix (%)	β Sheet (%)
30	67.84	9.46
35	69.24	8.63
40	70.50	8.40
45	69.48	8.63
50	62.57	9.60
55	4.23	29.74
60	3.80	31.70
65	1.53	33.80

Table 3 Effect of metal ions on the activity of the purified yellow laccase from *Lentinus squarrosulus* (MR13)

Metal ions (1 mM)	Relative activity (%)
Control	100.00
$CaCl_2$	84.82
$CuSO_4$	105.78
$CoCl_2$	90.57
$(NH_4)_2SO_4$	100.22
$FeCl_3$	72.70
$FeCl_2$	89.13
K_2SO_4	97.35
$NiSO_4$	98.99
KCl	95.91
$MnCl_2$	89.34
$MnSO_4$	93.04
NaCl	83.79
$MgSO_4$	88.31
Na_2SO_4	90.78
Na_2S	10.27
Hg_2Cl_2	44.57
$HgCl_2$	61.82

Kinetic properties of purified yellow laccase

ABTS was used as the substrate to determined two main kinetic parameters V_{max} and K_m. Lineweaver–Burk plot confirmed that the K_m and V_{max} values of the purified yellow laccase were 0.0714 mM and 0.0091 mM min^{-1}. The K_{cat} of the purified yellow laccase was found to be 303 s^{-1}. Catalytic efficiency (K_{cat}/K_m) of the enzyme was found as 4,089 s^{-1} mM^{-1}.

Lignin degradation studies of lignocellulosic substrate

The capability of white rot fungi to degrade polyaromatic compounds, dyes and lignin has been widely studied. In this present work, crude yellow laccase was used to degrade the lignin from *B. bamboos*. The degradation studies were conducted at the conditions most suitable for the enzymatic action. Lignin degradation rate has been described in Fig. 6. It was observed that the degree of lignin degradation attained was 65.3 % after 4 h of incubation. Subsequent lignin degradation was slow and reached 73.9 % after 8 h of incubation.

Discussion

The yellow laccase obtained from *L. squarrosulus* (MR13) was purified using ammonium sulfate precipitation, ion-exchange chromatography on DEAE-cellulose followed by

Table 4 Effect of Inhibitors on purified yellow laccase from *Lentinus squarrosulus* (MR13)

Inhibitors	Concentration (mM)	Inhibition (%)
Control (no inhibitor)	–	0.00
Sodium azide	0.1	33.33
	1.0	100.00
Dithiothreitol	0.1	100.00
L-Cysteine	0.1	32.22
	1.0	100.00
EDTA	0.1	6.67
	1.0	72.38
	5.0	100.00
Sodium thioglycolate	0.1	58.64
	1.0	100.00
H_2O_2	0.1	57.00
	1.0	84.63
	2.0	100.00

Table 5 Effect of different additives on purified yellow laccase from *Lentinus squarrosulus* (MR13)

Additives	Concentration (mM)	Inhibition (%)
Control	–	0.00
β-Mercaptoethanol	0.1	14.29
	1.0	42.86
	10.0	67.86
	100.0	100.00
SDS	0.01	7.14
	0.1	31.67
	1.0	69.64
	10.0	100.00
Urea	0.1	12.50
	1.0	20.28
	5.0	55.19
	10.0	100.00

Table 6 Substrate specificity of purified yellow laccase from *Lentinus squarrosulus* (MR13)

Substrate (1 mM)	Wavelength (nm)	Absorption coefficient (M^{-1} cm^{-1})	Relative activity (%)
ABTS	436	29,300	100.00
Resorcinol	500	32,800	73.68
Gallic acid	250	4,910	78.29
Catechol	450	2,211	81.65
Pyrogallol	450	4,400	53.23
Syringaldazine	530	65,000	116.87
Ferulic acid	287	12,483	84.92
Vanillic acid	316	2,340	83.98
Caffeic acid	311	18,500	85.69

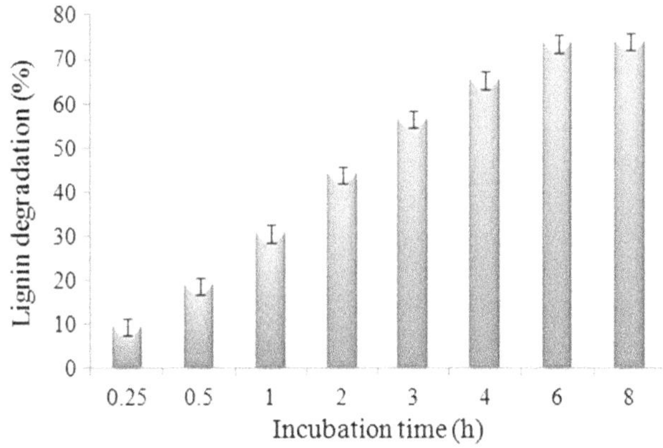

Fig. 6 Lignin degradation profile of yellow laccase in *Bambusa bamboos* against time

gel filtration chromatography. A laccase from monkey head mushroom *H. erinaceum* was purified up to a 15-fold purification factor. Its chromatographic steps included ion-exchange chromatography on DEAE-cellulose, CM-cellulose, Q-Sepharose and FPLC-gel filtration on Superdex 75 (Wang and Ng 2004). A blue laccase from *Ganoderma lucidum* was purified with a purification factor of 25.4-fold using DEAE-cellulose, Aff-gel blue gel, Con A-Sepharose and FPLC-gel filtration on Superdex 75 by Wang and Ng (2004). Similar yellow laccase from the fungi *Pleurotus ostreatus* D1 was purified from its crude solution by employing similar kind of purification steps (Pozdnyakova et al. 2006). In this present study 12.67-fold purification was achieved.

The concentrated purified enzyme solution was yellowish brown in colour. The absorption spectrum of purified laccase does not have the characteristic absorption at 610 nm, which is a typical characteristic of blue laccase for the presence of type 1 copper atom (Thurston 1994). These all sets of combinations were formerly pragmatic in yellow laccase of the fungi *P. tigrinus*, *Phlebia radiata* and *Phlebia tremellosa* (Leontievsky et al. 1997). The A_{280}/A_{610} ratio of purified enzyme was found to be 312.5, which was higher than typical blue laccases (Leontievsky et al. 1997; Bezalel et al. 1996; Leonowicz and Crzywnowicz 1981; Laemmli 1970; Niku-Paavola et al. 1988).

Lentinus squarrosulus (MR13) yellow laccase appeared as a single band with a molecular weight of 66 kDa in SDS-PAGE. The molecular mass found was within the range of molecular masses for most of the fungal laccases reported (50–90 kDa) (Baldrian 2006). Laccases from *Albatrella dispansus* (62 kDa) (Wang and Ng 2006), *Coriolus zonatus* (60 kDa) (Koroljova et al. 1999), *H. echinacea* (63 kDa) (Wang and Ng 2004), *Marasmius quercophilus* (60 kDa) (Farnet et al. 2000) and *Trametes sanguinea* (62 kDa) (Nishizawa et al. 1995) have a molecular mass very close to that of *L. squarrosulus* (MR13) laccase. On the other hand, *P. eryngii* laccase has a small molecular mass of 34 kDa (Wang and Ng 2006), and a laccase from *Podosporaa anserina* has a large molecular mass of 383 kDa (Ng and Wang 2004).

Shin and Kim (1998) reported that the optimal pH of laccase from the fungus *Coriolus hirsutus* was 4.0. Solano et al. (1997) reported an optimum pH of 6.5 for melanogenic bacteria MMB-1. Niladevi et al. (2008) reported optimal pH for laccase from *Streptomyces psammoticus* as 8.5. These findings were similar to the present finding where the laccase enzyme showed significantly broad range of pH activity (pH 4.5–8.5) and pH stability (pH 4.5–9.0).

The presently studied yellow laccase was active over a wide range of temperatures. The typical optimum temperature ranges for laccases are 35–45 °C. Significant activity was detected at lower temperature of 25 °C and ~75 % of activity was maintained at 50 °C. The wide functional range is similar to that observed by Wang and Ng (2004) for the *Albatrella* enzyme. The laccase from *P. sanguineus* was more thermostable than expected for typical laccases (Palonen et al. 2003; Shin and Kim 1998; Thakker et al. 1992).

CD spectroscopic studies at different temperatures can be used to provide information on the stability of proteins. CD spectra of the yellow laccase supported the data that the enzyme maintained stability at the temperature range of 30–50 °C. During further analysis of the CD spectrum, it was observed that the yellow laccase showed two negative troughs at 30–50 °C, one of which is in the region of 220 nm and the other at 208 nm. This is considered to be a characteristic features for all α, β or $\alpha + \beta$ proteins with varying intensities related to the extent of secondary structural elements. With the further increase in temperature these features tend to change towards increase in random structure signifying structural breakdown and loss in activity of the enzyme.

The activity of the presently investigated yellow laccase was not greatly influenced by any of the metal ion studied at a concentration of 1 mM. Nagai et al. (2002) reported that *L. edodes* laccase was inhibited in the presence of 1 mM Ca^{2+} (70 %) and Zn^{2+} (64 %) and was enhanced by 40 % in the presence of 10 mM Cu^{2+}. A laccase from *Coriolus versicolor* was strongly depressed by Ag^{+} (100 %), Al^{3+} (100 %), and Fe^{3+} (80 %), but was activated by Cu^{2+} (10 %) and Mg^{2+} (10 %) (Zhu and Ding 2003).

The enzyme was inhibited by all the inhibitors that were studied. However, the extent of inhibition varied greatly with the nature and concentration of the tested inhibitors. Sodium azide has been reported to prevent the substrate oxidation by laccase (Johannes and Majcherczyk 2000). The inhibitory effect of sodium azide on the *L. squarrosulus* was similar to other fungal laccases (Wu et al. 2011; Niladevi et al. 2008). EDTA, another well-known metal chelating agent fully inhibited the purified laccase at 2 and 5 mM concentrations. Similar results have also been reported from *S. cyaneus* (Arias et al. 2003). Most fungal laccases appear to be inhibited by this metal chelator (Baldrian 2006), although others, such as the laccase from *Pycnoporus sanguineus*, may be moderately resistant to EDTA (Lu et al. 2007). On the other hand, the inhibitory effect of thioglycolate on *L. squarrosulus* laccase is at par with previous reports for some other laccases (Niladevi et al. 2008; Sadhasivam et al. 2008).

The substrate specificity studies indicated that syringaldazine is the most suitable substrate for the presently studied yellow laccase (Table 6). Similar result has been reported by Calvo et al. (1998). Other methoxy-substituted compounds like ferulic acid and vanillic acid were not efficiently oxidized as compared to di-substituted compounds. These findings were in congruence with the previous reports (Palmieri et al. 1997). Among the substituted phenols such as catechol and resorcinol, the enzyme showed more affinity towards the ortho-substituted catechol than the meta-substituted resorcinol. Laccases have been reported to exhibit activity with para- and ortho-diphenols (Xu 1996) and very less reactivity has been observed generally with the meta-substituted phenols (Jolivalt et al. 1999).

In the presence of ABTS, the calculated K_m and V_{max} demonstrated the efficient oxidation ability of the yellow laccase toward the substrate. The K_m of presently studied yellow laccase was reletively lower than the previously mentioned reports. This feature implies that this yellow laccase have greater affinity towards the substrates. The K_m value of the laccases from *Trichoderma atroviride* and *Pycnoporus sanguineus* in the presence of the same substrate were determined to be 2,500 μM (Chakroun et al. 2010) and 77 μM (Lu et al. 2007), respectively. Laccase from *Chaetomium thermophilum* showed V_{max} value of 2.6 mmol min^{-1} mg^{-1} for the substrate ABTS (Chefetz et al. 1998). The values of the catalytic constants obtained for *L. squarrosulus* laccase were markedly higher than the above reported fungal laccases, which indicates the higher substrate specificity for laccase from *L. squarrosulus*.

In the present study it was found that,incubation time required for lignin degradation of *B. bamboos* was 6 h (Fig. 6) and further raise in incubation time does not demonstrate any significant effect. Lignin degradation ceased after attending upper limit, which might be due to the accumulation of lignin degradatory products which act as inhibitor for this yellow laccase. Incubation time for the present yellow laccase-mediated lignin degradation was shorter than any other reported values of enzyme mediated or whole cell mediated degradation of lignin or lignin model compounds (Cunha et al. 2010; Feng et al. 2010; Cho et al. 2009) but higher than the chemical or physical techniques (Yelle et al. 2008; Sun and Cheng 2007).

Conclusion

The present study deals with the purification and characterization of a pH stable yellow laccase from a newly isolated fungus *L. squarrosulus* MR13. The three-step purification process was effective in concentrating and purifying yellow laccase with a significant yield of 12.67 %. Characterization studies showed that the enzyme possesses a molecular mass of 66 kDa, a pH optimum at 4.5 and 40 °C. It demonstrates activity towards a range of phenolic as well as non-phenolic compounds. The effect of different inhibitors on the purified yellow laccase conformed to the common model of laccase inhibition.

Acknowledgments The authors would like to acknowledge "Petrotech Society", India for providing funds for this research.

Conflict of interest The authors declare that they have no conflict of interest.

References

Abadulla E, Tzanov T, Costa S, Robra K, Cavaco-Paulo A, Gubitz GM (2000) Decolorization and detoxification of textile dyes with a laccase from *Trametes hirsuta*. Appl Environ Microb 66:3357–3362

Arias ME, Arenas M, Rodriguez J, Soliveri J, Ball AS, Hernandez M (2003) Kraft pulp bioleaching and mediated oxidation of a non-phenolic substrate by laccase from *Streptomyces cyaneus* CECT3335. Appl Environ Microb 69:1953–1958

Baldrian P (2004) Increase of laccase activity during inter-specific interactions of white-rot fungi. FEMS Microbiol Ecol 50:245–253

Baldrian P (2006) Fungal laccases-occurrence and properties. FEMS Microbiol Rev 20:1–28

Bertrand G (1985) Sur la laccase et sur le pouvoir oxydant de cette diastase. CR Acad Sci (Paris) 120:266–269

Bezalel L, Hadar Y, Cerniglia CE (1996) Mineralization of polycyclic aromatic hydrocarbons by the white-rot fungus *Pleurotus ostreatus*. Appl Environ Microb 62:292–295

Bourbonnais R, Paice MG, Reid ID, Lanthier P, Yaguchi M (1995) Lignin oxidation by laccase isozymes from *Trametes versicolor* and role of the mediator 2,2′-azinobis(3-ethylbenzthiazoline-6-sulfonate) in kraft lignin depolymerization. Appl Environ Microb 61:1876–1880

Calvo AM, Copa-Patiño JL, Alonso O, González AE (1998) Studies of the production and characterization of laccase activity in the basidiomycete *Coriolopsis gallica*, an efficient decolorizer of alkaline effluents. Arch Microbiol 171:31–36

Chakroun H, Mechichi T, Martinez MJ, Dhouib A, Sayadi S (2010) Purification and characterization of a novel laccase from the ascomycete *Trichoderma atroviride*: application on bioremediation of phenolic compounds. Process Biochem 45:507–513

Chefetz B, Chen Y, Hadar Y (1998) Purification and characterization of laccase from *Chaetomium thermophilium* and its role in humification. Appl Environ Microbiol 64:3175–3179

Cho NS, Hatakka AI, Rogalski J, Cho HY, Ohga S (2009) Directional degradation of lignocellulose by *Phlebia radiata*. J Fac Agric Kyushu Univ 54:73–80

Crecchio C, Ruggiero P, Pizzigallo MDR (1995) Polyphenol oxidases immobilized in organic gels: properties and applications in the detoxification of aromatic compounds. Biotechnol Bioeng 48:585–591

Cunha GGS, Masarin F, Norambuena M, Freer J, Ferraz A (2010) Linoleic acid peroxidation and lignin degradation by enzymes produced by *Ceriporiopsis subvermispora* grown on wood or in submerged liquid cultures. Enzyme Microb Technol 46:262–267

Eggert C, Temp U, Eriksson KEL (1996) The ligninolytic system of the white rot fungus *Pycnoporus cinnabarinus*: purification and characterization of the laccase. Appl Environ Microb 62:1151–1158

Farnet AM, Criquet S, Tagger S, Gil G, Le Petit J (2000) Purification, partial characterisation and reactivity with aromatic compounds of two laccases from *Marasmius quercophilus* strain 17. Can J Microbiol 46:1–6

Feng Y, Simpson AJ, Schlesinger WH, Simpson MJ (2010) Altered microbial community structure and organic matter composition under elevated CO_2 and N fertilization in the Duke Forest. Glob Change Biol 16:2014–2116

Hussain MA, Huq ME, Rahman SM (2002) Estimation of lignin in jute by titration method. Pak J Biol Sci 5:521–522

Johannes C, Majcherczyk A (2000) Laccase activity and laccase inhibitors. J Biotechnol 78:193–199

Jolivalt C, Raynal A, Caminade E, Kokel B, Le Goffic F, Mougin C (1999) Transformation of *N′,N′*-dimethyl-*N*-(hydroxyphenyl) ureas by laccases from the white-rot fungus *Trametes versicolor*. Appl Microbiol Biotechnol 51:676–681

Koroljova OV, Stepanova EV, Gavrilova VP, Biniukov VI, Jaropolov AI, Varfolomeyev SD, Scheller F, Makower A, Otto A (1999) Laccase of *Coriolus zonatus*. Appl Biochem Biotech 76:115–127

Laemmli UK (1970) Cleavage of structural proteins during the assembly of the head of bacteriophage T4. Nature 227:680–685

Leonowicz A, Crzywnowicz K (1981) Quantitative estimation of laccase forms in some white-rot fungi using syringaldazine as a substrate. Enzyme Microb Technol 3:55–58

Leontievsky AA, Mysoedova NM, Pozdnyakova NN, Golovleva LA (1997) Yellow laccase of *Panus tigrinus* oxidizes non phenolic substrates without electron-transfer mediator. FEBS Lett 413:416–418

Lowry RJ, Rosebrough NJ, Farr AL, Randall RJ (1951) Protein measurement with the Folin phenol reagent. J Biol Chem 193:265–275

Lu L, Zhao M, Zhang BB, Yu SY, Bian XJ, Wang W (2007) Purification and characterization of laccase from *Pycnoporus sanguineus* and decolorization of an anthraquinone dye by the enzyme. Appl Environ Microbiol 74:1232–1239

Minussi RC, Pastore GM, Durany N (2002) Potential applications of laccase in the food industry. Trends Food Sci Technol 13:205–216

Nagai M, Sato T, Watanabe H, Saito K, Kawata M, Enei H (2002) Purification and characterization of an extracellular laccase from the edible mushroom *Lentinula edodes*, and decolorization of chemically different dyes. Appl Microbiol Biotechnol 60:327–335

Ng TB, Wang HX (2004) A homodimeric laccase with unique characteristics from the yellow mushroom *Cantharellus cibarius*. Biochem Biophys Res Commun 313:37–41

Niku-Paavola ML, Karhunen E, Salola P, Paunio V (1988) Ligninolytic enzymes of the white rot fungus *Phlebia radiata*. Biochem J 254:877–884

Niladevi KN, Jacob N, Prema P (2008) Evidence for a halotolerant-alkaline laccase in *Streptomyces psammoticus*: purification and characterization. Process Biochem 43:654–660

Nishizawa Y, Nakabayashi K, Shinagawa E (1995) Purification and characterization of laccase from white rot fungus *Trametes sanguinea* M85-2. J Ferment Bioeng 80:91–93

Palmieri G, Giardina P, Blanco C, Scaloni A, Capasso A, Sannia G (1997) A novel white laccase from *Pleurotus ostreatus*. J Biol Chem 272:31301–31307

Palonen H, Saloheimo M, Viikari L, Kruus K (2003) Purification, characterization and sequence analysis of a laccase from the ascomycete *Mauginiella* sp. Enzyme Microb Technol 33:854–862

Pozdnyakova NN, Rodakiewicz-Nowak J, Turkovskaya OV, Haber J (2006) Oxidative degradation of polyaromatic hydrocarbons catalyzed by blue laccase from *Pleurotus ostreatus* D1 in the presence of synthetic mediators. Enzyme Microb Technol 39:1242–1249

Rodakiewicz-Nowak J, Haber J, Pozdnyakova N, Leontievsky A, Goloveva LA (1999) Effect of ethanol on enzymatic activity of fungal laccases. Biosci Rep 19:589–600

Sadhasivam S, Savitha S, Swaminathan K, Lin FH (2008) Production, purification and characterization of mid-redox potential laccase from a newly isolated *Trichoderma harzianum* WL1. Process Biochem 43:736–742

Shin KS, Kim CJ (1998) Properties of laccase purified from nitrogen limited culture of white-rot fungus *Coriolus hirsutus*. Biotechnol Technol 12:101–104

Solano E, Garcia E, Pérez de Egea A, Sanchez-Amat A (1997) Isolation and characterization of strain MMB-1 (CECT 4803), a novel melanogenic marine bacterium. Appl Environ Microb 63:3499–3506

Sun FB, Cheng HZ (2007) Evaluation of enzymatic hydrolysis of wheat straw pretreated by atmospheric glycerol autocatalysis. J Chem Technol Biotechnol 82:1039–1044

Thakker GD, Evans CS, Rao KK (1992) Purification and characterization of laccase from *Monocillium indicum* Saxena. Appl Microbiol Biotechnol 37:321–323

Thurston CF (1994) The structure and function of fungal laccases. Microbiology 140:19–26

Wang HX, Ng TB (2004) A new laccase from dried fruiting bodies of the monkey head mushroom *Hericium erinaceum*. Biochem Biophys Res Commun 322:17–21

Wang HX, Ng TB (2006) Purification of a laccase from fruiting bodies of the mushroom *Pleurotus eryngii*. Appl Microbiol Biot 69:521–525

Wolfenden BS, Wilson RL (1982) Radical cations as reference chromogens in studies of one-electron transfer reactions; pulse radiolysis studies of 2,2′-azinobis-(3-ethylbenzothiazoline-6-sulfonate). J Chem Soc Perkin Trans 11:805–812

Wu XM, Liu MY, Xu Q, Guo WW (2011) Identification and characterization of miRNAs from citrus expressed sequence tags. Tree Genet Genomes 7:117–133

Xu F (1996) Oxidation of phenols, anilines and benzenethiols by fungal laccases: correlation between activity and redox potentials as well as halide inhibition. Biochemistry 35:7608–7614

Yang JT, Wu CS, Martinez HM (1986) Calculation of protein conformation from circular dichroism. Method Enzymol 130:208–271

Yelle DJ, Ralph J, Frihart CR (2008) Characterization of nonderivatized plant cell walls using high-resolution solution-state NMR spectroscopy. Magn Reson Chem 46:508–517

Yoshida H (1883) Chemistry of lacquer (*Urushi*) part 1. J Chem Soc 43:472–486

Zhu XF, Ding T (2003) Effect of metal ions on activity of two laccases. J Chem Res 14:50–62

27

Ecofriendly production of silver nanoparticles using *Candida utilis* and its mechanistic action against pathogenic microorganisms

Shailesh R. Waghmare · Mustopa N. Mulla · Suryakant R. Marathe · Kailas D. Sonawane

Abstract Silver nanoparticles (AgNPs) have attracted great interest due to their applications in various areas. In the present study ecofriendly biosynthesis of extracellular silver nanoparticles was carried out using *Candida utilis* NCIM 3469. Characterization of synthesized AgNPs was done by UV–visible spectroscopy, Scanning electron microscopy and antibacterial activity. AgNPs are found spherical in shape with size in the range of 20–80 nm. AgNPs showed antibacterial activity against pathogenic organisms such as *Pseudomonas aeruginosa, Staphylococcus aureus,* and *Escherichia coli*. The SEM analysis confirms the antibacterial activity of Ag nanoparticles due to damage of cytoplasmic membrane. AgNPs synthesized by *C. utilis* could be applicable in the development of antibacterial water filters for treatment of water.

Keywords Ecofriendly · Silver nanoparticles · *Candida utilis* · Antibacterial · *Pseudomonas aeruginosa*

Introduction

Silver nanoparticles have recently attracted a lot of interest due to their distinctive properties such as large surface areas, unique physical, chemical and biological properties (Sharma et al. 2009; Zhou et al. 2009). Silver nanoparticles are emerging as a new generation of antibacterial agent (Rai et al. 2009) medical applications (Atyeh et al. 2007; Chang and Weinstein 1975), and antibacterial water filter (Jain and Pradeep 2005). The outbreak of infectious diseases caused by different pathogenic bacteria as well as the development of antibiotic resistance (Hsueh 2010; Kumarasamy et al. 2010; Pitout 2010) has prompted pharmaceutical companies and researchers to search for new antibacterial agents. Among nanomaterials like copper, zinc, titanium, magnesium, silver, gold, and alginate, silver nanoparticles have proved to be the most effective as they have good antibacterial efficacy against bacteria, viruses and other eukaryotic microorganisms (Rai et al. 2009). Different studies have been performed to check the bactericidal effect of silver nanoparticles against Gram negative and Gram positive bacteria, but the bactericidal mechanism of this compound has not been clearly elucidated. Morones et al. (2005) studied the antibacterial activity of silver nanoparticles against four types of Gram negative bacteria, *Escherichia coli, V. cholera, Pseudomonas aeruginosa* and *S. typhus*. They have observed that these silver nanoparticles are attached to the surface of the cell membrane and disturb the function of cell membrane, penetrate bacteria, and release silver ions.

Water is the common breeding ground for many pathogens. In countries such as India, 80 % of the diseases are due to bacterial contamination in drinking water. The removal or inactivation of pathogenic microorganisms is the last step in the treatment of wastewater. During the past few decades, several investigations have been carried out concerning the use of metal ions (Ag, Cu, Zn, Hg, Ti, Ni, Co) as bactericides for water disinfection (Feng et al. 2000; Islam et al. 2003). The use of metal nanoparticles for water disinfection is relatively new (Stoimenov et al. 2002; Zhang et al. 2003). Because of their high reactivity due to

S. R. Waghmare (✉) · M. N. Mulla · S. R. Marathe · K. D. Sonawane
Department of Microbiology, Shivaji University, Kolhapur 416004, Maharashtra, India
e-mail: shlshwaghmare@gmail.com

K. D. Sonawane
Department of Biochemistry, Shivaji University, Kolhapur, Maharashtra, India

the large surface to volume ratio (Ichinose 1992), nanoparticles are expected to play a crucial role in water purification (Stoimenov et al. 2002; Zhang et al. 2003) when water becomes an important commodity (Barraque, 2003). Pesticide removal from drinking water with the help of nanoparticles was reported by Nair and Pradeep (2004).

Although, various chemical and biochemical methods are being explored for the AgNPs production, but microbes are exceedingly effective in this process (Narayanan and Sakthivel 2010). Biosynthesis of silver nanoparticles from bacteria, fungi, yeast, plants, and fruits have been reported (Kowshik et al. 2003). Based on their enormous biotechnological applications, microorganisms such as bacteria, fungi, and yeast are now regarded as possible ecofriendly "nano-factories" (Ahmad et al. 2002). The detail mechanism of microbial synthesis of Ag nanoparticles is yet to be elucidated, but several reports suggested role of certain enzymes such as NADH-dependent reductase from *Fusarium oxysporum* (Ahmad et al. 2002), proteins from *Phanerochaete chrysosporium* (Vigneshwaran et al. 2006) and *Plectonema boryanum* (Lengke et al. 2007) in the synthesis of nanoparticles. Here, we first time report the biofabrication of silver nanoparticles using *Candida utilis* and its application as antimicrobial agent against pathogenic microorganisms.

Materials and methods

Microorganisms

The cultures of *C. utilis* NCIM 3469, *E. coli* NCIM 2832, *Salmonella typhimurium* NCIM 2501, *Candida albicans* NCIM 3466 and *P. aeruginosa* NCIM 5032 used in this study were obtained from National Collection of Industrial Microorganisms (NCIM), Pune, Maharashtra, India.

Biosynthesis of silver nanoparticles

The *C. utilis* was grown on MGYP medium containing malt extract 3.0, glucose 10.0, yeast extract 3.0, peptone 5.0 and agar 20.0 in 1,000 ml distilled water. The fresh culture of *C. utilis* was inoculated in 100 ml mineral medium containing $NaNO_3$: 3, K_2HPO_4: 1, KCl: 0.5, $MgSO_4{\cdot}7H_2O$: 0.5, $FeSO_4$: 0.01, yeast extract 0.1, peptone: 0.5 per 1,000 ml distilled water and flask was incubated at 30 °C for 48 h. After incubation the cells were separated by centrifugation at 5,000 rpm for 15 min and 100 ml supernatant was collected in 250 ml Erlenmeyer flask containing 10 ml $AgNO_3$ solution (contain 37.18 mg $AgNO_3$) which makes the final concentration of $AgNO_3$ 2 mM. The flask was then incubated at 30 °C on shaking incubator at 120 rpm in dark. After 6 h, the sample was taken out for UV–visible (UV–Vis) spectrophotometric analysis. Another flask containing a mixture of 100 ml supernatant and 10 ml $AgNO_3$ solution (contain 37.18 mg $AgNO_3$) was autoclaved at 120 °C for 15 min and further analyzed by UV–Vis spectrophotometer.

In a separate experiment, culture of *C. utilis* was grown in 100 ml mineral medium as described above for 24 h. After incubation, cell biomass was collected and then added in 2 mM 100 ml $AgNO_3$ solution and incubated at shaking condition. The $AgNO_3$ reduced to AgNPs shows change in color of $AgNO_3$ from colorless to brown was monitored after 24 h for production of AgNPs by spectrophotometric analysis between the wavelength ranges 300 to 700 nm. The effect of different concentrations of $AgNO_3$ such as 2, 4, 6, 8, and 10 mM on synthesis of AgNPs was monitored for 24 h.

Characterization of AgNPs

The formation of AgNPs was monitored by UV–Vis spectroscopy by recording the spectra between 300 and 700 nm and simultaneously monitoring the appearance of the characteristic peak at 39–420 nm using Shimadzu U-1800 double beam spectrophotometer. The suspension obtained at the time point of maximum production of silver nanoparticles was air dried and then subjected to SEM, using JEOL JSM 6360 (Japan) Scanning Electron Microscope and micrographs were taken.

Antibacterial activity of AgNPs

AgNPs synthesized using whole cells of *C. utilis,* supernatant with heat treatment and 6 mM aqueous solution of silver nitrate as control were tested for antibacterial activity against pathogenic microorganisms such as *S. aureus, P. aeruginosa, and E. coli.* The antibacterial activity was tested by the agar disc diffusion method (Kim et al. 2007) and zone of inhibition was noted.

To examine bactericidal effect on Gram negative organism, *P. aeruginosa* was used as model organism. The fresh culture of *P. aeruginosa* was inoculated in the nutrient broth and then incubated at 37 °C for 24 h. 100 μl of nanoparticle solution having concentration 1 mg ml^{-1} was added in the nutrient broth containing *P. aeruginosa*. After 2 h of incubation, the sample was prepared according to method described by Galabova et al. (1996) for SEM analysis.

Statistical analysis

Results obtained were the mean of three or more determinants. Analysis of variance was carried out on all data at $p < 0.05$ using Graph Pad software (GraphPad InStat version 3.00, GraphPad Software, San Diego, CA, USA).

Results

Biosynthesis of Ag nanoparticles

The aqueous $AgNO_3$ was reduced to metallic silver. The synthesis of silver nanoparticles was monitored in three flasks: (1) in presence of supernatant at 30 °C; (2) in presence of supernatant at temperature 120 °C for 15 min; (3) in the presence of cell biomass of *C. utilis*. It has been observed that the silver nanoparticles were produced when $AgNO_3$ mixed with supernatant and heated at 120 °C for 15 min, whereas no brown color was seen in the flask which was kept at 30 °C. These results indicate the role of temperature in the synthesis of silver nanoparticles. In the case of flask containing cell biomass of *C. utilis* with $AgNO_3$, it was found that the intensity of brown color was increased as the incubation time increases at 30 °C. It confirms the synthesis of silver nanoparticles by *C. utilis*, as shown in Fig. 1.

Characterization of AgNPs

After 24 h of incubation the aqueous silver nitrate turned colorless to brownish yellow in the presence of *C. utilis* cell biomass. UV–Vis spectra of solution revealed peak between 401 and 410 nm, with maximum absorbance at 406 nm. In the present study we have studied the effect of $AgNO_3$ concentration on the synthesis of silver nanoparticles. As the concentration of $AgNO_3$ increases from 2 to 6 mM, synthesis of AgNPs increases, whereas by increasing the concentration of $AgNO_3$ up to 8 mM the synthesis of AgNPs sharply decreases as depicted in Fig. 2. The UV–Vis spectra of aqueous silver nitrate solution when monitored separately did not show any peak between 40 and 410 nm. SEM micrograph of synthesized silver nanoparticles revealed the formation of spherical nanoparticles with the size range of 20–100 nm along with some nanoclusters (Fig. 3).

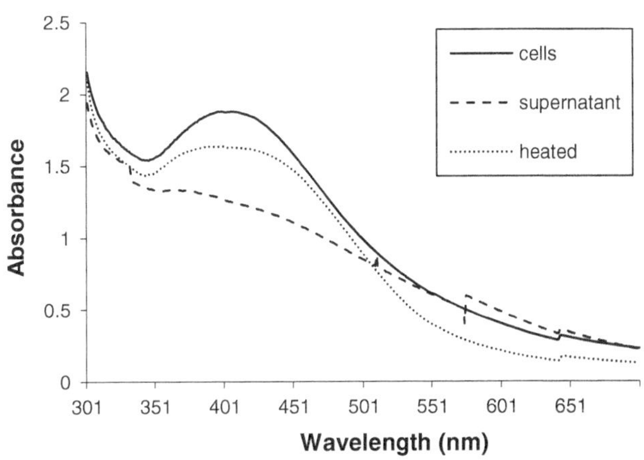

Fig. 1 UV–Vis spectroscopy of synthesized silver nanoparticles. Synthesis of silver nanoparticles was carried out using cells of *C. utilis*, supernatant and supernatant with heat treatment process

Antibacterial activity of synthesized AgNPs

The antibacterial activity of synthesized AgNPs by cell biomass of *C. utilis* (*B*), by the supernatant with heat treatment (*A*) and control as aqueous solution of silver nitrate (*C*) was tested on the Gram positive (*S. aureus*) and Gram negative (*P. aeruginosa*, *E. coli*) organism. AgNPs showed more antibacterial activity against Gram negative organisms such as *P. aeruginosa*, and *E. coli* as compared to Gram positive organism, i.e., *S. aureus* (Fig. 4).

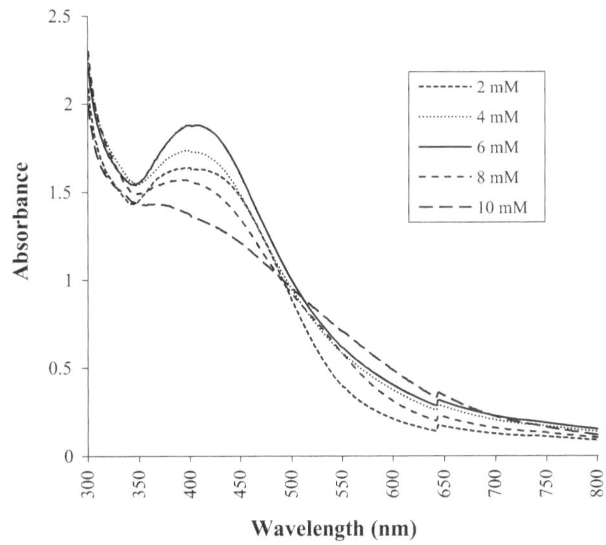

Fig. 2 Effect of silver nitrate concentration on synthesis of AgNPs. Synthesis of AgNPs was carried out at various concentration of silver nitrate such as 2, 4, 6, 8, and 10 mM using cells of *C. utilis*

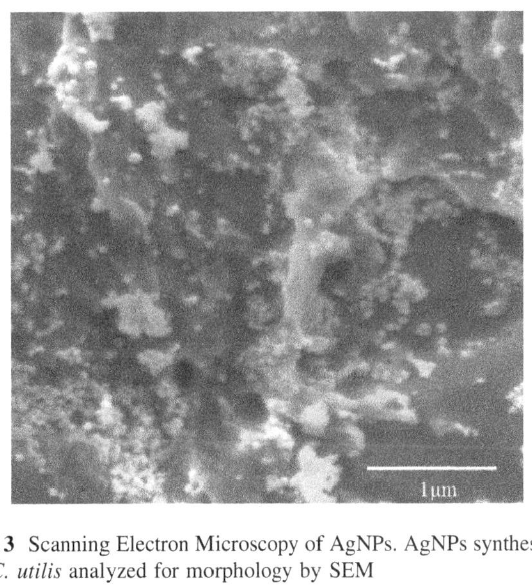

Fig. 3 Scanning Electron Microscopy of AgNPs. AgNPs synthesized by *C. utilis* analyzed for morphology by SEM

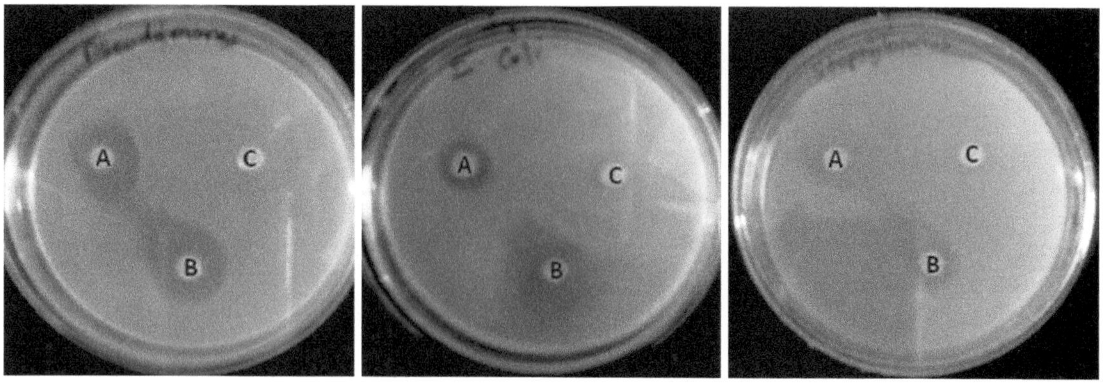

Fig. 4 Antibacterial activity of AgNPs against pathogens. Antibacterial activity of AgNPs synthesized by cells of *C. utilis* (**b**), Supernatant with heat treatment process (**a**) and control as aqueous solution of silver nitrate (**c**) were tested against pathogens such as *P.* aeruginosa, *E. coli*, and *S. aureus* by disc diffusion method

Table 1 Antibacterial activity of AgNPs against pathogens

Test organisms	Zone of inhibition in diameter (mm)	
	A	B
P. aeruginosa	13 ± 1.2	20 ± 1.1
E. coli	10 ± 1.0	18 ± 1.0
S. aureus	8 ± 0.8	12 ± 0.9

Each value represents the mean ± standard error values

A AgNPs synthesized heat treatment process with supernatant

B AgNPs synthesized using live *C.* utilis cells

However, AgNPs synthesized using cell biomass showed more antimicrobial activity than the supernatant (Table 1). SEM was used to evaluate the surface morphology of the both native (Fig. 5a) and treated (Fig. 5b) *P. aeruginosa.* The SEM micrograph revealed the damage of cell wall after the treatment of AgNPs (Fig. 5b) by releasing cytoplasmic content of cells of *P. aeruginosa.*

Discussion

In this study, we have for the first time reported the extracellular synthesis of silver nanoparticles using *C. utilis*. The extracellular synthesis of silver nanoparticles had been reported earlier from other yeast such as salt-tolerant yeast MKY3 (Kowshik et al. 2003), *C. guilliermondii* (Mishra et al. 2011), *Geotrichum* sp. (Jebali et al. 2011). The extracellular synthesis of AgNPs would be of great advantage to industry, since it would minimize the steps involved in the purification of AgNPs. The color of the reaction solution turned from colorless to brown indicating the formation of silver nanoparticles similar to that observed in earlier study. This color arises due to the excitation of surface plasmon vibrations in the metal nanoparticles (Narayanan and Sakthivel 2010).

The silver nanoparticles were synthesized using cells of *C. utilis*, which are spherical nanoparticles with the size range of 20–80 nm along with some nanoclusters (Fig. 3), whereas salt-tolerant yeast MKY3 synthesizes hexagonal silver nanoparticles with the size range of 2–5 nm (Kowshik et al. 2003). Antibacterial activity of AgNPs synthesized by *C. utilis* was found to be inhibitory to both Gram positive as well as Gram negative groups of bacteria, but it has more potential against Gram negative organisms such as *P. aeruginosa* and *E. coli* than the Gram positive, i.e., *S. aureus*. The Ag nanoparticles have shown more antimicrobial activity

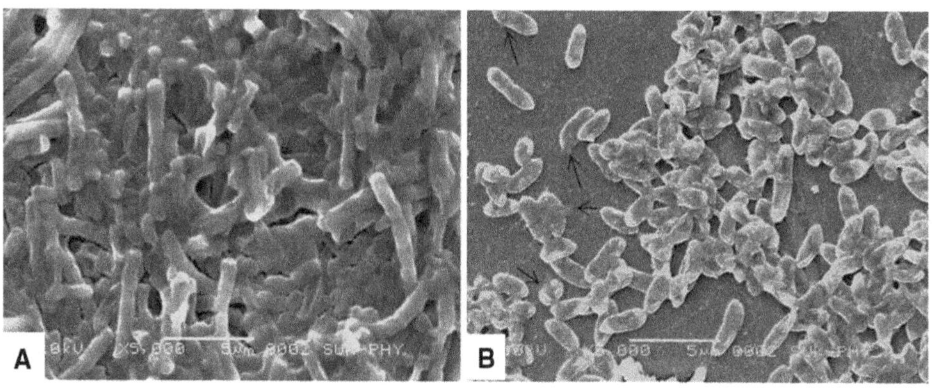

Fig. 5 SEM analysis of *P.* aeruginosa after the treatment with synthesized Ag nanoparticles (**b**) and native cells (**a**)

against Gram negative bacteria than the Gram positive bacteria. This could be because Gram negative bacterial cell wall contains thin layer of peptidoglycan whereas Gram positive bacterial cell wall contains thick layer (Le et al. 2010). The most of the disease causing Gram negative bacteria are transmitted through water, so the synthesized Ag nanoparticles impregnated with the polymer could be used for water purification filter systems (Jain and Pradeep, 2005).

The SEM micrograph of AgNPs treated with *P. aeruginosa* confirms the bactericidal activity of nanoparticles synthesized by *C. utilis*. The detailed mechanism of interaction and penetration of silver nanoparticles inside the cell has not been fully studied. The electrostatic attraction between negatively charged cell membranes and positively charged nanoparticles could be responsible for the interactions between nanoparticles and bacterial cells (Raffi et al (2008). The bactericidal activity of silver nanoparticles against pathogenic bacteria could be used in conjunction with advances in impregnation techniques and polymer technology to expand the range of applications of these nanoparticles in the preservation of food, disinfection of medical supplies and equipment, and decontamination of the surfaces of items such as toys and kitchenware (Matsumura et al. 2003). Besides their bactericidal activity and immediate antibacterial effect against a wide variety of pathogenic bacteria, silver nanoparticles have particular characteristics provided by the silver itself. It tends to induce low bacterial resistance and has low toxicity and minimal side effects after absorption by the body (Ip et al. 2006).

Conclusion

The present study concludes ecofriendly production of silver nanoparticles using yeast *C. utilis* NCIM 3469. It has more potential activity against Gram negative bacteria. Ag nanoparticles accumulate on the cell surface which causes cell membrane lysis as confirmed by SEM. The antibacterial activity of biofabricated AgNPs revealed applicability in the preparation of reactors for the treatment of waste water and in the development of antibacterial water filters for treatment of water.

Acknowledgments This work was supported by Department of Science and Technology, Government of India, New Delhi under the DST-PURSE scheme sanctioned to Department of Microbiology, Shivaji University, Kolhapur, Maharashtra, India.

Conflict of interest The authors declare that they have no conflict of interest.

References

Ahmad A, Mukherjee P, Senapati S, Mandal D, Khan MI, Kumar R, Sastry M (2002) Extracellular biosynthesis of silver nanoparticles using the fungus Fusarium oxysporum. Colloids Surface B 28:313–318

Atyeh BS, Costagliola M, Hayek SN, Dibo SA (2007) Effect of silver on burn wound infection control and healing: review of the literature. Burns 33:139–148

Barraque B (2003) Past and future sustainability of water policies in Europe. Nat Resour Forum 27:200–211

Chang TW, Weinstein L (1975) Prevention of Herpes Keratoconjunctivitis in rabbits by silver sulfadiazine. Antimicrob Agents Ch 8:677–678

Feng QL, Wu J, Chen GQ, Cui FZ, Kim TN, Kim JO (2000) A mechanistic study of the antibacterial effect of silver ions on *Escherichia coli* and *Staphylococcus aureus*. J Biomed Mater Res-A 52:662–668

Galabova D, Tuleva B, Spasova D (1996) Permeabilization of *Yarrowia lipolytica* cells by Triton X-100. Enzyme Microb Tech 18:18–22

Hsueh PR (2010) New Delhi metallo-lactamase-1 (NDM-1): an emerging threat among Enterobacteriaceae. J Formos Med Assoc 109:685–687

Ichinose N (1992) Superfine particle technology. Springer, Berlin

Ip M, Lui SL, Poon VK, Lung I, Burd A (2006) Antimicrobial activities of silver dressings: an in vitro comparison. J Med Microbiol 55:59–63

Islam MS, Motahar HM, Banu AL, Sultana C, Quadir MA (2003) Antibacterial and antifungal activity of mixed ligand complexes of oxovanadium (IV), titanium (III) and cadmium (II) metal ions. Oriental J Chem 19:547–554

Jain P, Pradeep T (2005) Potential of silver nanoparticle-coated polyurethane foam as an antibacterial water filter. Biotechnol Bioeng 90:59–63

Jebali A, Ramezani F, Kazemi B (2011) Biosynthesis of Silver Nanoparticles by *Geotrichum* sp. J Clust Sci 22:225–232

Kim JS, Kuk E, Yu KN, Kim JH, Park SJ, Lee HJ, Kim SH, Park YK, Park YH, Hwang CY, Kim YK, Lee YS, Jeong DH, Cho MH (2007) Antimicrobial effects of silver nanoparticles. Nanomed 3:95–101

Kowshik M, Ashtaputre S, Kharrazi S, Vogel W, Urban J, Kulkarni SK, Paknikar KM (2003) Extracellular synthesis of silver nanoparticles by a silver-tolerant yeast strain MKY3. Nanomed 14:95–100

Kumarasamy KK, Toleman MA, Walsh TR, Bagaria J, Butt F, Balakrishnan R, Chaudhary U, Doumith M, Giske CG, Irfan S, Krishnan P, Kumar AV, Maharjan S, Mushtaq S, Noorie T, Paterson DL, Pearson A, Perry C, Pike R, Rao B, Ray U, Sarma JB, Sharma M, Sheridan E, Thirunarayan MA, Turton J, Upadhyay S, Warner M, Welfare W, Livermore DM, Woodford N (2010) Emergence of a new antibiotic resistance mechanism in India, Pakistan, and the UK: a molecular, biological, and epidemiological study. Lancet Inf Dis 10:597–602

Le AT, Tam LT, Tam PD, Huy PT, Huy TQ, Hieu NV, Kudrinskiy AA, Krutyakov YA (2010) Synthesis of oleic acid-stabilized silver nanoparticles and analysis of their antibacterial activity. Mat Sci Eng C 30:910–916

Lengke MF, Fleet ME, Southam G (2007) Biosynthesis of silver nanoparticles by filamentous cyanobacteria from a silver(I) nitrate complex. Langmuir 23:2694–2699

Matsumura Y, Yoshikata K, Kunisaki S, Tsuchido T (2003) Mode of bactericidal action of silver zeolite and its comparison with that of silver nitrate. Appl Environ Microb 69:4278–4281

Mishra A, Tripathy SK, Yun SI (2011) Bio-synthesis of gold and silver nanoparticles from *Candida guilliermondii* and their antimicrobial effect against pathogenic bacteria. J Nanosci Nanotechnol 11:243–250

Morones JR, Elechiguerra JL, Camacho A, Holt K, Kouri JB, Tapia J, Yacaman MJ (2005) The bactericidal effect of silver nanoparticles. Nanotechnology 16:2346–2353

Nair AS, Pradeep T (2004) Reactivity of Au and Ag nanoparticles with halocarbons. Appl Nanosci 59–63

Narayanan KB, Sakthivel N (2010) Biological synthesis of silver nanoparticles. Adv Colloid Interfac 156:1–13

Pitout JDD (2010) The latest threat in the war on antibacterial resistance. Lancet Inf Dis 10:578–579

Raffi M, Hussain F, Bhatti T, Akhter J, Hameed A, Hasan M (2008) Antibacterial Characterization of Silver Nanoparticles against *E. Coli* ATCC-15224. J Mater Sci Technol 24:192–196

Rai M, Yadav A, Gade A (2009) Silver nanoparticles as a new generation of antibacterials. Biotechnol Adv 27:76–83

Sharma VK, Yngard RA, Lin Y (2009) Silver nanoparticles: green synthesis and their antimicrobial activities. Adv Colloid Interf 145:83–96

Stoimenov PK, Klinger RL, Marchin GL, Klabunde KJ (2002) Metal oxide nanoparticles as bactericidal agents. Langmuir 18: 6679–6686

Vigneshwaran A, Kathe AA, Varadarajan PV, Nachne RP, Balasubramanya RH (2006) Biomimetics of silver nanoparticles by white rot fungus, *Phaenerochaete chrysosporium*. Colloids Surface B 53:55–59

Zhang LZ, Yu JC, Yip HY, Li Q, Kwong KW, Xu AW, Wong PK (2003) Ambient light reduction strategy to synthesize silver nanoparticles and silver-coated TiO2 with enhanced photocatalytic and bactericidal activities. Langmuir 19:10372–10380

Zhou M, Wei Z, Qiao H, Zhu L, Yang H, Xia T (2009) Particle size and pore structure characterization of silver nanoparticles prepared by confined arc plasma. J Nanomater 2009:1–5

Permissions

All chapters in this book were first published in 3 Biotech, by Springer; hereby published with permission under the Creative Commons Attribution License or equivalent. Every chapter published in this book has been scrutinized by our experts. Their significance has been extensively debated. The topics covered herein carry significant findings which will fuel the growth of the discipline. They may even be implemented as practical applications or may be referred to as a beginning point for another development.

The contributors of this book come from diverse backgrounds, making this book a truly international effort. This book will bring forth new frontiers with its revolutionizing research information and detailed analysis of the nascent developments around the world.

We would like to thank all the contributing authors for lending their expertise to make the book truly unique. They have played a crucial role in the development of this book. Without their invaluable contributions this book wouldn't have been possible. They have made vital efforts to compile up to date information on the varied aspects of this subject to make this book a valuable addition to the collection of many professionals and students.

This book was conceptualized with the vision of imparting up-to-date information and advanced data in this field. To ensure the same, a matchless editorial board was set up. Every individual on the board went through rigorous rounds of assessment to prove their worth. After which they invested a large part of their time researching and compiling the most relevant data for our readers.

The editorial board has been involved in producing this book since its inception. They have spent rigorous hours researching and exploring the diverse topics which have resulted in the successful publishing of this book. They have passed on their knowledge of decades through this book. To expedite this challenging task, the publisher supported the team at every step. A small team of assistant editors was also appointed to further simplify the editing procedure and attain best results for the readers.

Apart from the editorial board, the designing team has also invested a significant amount of their time in understanding the subject and creating the most relevant covers. They scrutinized every image to scout for the most suitable representation of the subject and create an appropriate cover for the book.

The publishing team has been an ardent support to the editorial, designing and production team. Their endless efforts to recruit the best for this project, has resulted in the accomplishment of this book. They are a veteran in the field of academics and their pool of knowledge is as vast as their experience in printing. Their expertise and guidance has proved useful at every step. Their uncompromising quality standards have made this book an exceptional effort. Their encouragement from time to time has been an inspiration for everyone.

The publisher and the editorial board hope that this book will prove to be a valuable piece of knowledge for researchers, students, practitioners and scholars across the globe.

List of Contributors

Jeke Kullu
Department of Biotechnology, National Institute of Technology, Durgapur 713209, India

Abhishek Dutta
Department of Chemical Engineering, Groep T-Leuven Engineering College (Associatie KU Leuven), A. Vesaliusstraat 13, 3000 Leuven, Belgium

Denis Constales
Vakgroep Wiskundige Analyse, Ghent University, Galglaan 2, Blok S22, 9000 Ghent, Belgium

Surabhi Chaudhuri
Department of Biotechnology, National Institute of Technology, Durgapur 713209, India

Debjani Dutta
Department of Biotechnology, National Institute of Technology, Durgapur 713209, India

Khyati V. Pathak
BRD School of Biosciences, Sardar Patel University, Sardar Patel Maidan, Satellite Campus, Vadtal Road, P.O. Box 39, Vallabh Vidyangar 388120, Gujarat, India

Hareshkumar Keharia
BRD School of Biosciences, Sardar Patel University, Sardar Patel Maidan, Satellite Campus, Vadtal Road, P.O. Box 39, Vallabh Vidyangar 388120, Gujarat, India

K. Saravanakumar
Faculty of Marine Sciences, Centre of Advanced Study in Marine Biology, Annamalai University, Parangipettai 608502, Tamil Nadu, India

K. Kathiresan
Faculty of Marine Sciences, Centre of Advanced Study in Marine Biology, Annamalai University, Parangipettai 608502, Tamil Nadu, India

Komal Gupta
NIT, Surathkal, Karnataka, India

Shashank Garg
Department of Biotechnology, Lovely Professional University, Punjab, India

Joginder Singh
School of Biotechnology and Bio sciences, Lovely Professional University, Punjab, India

Manoj Kumar
Department of Biotechnology, Lovely Professional University, Punjab, India

J. Anderson
Department of Plant, Soil and Agricultural Systems, Southern Illinois University, Carbondale, IL 62901-4415, USA

M. Akond
Plant Genomics and Biotechnology Lab, Department of Biological Sciences, Fayetteville State University, Fayetteville, NC 28301-4298, USA

M. A. Kassem
Plant Genomics and Biotechnology Lab, Department of Biological Sciences, Fayetteville State University, Fayetteville, NC 28301-4298, USA

K. Meksem
Department of Plant, Soil and Agricultural Systems, Southern Illinois University, Carbondale, IL 62901-4415, USA

S. K. Kantartzi
Department of Plant, Soil and Agricultural Systems, Southern Illinois University, Carbondale, IL 62901-4415, USA

Sonica Sondhi
Department of Microbiology, BMS Block, Panjab University Chandigarh, India

Prince Sharma
Department of Microbiology, BMS Block, Panjab University, Chandigarh, India

Nancy George
Department of Microbiology, BMS Block, Panjab University, Chandigarh, India

Prakram Singh Chauhan
Department of Microbiology, BMS Block, Panjab University, Chandigarh, India

Neena Puri
Department of Industrial Microbiology, Guru Nanak Khalsa College, City Centre Road, Yamunanagar 135001, Haryana, India

Naveen Gupta
Department of Microbiology, BMS Block, Panjab University

C. Subathra Devi
Industrial Biotechnology Division, School of Biosciences and Technology, VIT University, Vellore 632014, Tamil Nadu, India

Anuj Saini
Industrial Biotechnology Division, School of Biosciences and Technology, VIT University, Vellore 632014, Tamil Nadu, India

Shubham Rastogi
Industrial Biotechnology Division, School of Biosciences and Technology, VIT University, Vellore 632014, Tamil Nadu, India

S. Jemimah Naine
Industrial Biotechnology Division, School of Biosciences and Technology, VIT University, Vellore 632014, Tamil Nadu, India

V. Mohanasrinivasan
Industrial Biotechnology Division, School of Biosciences and Technology, VIT University, Vellore 632014, Tamil Nadu, India

Fatma F. Abdel-Motaal
Department of Botany, Faculty of Science, Aswan University, Aswaˆn 81528, Egypt

Magdi A. El-Sayed
Department of Botany, Faculty of Science, Aswan University, Aswaˆn 81528, Egypt

Soad A. El-Zayat
Department of Botany, Faculty of Science, Aswan University, Aswaˆn 81528, Egypt

Shin-ichi Ito
Department of Biological and Environmental Sciences, Faculty of Agriculture, Yamaguchi University, Yamaguchi 753-8515, Japan

Hamish Cameron
Faculty of Science and Engineering, University of Waikato, Private Bag 3105, Hamilton, New Zealand

Sylke H. Campion
Scion, Te Papa Tipu Innovation Park, Private Bag 3020,Rotorua 3046, New Zealand

Tripti Singh
Scion, Te Papa Tipu Innovation Park, Private Bag 3020,Rotorua 3046, New Zealand

Alankar A. Vaidya
Scion, Te Papa Tipu Innovation Park, Private Bag 3020,Rotorua 3046, New Zealand

B. Gowramma
Department of Biotechnology, Sri Krishnadevaraya University, Anantapuram 515003, Andhra Pradesh, India

U. Keerthi
Department of Biotechnology, Sri Krishnadevaraya University, Anantapuram 515003, Andhra Pradesh, India

Mokula Rafi
Department of Biotechnology, Sri Krishnadevaraya University, Anantapuram 515003, Andhra Pradesh, India

D. Muralidhara Rao
Department of Biotechnology, Sri Krishnadevaraya University, Anantapuram 515003, Andhra Pradesh, India

Monika Kayasth
Department of Microbiology, CCS Haryana Agricultural University, Hisar 125004, India

Varun Kumar
Department of Biotechnology and Bioinformatics, Jaypee University of Information Technology, Waknaghat 173234, Himachal Pradesh, India

Rajesh Gera
Department of Microbiology, CCS Haryana Agricultural University, Hisar 125004, India

Mohd. Shahbaaz
Department of Computer Science, Jamia Millia Islamia, New Delhi 110025, India

Faizan Ahmad
Center for Interdisciplinary Research in Basic Sciences, Jamia Millia Islamia, Jamia Nagar, New Delhi 110025, India Md. Imtaiyaz Hassan

Priya Vashisth
Department of Biotechnology, Indian Institute of Technology Roorkee, Roorkee 247667, India

Mohit Sharma
Department of Biotechnology, Indian Institute of Technology Roorkee, Roorkee 247667, India

Kumar Nikhil
Department of Biotechnology, Indian Institute of Technology Roorkee, Roorkee 247667, India

Harmeet Singh
Department of Biotechnology, Indian Institute of Technology Roorkee, Roorkee 247667, India

Richa Panwar
Department of Biotechnology, Indian Institute of Technology Roorkee, Roorkee 247667, India

Parul A. Pruthi
Department of Biotechnology, Indian Institute of Technology Roorkee, Roorkee 247667, India

Vikas Pruthi
Department of Biotechnology, Indian Institute of Technology Roorkee, Roorkee 247667, India

Arijit Das
Department of Microbiology, Karpagam University, Coimbatore 641021, Tamil Nadu, India
Department of Microbiology, Centre for Advanced Studies in Biosciences, Jain University, Bangalore 560019, Karnataka, India

Sourav Bhattacharya
Department of Microbiology, Karpagam University, Coimbatore 641021, Tamil Nadu, India
Department of Microbiology, Centre for Advanced Studies in Biosciences, Jain University, Bangalore 560019, Karnataka, India

Muthusamy Palaniswamy
Department of Microbiology, Karpagam University, Coimbatore 641021, Tamil Nadu, India

Jayaraman Angayarkanni
Department of Microbial Biotechnology, Bharathiar University, Coimbatore 641046, Tamil Nadu, India

Jigisha Anand
Department of Biotechnology, Graphic Era University, Dehradun, India

Bhagwati Upadhyaya
Department of Microbiology, Guru Nanak Dev University, Amritsar, Punjab, India

Pramod Rawat
Department of Biotechnology, Graphic Era University, Dehradun, India

Nishant Rai
Department of Biotechnology, Graphic Era University, Dehradun, India

S. K. Mohan
Department of Chemical Engineering, Annamalai University, Annamalainagar 608002, Tamilnadu, India

T. Viruthagiri
Department of Chemical Engineering, Annamalai University, Annamalainagar 608002, Tamilnadu, India

C. Arunkumar
Department of Chemical Engineering, Annamalai University, Annamalainagar 608002, Tamilnadu, India

Pavan Kumar Pindi
Department of Microbiology, Palamuru University, Mahabubnagar 509001, Andhra Pradesh, India

Tasleem Sultana
Department of Microbiology, Palamuru University, Mahabubnagar 509001, Andhra Pradesh, India

Praveen Kumar Vootla
Department of Microbiology, Palamuru University, Mahabubnagar 509001, Andhra Pradesh, India

S. K. Bhatia
Department of Biotechnology, Himachal Pradesh University, Summer Hill, Shimla 171005, India

P. K. Mehta
Department of Biotechnology, Himachal Pradesh University, Summer Hill, Shimla 171005, India

R. K. Bhatia
Department of Biotechnology, Himachal Pradesh University, Summer Hill, Shimla 171005, India

T. C. Bhalla
Department of Biotechnology, Himachal Pradesh University, Summer Hill, Shimla 171005, India

Seyedeh Mahsan Hoseini Alfatemi
Department of Bacteriology and Virology, Shiraz Medical School, Shiraz University of Medical Sciences, P.O. Box 71455-119, Shiraz, Iran

Javad Sharifi Rad
Zabol Medicinal Plants Research Center, Zabol University of Medical Sciences, P.O. Box 61615-585, Zabol, Iran
Department of Pharmacognosy, Faculty of Pharmacy,Zabol University of Medical Sciences, P.O. Box 61615-585, Zabol, Iran
Cereal Health Research Center of Zabol, Zabol University of Medical Sciences, P.O. Box 61615-585, Zabol, Iran

Majid Sharifi Rad
Department of Range and Watershed Management, Faculty of Natural Resources, University of Zabol, Zabol, Iran
Department of Rangeland Science, Gorgan University of Agricultural Sciences and Natural Resources, 49138-15739 Gorgan, Iran

Sasan Mohsenzadeh
Department of Biology, College of Sciences, Shiraz University, 71454 Shiraz, Iran

Jaime A. Teixeira da Silva
P.O. Box 7, Miki Cho Post Office, 3011-2, Ikenobe, Kagawa-Ken 761-0799, Japan

Lalit Kumar
Department of Biochemistry, Kurukshetra University, Kurukshetra 136119, Haryana, India

Davender Kumar
Department of Biochemistry, Kurukshetra University, Kurukshetra 136119, Haryana, India

Sushil Nagar
Department of Biochemistry, Kurukshetra University, Kurukshetra 136119, Haryana, India

Rishi Gupta
Lignocellulose Biotechnology Laboratory, Departmentof Microbiology, University of Delhi South Campus,Benito Juarez Road, New Delhi 110021, India

Neelam Garg
Department of Microbiology, Kurukshetra University, Kurukshetra 136119, Haryana, India

Ramesh Chander Kuhad
Lignocellulose Biotechnology Laboratory, Departmentof Microbiology, University of Delhi South Campus,Benito Juarez Road, New Delhi 110021, India

Vijay Kumar Gupta
Department of Biochemistry, Kurukshetra University, Kurukshetra 136119, Haryana, India

Abrar Ahmad
Environmental Biotechnology Division, CSIR-Indian Institute of Toxicology Research, M.G. Marg, Lucknow 226001, India

Ashok Kumar Chauhan
Environmental Biotechnology Division, CSIR-Indian Institute of Toxicology Research, M.G. Marg, Lucknow 226001, India

Hari Narayan Kushwaha
Pharmacokinetics and Metabolism Division, Central Drug Research Institute, Lucknow 226001, India

Saleem Javed
Department of Biochemistry, Faculty of Science, Jamia Hamdard, New Delhi 110062, India

Ashwani Kumar
Environmental Biotechnology Division, CSIR-Indian Institute of Toxicology Research, M.G. Marg, Lucknow 226001, India

Manjit Jaiswal
Department of Pharmacy, School of Medical and Allied Sciences, Galgotias University, G. B. Nagar, Uttar Pradesh, India

Rupesh Dudhe
Department of Pharmacy, School of Medical and Allied Sciences, Galgotias University, G. B. Nagar, Uttar Pradesh, India

P. K. Sharma
Department of Pharmacy, School of Medical and Allied Sciences, Galgotias University, G. B. Nagar, Uttar Pradesh, India

Munees Ahemad
Department of Agricultural Microbiology, Facultyof Agricultural Sciences, Aligarh Muslim University,Aligarh 202002, UP, India

Suki Roy
Molecular and Microbiology Research Laboratory, School of Bio Sciences and Technology, VIT University, Vellore 632014, Tamil Nadu, India

Shreta Chandni
Molecular and Microbiology Research Laboratory, School of Bio Sciences and Technology, VIT University, Vellore 632014, Tamil Nadu, India

Ishita Das
Molecular and Microbiology Research Laboratory, School of Bio Sciences and Technology, VIT University, Vellore 632014, Tamil Nadu, India

Loganathan Karthik
Department of Biotechnology, Shri JJT University, Jhunjhunu, Rajasthan, India

Gaurav Kumar
Molecular and Microbiology Research Laboratory, School of Bio Sciences and Technology, VIT University, Vellore 632014, Tamil Nadu, India

Kokati Venkata Bhaskara Rao
Molecular and Microbiology Research Laboratory, School of Bio Sciences and Technology, VIT University, Vellore 632014, Tamil Nadu, India

Dominik Winckelmann
School of Engineering and Science, Jacobs-University Bremen, Campus Ring 1, 28759 Bremen, Germany

Franziska Bleeke
School of Engineering and Science, Jacobs-University Bremen, Campus Ring 1, 28759 Bremen, Germany

Peter Bergmann
University of Applied Sciences Bremen, Am Neustadtswall 30,28199 Bremen, Germany

Gerd Klöck
University of Applied Sciences Bremen, Am Neustadtswall 30,28199 Bremen, Germany

Mainak Mukhopadhyay
Microbial Biotechnology and Downstream Processing Laboratory, Agricultural and Food Engineering Department, Indian Institute of Technology, Kharagpur 721 302, India

Rintu Banerjee
Microbial Biotechnology and Downstream Processing Laboratory, Agricultural and Food Engineering Department, Indian Institute of Technology, Kharagpur 721 302, India

Shailesh R. Waghmare
Department of Microbiology, Shivaji University, Kolhapur 416004, Maharashtra, India

Mustopa N. Mulla
Department of Microbiology, Shivaji University, Kolhapur 416004, Maharashtra, India

Suryakant R. Marathe
Department of Microbiology, Shivaji University, Kolhapur 416004, Maharashtra, India

Kailas D. Sonawane
Department of Microbiology, Shivaji University, Kolhapur 416004, Maharashtra, India
Department of Biochemistry, Shivaji University, Kolhapur, Maharashtra, India